Topics in Geometry

Topics in Geometry

Robert Bix

Department of Mathematics
University of Michigan–Flint
Flint, Michigan

ACADEMIC PRESS, INC.
Harcourt Brace & Company, Publishers
Boston   San Diego   New York
London   Sydney   Tokyo   Toronto

ACADEMIC PRESS, INC.
1250 Sixth Avenue, San Diego, CA 92101-4311

United Kingdom Edition published by
ACADEMIC PRESS LIMITED
24–28 Oval Road, London NW1 7DX

Library of Congress Cataloging-in-Publication Data

Bix, Robert, 1953–
   Topics in geometry/Robert Bix.
    p. cm.
   Includes bibliographical references and index.
   ISBN 0-12-102740-6 (acid-free)
   1. Geometry. I. Title.
  QA445.B53 1993
  516—dc20   93-1091
                                       CIP

PRINTED IN THE UNITED STATES OF AMERICA

93 94 95 96 BC 9 8 7 6 5 4 3 2 1

To Peggy
and Jonathan

Table of Contents

Preface

This book is a text for a junior-level College Geometry course designed for prospective secondary school teachers and other mathematics majors. The only prerequisite is first-year calculus. Chapter I starts with a brief, self-contained review of elementary Euclidean geometry before proceeding to more advanced material.

The book includes material for a number of different courses. Topics include advanced Euclidean geometry, transformation geometry, projective geometry, conic sections, and hyperbolic and absolute geometry. Substantial results are obtained in each topic.

Considerable variables in coverage are possible. The only restrictions are that Chapter I must be covered before any other chapter and that Chapter III must be covered before Chapter IV.

One of the most distinctive features of this book is its emphasis on the connections between Euclidean and non-Euclidean geometry. This emphasis unifies much of the book. For example, the Euclidean results on division ratios in Chapter I are used in Chapter III to prove theorems in projective geometry by projecting between planes. Striking results in Euclidean geometry are then deduced as special cases of theorems in projective geometry. Similarly, the Euclidean and projective results of Chapters I and III are combined to study conic sections in Chapter IV.

Euclidean results from Chapter I are used to study inversion in circles
and construct the hyperbolic plane in Chapter V. Chapter V also includes
a comparative study of axiom systems for hyperbolic, absolute, and
Euclidean geometries.

The text is intended to be as complete and as readable as possible.
Because full details of proofs are presented in the book, they can be omitted
in class, where time is scarce and informal presentations are most effective.
Accordingly, instructors can use class time to present an overview of
material and to answer questions or discuss exercises.

The book contains many exercises of varying difficulty so that students
can become actively involved in doing mathematics. The more challenging
exercises are divided into several parts and provided with hints. Nowhere
does the body of the text depend on the exercises.

The following books are excellent sources of further reading about the
history of geometry, and they provided much of the historical information
in the chapter introductions.

Boyer, Carl B., and Merzbach, Uta C., *A History of Mathematics*, Second
Edition, Wiley, New York, 1989.

Eves, Howard, *An Introduction to the History of Mathematics*, Fifth
Edition, Saunders College Publishing, Philadelphia, 1983.

Kline, Morris, *Mathematical Thought from Ancient to Modern Times*,
Oxford University Press, New York, 1972.

Rosenfeld, B. A., *A History of Non-Euclidean Geometry* (translated by
Abe Shenitzer), Springer-Verlag, New York, 1988.

Other references appear at the end of the chapter introductions.

I would like to thank Steven C. Althoen, Robert J. Bumcrot, Harry J.
D'Souza, Paul K. Garlick, Lawrence D. Kugler, Carole B. Lacampagne,
James R. Shivlie, and Kenneth E. Schilling for their generous assistance.
Special thanks go to Ira N. Bix for his exceptional work drawing the figures
and to Michael C. Bix for his extraordinary help in reviewing the manuscript. I am also grateful to the students who tried out the manuscript in
classes at the University of Michigan-Flint. The research for this project
was supported in part by a grant from the Faculty Development Fund of the
University of Michigan-Flint.

Chapter I
Division Ratios and Triangles

INTRODUCTION AND HISTORY

Mathematics has often advanced over the past 2500 years by increasing the scope of geometric ideas. Mathematical research has focused on geometry in varying degrees over time, but geometry has always been a key source of mathematical intuition.

From the sixth through the fourth centuries B.C., Greek scholars transformed the empirical geometry of the Babylonians and Egyptians into a masterpiece of logical reasoning. Such mathematicians and philosophers as Thales, Pythagoras, Plato, Eudoxus, and Aristotle sought irrefutable demonstrations of abstract geometric truths. Euclid presented their fundamental results in definitive form in the *Elements* about 300 B.C. Ever since its appearance, Euclid's work has inspired mathematicians with the brilliance of its construction and the elegance of its reasoning. It includes the results on triangles and parallel lines reviewed in Section 0, the theorems on angle bisectors in Section 5, and the results on inscribed angles and chords of circles in Section 6.

Euclid lived in Alexandria, the center of mathematics from approximately 300 B.C. to 500 A.D. An extraordinary university was founded there about 290 B.C. by Ptolemy Soter, who ruled Egypt as the successor to a third of the empire of Alexander the Great. Spurred by intercontinental trade, the needs of navigation, and remarkable advances in mechanics, optics, and other sciences, geometry became far more quantitative. In the middle of the third century B.C., Archimedes used physical reasoning to deduce remarkable results on areas and volumes, results that he then proved rigorously. At the end of the third century B.C., Apollonius derived extraordinary theorems on conic sections by considering the distances from a variable point on a conic section to two fixed lines. The study of astronomy led to a geometric development of trigonometry and to a detailed study of triangles on the surface of a sphere. This work was brought to a high point in Alexandria by Menelaus and Claudius Ptolemy in the second century A.D. Section 2 centers around a theorem that Menelaus generalized to the sphere; he apparently considered the planar version in Section 2 to be too well-known to prove. Ptolemy proved the theorem in Exercise 6.16 for use in trigonometric calculations. The last great classical geometer was Pappus of Alexandria, who, at the start of the fourth century A.D., anticipated results in projective geometry by more than 1300 years.

Arabic mathematicians played a key role in preserving classical geometry from the eighth through the thirteenth centuries. They significantly advanced the study of Euclid's Fifth Postulate, which we take up in Chapter V. European mathematicians from the twelfth through the seventeenth centuries continued the work of restoring classical mathematics.

The interest of Renaissance painters in perspective led to the creation of projective geometry in the seventeenth century, as we discuss in the introduction to Chapter III. This work was overshadowed, however, by the explosive development of analytic geometry and calculus in the seventeenth and eighteenth centuries through the work of such extraordinary mathematicians as René Descartes, Pierre de Fermat, Isaac Newton, Gottfried Wilhelm Leibniz, and Leonhard Euler. It became routine to study curves defined by algebraic equations rather than curves constructed geometrically.

A limited amount of work continued in the seventeenth and eighteenth centuries on synthetic—that is, nonanalytic—geometry. Giovanni Ceva rediscovered Menelaus' Theorem and published his own closely related theorem, which appears in Section 3. The Simson line, which we consider in Exercises 6.23 and 6.24, is named after Robert Simson, who worked to revive classical geometry in Britain. Girolamo Saccheri and Johann Heinrich Lambert continued work on Euclid's Fifth Postulate.

Synthetic geometry was revived strongly in the nineteenth century by mathematicians unhappy with the dominant role of algebraic and analytic computations in geometry. Lazare Carnot gave a synthetic derivation of the Euler line—the subject of Section 4—which Euler had constructed by analytic methods. The nine-point circle, which we discuss in Section 5, was discovered by Jean-Victor Poncelet, Charles Jules Brianchon, and Karl Wilhelm Feuerbach. The study of projective geometry, which had lagged for 150 years, flourished in the work of Carnot, Poncelet, Brianchon, Jacob Steiner, Michel Chasles, Karl Georg Christian von Staudt, and others.

Carnot based geometric results on directed distances, which we introduce in Section 1 and which play a key role in much of this book. Attaching a sign to the distance between two points on a line automatically keeps track of their relative positions. We use directed distances anachronistically in many theorems, such as those of Menelaus and Ceva, to obtain general results without considering special cases. Directed distances were generalized to barycentric coordinates by Augustus Ferdinand Möbius and to homogeneous coordinates by Julius Plücker, who developed powerful techniques of analytic projective geometry in the nineteenth century.

Spherical geometry and projective geometry are examples of non-Euclidean geometries dating from the first century B.C. and the seventeenth century A.D., respectively. The idea that there could be geometries analogous to but different from Euclidean geometry, however, didn't take hold until hyperbolic geometry, which we introduce in Chapter V, developed in the nineteenth century. This breakthrough led in the twentieth century to new types of geometric spaces linked to abstract algebra and analysis. This linkage can be seen either as extending geometric ideas throughout mathematics or as abandoning classical geometry. On the other hand, current issues of the *American Mathematical Monthly* and *Mathematics Magazine* show that many of the aspects of synthetic geometry that were introduced in the nineteenth century are still being explored. Moreover, topics in pure geometry such as tilings and combinatorial geometry provide new focuses of research and are rich sources of open problems.

The following books are excellent sources of further reading. They provided much of the material in this chapter.

Altshiller-Court, Nathan, *College Geometry*, Second Edition, Barnes and Noble, New York, 1952.

Coolidge, Julian Lowell, *A Treatise on the Circle and the Sphere*, Chelsea, New York, 1971.

Coxeter, H. S. M., and Greitzer, S. L., *Geometry Revisited*, The Mathematical Association of America, Washington, D.C., 1967.

Davis, David R., *Modern College Geometry*, Addison-Wesley, Cambridge, Mass., 1954.

Eves, Howard, *A Survey of Geometry*, Revised Edition, Allyn and Bacon, Boston, 1972.

Section 0.

Elementary Euclidean Geometry

The primary aim of this book is to develop striking geometric results as simply as possible by adding a few new ideas to elementary Euclidean geometry. In this section, we review the basic facts that we need from Euclidean geometry. This review serves two main purposes. First, it emphasizes how little Euclidean geometry we need to obtain remarkable results. Second, it should enable readers, regardless of their preparation in geometry, to proceed comfortably to the more advanced material in the rest of the book. Readers who are well prepared in geometry may prefer either to skim or omit this section.

We do not develop Euclidean geometry from formal axioms until Chapter V. Until then, we rely on the common understanding of points, lines, distances, angles, and areas in the Euclidean plane. This understanding is informal but clear and unambiguous.

There are several advantages to postponing an axiomatic development of Euclidean geometry until Chapter V. By minimizing formalism, we emphasize the role of physical intuition; the use of physical intuition provides much of the appeal of geometry and distinguishes it from other mathematics. An informal approach lets us move quickly through elementary geometry to more intesting material. Finally, by postponing an axiomatic treatment of geometry until Chapter V, we can demonstrate its usefulness by developing hyperbolic and Euclidean geometry simultaneously.

In fact, the Euclidean plane can be defined formally without axioms. The points are the ordered pairs (x, y) of real numbers. The lines are the graphs of linear equations $ax + by = c$, where a, b, c are real numbers such that a and b are not both zero. The distance between points (x_1, y_1) and (x_2, y_2) is

$$\sqrt{(x_1 - x_2)^2 + (y_1 - y_2)^2}.$$

Areas and angles can be defined using calculus. This approach, however, has the same disadvantages as the axiomatic approach: it is cumbersome and time-consuming, and it minimizes the role of geometric intuition. Accordingly, until Chapter V, we approach the Euclidean plane informally.

In reviewing basic Euclidean geometry, we state without proof Properties 0.1–0.5, which we consider to be intuitively obvious. We then use these properties to derive results that are less immediately evident. We include the words "in the Euclidean plane" in stating all results in Chapters I and II because we move beyond beyond the Euclidean plane in Chapters III–V.

In the Euclidean plane, a *triangle* consists of three points that don't lie on one line and the three segments that have pairs of these points as endpoints

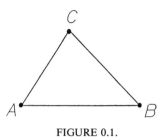

FIGURE 0.1.

(Figure 0.1). The three points are called the *vertices* of the triangle, and the three segments are called the *sides*. A side is named by its endpoints; for example, in Figure 0.1, side *BC* is the segment with endpoints *B* and *C*. A vertex and a side are called *opposite* if one doesn't lie on the other; in Figure 0.1, vertex *A* and side *BC* are opposite, vertex *B* and side *AC* are opposite, and vertex *C* and side *AB* are opposite. A triangle is named by its vertices, and so Figure 0.1 shows triangle *ABC*.

Triangles *ABC* and *A'B'C'* are called *congruent* if they have sides *AB* and *A'B'* of equal length, sides *AC* and *A'C'* of equal length, sides *BC* and *B'C'* of equal length, equal angles at *A* and *A'*, equal angles at *B* and *B'*, and equal angles at *C* and *C'* (Figure 0.2). In short, two triangles are congruent if corresponding sides have equal lengths and if corresponding angles are equal. Intuitively, two triangles are congruent when they have the same size and shape.

Property 0.1 states that two triangles are congruent if corresponding sides have equal lengths. In other words, when the sides of a triangle are given, there is essentially at most one way to form the triangle. We consider this result intuitively obvious, and we assume without proof that it holds.

PROPERTY 0.1 (SSS Criterion). In the Euclidean plane, let *ABC* and *A'B'C'* be triangles. Assume that sides *AB* and *A'B'* have equal length, sides *AC* and *A'C'* have equal length, and sides *BC* and *B'C'* have equal length. Then triangles *ABC* and *A'B'C'* are congruent (Figure 0.2). □

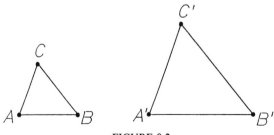

FIGURE 0.2.

We call this property the *SSS Criterion* because it states that a triangle is determined by the lengths of its three sides. Similarly, the *SAS Criterion*, which is the next property we assume, states that a triangle is determined by the lengths of two sides and by the angle between these sides.

PROPERTY 0.2 (SAS Criterion). In the Euclidean plane, let ABC and $A'B'C'$ be triangles that have sides AB and $A'B'$ of equal length, sides AC and $A'C'$ of equal length, and equal angles at A and A'. Then the triangles are congruent (Figure 0.2). ☐

In other words, if we're given the angle at A and the distances from A to B and C, there is essentially only one way to form triangle ABC: choose B and C to be the points at the required distances from A along the sides of the angle at A.

The next property we assume states that a triangle is determined by two angles and the length of their common side.

PROPERTY 0.3 (ASA Criterion). In the Euclidean plane, let ABC and $A'B'C'$ be triangles that have equal angles at A and A', equal angles at B and B', and sides AB and $A'B'$ of equal length. Then the triangles are congruent (Figure 0.2). ☐

In other words, if we're given side AB and the angles at A and B, there is essentially only one way to form triangle ABC: side AB and the angles at A and B determine the three sides of the triangle, and the two sides of the triangle other than side AB intersect at the third vertex.

We say that triangles ABC and $A'B'C'$ are *similar* if they have equal angles at A and A', at B and B', and at C and C' (Figure 0.3). Similar triangles have the same shape, but may have different sizes. The next property we assume states that corresponding sides of similar triangles are proportional, a fact that corresponds to the idea that similar triangles have the same shape.

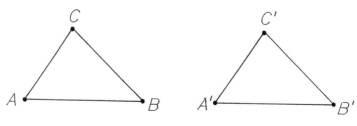

FIGURE 0.3.

FIGURE 0.4a.   FIGURE 0.4b.

PROPERTY 0.4. In the Euclidean plane, let ABC and $A'B'C'$ be similar triangles (Figure 0.3). If XY denotes the distance between points X and Y, then we have

$$AB/A'B' = AC/A'C' = BC/B'C'. \quad \square$$

If A and B are two points, *line AB* is the unique line through A and B (Figure 0.4a). The *ray $\overrightarrow{AB}$* consists of A together with all points on line AB that lie on the same side of A as B (Figure 0.4b). We say that the ray *originates* at A.

If A, B, C are three points, $\angle BAC$ is the angle θ formed by the rays $\overrightarrow{AB}$ and $\overrightarrow{AC}$, where $0° \le \theta \le 180°$ (Figure 0.5). We call $\angle BAC$ a *straight angle* when $\theta = 180°$, which means that A lies on line BC between B and C (Figure 0.6). When $\theta = 90°$, we call $\angle BAC$ a *right angle* and say that lines AB and AC are *perpendicular* (Figure 0.7).

In the Euclidean plane, we call two lines *parallel* if they either have no points in common or are equal. Note that *we consider a line to be parallel to itself*; this convention simplifies many of our statements. Intuitively, lines are parallel when they point in the same direction. This idea makes it natural to assume the following property.

PROPERTY 0.5. In the Euclidean plane, let l, m, n be three lines. Assume that l and n intersect at a point A and that m and n intersect at a point B. Let C be a point on l other than A, and let D be a point on m other than B that lies on the same side of n as C. Let E be a point on n other than A and B that doesn't lie between A and B. Then l and m are parallel if and only if $\angle CAE$ and $\angle DBE$ are equal (Figure 0.8). $\quad \square$

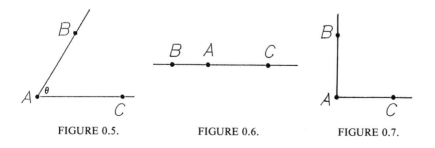

FIGURE 0.5.   FIGURE 0.6.   FIGURE 0.7.

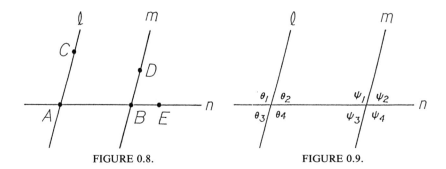

FIGURE 0.8.   FIGURE 0.9.

In the Euclidean plane, let l and m be two parallel lines (Figure 0.9). Let n be a line that intersects l and m. We say that an angle determined by l and n and an angle determined by m and n *correspond* to each other if one of them contains the other. For example, in Figure 0.9, θ_i and ψ_i are corresponding angles for $i = 1, 2, 3, 4$. Informally, the pairs of corresponding angles are those that lie in the same position relative to n and either l and m. Property 0.5 shows that, *when two parallel lines intersect a third line, corresponding angles are equal.*

We call two angles *supplementary* when their sum is 180°.

Two intersecting lines form four angles (Figure 0.10). Consider two of these angles. If the angles share a common side, they are called *adjacent*; if the angles don't share a common side, they are called *vertical*. In Figure 0.10, θ_i and ψ_j are adjacent angles for all i and j, and the two pairs of vertical angles are $\{\theta_1, \theta_2\}$ and $\{\psi_1, \psi_2\}$. Adjacent angles are supplementary because they combine to form a straight angle. Vertical angles are equal because they are each supplementary to an angle adjacent to them both. For example, in Figure 0.10, the vertical angles θ_1 and θ_2 each form a straight angle with ψ_1, and the equations $\theta_1 + \psi_1 = 180°$ and $\theta_2 + \psi_1 = 180°$ imply that θ_1 equals θ_2. Thus, we've proved the following result.

THEOREM 0.6. In the Euclidean plane, when two lines intersect, adjacent angles are supplementary, and vertical angles are equal (Figure 0.10). □

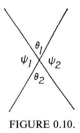

FIGURE 0.10.

In particular, Theorem 0.6 shows that two perpendicular lines form four right angles.

Let l and m be two parallel lines that intersect a line n (Figure 0.9). An angle formed by l and n and an angle formed by m and n are called *alternate interior angles* if they lie between l and m on opposite sides of n. In Figure 0.9, θ_2 and ψ_3 are alternate interior angles, and so are θ_4 and ψ_1. Given alternate interior angles α and β, the angle that forms a pair of vertical angles with α also forms a pair of corresponding angles with β. For example, given the alternate interior angles θ_2 and ψ_3 in Figure 0.9, the angle θ_3, which forms a pair of vertical angles with θ_2, also forms a pair of corresponding angles with ψ_3. Since vertical angles are equal (by Theorem 0.6), and since corresponding angles are equal (by Property 0.5), it follows that alternate interior angles are equal. Thus, we've proved the following result.

THEOREM 0.7. In the Euclidean plane, let two parallel lines be intersected by a third line. Then alternate interior angles are equal. □

We can use the previous result to prove that the sum of the angles of a triangle is 180°, the measure of a straight angle.

THEOREM 0.8. In the Euclidean plane, the sum of the angles of a triangle is 180°.

Proof: Let A, B, C be the vertices of the triangle (Figure 0.11). Consider the line through C parallel to the line through A and B. Let θ_1 be the angle at C that forms a pair of alternate interior angles with $\angle BAC$, and let θ_2 be the angle at C that forms a pair of alternate interior angles with $\angle ABC$. We have

$$\theta_1 = \angle BAC \quad \text{and} \quad \theta_2 = \angle ABC$$

(by Theorem 0.7). We also have

$$\theta_1 + \theta_2 + \angle ACB = 180°,$$

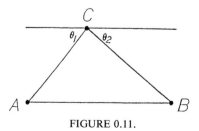

FIGURE 0.11.

since the angles on the left side of this equation comprise a straight angle. Combining the equations in the last two sentences shows that

$$\angle BAC + \angle ABC + \angle ACB = 180°,$$

and so the sum of the angles of triangle ABC is $180°$. ☐

We can combine the previous result with Property 0.4 of similar triangles to prove the Pythagorean Theorem. A *right triangle* is a triangle that has a right angle as one of its angles. The side opposite the right angle is the *hypotenuse*, and the other two sides are the *legs*.

THEOREM 0.9 (Pythagorean Theorem). In the Euclidean plane, let a right triangle have hypotenuse of length c and legs of lengths a and b. Then we have

$$a^2 + b^2 = c^2.$$

Proof: Let the triangle have vertices A, B, C, where the right angle is at C, a is the length of side BC, and b is the length of side AC (Figure 0.12). Let F be the point where line AB intersects the line through C perpendicular to line AB. Let d be the distance from B to F, let e be the distance from A to F, and let h be the distance from C to F.

Triangle ABC has a $90°$ angle at C, and so the angles at A and B are each less than $90°$ (by Theorem 0.8). It follows that F lies between A and B, and so we have

$$d + e = c. \tag{1}$$

Triangles ABC and CBF have the same angle at B (since F lies between A and B) and have right angles $\angle ACB$ and $\angle CFB$. Thus, the triangles also have equal angles $\angle BAC$ and $\angle BCF$ (by Theorem 0.8) and are similar, and so Property 0.4 shows that

$$c/a = a/d. \tag{2}$$

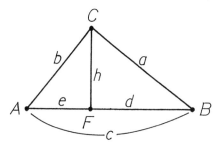

FIGURE 0.12.

Similarly, triangles ABC and ACF have the same angle at A (since F lies between A and B) and have right angles $\angle ACB$ and $\angle AFC$. Thus, the triangles also have equal angles $\angle ABC$ and $\angle ACF$ (by Theorem 0.8) and are similar, and so Property 0.4 gives

$$c/b = b/e. \tag{3}$$

Combining Equations 1–3 shows that

$$a^2 + b^2 = cd + ce = c(d + e) = c^2. \quad \square$$

The ASA and SAS criteria imply that a triangle has two equal angles if and only if it has two sides of equal length.

THEOREM 0.10. In the Euclidean plane, two angles of a triangle are equal if and only if the opposite sides have equal length.

Proof: Label the vertices of the triangle A, B, C so that the two angles under consideration are those at A and B (Figure 0.13). The sides opposite these angles are BC and AC. Triangles ABC and BAC have a common side AB, and so the triangles are congruent if and only if the angles at A and B are equal, by the ASA Criterion (Property 0.3). Triangles ABC and BAC have the same angle at C, and so the triangles are congruent if and only if the sides AC and BC have equal length, by the SAS Criterion (Property 0.2). The last two sentences show that the angles at A and B are equal if and only if the sides AC and BC have equal length, since each of these conditions holds if and only if triangles ABC and BAC are congruent. $\square$

The next result states that *two lines that are each parallel to a third line are parallel to each other*. This follows informally from the idea that lines are parallel when they point in the same direction. We can use Property 0.5 of corresponding angles to give a formal proof.

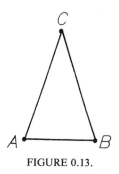

FIGURE 0.13.

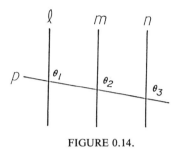

FIGURE 0.14.

THEOREM 0.11. In the Euclidean plane, let l, m, n be three lines. If l and m are parallel, and if m and n are parallel, then l and n are parallel.

Proof: Let p be a line that intersects l, m, and n (Figure 0.14). Let θ_1 be an angle formed by l and p. Since l and m are parallel, θ_1 equals the angle θ_2 formed by m and p that lies in corresponding position (by Property 0.5). Since m and n are parallel, θ_2 equals the angle θ_3 formed by n and p that lies in corresponding position (by Property 0.5). The last two sentences show that θ_1 equals θ_3, and so l and n are parallel (by Property 0.5). $\square$

A *parallelogram ABCD* consists of four points A, B, C, D and the four segments with endpoints A and B, B and C, C and D, and D and A, where the lines AB and CD are parallel and distinct, and where the lines BC and DA are parallel and distinct (Figure 0.15). The four segments are called *sides* and are named by their endpoints. Sides AB and CD and sides BC and DA are the two pairs of *opposite sides*. Combining the ASA Criterion with Theorem 0.7 on alternate interior angles gives the following result.

THEOREM 0.12. In the Euclidean plane, opposite sides of a parallelogram have equal length (Figure 0.15).

Proof: Label the vertices so that we're considering parallelogram $ABCD$. Draw the diagonal line AC (Figure 0.16). We have

$$\angle BAC = \angle DCA \quad \text{and} \quad \angle BCA = \angle DAC$$

(by Theorem 0.7 on alternate interior angles, since opposite sides of a parallelogram lie on parallel lines). Triangles ABC and CDA share a common side, AC. The last two sentences show that triangles ABC and CDA are congruent, by the ASA Criterion (Property 0.3). Thus, sides AB and CD have equal length, and sides BC and DA have equal length. $\square$

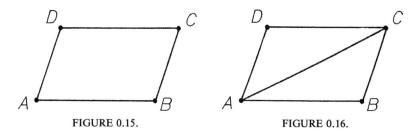

FIGURE 0.15.   FIGURE 0.16.

We end this introductory section by considering areas of triangles and sines of angles. We start by reviewing the fact that *the area of a triangle is one-half of the base times the height*. A *rectangle* is a parallelogram such that intersecting sides lie on perpendicular lines. We take it for granted that the area of a rectangle is the product of the lengths of two intersecting sides. We let area($\triangle ABC$) denote the area of triangle ABC. In the next result and its proof, we let XY denote the distance between points X and Y except when we write "line XY" or "segment XY."

THEOREM 0.13. In the Euclidean plane, let ABC be a triangle. Let F be the point where line AB intersects the line through C perpendicular to line AB. Then we have

$$\text{area}(\triangle ABC) = \tfrac{1}{2}AB \cdot CF$$

(Figures 0.17–0.19).

Proof: There are three cases, depending on the relative positions of A, B, and F.

Case 1: F equals either A or B. By symmetry, we can assume that F equals B, which is equivalent to the condition that $\angle ABC = 90°$ (Figure 0.18). A rectangle of base AB and height CF can be divided into two

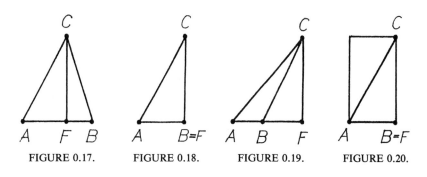

FIGURE 0.17.   FIGURE 0.18.   FIGURE 0.19.   FIGURE 0.20.

triangles that are each congruent to triangle ABC (by the proof of Theorem 0.12) (Figure 0.20). The rectangle has area $AB \cdot CF$, and so triangle ABC has area $\frac{1}{2}AB \cdot CF$.

Case 2: F lies between A and B (Figure 0.17). Applying Case 1 to the right triangles AFC and BFC shows that

$$\begin{aligned}
\text{area}(\triangle ABC) &= \text{area}(\triangle AFC) + \text{area}(\triangle BFC) \\
&= \tfrac{1}{2}AF \cdot CF + \tfrac{1}{2}BF \cdot CF \\
&= \tfrac{1}{2}(AF + BF)CF \\
&= \tfrac{1}{2}AB \cdot CF.
\end{aligned}$$

Case 3: F lies outside segment AB. By symmetry, we can assume that F lies on the side of B opposite A (Figure 0.19). Applying Case 1 to the right triangles AFC and BFC shows that

$$\begin{aligned}
\text{area}(\triangle ABC) &= \text{area}(\triangle AFC) - \text{area}(\triangle BFC) \\
&= \tfrac{1}{2}AF \cdot CF - \tfrac{1}{2}BF \cdot CF \\
&= \tfrac{1}{2}(AF - BF)CF \\
&= \tfrac{1}{2}AB \cdot CF. \quad \square
\end{aligned}$$

Next we review sines of angles. Let θ be an angle such that $0° < \theta < 180°$. Set up x and y axes in the Euclidean plane. Let $P(\theta)$ be the point such that the line segment from the origin $(0,0)$ to $P(\theta)$ has length 1, lies above the x axis, and forms angle θ with the positive x axis. We define the *sine* of θ, $\sin(\theta)$, to be the y coordinate of $P(\theta)$. Figure 0.21a shows the case $0° < \theta < 90°$, Figure 0.21b shows the case $\theta = 90°$, and Figure 0.21c shows the case $90° < \theta < 180°$. In each case, $\sin(\theta)$ is positive because $P(\theta)$ lies above the x axis. Throughout this text, we only consider sines of angles between $0°$ and $180°$, and so these sines are always positive.

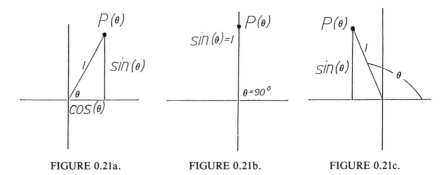

FIGURE 0.21a.   FIGURE 0.21b.   FIGURE 0.21c.

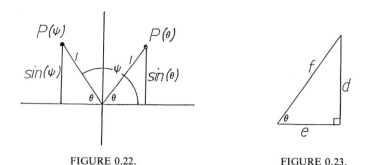

FIGURE 0.22.   FIGURE 0.23.

If we set $\psi = 180° - \theta$, where $0° < \theta < 180°$, then we also have $0° < \psi < 180°$. The segment from the origin to $P(\psi)$ makes angle ψ with the positive x axis, and so it makes angle θ with the negative x axis, since $\theta + \psi = 180°$ (Figure 0.22). Thus, the points $P(\theta)$ and $P(\psi)$ have equal y coordinates, and so we have $\sin(\psi) = \sin(\theta)$. In other words, the relation

$$\sin(180° - \theta) = \sin(\theta) \tag{4}$$

holds for all angles θ between $0°$ and $180°$. (Of course, trigonometric identities show that this equation actually holds for all angles θ.)

In a right triangle, let θ be an angle other than the right angle (Figure 0.23). Let f be the length of the hypotenuse, and let d be the length of the leg opposite θ. Theorem 0.8 implies that $\theta < 90°$ and that the triangles in Figures 0.21a and 0.23 have equal angles θ, $90°$, and $90° - \theta$. Thus, these triangles are similar, and so corresponding sides are proportional (by Property 0.4). This shows that $\sin(\theta)/1 = d/f$, and so we have

$$\sin(\theta) = d/f. \tag{5}$$

In other words, the sine of an angle in a right triangle is the ratio of the length of the opposite side to the length of the hypotenuse.

Similarly, if $0° < \theta < 90°$, we define the *cosine* of θ, $\cos(\theta)$, to be the x coordinate of the point $P(\theta)$ in Figure 0.21a. As in the previous paragraph, if θ is an angle in a right triangle and $\theta \neq 90°$, Property 0.4 of similar triangles implies that

$$\cos(\theta) = e/f, \tag{6}$$

where e is the length of the leg adjacent to θ and f is the length of the hypotenuse (Figure 0.23).

One way to find the area of a triangle is to take one-half the base times the height (by Theorem 0.13). The next result gives another way: the area of a triangle is one-half the product of the lengths of two sides times the sine of the included angle.

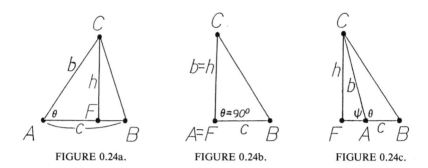

FIGURE 0.24a.   FIGURE 0.24b.   FIGURE 0.24c.

THEOREM 0.14. In the Euclidean plane, let ABC be a triangle. Let b be the distance from A to C, let c be the distance from A to B, and let θ be $\angle BAC$. Then the relation

$$\text{area}(\triangle ABC) = \tfrac{1}{2}bc \sin(\theta) \tag{7}$$

holds (Figures 0.24a–c).

Proof: Let F be the point where line AB intersects the line through C perpendicular to line AB. Let h be the distance from C to F. We claim that

$$h = b \sin(\theta). \tag{8}$$

In fact, if $0° < \theta < 90°$ (as in Figure 0.24a), h is the length of the side opposite θ in a right triangle with hypotenuse of length b, and so Equation 8 follows from Equation 5. If $\theta = 90°$ (as in Figure 0.24b), we have $h = b$ and $\sin(\theta) = 1$ (as in Figure 0.21b), and so Equation 8 holds. If $90° < \theta < 180°$ (as in Figure 0.24c), $\psi = 180° - \theta$ is an angle in a right triangle where the side opposite ψ has length h and the hypotenuse has length b. Then we have $h = b \sin(\psi)$ (by Equation 5) and $\sin(\psi) = \sin(\theta)$ (by Equation 4), and so Equation 8 holds.

We've shown that Equation 8 holds in every case. Triangle ABC has base c and height h, and so we have

$$\text{area}(\triangle ABC) = \tfrac{1}{2}ch \tag{9}$$

(by Theorem 0.13). Substituting Equation 8 into Equation 9 establishes Equation 7. □

We can now derive the Law of Sines by using Theorem 0.14 to compute the area of a triangle in two ways and by setting the two expressions equal to each other. In the next theorem and its proof, let XY be the distance between points X and Y.

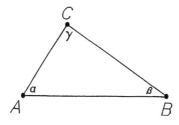

FIGURE 0.25.

THEOREM 0.15 (Law of Sines). In the Euclidean plane, let ABC be a triangle. Let α, β, γ be the angles at A, B, C, respectively. Then the equations

$$\frac{\sin \alpha}{BC} = \frac{\sin \beta}{CA} = \frac{\sin \gamma}{AB}$$ (10)

hold (Figure 0.25).

Proof: Theorem 0.14 shows that

$$\text{area}(\triangle ABC) = \tfrac{1}{2} AB \cdot CA \sin \alpha$$

and

$$\text{area}(\triangle ABC) = \tfrac{1}{2} AB \cdot BC \sin \beta.$$

Equating these two expressions for the area of triangle ABC and cancelling $\tfrac{1}{2} AB$ gives

$$CA \sin \alpha = BC \sin \beta,$$

which yields

$$\frac{\sin \alpha}{BC} = \frac{\sin \beta}{CA}.$$

The rest of Equation 10 follows by symmetry. $\square$

Thus, for any triangle in the Euclidean plane, the ratio of the sine of an angle to the length of the opposite side is the same for all three vertices.

EXERCISES

0.1. In the Euclidean plane, let ABC be a triangle. Let D be a point on line BC that lies on the side of C opposite B (Figure 0.26).

(a) Prove that $\angle ACD = \angle BAC + \angle ABC$. (*Thus, an exterior angle of a triangle equals the sum of the two remote interior angles.*)
(b) Prove that $\angle ABC < \angle ACD$.

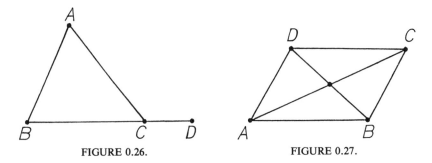

FIGURE 0.26.   FIGURE 0.27.

Exercises 0.2–0.7 use the following terminology. In the Euclidean plane, a *quadrilateral ABCD* consists of four points A, B, C, D and the four segments (called *sides*) from A to B, B to C, C to D, and D to A, where the points A–D don't lie on one line. A *rhombus* is a quadrilateral whose four sides are of equal length. The *diagonals* of quadrilateral $ABCD$ are the segments from A to C and from B to D.

0.2. In the Euclidean plane, prove that a quadrilateral is a parallelogram if and only if the two diagonals have the same midpoint (Figure 0.27).

0.3. In the Euclidean plane, let $ABCD$ be a parallelogram. Let M be the point of intersection of the lines AC and BD. Let l be a line through M that intersects line AB at a point E and that intersects line CD at a point F. Prove that M is the midpoint of E and F. (See Exercise 0.2.)

0.4. In the Euclidean plane, prove that a parallelogram is a rectangle if and only if the two diagonals are of equal length (Figure 0.28).

0.5. In the Euclidean plane, prove that every rhombus is a parallelogram (Figure 0.29).

0.6. In the Euclidean plane, prove that a parallelogram is a rhombus if and only if the diagonals are perpendicular (Figure 0.30).

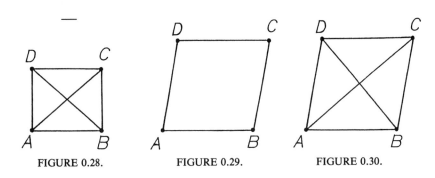

FIGURE 0.28.   FIGURE 0.29.   FIGURE 0.30.

0.7. In the Euclidean plane, let $ABCD$ be a parallelogram. Prove that the following conditions are equivalent:

(i) $ABCD$ is a rhombus.
(ii) $\angle CAB = \angle CAD$, $\angle DBA = \angle DBC$, $\angle ACB = \angle ACD$, $\angle BDA = \angle BDC$.
(iii) One of the equalities in part (ii) holds.

(The statement that conditions (i)–(iii) are equivalent means that each of these conditions implies all of the others. Thus, in any particular case, either all of the conditions hold (Figure 0.30) or none of the conditions holds (Figure 0.27). One convenient way to prove the equivalence of (i)–(iii) is to prove that the conditions imply each other in the cyclic order

$$(i) \Rightarrow (ii) \Rightarrow (iii) \Rightarrow (i).$$

Another way is to prove that one of the conditions holds if and only if each of the other conditions holds.)

0.8. In the Euclidean plane, let A, B, C, D be four points. Assume that the distance from A to B equals the distance from C to D and that the distance from A to D equals the distance from B to C.

(a) Are the lines AB and CD necessarily parallel? If so, prove it. If not, give a counterexample.
(b) If B and D lie on opposite sides of line AC, are the lines AB and CD necessarily parallel? If so, prove it. If not, give a counterexample.

0.9. In the Euclidean plane, let A and C be two points on a line l. Let B and D be two points that don't lie on l. Assume that $\angle BAC = \angle DCA$.

(a) Are the lines AB and CD necessarily parallel? If so, prove it. If not, give a counterexample.
(b) If B and D lie on opposite sides of l, are the lines AB and CD necessarily parallel? If so, prove it. If not, give a counterexample.

0.10. In the Euclidean plane, let ABC be a triangle. Let M be the midpoint of A and B. Prove that the sides AC and BC have equal length if and only if the lines AB and CM are perpendicular. Illustrate this result with a figure.

0.11. An equilateral triangle is a triangle whose three sides have equal length. The distance from a point X to a line l is the distance from X to the point Y where l intersects the line through X perpendicular to l (Figure 0.31). The distance from a point X to a side of a triangle is the distance from X to the line that contains the side of the triangle.

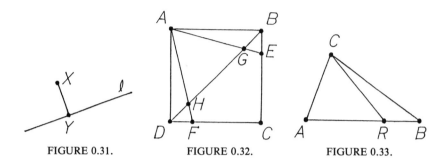

FIGURE 0.31.   FIGURE 0.32.   FIGURE 0.33.

(a) In the Euclidean plane, let P be any point that lies on or inside of an equilateral triangle. Prove that the sum of the distances from P to the three sides of the triangle equals the distance from a vertex of the triangle to the opposite side. (Thus, the sum of the distances from P to the sides of an equilateral triangle is the same for every point P on or inside of the triangle. One possible approach to this exercise is as follows. Let A, B, C be the vertices of the triangle. If P lies inside the triangle, express the area of triangle ABC in terms of the areas of triangles PBC, APC, and ABP, and then apply Theorem 0.13. The case where P lies on a side of the triangle can be handled similarly.)

(b) State and prove results analogous to those in part (a) when P is a point that lies outside an equilateral triangle.

0.12. In the Euclidean plane, let $ABCD$ be consecutive vertices of a square (Figure 0.32). Let r be a positive number. Let E be the point on side BC that lies r times as far from C as from B, and let F be the point on side CD that lies r times as far from C as from D. Let G and H be the points where the line BD intersects the lines AE and AF, respectively. For what value of r is the distance from G to H two-thirds of the distance from B to D?

0.13. In the Euclidean plane, let ABC be a triangle (Figure 0.33). Let R be a point on line AB other than A and B. Prove that

$$\frac{RA}{RB} = \frac{\sin(\angle ABC)\sin(\angle ACR)}{\sin(\angle BAC)\sin(\angle BCR)},$$

where RA and RB are the distances from R to A and B. (*Hint:* Remember to take into account the fact that R may or may not lie between A and B.)

0.14. In the notation of Exercise 0.13, prove that

$$\frac{RA}{RB} = \frac{AC \cdot \sin(\angle ACR)}{BC \cdot \sin(\angle BCR)},$$

where XY denotes the distance between points X and Y.

0.15. In the Euclidean plane, let ABC be a triangle. Let D be a point on line AB other than A, and let E be a point on line AC other than A. If A lies r times as far from D as from B, and if A lies s times as far from E as from C, prove that the area of triangle ADE is rs times the area of triangle ABC.

0.16. In the Euclidean plane, let quadrilaterals and their sides be defined as before Exercise 0.2. Let $A_1A_2A_3A_4$ be a quadrilateral such that A_2 and A_4 lie on opposite sides of line A_1A_3 and such that A_1 and A_3 lie on opposite sides of line A_2A_4. Let r be a positive number. For $i = 1, 2, 3, 4$, let B_i be the point on the segment with endpoints A_i and A_{i+1} that lies r times as far from A_i as from A_{i+1}, where we set $A_5 = A_1$. Prove that the area enclosed by the sides of quadilateral $B_1B_2B_3B_4$ is $(r^2 + 1)/(r + 1)^2$ times the area enclosed by the sides of quadrilateral $A_1A_2A_3A_4$. (See Exercise 0.15.)

Section 1.

Division Ratios

We now add a new idea to the familiar concepts reviewed in Section 0. The idea is to describe the relative positions of three points on a line by a single number, their division ratio. Division ratios play a key role in all our later studies.

Throughout this book, when we say that certain objects are *distinct*, we mean that no two of the objects are equal. Whenever we say "let $A_1, ..., A_n$ be n points" or "let $l_1, ..., l_n$ be n lines," we mean that these points or lines are distinct. For example, if we say "let A, B, C be three points," we mean that no two of these points are equal, and so we have $A \neq B$, $A \neq C$, and $B \neq C$.

A line l in the Euclidean plane has two ends. We pick one of these ends to consider positive and the other to consider negative. Let A and B be two points on l. We define the *directed distance* $\overline{AB}$ *from A to B* by

$$\overline{AB} = \pm d, \tag{1}$$

where d is the distance from A to B, and where we use the plus sign if motion from A to B is toward the positive end of l (Figure 1.1a) and we use the negative sign if motion from A to B is toward the negative end of l

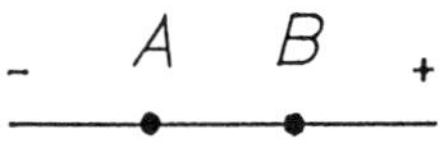

FIGURE 1.1a. 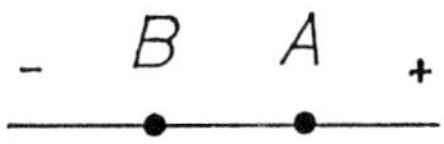   FIGURE 1.1b.

(Figure 1.1b). In particular, Equation 1 implies that

$$|\overline{AB}| = d, \tag{2}$$

and so the absolute value of the directed distance $\overline{AB}$ is the distance from A to B. The sign of $\overline{AB}$ shows the direction of motion from A to B. Accordingly, the sign of $\overline{AB}$ depends on which end of l we consider positive, as part (ii) of the next result illustrates. Part (i) of this result shows that we can specify the locations of points on a line by their directed distances from a fixed point. As noted in Section 0, we use the phrase "in the Euclidean plane" in theorems because we move beyond the Euclidean plane to Euclidean space and the extended plane from Chapter III on.

THEOREM 1.1. In the Euclidean plane, let A, B, C be three points on a line l.

(i) If we choose a positive end of l for directed distances, then we have $\overline{CA} \neq \overline{CB}$.

(ii) If we interchange the positive and negative ends of l, then the sign of $\overline{AB}$ is reversed.

Proof:

(i) If we had $\overline{CA} = \overline{CB}$, then A and B would lie the same distance away from C in the same direction along l. This would imply that $A = B$, contradicting the assumption that A and B are distinct. Thus, we have $\overline{CA} \neq \overline{CB}$.

(ii) Interchanging the positive and negative ends of l changes motion toward the positive end of l into motion toward the negative end, and vice versa. Thus, $\overline{AB}$ changes sign when the ends of l are relabeled. Relabeling the ends of l doesn't affect the distance from A to B, which equals the absolute value of $\overline{AB}$ (by Equation 2). Hence, relabeling sends $\overline{AB}$ to its negative. $\square$

Theorem 1.1(ii) shows that the sign of $\overline{AB}$ depends on the choice of a positive end on l. This is awkward, since it means that the directed distance between two points depends on something other than the points themselves. This problem disappears, however, if we consider a ratio of two directed distances on the same line. Specifically, we consider the ratio of the directed distances from one point on a line to two others.

THEOREM 1.2. In the Euclidean plane, let A, B, C be three points on a line l.

(i) Then the quotient $\overline{CA}/\overline{CB}$ is a real number other than 0 and 1.

(ii) The value of $\overline{CA}/\overline{CB}$ doesn't depend on the choice of a positive end of l.

Proof:

(i) Since $A \neq C$ and $B \neq C$, we have $\overline{CA} \neq 0$ and $\overline{CB} \neq 0$, and so $\overline{CA}/\overline{CB}$ is a nonzero real number. Since $A \neq B$, we have $\overline{CA} \neq \overline{CB}$ (by Theorem 1.1i), which shows that $\overline{CA}/\overline{CB} \neq 1$.

(ii) If the positive and negative ends of l are interchanged, Theorem 1.1(ii) shows that $\overline{CA}/\overline{CB}$ is replaced by $(-\overline{CA})/(-\overline{CB}) = \overline{CA}/\overline{CB}$, and so the value of $\overline{CA}/\overline{CB}$ remains unchanged. □

If A, B, C are three points on a line, Theorem 1.2 shows that the value of the quotient $\overline{CA}/\overline{CB}$ depends only on the points A, B, C. This quotient plays a key role in our work, and we call it the division ratio of A, B by C.

DEFINITION 1.3. In the Euclidean plane, let A, B, C be three points on a line. The *division ratio of A, B by C* is the value of the quotient $\overline{CA}/\overline{CB}$. We also say that C divides A, B in ratio $\overline{CA}/\overline{CB}$. □

A division ratio is the quotient of the directed distances from a point on a line to two other points. The next result states that the sign of a division ratio shows whether the first point lies between the other two.

THEOREM 1.4. In the Euclidean plane, let A, B, C be three points on a line. The division ratio $\overline{CA}/\overline{CB}$ is negative if and only if C lies between A and B.

Proof: The quotient $\overline{CA}/\overline{CB}$ is negative if and only if $\overline{CA}$ and $\overline{CB}$ have opposite signs. This happens if and only if exactly one of the points A and B lies farther toward the positive end than C does. This occurs if and only if C lies between A and B. □

To illustrate Theorem 1.4, Figure 1.2a shows an example where $\overline{CA}$ is positive and $\overline{CB}$ is negative. Then the division ratio $\overline{CA}/\overline{CB}$ is negative, and C lies between A and B. Figure 1.2b shows an example where $\overline{CA}$ and $\overline{CB}$ are both negative. Here $\overline{CA}/\overline{CB}$ is positive, and C doesn't lie between A and B.

FIGURE 1.2a.   FIGURE 1.2b.

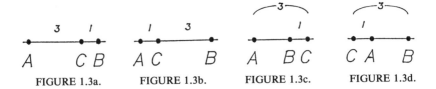

FIGURE 1.3a.   FIGURE 1.3b.   FIGURE 1.3c.   FIGURE 1.3d.

Equation 2 implies that the absolute value of the division ratio $\overline{CA}/\overline{CB}$ is the ratio of the distances from C to A and B. By Theorem 1.4, the sign of the division ratio shows whether C lies between A and B. For example, if C lies between A and B three times as far from A as from B, then $\overline{CA}/\overline{CB} = -3$ (Figure 1.3a). If C lies between A and B three times as far from B as from A, then $\overline{CA}/\overline{CB} = -1/3$ (Figure 1.3b). If C doesn't lie between A and B and is three times as far from A as from B, then $\overline{CA}/\overline{CB} = 3$ (Figure 1.3c). If C doesn't lie between A and B and is three times as far from B as from A, then $\overline{CA}/\overline{CB} = 1/3$.

We can use division ratios to characterize midpoints. Specifically, the midpoint of two points is determined by the property that it divides them in ratio -1.

THEOREM 1.5. In the Euclidean plane, let A, B, C be three points on a line. Then C is the midpoint of A and B if and only if $\overline{CA}/\overline{CB} = -1$ (Figure 1.4).

Proof: First assume that C is the midpoint of A and B. Then C is equidistant from A and B, and so we have $|\overline{CA}| = |\overline{CB}|$ (by Equation 2). It follows that $|\overline{CA}/\overline{CB}| = |\overline{CA}|/|\overline{CB}| = 1$. Moreover, $\overline{CA}/\overline{CB}$ is negative (by Theorem 1.4), since the midpoint C of A and B lies between A and B. Hence, we have $\overline{CA}/\overline{CB} = -1$.

Conversely, assume that $\overline{CA}/\overline{CB} = -1$. It follows that $|\overline{CA}|/|\overline{CB}| = |\overline{CA}/\overline{CB}| = 1$, and so we have $|\overline{CA}| = |\overline{CB}|$. Then C is equidistant from A and B (by Equation 2); so C is the midpoint of A and B. □

In order to work with division ratios, it is often convenient to express them in terms of coordinates on a line. The next definition states exactly what we mean by coordinates on a line.

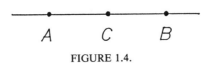

FIGURE 1.4.

DEFINITION 1.6. In the Euclidean plane, a *coordinate system* on a line *l* is an assignment of a real number x to each point X on *l* such that $|b - a|$ is the distance from A to B for every pair of points A and B on *l*. $\square$

This definition says that coordinates on a line are scaled by distance. The x coordinates of points on the x axis provide an example of Definition 1.6, since the distance between the points $(a, 0)$ and $(b, 0)$ is $|b - a|$. Likewise, the y coordinates of points on the y axis give an example of Definition 1.6, since the distance between the points $(0, a)$ and $(0, b)$ is $|b - a|$. In general, a coordinate system on a line *l* makes *l* look the x or y axis.

To set up a coordinate system on a line *l*, we assign coordinate 0 to any point Q on *l*, and we pick one end of *l* to be positive and the other to be negative (Figure 1.5). Once these choices are made, there is exactly one way to assign coordinates: each real number $r > 0$ is assigned to a point r units from Q toward the positive end of *l*, and each real number $r < 0$ is assigned to the point $|r|$ units from Q toward the negative end of *l*. *Thus, a coordinate system on a line l matches up the points on l with the real numbers.*

The next result shows that we can compute directed distances by subtracting coordinates. When we say that $X \to x$ is a coordinate system on a line *l*, we mean that points A, B, ..., Z on *l* have coordinates $a, b, ..., z$, respectively, in the sense of Definition 1.6.

THEOREM 1.7. In the Euclidean plane, let $X \to x$ be a coordinate system on a line *l*. Assume that the end of *l* with positive coordinates is chosen as the positive end for directed distances. Then the relation

$$\overline{AB} = b - a \tag{3}$$

holds for every pair of points A and B on *l*.

Proof: Equation 2 and Definition 1.6 show that $|\overline{AB}|$ and $|b - a|$ both equal the distance from A to B, and so we have $|\overline{AB}| = |b - a|$. By assumption, motion from A to B goes toward the positive end of *l* if and only if a is less than b. Thus, $\overline{AB}$ is positive if and only if $b - a$ is positive. Hence, both sides of Equation 3 have the same sign as well as the same absolute value, and so Equation 3 holds. $\square$

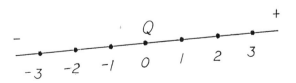

FIGURE 1.5.

Theorems 1.2 and 1.7 imply that we can use any coordinate system on a line to evaluate division ratios.

THEOREM 1.8. In the Euclidean plane, let $X \to x$ be a coordinate system on a line l. Let A, B, C be three points on l. Then we have

$$\frac{\overline{CA}}{\overline{CB}} = \frac{a - c}{b - c}. \tag{4}$$

Proof: By Theorem 1.2, the value of the division ratio $\overline{CA}/\overline{CB}$ doesn't depend on the end of l we consider positive for directed distances. Thus, we're freee to choose the end of l that has positive coordinates to be the positive end for directed distances. Then Theorem 1.7 shows that $\overline{CA} = a - c$ and $\overline{CB} = b - c$, and so Equation 4 holds. $\square$

Theorem 1.5 shows that the midpoint of two points is characterized by the fact that it divides them in ratio -1. We use Theorem 1.8 in the next result to show that every point on a line is characterized by the ratio in which it divides two fixed points.

THEOREM 1.9. In the Euclidean plane, let A and B be two points on a line l. Then the map

$$C \to \frac{\overline{CA}}{\overline{CB}}$$

matches up the points C on l other than A and B with the real numbers other than 0 and 1.

Proof: If C is a point on l other than A and B, then $\overline{CA}/\overline{CB}$ is a real number other than 0 and 1, by Theorem 1.2. Conversely, let r be a real number other than 0 and 1. We must prove that there is a unique point C on l other than A and B such that

$$\overline{CA}/\overline{CB} = r. \tag{5}$$

Consider any coordinate system $X \to x$ on l. Equation 5 implies that

$$\frac{a - c}{b - c} = r, \tag{6}$$

by Theorem 1.8, which in turn implies that

$$a - c = r(b - c).$$

Since $r \neq 1$, the following equations are all equivalent:

$$a - c = r(b - c); \tag{7}$$

$$a - c = rb - rc;$$

$$rc - c = rb - a;$$

$$(r - 1)c = rb - a;$$

$$c = \frac{rb - a}{r - 1}. \tag{8}$$

Equation 8 shows that there is a unique real number c satisfying Equation 7. We have $c \neq a$, since replacing c by a in Equation 7 gives $0 = r(b - a)$, contradicting the facts that $r \neq 0$ and $a \neq b$ (since $A \neq B$). Then Equation 7 shows that $r(b - c) = a - c \neq 0$, and so we have $c \neq b$. There is a unique point C on l with coordinate c. We have $C \neq A$ and $C \neq B$, since we have $c \neq a$ and $c \neq b$. Equation 6 follows from Equation 7, and Equation 5 follows from Equation 6 (by Theorem 1.8). Thus, there is a unique point C on l other than A and B such that $\overline{CA}/\overline{CB} = r$, as desired. $\square$

Theorem 1.9 shows that a point on a line is uniquely determined by the ratio in which it divides two fixed points. That is, given two points A and B on a line l, there is a unique point C on l such that the division ratio $\overline{CA}/\overline{CB}$ takes a particular value. As C varies once over the points of l other than A and B, the division ratio $\overline{CA}/\overline{CB}$ varies once over the real numbers other than 0 and 1. For example, let A and B be the points $(-1, 0)$ and $(5, 0)$ on the x axis, and let $C = (c, 0)$ be a point on the x axis other than A and B. The x coordinates of points on the x axis give a coordinate system in the sense of Definition 1.6. Thus, Theorem 1.8 shows that we can use x coordinates as follows to compute division ratios:

$$\frac{\overline{CA}}{\overline{CB}} = \frac{-1 - c}{5 - c}. \tag{9}$$

Figure 1.6 shows various points C other than A and B on the x axis. Each point C is labeled beneath the x axis with its x coordinate c and is labeled above the x axis with the value of the division ratio $\overline{CA}/\overline{CB}$ computed

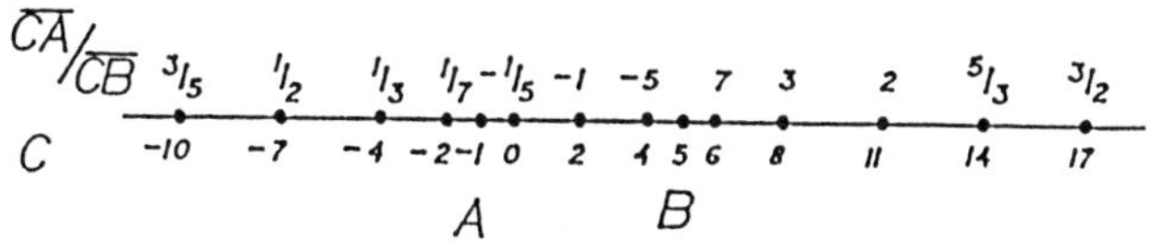

FIGURE 1.6.

from Equation 9. For example, setting $c = -10$ in Equation 9 gives $\overline{CA}/\overline{CB} = 9/15 = 3/5$, as is shown by the leftmost point in Figure 1.6. As C moves from the far left of the x axis toward A, the division ratio decreases from values slightly less than 1 through values approaching 0 from above. As C moves from A toward B, the division ratio decreases from values slightly less than 0 through arbitrarily negative values. As C moves from B toward the far right of the x axis, the division ratio decreases from arbitrarily large positive values through values approaching 1 from above. Thus, the division ratio takes each value except 0 and 1 exactly once, as Theorem 1.9 states. The correspondence $C \to \overline{CA}/\overline{CB}$ is also illustrated by Figure 1.7, which shows the graph of $y = (-1 - c)/(5 - c)$ for all $c \neq -1, 5$. Each horizontal line except $y = 0$ and $y = 1$ crosses the graph exactly once, again showing that the division ratio assumes each value except 0 and 1 exactly once.

Figures 1.6 and 1.7 illustrate several other general properties of division ratios. For example, the division ratio is negative exactly when C is between A and B, as Theorem 1.4 states. When C approaches A, the directed distance $\overline{CA}$ approaches 0, and so the division ratio $\overline{CA}/\overline{CB}$ approaches 0. When C approaches B, the directed distance $\overline{CB}$ approaches 0, and so the division ratio CA/CB approaches ∞. (We don't distinguish between $+\infty$ and $-\infty$ anywhere in the text.) Dividing the numerator and denominator of the right side of Equation 9 by $-c$ gives

$$\frac{\overline{CA}}{\overline{CB}} = \frac{-1 - c}{5 - c} = \frac{1/c + 1}{-5/c + 1},$$

and so $\overline{CA}/\overline{CB}$ approaches 1 as c approaches ∞ (since $1/c$ and $-5/c$ approach 0 as c approaches ∞). Thus, as Figures 1.6 and 1.7 show, $\overline{CA}/\overline{CB}$ approaches 1 as C moves far away from A and B toward either end of the x axis.

Like the proof of Theorem 1.9, the proof of the next result illustrates the fact that it's often convenient to use coordinate systems when working with directed distances.

THEOREM 1.10. In the Euclidean plane, let A, B, C be three points on a line l. Choose a positive end of l for directed distances. Then we have

$$\overline{BA} = -\overline{AB} \tag{10}$$

and

$$\overline{AC} = \overline{AB} + \overline{BC}. \tag{11}$$

Proof: Choose a coordinate system $X \to x$ on l such that positive coordinates are assigned to points on the end of l considered positive for

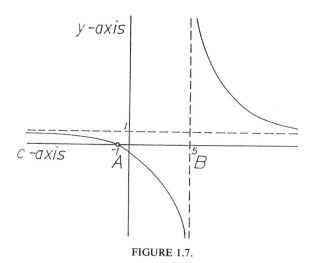

FIGURE 1.7.

directed distances. Then Theorem 1.7 implies that

$$\overline{BA} = a - b = -(b - a) = -\overline{AB}$$

and

$$\overline{AC} = c - a = (c - b) + (b - a) = \overline{AB} + \overline{BC}. \quad \square$$

Theorem 1.10 also follows geometrically if we think of the directed distance $\overline{AB}$ as showing both the distance and the direction of motion from A to B. Then Equation 10 states that motion from B to A covers the same distance in the opposite direction as motion from A to B. Equation 11 states that we can travel from A to C by traveling first from A to B and then from B to C. The signs of the directed distances ensure that Equation 11 holds regardless of the relative positions of the points A, B, C: if B lies between A and C, Equation 11 corresponds to the fact that $|\overline{AC}| = |\overline{AB}| + |\overline{BC}|$ (Figure 1.8a); if C lies between A and B, Equation 11 corresponds to the fact that $|\overline{AC}| = |\overline{AB}| - |\overline{BC}|$ (Figure 1.8b); and if A lies between B and C, Equation 11 corresponds to the fact that $|\overline{AC}| = -|\overline{AB}| + |\overline{BC}|$.

By Theorem 1.2(ii), it's unnecessary to specify a positive end for directed distances on a line where we consider division ratios. On the other hand, by Theorem 1.1(ii), we need to choose a positive end on a line where we

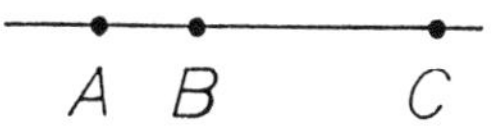

A B   C   A C   B   B A   C

FIGURE 1.8a. 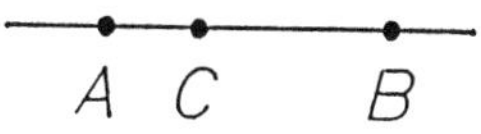   FIGURE 1.8b.    FIGURE 1.8c.

consider directed distances outside of division ratios. Thus, for example, we didn't specify the choice of a positive end on the line in Theorem 1.9, but we did choose a positive end on the line in Theorem 1.10.

The next result shows how division ratios characterize the points where intersecting lines are crossed by parallel lines. This characterization is one of the key ways we use division ratios. The result is essentially just a restatement of the fact that corresponding sides of similar triangles are proportional (by Property 0.4). As we do throughout the text, we let *XY be the unique line through two points X and Y* unless we state otherwise.

THEOREM 1.11. In the Euclidean plane, let *l* and *m* be two lines on a point *C*. Let *A* and *D* be two points on *l* other than *C*, and let *B* and *E* be two points on *m* other than *C*. Then the lines *AB* and *DE* are parallel if and only if the equality

$$\overline{CA}/\overline{CD} = \overline{CB}/\overline{CE} \tag{12}$$

holds (Figures 1.9a and b).

Proof: *B* lies on *m* and doesn't equal the point *C* where *l* intersects *m*, and so *B* doesn't lie on *l*. Thus, we have $A \neq B$ (since *l* contains *A* but not *B*), and so line *AB* exists. Moreover, we have $AB \neq l$ (since *B* lies on *AB* but not *l*), and so *AB* and *l* intersect at the unique point *A*. By symmetry, line *DE* exists and intersects *l* at the unique point *D*. In particular, *AB* and *DE* are distinct lines, since they intersect *l* at distinct points *A* and *D*.

First, assume that the lines *AB* and *DE* are parallel; we must prove that Equation 12 holds. Neither *AB* nor *DE* contains *C*, by the previous paragraph. Thus, either *C* lies on the same side of *AB* and *DE* (as in Figure 1.9a) or *C* lies between *AB* and *DE* (as in Figure 1.9b). We consider these two cases separately. (Of course it makes sense to consider whether *C* lies on the same side of *AB* and *DE* only because we're assuming that *AB* and *DE* are parallel.)

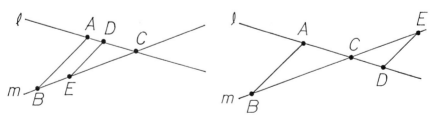

FIGURE 1.9a.   FIGURE 1.9b.

Case 1: C lies on the same side of AB and DE (Figure 1.9a). The assumption that AB and DE are parallel implies that the corresponding angles where they intersect l are equal and that the corresponding angles where they intersect m are equal (by Property 0.5). Thus, we have $\angle BAC = \angle EDC$ and $\angle ABC = \angle DEC$. Hence, triangles ABC and DEC have three pairs of equal angles (since they also have the same angle at C). Then ABC and DEC are similar triangles, and so corresponding sides are proportional (by Property 0.4). Thus, we have

$$|\overline{CA}/\overline{CD}| = |\overline{CB}/\overline{CE}|. \tag{13}$$

C doesn't lie between the lines AB and DE in this case, and so C doesn't lie between A and D on l, and it doesn't lie between B and E on m. Thus, both of the division ratios $\overline{CA}/\overline{CD}$ and $\overline{CB}/\overline{CE}$ are positive in this case (by Theorem 1.4). Hence, these division ratios are equal, since they have the same absolute value (by Equation 13) as well as the same sign.

Case 2: C lies between AB and DE (Figure 1.9b). The assumption that AB and DE are parallel implies that the alternate interior angles where they intersect l are equal and that the alternate interior angles where they intersect m are equal (by Theorem 0.7). Thus, we have $\angle BAC = \angle EDC$ and $\angle ABC = \angle DEC$. Hence, triangles ABC and DEC have three pairs of equal angles (since they also have the same angle at C, by either Theorem 0.6 or Theorem 0.8). Then triangles ABC and DEC are similar, and so corresponding sides are proportional (by Property 0.4). Thus, we have

$$|\overline{CA}/\overline{CD}| = |\overline{CB}/\overline{CE}|. \tag{14}$$

C lies between the lines AB and DE in this case, and so it lies between A and D on l and between B and E on m. Thus, both of the division ratios $\overline{CA}/\overline{CD}$ and $\overline{CB}/\overline{CE}$ are negative in this case (by Theorem 1.4). Together with Equation 14, this shows that Equation 12 holds.

We've proved that Equation 12 holds whenever AB and DE are parallel. Conversely, assume that Equation 12 holds; we must show that AB and DE are parallel. Let n be the line on D parallel to AB (Figure 1.10). The parallel lines n and AB are distinct (since D lies on n but not AB, by the first paragraph of the proof), and so they have no points in common. Thus, n doesn't contain B. Neither l nor m is parallel to AB, and AB is parallel to n, and so neither l nor m is parallel to n (by Theorem 0.11). Thus, n intersects l at the unique point D and intersects m at a unique point F. We have $F \neq B$ (since n doesn't contain B) and $F \neq C$ (since n doesn't contain C because n intersects l at the unique point $D \neq C$). In short, F is a point on m other than B and C such that AB and DF are parallel lines.

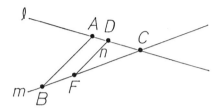

FIGURE 1.10.

Then we have

$$\overline{CA}/\overline{CD} = \overline{CB}/\overline{CF}, \tag{15}$$

by the preceding paragraphs of the proof. Equation 12 also holds, by assumption. Combining Equations 12 and 15 shows that

$$\overline{CB}/\overline{CE} = \overline{CB}/\overline{CF}. \tag{16}$$

If we choose a positive end on m for directed distances, Equation 16 implies that $\overline{CE} = \overline{CF}$. It follows that $E = F$ (by Theorem 1.1i). Therefore, the line $DE = DF = n$ is parallel to AB, as desired. □

In Theorem 1.11, the lines AB and DE are parallel if and only if the ratio in which C divides A, D equals the ratio in which C divides B, E (Figures 1.11a and b). By using the properties of directed distances in Theorem 1.10, we can restate Theorem 1.11 in terms of the ratio in which D divides A, C and the ratio in which E divides B, C. Once again we obtain a result that is essentially just a restatement of the fact that similar triangles have proportional sides.

THEOREM 1.12. In the Euclidean plane, let l and m be two lines on a point C. Let A and D be two points on l other than C, and let B and E be two points on m other than C. Then the lines AB and DE are parallel if and

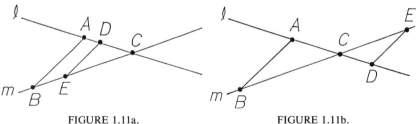

FIGURE 1.11a.   FIGURE 1.11b.

only if the equality

$$\overline{DA}/\overline{DC} = \overline{EB}/\overline{EC}$$

holds (Figures 1.11a and b).

Proof: Choose positive ends for directed distances on l and m. Then the following conditions are equivalent:

$$\overline{CA}/\overline{CD} = \overline{CB}/\overline{CE}; \tag{17}$$

$$\frac{\overline{CD} + \overline{DA}}{\overline{CD}} = \frac{\overline{CE} + \overline{EB}}{\overline{CE}} \quad \text{(by Equation 11)};$$

$$1 - \overline{DA}/\overline{DC} = 1 - \overline{EB}/\overline{EC} \quad \text{(by Equation 10)};$$

$$\overline{DA}/\overline{DC} = \overline{EB}/\overline{EC}. \tag{18}$$

By Theorem 1.11, the lines AB and DE are parallel if and only if Equation 17 holds. Since Equations 17 and 18 are equivalent, the theorem follows. □

As an example of how Theorems 1.11 and 1.12 can be used, we note that Theorem 1.12 immediately implies the following result: The midpoints of two sides of a triangle determine a line parallel to the third side.

THEOREM 1.13. In the Euclidean plane, let ABC be a triangle. Let D be the midpoint of A and C, and let E be the midpoint of B and C. Then the lines AB and DE are parallel (Figure 1.12).

Proof: Since D is the midpoint of A and C, and since E is the midpoint of B and C, we have

$$\overline{DA}/\overline{DC} = -1 = \overline{EB}/\overline{EC}$$

(by Theorem 1.5). Hence, the lines AB and DE are parallel, by Theorem 1.12. □

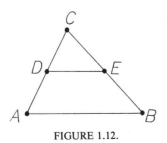

FIGURE 1.12.

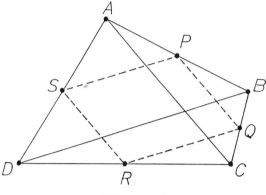

FIGURE 1.13.

The next result is an interesting consequence of the previous theorem. As mentioned in Section 0, we use the following conventions throughout the book: We consider a line to be parallel to itself, and we require opposite sides of parallelogram to be distinct parallel lines.

THEOREM 1.14. In the Euclidean plane, let A, B, C, D be four points such that the lines AC and BD aren't parallel. Let P be the midpoint of A and B, let Q be the midpoint of B and C, let R be the midpoint of C and D, and let S be the midpoint of D and A. Then P, Q, R, S are the vertices of a parallelogram, and the sides of the parallelogram are parallel to AC and BD (Figure 1.13).

Proof: PQ and AC are parallel lines; this holds by Theorem 1.13 if B doesn't lie on AC, and it holds because $PQ = AC$ if B lies on AC. By symmetry, RS is also parallel to AC, and so PQ, RS, AC are all parallel (by Theorem 0.11). Then QR, PS, BD are all also parallel, by symmetry.

Since PQ and QR are parallel to AC and BD, respectively, and since AC and BD aren't parallel, neither are PQ and QR (by Theorem 0.11). Thus, the points P, Q, R, S don't lie on one line. Accordingly, the parallel lines PQ and RS are distinct, and the parallel lines QR and PS are distinct. Hence, $PQRS$ is a parallelogram. We've also seen that its sides are parallel to AC and BD. □

In Figure 1.13, quadrilateral $ABCD$ is the figure that has four vertices A, B, C, D and four sides AB, BC, CD, DA. P, Q, R, S are the midpoints of the four sides. AC and BD are the two diagonals of the quadrilateral. Thus, we can restate Theorem 1.14 as follows:

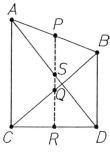

FIGURE 1.14.

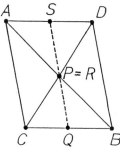

FIGURE 1.15.

In the Euclidean plane, consider a quadrilateral whose vertices are distinct and whose diagonals aren't parallel. Then the midpoints of the four sides of the quadrilateral are the vertices of a parallelogram, and the sides of the parallelogram are parallel to the diagonals of the quadrilateral.

It's necessary to assume that the diagonals AC and BD aren't parallel in order for $PQRS$ to be a parallelogram, as Figures 1.14 and 1.15 show. On the other hand, the assumption that the diagonals AC and BD aren't parallel allows the following possibilities: the sides of quadrilateral $ABCD$ can cross (Figure 1.16); one of the vertices of the quadrilateral can lie inside the triangle determined by the other three (Figure 1.17); and three of the vertices of the quadrilateral can lie on a line (Figure 1.18). In each of these cases, so long as A, B, C, D are distinct points and AC isn't parallel to BD, Theorem 1.14 shows that $PQRS$ is a parallelogram.

We end this chapter with a companion result to Theorem 1.11. Theorem 1.11 uses division ratios to characterize the points where two parallel lines cross two intersecting lines. The next theorem uses directed distances to characterize the points where two parallel lines cross two parallel lines. Just as Theorem 1.11 is based on the fact that similar triangles have proportional sides, the next theorem is based on the fact that opposite sides of a parallelogram have equal length (by Theorem 0.12). We begin with a preliminary definition.

DEFINITION 1.15. In the Euclidean plane, let l and m be parallel lines on which positive ends are chosen for directed distances. We say that the positive ends *correspond* if the positive ends match up when one line is moved parallel to itself until it coincides with the other line. □

For example, the lines in Figure 1.19a have corresponding positive ends, while those in Figure 1.19b don't. We can now state the companion result to Theorem 1.11.

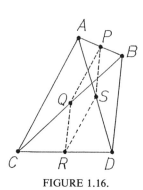

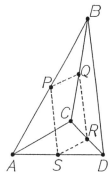

FIGURE 1.16.

FIGURE 1.17.

THEOREM 1.16. In the Euclidean plane, let l and m be two parallel lines. Let A and D be two points on l, and let B and E be two points on m. Choose corresponding positive ends on l and m for directed distances. Then the lines AB and DE are parallel if and only if $\overline{AD} = \overline{BE}$ (Figure 1.20).

Proof: First assume that the lines AB and DE are parallel. Then $ADEB$ is a parallelogram, and so its opposite sides are equal (by Theorem 0.12). Thus, we have $|\overline{AD}| = |\overline{BE}|$ (by Equation 2). Since A, D, E, B are consecutive vertices of a parallelogram, D and E lie on the same side of line AB. Thus, motions from A to D and from B to E are in the same direction, and so $\overline{AD}$ and $\overline{BE}$ have the same sign (since the positive ends of l and m correspond). Hence, we have $\overline{AD} = \overline{BE}$, since both directed distances have the same absolute value and the same sign.

Conversely, assume that $\overline{AD} = \overline{BE}$. B doesn't lie on l (since l and m are distinct parallel lines), and so line AB exists and doesn't equal l. Thus, AB and l intersect at the unique point A. Let n be the line on D parallel to AB (Figure 1.21). AB and n are parallel, l and m are parallel, and AB and l aren't parallel; thus, neither AB nor n is parallel to either l or m

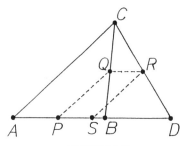

FIGURE 1.18.

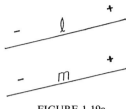

FIGURE 1.19a.

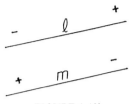

FIGURE 1.19b.

(by Theorem 0.11). Hence, AB intersects m at the unique point B, and n intersects m at a unique point F. The parallel lines AB and n are distinct (since AB intersects l at the unique point A, while n contains D). Thus, the points B and F are distinct (since they lie on the distinct parallel lines AB and n). In short, B and F are two points on m such that the lines AB and DF are parallel. Then the first paragraph of the proof shows that $\overline{AD} = \overline{BF}$. The equality $\overline{AD} = \overline{BE}$ also holds, by assumption, and so we have $\overline{BE} = \overline{BF}$. Then E equals F (by Theorem 1.1i), and so $DE = DF = n$ is parallel to AB, as desired. □

EXERCISES

1.1. In the Euclidean plane, let A, B, C, D be four points on a line l. Choose a positive end of l for directed distances.

(a) Prove that $\overline{AC} + \overline{BD} = \overline{AD} + \overline{BC}$. (See Theorem 1.7.)
(b) Prove that $\overline{AC} = \overline{DB}$ if and only if $\overline{AD} = \overline{CB}$.

1.2. In the Euclidean plane, let A and B be two points on a line l. Let $X \to x$ be a coordinate system on l.

(a) Use Theorems 1.5 and 1.8 to prove that the midpoint of A and B has coordinate $\frac{1}{2}(a + b)$.
(b) Use Definition 1.6 to check that the point on l with coordinate $\frac{1}{2}(a + b)$ is equidistant from A and B.

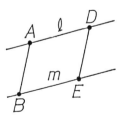

FIGURE 1.20.

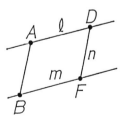

FIGURE 1.21.

1.3. In the Euclidean plane, let A, B, C, D be four points on a line l. Let M be the midpoint of A and B, and let N be the midpoint of C and D. Choose a positive end of l for directed distances.

(a) Prove that $M = N$ if and only if $\overline{AC} = \overline{DB}$.
(b) If $M \neq N$, prove that $\overline{MN} = \frac{1}{2}(\overline{AC} + \overline{BD})$.

(See Exercise 1.2 and Theorem 1.7.)

1.4. In the Euclidean plane, choose a positive end on a line l for directed distances. If A, B, C, D are four points on l, use Theorem 1.7 to prove that

$$\overline{AB} \cdot \overline{CD} + \overline{AC} \cdot \overline{DB} + \overline{AD} \cdot \overline{BC} = 0.$$

1.5. In the Euclidean plane, let B, C, D be three points on a line l. Let F be a point on l that isn't necessarily distinct from B, C, D. Choose a positive end of l for directed distances. Use Theorem 1.7 to prove that

$$BF^2 \cdot \overline{CD} + CF^2 \cdot \overline{DB} + DF^2 \cdot \overline{BC} + \overline{CD} \cdot \overline{DB} \cdot \overline{BC} = 0,$$

where XY denotes the distance between points X and Y. (As always, we take the distance from a point to itself to be zero.)

1.6. Consider the following result:

Stewart's Theorem. In the Euclidean plane, let A be any point, and let l be any line. Let B, C, D be three points on l. Choose a positive end of l for directed distances. Then we have

$$AB^2 \cdot \overline{CD} + AC^2 \cdot \overline{DB} + AD^2 \cdot \overline{BC} + \overline{CD} \cdot \overline{DB} \cdot \overline{BC} = 0,$$

where XY denotes the distance between any points X and Y.

Let F be the point where l intersects the line through A perpendicular to l. Prove Stewart's Theorem by combining the Pythagorean Theorem 0.9 with Exercise 1.5.

1.7. In the Euclidean plane, let B and C be two points, and let A' be their midpoint. For any point A in the plane, prove that

$$AA'^2 = \tfrac{1}{2}AB^2 + \tfrac{1}{2}AC^2 - \tfrac{1}{4}BC^2,$$

where XY denotes the distance between any points X and Y.

(*Hint:* One possible approach is to use Stewart's Theorem from Exercise 1.6. Another possible approach is to adapt the method of proving Stewart's Theorem outlined in Exercises 1.5 and 1.6.)

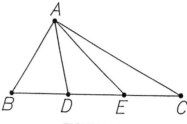

FIGURE 1.22.

1.8. In the Euclidean plane, let ABC be a triangle. Let A' be the midpoint of B and C, let B' be the midpoint of C and A, and let C' be the midpoint of A and B. Use Exercise 1.7 to prove that

$$AA'^2 + BB'^2 + CC'^2 = \tfrac{3}{4}(BC^2 + CA^2 + AB^2),$$

where XY denotes the distance between any points X and Y.

1.9. In the Euclidean plane, let ABC be a triangle. Let D and E be the two points that divide the segment with endpoints B and C into thirds (Figure 1.22). Prove that

$$AD^2 + AE^2 = AB^2 + AC^2 - \tfrac{4}{9}BC^2,$$

where XY denotes the distance between any two points X and Y.
    (See the hint to Exercise 1.7.)

1.10. In the Euclidean plane, let B and C be two points, and let M be their midpoint. Let k be a real number, and let XY denote the distance between any points X and Y. Let $\mathcal{S}$ be the set of all points A in the plane such that $AB^2 + AC^2 = k$. Use Exercise 1.7 to prove that $\mathcal{S}$ is either the empty set, $\{M\}$, or a circle with center M. (A circle with center M is the set of points at a fixed positive distance from M.)

1.11. Consider the following result (Figure 1.13):

Theorem. In the Euclidean plane, let A, B, C, D be four points. Let P be the midpoint of A and B, and let R be the midpoint of C and D. Then the equality

$$AC^2 + BD^2 + AD^2 + BC^2 = AB^2 + CD^2 + 4PR^2$$

holds, where XY denotes the distance from X to Y.

(a) Use Exercise 1.7 to express PR^2 in terms of AR^2, BR^2, and AB^2.
(b) Use Exercise 1.7 to express AR^2 and BR^2 in terms of AC^2, AD^2, BC^2, BD^2, and CD^2.
(c) Prove the theorem by substituting the expressions in (b) into the expression in (a) and simplifying the result.

1.12. Use the theorem in Exercise 1.11 to prove the following result (Figure 1.13):

Theorem. In the Euclidean plane, let A, B, C, D be four points. Let P be the midpoint of A and B, let Q be the midpoint of B and C, let R be the midpoint of C and D, and let S be the midpoint of D and A. Then the equality

$$AC^2 + BD^2 = 2PR^2 + 2QS^2$$

holds, where XY denotes the distance from X to Y.

1.13. In the Euclidean plane, let A, B, C be three points on a line l. Let the division ratio of A, B by C equal r. Use Theorem 1.10 to express each of the following division ratios in terms of r.

   (a) The division ratio of B, A by C.
   (b) The division ratio of A, C by B.
   (c) The division ratio of C, A by B. (*Hint:* Use part (b).)
   (d) The division ratio of B, C by A.
   (e) The division ratio of C, B by A.

1.14. In the Euclidean plane, let B and C be two points on a line l. Let F be a point on l that isn't necessarily distinct from B and C. Use Theorem 1.7 to prove that

$$2\overline{CF} \cdot \overline{CB} = BC^2 + CF^2 - BF^2,$$

where XY denotes the distance between any points X and Y and where we set $\overline{CF} = 0$ if $C = F$.

1.15. In the Euclidean plane, let ABC be a triangle (Figure 1.23). Let a, b, c be the respective lengths of sides BC, CA, AB. Let F be the point where line BC intersects the line through A perpendicular to line BC. Use the Pythagorean Theorem 0.9 and Exercise 1.14 to prove that

$$2\overline{CF} \cdot \overline{CB} = a^2 + b^2 - c^2,$$

where we set $\overline{CF} = 0$ if $C = F$.

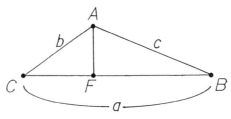

FIGURE 1.23.

(This result is a restatement of the Law of Cosines. Do not use the Law of Cosines in the proof.)

1.16. In the Euclidean plane, let ABC be a triangle (Figure 1.23). Let a, b, c be the respective lengths of sides BC, CA, AB.

(a) Prove that

$$(\text{area } \triangle ABC)^2 = \tfrac{1}{8}(a^2b^2 + a^2c^2 + b^2c^2) - \tfrac{1}{16}(a^4 + b^4 + c^4). \quad (19)$$

(*Hint:* One possible approach is as follows. In the notation of Exercise 1.15, if $\angle ACF \neq 90°$, apply the Pythagorean Theorem 0.9 to triangle AFC, eliminate $\overline{CF}^2$ from this expression by using Exercise 1.15, and combine the result with Theorem 0.13 to deduce Equation 19. Then consider the case when $\angle ACF = 90°$.)

(b) If we set $s = \tfrac{1}{2}(a + b + c)$, prove that

$$\text{area } \triangle ABC = [s(s - a)(s - b)(s - c)]^{1/2}.$$

(This equation is called *Heron's Formula.* It expresses the area of a triangle in terms of the lengths of the sides. The quantity s is called the *semiperimeter* of triangle ABC.)

1.17. In the Euclidean plane, let a, b, c be three parallel lines (Figure 1.24). Let l be a line that intersects a, b, c at points A, B, C, and let l' be a line that intersects a, b, c at points A', B', C'. Prove that $\overline{AB}/\overline{AC} = \overline{A'B'}/\overline{A'C'}$.

1.18. In the Euclidean plane, let A, B, C, D be four points such that the lines AC and BD are parallel. Let P be the midpoint of A and B, let Q be the midpoint of B and C, let R be the midpoint of C and D, and let S be the midpoint of D and A.

(a) Prove that P, Q, R, S lie on a line (Figures 1.14 and 1.15).

(b) Assume that $AC \neq BD$. Prove that $P = R$ if and only if AD is parallel to BC (Figure 1.15).

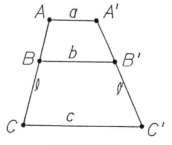

FIGURE 1.24.

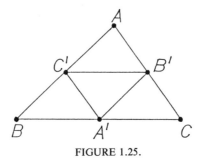

FIGURE 1.25.

1.19. In the Euclidean plane, let ABC be a triangle. Let A' be the midpoint of B and C, let B' be the midpoint of C and A, and let C' be the midpoint of A and B (Figure 1.25).

(a) Prove that $AB'A'C'$, $BC'B'A'$, and $CA'C'B'$ are parallelograms.

(b) Prove that $AB'C'$, $A'BC'$, $A'B'C$, $A'B'C'$, are congruent triangles similar to triangle ABC.

1.20. In the Euclidean plane, let ABC be a triangle (Figure 1.25). Let A' be a point on line BC other than B and C, let B' be a point on line CA other than C and A, and let C' be a point on line AB other than A and B. Assume that $B'C'$ is parallel to BC, $C'A'$ is parallel to CA, and $A'B'$ is parallel to AB. Prove that A', B', C' are the midpoints of the sides of triangle ABC.

1.21. In the Euclidean plane, let $ABCD$ be a parallelogram (Figure 0.30). In each part of this exercise, use the results listed to prove that

$$AC^2 + BD^2 = 2AB^2 + 2BC^2.$$

(a) Exercises 0.2 and 1.7

(b) Exercise 1.12

(c) Exercises 1.15 and 0.2.

Section 2.

Menelaus' Theorem

Having introduced division ratios in the last section, we now use them in Menelaus' Theorem. We consider three points, one on each side of a triangle. We prove that the three points lie on a line if and only if the product of the ratios in which they divide the sides of the triangle is 1.

We start by using Theorem 1.12 to prove one direction of Menelaus' Theorem.

THEOREM 2.1. In the Euclidean plane, let ABC be a triangle. Let A' be a point on BC other than B and C, let B' be a point on CA other than C and A, and let C' be a point on AB other than A and B. If the points A', B', C' lie on a line l, then the equation

$$\frac{\overline{A'B}}{\overline{A'C}} \cdot \frac{\overline{B'C}}{\overline{B'A}} \cdot \frac{\overline{C'A}}{\overline{C'B}} = 1 \tag{1}$$

holds (Figure 2.1).

Proof: A' doesn't equal the point B where BC intersects AB, and so A' doesn't lie on AB. Thus, l doesn't equal AB. Accordingly, since l and AB both contain C', they aren't parallel. By symmetry, l isn't parallel to BC or CA, either.

Let m be the line through B parallel to l (Figure 2.2). Since l isn't parallel to CA, neither is m, and so m intersects CA at a unique point B''. Since l isn't parallel to AB or BC, B'' doesn't equal A or C. Since l is parallel to m but not to BC, the fact that $A' \neq B$ implies that $l \neq m$ and $B' \neq B''$. Thus, the fact that l and m are parallel implies that

$$\frac{\overline{A'B}}{\overline{A'C}} \cdot \frac{\overline{B'C}}{\overline{B'A}} = \frac{\overline{B'B''}}{\overline{B'C}} \cdot \frac{\overline{B'C}}{\overline{B'A}} \qquad \text{(by Theorem 1.12)}$$

$$= \frac{\overline{B'B''}}{\overline{B'A}}$$

$$= \frac{\overline{C'B}}{\overline{C'A}} \qquad \text{(by Theorem 1.12),}$$

and Equation 1 follows. $\square$

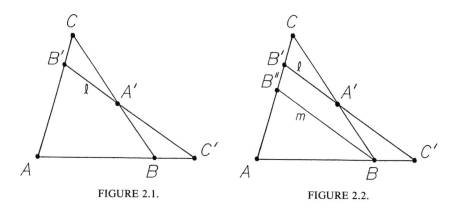

FIGURE 2.1. FIGURE 2.2.

We note that the three division ratios in Equation 1 are obtained from each other by substituting B for A, C for B, and A for C. In other words, we substitute by moving in one direction around triangle ABC in Figure 2.1.

The previous theorem states that the product of the three division ratios $\overline{A'B}/\overline{A'C}$, $\overline{B'C}/\overline{B'A}$, $\overline{C'A}/\overline{C'B}$ is 1. These are the ratios in which the sides of triangle ABC are divided by a line l that intersects each side at a point other than a vertex. In particular, Equation 1 shows that the product of these three division ratios is positive, and so either 0 or 2 of them are negative. Thus, either 0 or 2 of the points A', B', C' lie between the vertices of triangle ABC (by Theorem 1.4). In fact, none of the points A', B', C' lie between the vertices of triangle ABC when l doesn't pass through the interior of the triangle (Figure 2.3). Two of the points A', B', C' lie between the vertices of triangle ABC when l passes through the interior of the triangle: the two points are those where l intersects the sides of the triangle as it leaves the interior (Figure 2.1).

We would like to prove the converse of Theorem 2.1 and show that A', B', C' lie on a line when Equation 1 holds. We can deduce this fact from Theorem 2.1 itself once we consider another possibility that can arise: the line $l = A'B'$ could be parallel to side AB of the triangle (Figure 2.4). In this case, the point C' where l intersects AB vanishes, and so we might guess that the division ratio involving C' vanishes from Equation 1, leaving the equation

$$\frac{\overline{A'B}}{\overline{A'C}} \cdot \frac{\overline{B'C}}{\overline{B'A}} = 1. \tag{2}$$

We can further justify Equation 2 as follows. Imagine that the line l in Figure 2.1 revolves around A' until it becomes parallel to AB. As l becomes

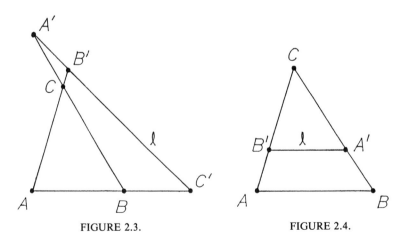

FIGURE 2.3.   FIGURE 2.4.

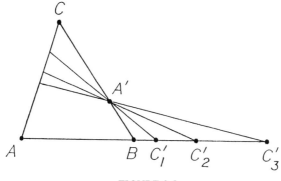

FIGURE 2.5.

parallel to AB, the point C' where l intersects AB moves off to infinity along either end of AB. (See Figures 2.5 and 2.6, which show various lines on A' and the the points C_1', C_2', C_3' where they intersect AB.) If $X \to x$ is a coordinate system on AB, then we have

$$\frac{\overline{C'A}}{\overline{C'B}} = \frac{a - c'}{b - c'}$$

(by Theorem 1.8). Dividing the numerator and denominator by $-c'$ shows that

$$\frac{\overline{C'A}}{\overline{C'B}} = \frac{-a/c' + 1}{-b/c' + 1}.$$

The right-hand side of this equation approaches 1 as c' approaches ∞, since $-a/c'$ and $-b/c'$ both approach 0 as c' approaches ∞. (Here, as always, we don't distinguish between $\pm\infty$.) We observed in Figures 2.5

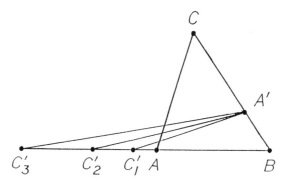

FIGURE 2.6.

and 2.6 that c' approaches ∞ as l becomes parallel to AB, and so the previous sentence suggests that $\overline{C'A}/\overline{C'B}$ should be replaced by 1 when l is parallel to AB. Thus, we expect that Equation 2 is the version of Equation 1 that holds when l is parallel to AB. In fact, this result is just a restatement of Theorem 1.12, as the proof of the next theorem shows.

THEOREM 2.2. In the Euclidean plane, let ABC be a triangle. Let A' be a point on BC other than B and C, and let B' be a point on CA other than C and A. Then the lines AB and $A'B'$ are parallel if and only if the equality

$$\frac{\overline{A'B}}{\overline{A'C}} \cdot \frac{\overline{B'C}}{\overline{B'A}} = 1 \tag{3}$$

holds (Figure 2.4).

Proof: Theorem 1.12 shows that AB and $A'B'$ are parallel if and only if the equality

$$\overline{A'B}/\overline{A'C} = \overline{B'A}/\overline{B'C} \tag{4}$$

holds. Moreover, Equations 3 and 4 are equivalent: since division ratios are nonzero real numbers (by Theorem 1.2i), we can obtain Equation 4 by multiplying Equation 3 by $\overline{B'A}/\overline{B'C}$, and we can obtain Equation 3 by multiplying Equation 4 by $\overline{B'C}/\overline{B'A}$. Thus, AB and $A'B'$ are parallel if and only if Equation 3 holds. $\square$

Theorem 2.2 gives a criterion for AB and $A'B'$ to be parallel. The next theorem shows that Theorem 2.1 applies whenever AB and $A'B'$ aren't parallel.

THEOREM 2.3. In the Euclidean plane, let ABC be a triangle. Let A' be a point on BC other than B and C, and let B' be a point on CA other than C and A. Assume that AB and $A'B'$ aren't parallel. Then AB and $A'B'$ intersect at a point C' other than A and B (Figure 2.7).

Proof: A' doesn't equal the point C where CA and BC intersect, and so A' doesn't lie on CA. (This shows once again that $A' \neq B'$, and so $A'B'$ exists.) It follows that $A'B' \neq CA$, and so these lines intersect at the unique point B'. Thus, $A'B'$ doesn't contain A. Then $A'B'$ doesn't contain B either, by symmetry. Since AB and $A'B'$ aren't parallel, by assumption, they intersect at a unique point C'. We've seen that neither A nor B lies on $A'B'$, and so neither of these points equals C'. $\square$

We can now prove Menelaus' Theorem, which combines Theorem 2.1 with its converse.

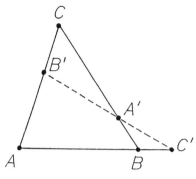

FIGURE 2.7.

THEOREM 2.4 (Menelaus' Theorem). In the Euclidean plane, let ABC be a triangle. Let A' be a point on BC other than B and C, let B' be a point on CA other than C and A, and let C' be a point on AB other than A and B. Then the points A', B', C' lie on a line if and only if the equation

$$\frac{\overline{A'B}}{\overline{A'C}} \cdot \frac{\overline{B'C}}{\overline{B'A}} \cdot \frac{\overline{C'A}}{\overline{C'B}} = 1 \qquad (5)$$

holds (Figure 2.7).

Proof: If A', B', C' lie on a line, Theorem 2.1 shows that Equation 5 holds. Conversely, assume that Equation 5 holds. We have $\overline{C'A}/\overline{C'B} \neq 1$ (by Theorem 1.2i), and so Equation 5 implies that

$$\frac{\overline{A'B}}{\overline{A'C}} \cdot \frac{\overline{B'C}}{\overline{B'A}} \neq 1.$$

Then the lines AB and $A'B'$ aren't parallel (by Theorem 2.2), and so they intersect at a point C'' other than A and B, by Theorem 2.3 (Figure 2.8). Since A', B', C'' lie on a line, Theorem 2.1 shows that

$$\frac{\overline{A'B}}{\overline{A'C}} \cdot \frac{\overline{B'C}}{\overline{B'A}} \cdot \frac{\overline{C''A}}{\overline{C''B}} = 1. \qquad (6)$$

Equation 5 also holds, by assumption. Combining Equations 5 and 6 with the fact that division ratios are nonzero (by Theorem 1.2i) shows that

$$\overline{C'A}/\overline{C'B} = \overline{C''A}/\overline{C''B}.$$

It follows that $C' = C''$ (by Theorem 1.9). Thus, since A', B', C'' lie on a line, so do A', B', C'. $\square$

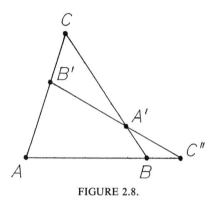

FIGURE 2.8.

We note that Theorem 1.9 is the key in the previous proof to using Theorem 2.1 to obtain its converse. Theorem 1.9 shows that points on a line are determined by the ratios in which they divide two fixed points on the line.

Menelaus' Theorem uses division ratios to characterize when three points, one on each side of a triangle, lie on a line. The next example illustrates how Menelaus' Theorem can be applied in a particular case.

EXAMPLE 2.5. In the Euclidean plane, let ABC be a triangle whose sides have lengths $|\overline{AB}| = 2$, $|\overline{BC}| = 5$, and $|\overline{CA}| = 4$ (Figure 2.9). Let B' be the point between C and A that lies 1 unit from C, and let C' be the point on line AB that lies 2 units from B on the side of B opposite A. Prove that the lines BC and $B'C'$ intersect at a point A', and find the distances from A' to B and C.

Proof: We have $\overline{B'C}/\overline{B'A} = -1/3$ (by Theorem 1.4), because B' lies between C and A, 1 unit from C and 3 units from A (since $|\overline{CA}| = 4$). We also have $\overline{C'A}/\overline{C'B} = 4/2 = 2$ (by Theorem 1.4), because C' doesn't lie between A and B and is 2 units from B and 4 units from A (since $|\overline{AB}| = 2$). Replacing A by B, B by C, and C by A in Theorem 2.2 shows that BC and $B'C'$ are parallel if and only if the equation

$$\frac{\overline{B'C}}{\overline{B'A}} \cdot \frac{\overline{C'A}}{\overline{C'B}} = 1$$

holds. In fact, we have

$$\frac{\overline{B'C}}{\overline{B'A}} \cdot \frac{\overline{C'A}}{\overline{C'B}} = \frac{-1}{3} \cdot 2 = \frac{-2}{3}, \tag{7}$$

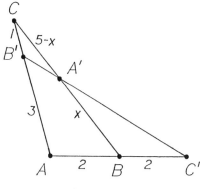

FIGURE 2.9.

and so BC and $B'C'$ aren't parallel. Thus, BC and $B'C'$ intersect at a point A' other than B and C (by Theorem 2.3, where we replace A by B, B by C, and C by A). Hence, Menelaus' Theorem 2.4 shows that

$$\frac{\overline{A'B}}{\overline{A'C}} \cdot \frac{\overline{B'C}}{\overline{B'A}} \cdot \frac{\overline{C'A}}{\overline{C'B}} = 1. \tag{8}$$

Substituting Equation 7 into Equation 8 shows that

$$\frac{\overline{A'B}}{\overline{A'C}} \cdot \frac{-2}{3} = 1,$$

and so we have

$$\overline{A'B}/\overline{A'C} = -3/2. \tag{9}$$

Set up a coordinate system on BC such that B has coordinate 0 and C has coordinate 5 (since $|\overline{BC}| = 5$). If A' has coordinate x, then Equation 9 becomes

$$\frac{0 - x}{5 - x} = \frac{-3}{2}$$

(by Theorem 1.8). It follows that $-2x = -15 + 3x$, $-5x = -15$, and $x = 3$. Thus, since B has coordinate 0 and C has coordinate 5, A' lies between B and C, 3 units from B and 2 units from C. $\square$

An interesting consequence of Menelaus' Theorem concerns points symmetric about the midpoints of the sides of a triangle. In order to apply Menelaus' Theorem, we begin with a result describing the ratios in which an interval is divided by two points symmetric about the midpoint of the interval.

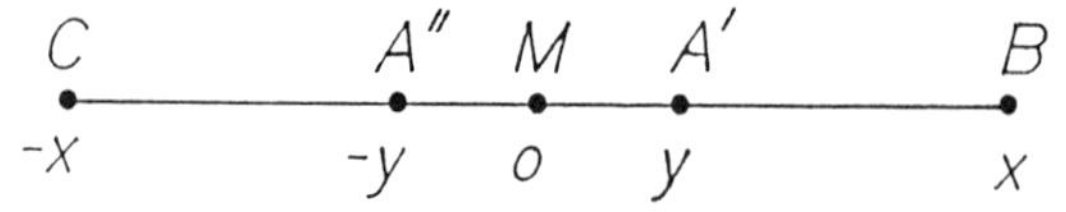

FIGURE 2.10.

THEOREM 2.6. In the Euclidean plane, let B and C be two points, and let M be their midpoint. Let A' and A'' be two points on BC other than B and C that have M as their midpoint. Then the equation

$$\overline{A'B}/\overline{A'C} = \overline{A''C}/\overline{A''B} \tag{10}$$

holds, and so the ratios in which A' and A'' divide B, C are reciprocals (Figure 2.10).

Proof: Choose a coordinate system on line BC such that M has coordinate 0. Let B and A' have respective coordinates x and y. Then C and A'' have respective coordinates $-x$ and $-y$, since the point M with coordinate 0 is the midpoint of B and C and of A' and A''. Then we have

$$\frac{\overline{A'B}}{\overline{A'C}} = \frac{x-y}{-x-y} = \frac{-(-x+y)}{-(x+y)} = \frac{-x+y}{x+y}$$

and

$$\frac{\overline{A''C}}{\overline{A''B}} = \frac{-x-(-y)}{x-(-y)} = \frac{-x+y}{x+y}$$

(By Theorem 1.8), and so Equation 10 holds. □

We can now use Menelaus' Theorem to prove the following result.

THEOREM 2.7. In the Euclidean plane, let ABC be a triangle. Let M be the midpoint of B and C, let N be the midpoint of C and A, and let O be the midpoint of A and B. Let A' and A'' be two points on BC other than B and C that have M as their midpoint, let B' and B'' be two points on CA other than C and A that have N as their midpoint, and let C' and C'' be two points on AB other than A and B that have O as their midpoint. If A', B', C' lie on a line, then so do A'', B'', C'' (Figure 2.11).

Proof: Since A', B', C' lie on a line, we have

$$\frac{\overline{A'B}}{\overline{A'C}} \cdot \frac{\overline{B'C}}{\overline{B'A}} \cdot \frac{\overline{C'A}}{\overline{C'B}} = 1, \tag{11}$$

by Menelaus' Theorem 2.4. Since the pairs of points $\{A', A''\}$, $\{B', B''\}$, $\{C', C''\}$ are symmetric about the midpoints of the sides of the triangle,

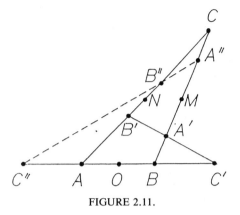

FIGURE 2.11.

we have

$$\frac{\overline{A'B}}{\overline{A'C}} = \frac{\overline{A''C}}{\overline{A''B}}, \qquad \frac{\overline{B'C}}{\overline{B'A}} = \frac{\overline{B''A}}{\overline{B''C}}, \qquad \frac{\overline{C'A}}{\overline{C'B}} = \frac{\overline{C''B}}{\overline{C''A}}, \qquad (12)$$

by Theorem 2.6. Substituting the equations in (12) into Equation 11 gives

$$\frac{\overline{A''C}}{\overline{A''B}} \cdot \frac{\overline{B''A}}{\overline{B''C}} \cdot \frac{\overline{C''B}}{\overline{C''A}} = 1.$$

Taking reciprocals of both sides of this equation gives

$$\frac{\overline{A''B}}{\overline{A''C}} \cdot \frac{\overline{B''C}}{\overline{B''A}} \cdot \frac{\overline{C''A}}{\overline{C''B}} = 1.$$

Thus, A'', B'', C'' lie on a line, by Menelaus' Theorem 2.4. □

We can restate this theorem as follows: *On each of the three sides of a triangle, take two points, other than vertices, that are symmetric about the midpoint of the side. If three of these six points lie on a line, then so do the three remaining points* (Figure 2.11).

EXERCISES

2.1. In the Euclidean plane, let ABC be a triangle whose sides have lengths $|\overline{AB}| = 6$, $|\overline{BC}| = 5$, and $|\overline{AC}| = 2$. Let A' be the point on line BC that lies 1 unit from C on the side of C opposite B, and let B' be the point on line AC that lies 1 unit from C on the side of C opposite A. Prove that the lines AB and $A'B'$ intersect at a point C', and find the distances from C' to A and B. Illustrate your answer with a figure.

2.2. In the Euclidean plane, let ABC be a triangle whose sides have lengths $|\overline{AB}| = 4$, $|\overline{BC}| = 3$, and $|\overline{AC}| = 2$. Let A' be the point on line BC that lies 1 unit from B between B and C, and let C' be the point on line AB that lies 2 units from B on the side of B opposite A. Prove that the lines AC and $A'C'$ intersect at a point B', and find the distances from B' to A and C. Illustrate your answer with a figure.

2.3. In the Euclidean plane, let ABC be a triangle whose sides have lengths $|\overline{AB}| = 8$, $|\overline{BC}| = 3$, and $|\overline{AC}| = 6$. Let B' be the point on line AC that lies 4 units from A between A and C, and let C' be the point on line AB that lies 2 units from A between A and B. Prove that the lines BC and $B'C'$ intersect at a point A', and find the distances from A' to B and C. Illustrate your answer with a figure.

2.4. In the Euclidean plane, let ABC be a triangle whose sides have lengths $|\overline{AB}| = 5$, $|\overline{BC}| = 4$, and $|\overline{AC}| = 2$. Let A' be the point on line BC that lies 3 units from C between B and C, and let B' be the point on line AC that lies 2 units from A on the side of A opposite C. Prove that the lines AB and $A'B'$ intersect at a point C', and find the distances from C' to A and B. Illustrate your answer with a figure.

2.5. In the Euclidean plane, let ABC be a triangle whose sides have lengths $|\overline{AB}| = 7$, $|\overline{BC}| = 9$, and $|\overline{AC}| = 8$. Let A' be the point on line BC that lies 3 units from C between B and C. Let B' be the point on AC such that $A'B'$ is parallel to AB. Let C' be the point on AB such that $A'C'$ is parallel to AC. Find the distances from B' to A and C, and from C' to A and B. Illustrate your answer with a figure.

2.6. In the Euclidean plane, let ABC be a triangle whose sides have lengths $|\overline{AB}| = 4$, $|\overline{BC}| = 7$, and $|\overline{AC}| = 8$. Let C' be the point on line AB that lies 5 units from B on the side of B opposite A. Let A' be the point on BC such that $A'C'$ is parallel to AC. Let B' be the point on AC such that $B'C'$ is parallel to BC. Find the distances from A' to B and C and from B' to A and C. Illustrate your answer with a figure.

2.7. In the Euclidean plane, let ABC be a triangle whose sides have lengths $|\overline{AB}| = 3$, $|\overline{BC}| = 5$, and $|\overline{AC}| = 4$. Let B' be the point on line AC that lies 2 units from A on the side of A opposite C. Let A' be the point on BC such that $A'B'$ is parallel to AB. Let C' be the point on AB such that $B'C'$ is parallel to BC. Find the distances from A' to B and C and from C' to A and B. Illustrate your answer with a figure.

2.8. In the Euclidean plane, let ABC be a triangle. Let M be the midpoint of B and C, and let N be the midpoint of C and A. Let A' and A'' be two points on BC other than B and C that have M as their midpoint, and let B' and B'' be two points on CA other than C and A that have N as their midpoint. Prove or disprove that $A'B'$ is parallel to AB if and only if $A''B''$ is parallel to AB. Illustrate your answer with a figure.

2.9. In the Euclidean plane, let ABC be a triangle. Let M be the midpoint of B and C, let N be the midpoint of C and A, and let O be the midpoint of A and B. Let A' and A'' be two points on BC other than B and C that have M as their midpoint, and let B' and B'' be two points on CA other than C and A that have N as their midpoint. Prove or disprove that $A'B'$ contains O if and only if $A''B''$ contains O. Illustrate your answer with a figure.

2.10. In the Euclidean plane, let ABC be a triangle. Let A' be a point on BC other than B and C, let B' be a point on CA other than C and A, and let C' be a point on AB other than A and B. If the lines AA', BB', CC' are parallel, prove that

$$\frac{\overline{A'B}}{\overline{A'C}} \cdot \frac{\overline{B'C}}{\overline{B'A}} \cdot \frac{\overline{C'A}}{\overline{C'B}} = -1.$$

(See Figure 2.12. One possible approach is to combine the two equations that result from using either Theorem 1.12 or Theorem 2.2 together with the fact that $\{AA', BB'\}$ and $\{BB', CC'\}$ are two pairs of parallel lines.)

2.11. In the Euclidean plane, let p and q be two parallel lines (Figure 2.13). Let A be a point on p, and let B be a point on q. Let A' be a point on q other than B, let B' be a point on p other than A, and let C' be a point on AB other than A and B. Choose corresponding positive ends for directed

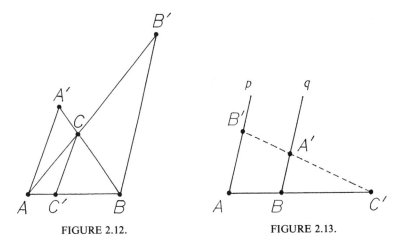

FIGURE 2.12.   FIGURE 2.13.

distances on the parallel lines p and q (as in Definition 1.15). Prove that A', B', C' lie on a line if and only if the equality

$$\frac{\overline{A'B}}{\overline{B'A}} \cdot \frac{\overline{C'A}}{\overline{C'B}} = 1$$

holds. (This is the version of Menelaus' Theorem 2.4 where C has vanished, just as Theorem 2.2 is the version of Theorem 2.4 where C' has vanished. One possible approach is to use Properties 0.4 and 0.5 and Theorem 1.4, taking into account the fact that C' may or may not lie between A and B.)

2.12. In the Euclidean plane, let l and l' be two parallel lines. Let A, B, C be three points on l, and let A', B', C' be three points on l'. Prove that the lines AA', BB', CC' either lie on a common point or are parallel if and only if the equation $\overline{AB}/\overline{AC} = \overline{A'B'}/\overline{A'C'}$ holds. Illustrate this result with a figure in each of the cases where the lines AA', BB', CC' lie on a common point or are parallel.

2.13. In the Euclidean plane, let l and m be two parallel lines. Let C and D be two points of l, and let A be their midpoint. Let E and F be two points of m, and let B be their midpoint.

(a) Prove that CE is parallel to AB if and only if DF is parallel to AB. Illustrate this result with a figure.
(b) If neither CE nor DF is parallel to AB, prove that they intersect AB at the same point. Illustrate this result with a figure.

2.14. In the Euclidean plane, let l and m be two parallel lines. Let C and D be two points on l, and let E and F be two points on m. Assume that CE and DF intersect at a point P. Let n be a line on P. Prove that the following conditions are equivalent.

(i) The lines CF, DE, n either lie on a common point or are parallel.
(ii) The lines l and n intersect at the midpoint of C and D.
(iii) The lines m and n intersect at the midpoint of E and F.

(See Exercise 2.13. Figure 2.14 illustrates the case where CF, DE, and n lie on a common point, and Figure 2.15 illustrates the case where these lines are parallel. See the parenthetical note after Exercise 0.7.)

2.15. In the Euclidean plane, let l, m, n be three lines on a point A. Let p and q be two lines parallel to and distinct from l (Figure 2.16). Neither m nor n is parallel to either p or q, by Theorem 0.11. Let B be the point where m intersects p, let C be the point where m intersects q, let D be the point

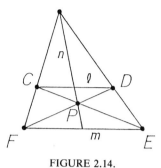

FIGURE 2.14.

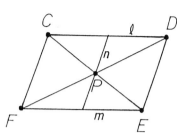

FIGURE 2.15.

where n intersects p, and let E be the point where n intersects q.

(a) Why does line BE exist and intersect l at a point F? Why does line CD exist and intersect l at a point G?

(b) Prove that $F \neq G$ and that A is the midpoint of F and G.

(See Exercises 2.13 and 2.14.)

2.16. In the notation of Exercise 2.15, prove that the following conditions are equivalent (Figure 2.17).

(i) F is the midpoint of B and E.
(ii) A is the midpoint of B and C.
(iii) A is the midpoint of D and E.
(iv) G is the midpoint of C and D.
(v) The lines BE and CD are parallel.

(See the parenthetical note after Exercise 0.7.)

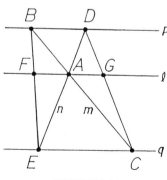

FIGURE 2.16.

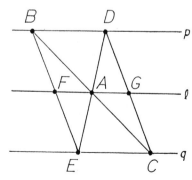

FIGURE 2.17.

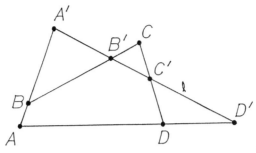

FIGURE 2.18.

2.17. In the Euclidean plane, let A, B, C, D be four points, no three of which lie on any line. Let A' be a point on AB other than A and B, let B' be a point on BC other than B and C, let C' be a point on CD other than C and D, and let D' be a point on DA other than D and A.

(a) If A', B', C', D' lie on a line, prove that

$$\frac{\overline{A'A}}{\overline{A'B}} \cdot \frac{\overline{B'B}}{\overline{B'C}} \cdot \frac{\overline{C'C}}{\overline{C'D}} \cdot \frac{\overline{D'D}}{\overline{D'A}} = 1. \tag{13}$$

(See Figure 2.18. One possible approach is to consider whether the line through A', B', C', D' intersects AC and to apply Theorems 2.1 and 2.2 to triangles ABC and CDA.)

(b) If Equation 13 holds, does it follow that A', B', C', D' lie on a line? Justify your answer completely.

Section 3.

Ceva's Theorem

We've proved Meneleus' Theorem, which uses division ratios to describe when three points, one on each side of a triangle, lie on a line. We now prove an analogous result called Ceva's Theorem. Ceva's Theorem uses division ratios to characterize when three lines, one on each vertex of a triangle, lie on a common point. The key step in proving Ceva's Theorem is part (iii) of the next result, which is an application of Meneleus' Theorem.

THEOREM 3.1. In the Euclidean plane, let ABC be a triangle. Let A' be a point on BC other than B and C, and let B' be a point on CA other than C and A. Assume that there is a point P that lies on both AA' and BB' (Figure 3.1).

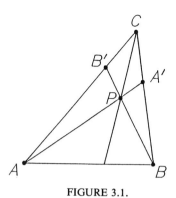

FIGURE 3.1.

(i) Then P is the unique point where AA' and BB' intersect.

(ii) P doesn't lie on any of the lines AB, BC, CA.

(ii) The following equation holds:

$$\frac{\overline{A'B}}{\overline{A'C}} \cdot \frac{\overline{AC}}{\overline{AB'}} \cdot \frac{\overline{PB'}}{\overline{PB}} = 1.$$

Proof: A doesn't lie on BC (since ABC is a triangle), and so line AA' exists and doesn't equal BC. Thus, AA' intersects BC at the unique point A', and so neither B nor C lies on AA'. By symmetry, neither A nor C lies on BB'.

(i) We have $AA' \neq BB'$ (since AA' doesn't contain B, by the previous paragraph). Thus, P is the unique point where AA' and BB' intersect.

(ii) Neither B nor C lies on AA' (by the first paragraph of the proof), and so neither AB nor CA equals AA'. Thus, A is the unique point where each of the lines AB and CA intersects AA'. P doesn't equal A (since BB' contains P but not A, by the first paragraph of the proof). Since P lies on AA', the two previous sentences imply that neither AB nor CA contains P. Then CB doesn't contain P either, since we can interchange A with B and A' with B' without affecting the assumptions of the theorem.

(iii) $B'BC$ is a triangle (since C doesn't lie on BB', by the first paragraph of the proof). A' is a point on BC other than B and C, A is a point on CB' other than C and B', and P is a point on $B'B$ other than B' and B (by part ii). Since A', A, P lie on a line, Menelaus' Theorem 2.4 shows that

$$\frac{\overline{A'B}}{\overline{A'C}} \cdot \frac{\overline{AC}}{\overline{AB'}} \cdot \frac{\overline{PB'}}{\overline{PB}} = 1. \quad \square$$

We can now prove one direction of Ceva's Theorem.

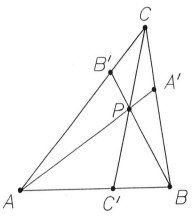

FIGURE 3.2.

THEOREM 3.2. In the Euclidean plane, let ABC be a triangle. Let A' be a point on BC other than B and C, let B' be a point on CA other then C and A, and let C' be a point on AB other than A and B. If the lines AA', BB', CC' all lie on a point P, then the equation

$$\frac{\overline{A'B}}{\overline{A'C}} \cdot \frac{\overline{B'C}}{\overline{B'A}} \cdot \frac{\overline{C'A}}{\overline{C'B}} = -1 \tag{1}$$

holds (Figure 3.2).

Proof: Theorem 3.1(iii) shows that

$$\frac{\overline{A'B}}{\overline{A'C}} \cdot \frac{\overline{AC}}{\overline{AB'}} \cdot \frac{\overline{PB'}}{\overline{PB}} = 1. \tag{2}$$

The hypotheses of Theorem 3.2 are unaffected if we interchange A with C and A' with C', and so we can interchange A with C and substitute C' for A' in Equation 2. Thus, we obtain

$$\frac{\overline{C'B}}{\overline{C'A}} \cdot \frac{\overline{CA}}{\overline{CB'}} \cdot \frac{\overline{PB'}}{\overline{PB}} = 1,$$

which we can rewrite as

$$\frac{\overline{PB'}}{\overline{PB}} = \frac{\overline{CB'}}{\overline{CA}} \cdot \frac{\overline{C'A}}{\overline{C'B}}. \tag{3}$$

Choose a positive end for directed distances on line CA. Substituting Equation 3 into Equation 2 gives

$$1 = \frac{\overline{A'B}}{\overline{A'C}} \cdot \frac{\overline{AC}}{\overline{AB'}} \cdot \frac{\overline{CB'}}{\overline{CA}} \cdot \frac{\overline{C'A}}{\overline{C'B}}$$

$$= \frac{\overline{A'B}}{\overline{A'C}} \cdot \frac{-\overline{CA}}{-\overline{B'A}} \cdot \frac{-\overline{B'C}}{\overline{CA}} \cdot \frac{\overline{C'A}}{\overline{C'B}} \qquad \text{(by Theorem 1.10)}$$

$$= -\frac{\overline{A'B}}{\overline{A'C}} \cdot \frac{\overline{B'C}}{\overline{B'A}} \cdot \frac{\overline{C'A}}{\overline{C'B}} \cdot \qquad \text{(by canceling } -\overline{CA}\text{)}.$$

Multiplying both sides of this equation by -1 gives Equation 1. $\quad\square$

The left side of Equation 1 is the product of the ratios in which A', B', C' divide the sides of triangle ABC. In particular, Theorem 3.2 shows that the product of these division ratios is negative, and so either all three ratios are negative or exactly one of them is negative. Thus, either all three of the points A', B', C' lie between the vertices of triangle ABC, or exactly one of them does so (by Theorem 1.4). In fact, all three of the points A', B', C' lie between the vertices of triangle ABC when P lies inside the triangle (Figure 3.2), and exactly one of the points A', B', C' lies between the vertices of triangles ABC when P lies outside the triangle (Figures 3.3–3.5).

Theorem 2.1 implied its converse once we examined in Theorem 2.2 an alternative arrangement involving parallel lines. Similarly, Theorem 3.2 implies its converse once we consider an alternative arrangement involving parallel lines. Specifically, suppose that the point P in Theorem 3.2 is repositioned so that CP and AB are parallel (Figure 3.6). Then the point C' where CP intersects AB vanishes. As in the discussion before Theorem 2.2, we expect that the division ratio involving C' disappears from Equation 1 in this case. Thus, we expect the following result to hold.

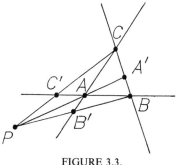

FIGURE 3.3.

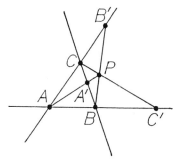

FIGURE 3.4.

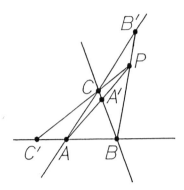

FIGURE 3.5.

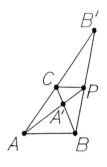

FIGURE 3.6.

THEOREM 3.3. In the Euclidean plane, let ABC be a triangle. Let A' be a point on BC other than B and C, and let B' be a point on CA other than C and A. Assume that AA' and BB' both contain a point P such that CP and AB are parallel. Then the equation

$$\frac{\overline{A'B}}{\overline{A'C}} \cdot \frac{\overline{B'C}}{\overline{B'A}} = -1 \tag{4}$$

holds (Figure 3.6).

Proof: Theorem 3.1(iii) shows that

$$\frac{\overline{A'B}}{\overline{A'C}} \cdot \frac{\overline{AC}}{\overline{AB'}} \cdot \frac{\overline{PB'}}{\overline{PB}} = 1. \tag{5}$$

AC and BP are distinct lines (by Theorem 3.1ii), and they both contain B'. The points A, C, B' are distinct, and the points B, P, B' are distinct (by Theorem 3.1ii). The lines AB and CP are parallel, and so Theorem 1.12 shows that

$$\overline{PB}/\overline{PB'} = \overline{CA}/\overline{CB'}.$$

We can take reciprocals of both sides of this equation (since division ratios are nonzero, by Theorem 1.2i), and so we have

$$\overline{PB'}/\overline{PB} = \overline{CB'}/\overline{CA}. \tag{6}$$

We choose a positive end on line CA for directed distances. Then substituting Equation 6 into Equation 5 gives

$$1 = \frac{\overline{A'B}}{\overline{A'C}} \cdot \frac{\overline{AC}}{\overline{AB'}} \cdot \frac{\overline{CB'}}{\overline{CA}}$$

$$= \frac{\overline{A'B}}{\overline{A'C}} \cdot \frac{-\overline{CA}}{-\overline{B'A}} \cdot \frac{-\overline{B'C}}{\overline{CA}} \qquad \text{(by Theorem 1.10)}$$

$$= -\frac{\overline{A'B}}{\overline{A'C}} \cdot \frac{\overline{B'C}}{\overline{B'A}} \qquad \text{(by canceling } -\overline{CA}\text{)}.$$

Multiplying this equation by -1 gives Equation 4. □

We can now prove Ceva's Theorem, which combines Theorem 3.2 and its converse.

THEOREM 3.4 (Ceva's Theorem). In the Euclidean plane, let ABC be a triangle. Let A' be a point on BC other than B and C, let B' be a point on CA other than C and A, and let C' be a point on AB other than A and B. Assume that the lines AA', BB', CC' are not all parallel to each other. Then these lines have a point in common if and only if the equation

$$\frac{\overline{A'B}}{\overline{A'C}} \cdot \frac{\overline{B'C}}{\overline{B'A}} \cdot \frac{\overline{C'A}}{\overline{C'B}} = -1 \qquad (7)$$

holds (Figure 3.7).

Proof: We note that $A \neq A'$ (since BC contains A' but not A), and so line AA' exists. Likewise, the lines BB' and CC' exist. If AA', BB', CC' all contain the same point P, then Equation 7 holds (by Theorem 3.2).

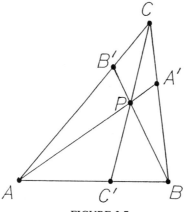

FIGURE 3.7.

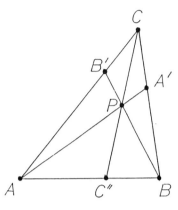

FIGURE 3.8.

Conversely, assume that Equation 7 holds. By assumption, the lines AA', BB', CC' aren't all parallel, and so at least two of them intersect. By symmetry, we can assume that AA' and BB' intersect at a point P. Line CP exists, by Theorem 3.1(ii). Equation 7 holds by assumption, and we have $\overline{C'A}/\overline{C'B} \neq 1$ (by Theorem 1.2i), and so it follows that

$$\frac{\overline{A'B}}{\overline{A'C}} \cdot \frac{\overline{B'C}}{\overline{B'A}} \neq -1.$$

Thus, CP and AB aren't parallel (by Theorem 3.3), and so they intersect at a point C'' (Figure 3.8). Neither A nor B equals C'' (since A and B don't lie on CP, by Theorem 3.1ii). Thus, Theorem 3.2 shows that

$$\frac{\overline{A'B}}{\overline{A'C}} \cdot \frac{\overline{B'C}}{\overline{B'A}} \cdot \frac{\overline{C''A}}{\overline{C''B}} = -1. \tag{8}$$

Equation 7 also holds, by assumption. Combining Equations 7 and 8 with the fact that division ratios are nonzero (by Theorem 1.2i) shows that

$$\overline{C'A}/\overline{C'B} = \overline{C''A}/\overline{C''B}.$$

It follows that $C' = C''$ (by Theorem 1.9). Thus, since the lines AA', BB', CC'' all contain P, so do the lines AA', BB', CC'. □

Ceva's Theorem and Menelaus' Theorem are analogous. Ceva's Theorem characterizes when three lines, one on each vertex of a triangle, contain the same point. Menelaus' Theorem characterizes when three points, one on each side of a triangle, lie on the same line. The product of the division ratios $\overline{A'B}/\overline{A'C}$, $\overline{B'C}/\overline{B'A}$, $\overline{C'A}/\overline{C'B}$ on the three sides of the triangle is -1 in Ceva's Theorem and 1 in Menelaus' Theorem.

In order to apply Ceva's Theorem, we must know that AA', BB', CC' aren't all parallel. As the next result shows, we can use Menelaus' Theorem to determine when this happens in terms of the division ratios $\overline{A'B}/\overline{A'C}$, $\overline{B'C}/\overline{B'A}$, $\overline{C'A}/\overline{C'B}$ on the sides of the triangle. Moreover, when AA', BB', CC' meet at a point P, the next result determines the ratios in which P divides each of the pairs $\{A, A'\}$, $\{B, B'\}$, $\{C, C'\}$.

THEOREM 3.5. In the Euclidean plane, let ABC be a triangle. Let A' be a point on BC other than B and C, and let B' be a point on CA other than C and A.

(i) The lines AA' and BB' are parallel if and only if the equation

$$\frac{\overline{A'B}}{\overline{A'C}}\left(1 - \frac{\overline{B'C}}{\overline{B'A}}\right) = 1$$

holds (Figure 3.9).

(ii) If the lines AA' and BB' both contain a point P, then the equation

$$\frac{\overline{PB}}{\overline{PB'}} = \frac{\overline{A'B}}{\overline{A'C}}\left(1 - \frac{\overline{B'C}}{\overline{B'A}}\right)$$

holds (Figure 3.10).

Proof: We use the following observation to prove both parts of the theorem: If we choose a positive end for directed distances on line AB, then we have

$$\frac{\overline{AC}}{\overline{AB'}} = \frac{\overline{AB'} + \overline{B'C}}{\overline{AB'}} \qquad \text{(by Theorem 1.10)}$$

$$= 1 + \frac{\overline{B'C}}{\overline{AB'}}$$

$$= 1 - \frac{\overline{B'C}}{\overline{B'A}} \qquad \text{(by Theorem 1.10).} \qquad (9)$$

(i) By Theorem 1.12, AA' and BB' are parallel if and only if the equation

$$\overline{A'B}/\overline{A'C} = \overline{AB'}/\overline{AC}$$

holds (Figure 3.9). This equation is equivalent to the equation

$$\frac{\overline{A'B}}{\overline{A'C}} \cdot \frac{\overline{AC}}{\overline{AB'}} = 1,$$

since division ratios are nonzero (by Theorem 1.2i). Combining the last two sentences with Equation 9 establishes part (i).

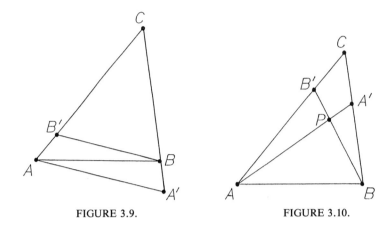

FIGURE 3.9.   FIGURE 3.10.

(ii) Theorem 3.1(iii) shows that

$$\frac{\overline{A'B}}{\overline{A'C}} \cdot \frac{\overline{AC}}{\overline{AB'}} \cdot \frac{\overline{PB'}}{\overline{PB}} = 1$$

(Figure 3.10). We can rewrite this equation as

$$\frac{\overline{PB}}{\overline{PB'}} = \frac{\overline{A'B}}{\overline{A'C}} \cdot \frac{\overline{AC}}{\overline{AB'}}. \tag{10}$$

Substituting Equation 9 into Equation 10 establishes part (ii). □

The next example illustrates how Ceva's Theorem can be applied in a particular case.

EXAMPLE 3.6. In the Euclidean plane, let ABC be a triangle whose sides have lengths $|\overline{AB}| = 6$, $|\overline{BC}| = 5$, and $|\overline{CA}| = 4$ (Figure 3.11). Let A' be the point on BC between B and C that lies 3 units from B, and let B' be the point on CA that lies 2 units from C on the side of C opposite A.

(a) Show that there is a unique point C' on AB such that the lines AA', BB', CC' contain the same point P. Find the distances from C' to A and B.
(b) Find the values of the division ratios $\overline{PA}/\overline{PA'}$, $\overline{PB}/\overline{PB'}$, $\overline{PC}/\overline{PC'}$.

Solution: (a) It follows from Theorem 1.4 that

$$\overline{A'B}/\overline{A'C} = -3/2, \tag{11}$$

because A' lies between B and C and is 3 units from B and 2 units from C (since $|\overline{BC}| = 5$). Likewise, Theorem, 1.4 implies that

$$\overline{B'C}/\overline{B'A} = 2/6 = 1/3, \tag{12}$$

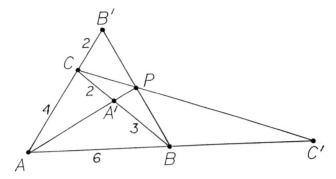

FIGURE 3.11.

because B' doesn't lie between C and A and is 2 units from C and 6 units from A (since $|\overline{CA}| = 4$). Thus, we have

$$\frac{\overline{A'B}}{\overline{A'C}}\left(1 - \frac{\overline{B'C}}{\overline{B'A}}\right) = \frac{-3}{2}\left(1 - \frac{1}{3}\right) = \frac{-3}{2}\cdot\frac{2}{3} = -1, \tag{13}$$

and so AA' and BB' aren't parallel (by Theorem 3.5i). Then AA' and BB' intersect at a point P. Line CP exists and doesn't equal AB (by Theorem 3.1ii). Equations 11 and 12 show that

$$\frac{\overline{A'B}}{\overline{A'C}}\cdot\frac{\overline{B'C}}{\overline{B'A}} = \frac{-3}{2}\cdot\frac{1}{3} = \frac{-1}{2}, \tag{14}$$

and so CP isn't parallel to AB (by Theorem 3.3). Thus, CP intersects AB at a unique point C'. Therefore, since the point P where AA' and BB' intersect is unique (by Theorem 3.1i), C' is the unique point on AB such that AA', BB', CC' lie on a common point. Neither A nor B equals C', since neither A nor B lies on CP (by Theorem 3.1ii). Thus, Ceva's Theorem 3.4 shows that

$$\frac{\overline{A'B}}{\overline{A'C}}\cdot\frac{\overline{B'C}}{\overline{B'A}}\cdot\frac{\overline{C'A}}{\overline{C'B}} = -1. \tag{15}$$

Substituting Equation 14 into Equation 15 shows that

$$\frac{-1}{2}\cdot\frac{\overline{C'A}}{\overline{C'B}} = -1,$$

and so we have

$$\overline{C'A}/\overline{C'B} = 2. \tag{16}$$

Set up a coordinate system on AB so that A has coordinate 0 and B has coordinate 6 (since $|\overline{AB}| = 6$). If C' has coordinate x, then Equation 16 shows that

$$\frac{0 - x}{6 - x} = 2$$

(by Theorem 1.8), and so it follows that $-x = 12 - 2x$ and $x = 12$. Thus, since A has coordinate 0 and B has coordinate 6, C' lies 12 units from A and 6 units from B on the side of B opposite A.

(b) Theorem 3.5(ii) shows that

$$\frac{\overline{PB}}{\overline{PB'}} = \frac{\overline{A'B}}{\overline{A'C}}\left(1 - \frac{\overline{B'C}}{\overline{B'A}}\right) = -1$$

(by Equation 13), and so P is the midpoint of B and B' (by Theorem 1.5). Replacing A by B, B by C, and C by A in Theorem 3.5(ii) shows that

$$\frac{\overline{PC}}{\overline{PC'}} = \frac{\overline{B'C}}{\overline{B'A}}\left(1 - \frac{\overline{C'A}}{\overline{C'B}}\right) = \frac{1}{3}(1 - 2) = -\frac{1}{3}$$

(by Equations 12 and 16), and so P lies between C and C' three times as far from C' as from C. Replacing A by C, B by A, and C by B in Theorem 3.5(ii) shows that

$$\frac{\overline{PA}}{\overline{PA'}} = \frac{\overline{C'A}}{\overline{C'B}}\left(1 - \frac{\overline{A'B}}{\overline{A'C}}\right) = 2\left(1 - \frac{-3}{2}\right) = 2\left(\frac{5}{2}\right) = 5$$

(by Equations 11 and 16), and so P lies 5 times as far from A as from A' and is on the side of A' opposite A. $\square$

In particular, the product of the three division ratios $\overline{A'B}/\overline{A'C}$, $\overline{B'C}/\overline{B'A}$, $\overline{C'A}/\overline{C'B}$ equals -1 when each of these ratios equals -1. This occurs when A', B', C' are the midpoints of the sides of triangle ABC (by Theorem 1.5). Thus, Ceva's Theorem shows that *the lines joining each vertex of a triangle to the midpoint of the opposite side lie on a common point G* (Figure 3.12). This fact is part (i) of the next theorem.

THEOREM 3.7. In the Euclidean plane, let ABC be a triangle. Let A' be the midpoint of B and C, let B' be the midpoint of C and A, and let C' be the midpoint of A and B (Figure 3.12).

(i) Then the lines AA', BB', CC' lie on a common point G.
(ii) G doesn't lie on any side of triangle ABC, and G is the unique point of intersection of each pair of the lines AA', BB', CC'.
(iii) Each of the division ratios $\overline{GA}/\overline{GA'}$, $\overline{GB}/\overline{GB'}$, $\overline{GC}/\overline{GC'}$ equals -2.

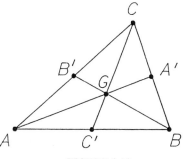

FIGURE 3.12.

Proof: Each of the division ratios $\overline{A'B}/\overline{A'C}$, $\overline{B'C}/\overline{B'A}$, $\overline{C'A}/\overline{C'B}$ equals -1, by Theorem 1.5. Thus, we have

$$\frac{\overline{A'B}}{\overline{A'C}} \cdot \frac{\overline{B'C}}{\overline{B'A}} \cdot \frac{\overline{C'A}}{\overline{C'B}} = (-1)^3 = -1.$$

Moreover, AA' and BB' aren't parallel, by Theorem 3.5(i), since

$$\frac{\overline{A'B}}{\overline{A'C}}\left(1 - \frac{\overline{B'C}}{\overline{B'A}}\right) = -1(1 - (-1)) = -1 \cdot 2 = -2. \qquad (17)$$

Thus, Ceva's Theorem 3.4 shows that the lines AA', BB', CC' lie on a common point G, and so part (i) holds. Part (ii) now follows from Theorem 3.1. Theorem 3.5(ii) shows that

$$\frac{\overline{GB}}{\overline{GB'}} = \frac{\overline{A'B}}{\overline{A'C}}\left(1 - \frac{\overline{B'C}}{\overline{B'A}}\right) = -2$$

(by Equation 17). It follows by symmetry that $\overline{GA}/\overline{GA'}$ and $\overline{GC}/\overline{GC'}$ also equal -2, and so part (iii) holds. □

Theorems 1.4 and 3.7(iii) show that *G lies between each vertex and the midpoint of the opposite side, and G is twice as far from the vertex as from the opposite midpoint. In other words, G lies one-third of the way from the midpoint of each side to the opposite vertex* (Figure 3.12). The next definition presents the language to state Theorem 3.7 economically.

DEFINITION 3.8.

   (i) In the Euclidean plane, a *median* of a triangle is the line through a vertex and the midpoint of the opposite side.
   (ii) The point G in Theorem 3.7 is called the *centroid* of triangle *ABC*. □

Parts (ii) and (iii) of Theorem 3.7 each show that every triangle in the Euclidean plane has a unique centroid. We can now restate Theorem 3.7 as follows.

THEOREM 3.9. Consider any triangle in the Euclidean plane (Figure 3.12).

(i) The three medians lie on a common point, the centroid.

(ii) The centroid doesn't lie on any side of the triangle, and the centroid is the unique point where each pair of medians intersects.

(iii) The centroid trisects each median, lying one-third of the way from the midpoint of each side to the opposite vertex. $\square$

We emphasize that the three medians of a triangle are not generally equal in length (Figure 3.12). Thus, although the centroid divides each median in the same ratio, it generally lies at a different distance from each vertex and at a different distance from the midpoint of each side.

Theorems 3.7 and 3.9 represent the case where the three division ratios in Ceva's Theorem each equal -1. Ceva's Theorem applies far more generally, whenever the product of these three ratios is -1. Thus, we can think of Ceva's Theorem as a generalization of the fact that the medians of a triangle lie on a common point, the centroid.

EXERCISES

3.1. Let A, B, C, A', B', be as in Exercise 2.1. Prove that there is a unique point C' on AB such that AA', BB', CC' lie on a common point P. Find the distances from C' to A and B. Find the values of the division ratios $\overline{PA}/\overline{PA'}$, $\overline{PB}/\overline{PB'}$, $\overline{PC}/\overline{PC'}$. Illustrate your answer with a figure.

3.2. Let A, B, C, A', C' be as in Exercise 2.2. Prove that there is a unique point B' on CA such that AA', BB', CC' lie on a common point P. Find the distances from B' to A and C. Find the values of the division ratios $\overline{PA}/\overline{PA'}$, $\overline{PB}/\overline{PB'}$, $\overline{PC}/\overline{PC'}$. Illustrate your answer with a figure.

3.3. Let A, B, C, B', C' be as in Exercise 2.3. Prove that there is a unique point A' on BC such that AA', BB', CC' lie on a common point P. Find the distances from A' to B and C. Find the values of the division ratios $\overline{PA}/\overline{PA'}$, $\overline{PB}/\overline{PB'}$, $\overline{PC}/\overline{PC'}$. Illustrate your answer with a figure.

3.4. Let A, B, C, A', B' be as in Exercise 2.4. Prove that there is a unique point C' on AB such that AA', BB', CC' lie on a common point P. Find the distances from C' to A and B. Find the values of the division ratios $\overline{PA}/\overline{PA'}$, $\overline{PB}/\overline{PB'}$, $\overline{PC}/\overline{PC'}$. Illustrate your answer with a figure.

3.5. Let A, B, C, A' be as in Exercise 2.5. Let B' be the point on AC and let C' be the point on AB such that AA', BB', CC' are parallel. Use Theorem 3.5(i) to find the distances from B' to A and C and the distances from C' to A and B. Illustrate your answer with a figure.

3.6. Let A, B, C, C' be as in Exercise 2.6. Let A' be the point on BC and let B' be the point on AC such that AA', BB', CC' are parallel. Use Theorem 3.5(i) to find the distances from A' to B and C and the distances from B' to A and C. Illustrate your answer with a figure.

3.7. Let A, B, C, A' be as in Exercise 2.5.

(a) Prove that there is a unique point B' on CA such that neither C nor A equals B', AA' and BB' lie on a point P, and CP is parallel to AB. Find the distances from B' to C and A. Illustrate your answer with a figure. (*Hint:* See Theorem 3.3.)

(b) Prove that there is a unique point C' on AB such that neither A nor B equals C', AA' and CC' lie on a point P, and BP is parallel to CA. Find the distances from C' to A and B. Illustrate your answer with a figure.

3.8. Let A, B, C, C' be as in Exercise 2.6.

(a) Prove that there is a unique point A' on BC such that neither B nor C equals A', AA' and CC' lie on a point P, and BP is parallel to CA. Find the distances from A' to B and C. Illustrate your answer with a figure.

(b) Prove that there is a unique point B' on CA such that neither C nor A equals B', BB' and CC' lie on a point P, and AP is parallel to BC. Find the distances from B' to C and A. Illustrate your answer with a figure.

3.9. In the Euclidean plane, let ABC be a triangle (Figure 3.13). Let A' be a point on BC other than B and C, and let B' be a point on CA other than C and A. Assume that AA' and BB' intersect at a point P.

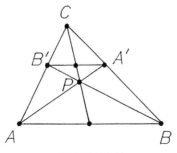

FIGURE 3.13.

Use Theorems 1.5, 2.2, and 3.4 to prove that the following conditions are equivalent.

  (i) $A'B'$ is parallel to AB.
  (ii) CP contains the midpoint of A and B.
  (iii) CP contains the midpoint of A' and B'.

(See the parenthetical note after Exercise 0.7.)

3.10. In the Euclidean plane, let ABC be a triangle. Let l be the line on C parallel to AB, and let P be a point on l other than C. Let M be the midpoint of B and C, and let N be the midpoint of C and A. Assume that AP intersects BC at a point A' other than M and that BP intersects CA at a point B' other than N. Let A'' be the point on BC such that M is the midpoint of A' and A'', and let B'' be the point on CA such that N is the midpoint of B' and B''. Prove that l, AA'', BB'' lie on a common point. Illustrate this result with a figure.

3.11. In the Euclidean plane, let ABC be a triangle (Figure 3.7). Let A' be a point on BC other than B and C, let B' be a point on CA other than C and A, and let C' be a point on AB other than A and B. Assume that the lines AA', BB', CC' are not all parallel. Prove that these lines all lie on the same point if and only if the equation

$$\frac{\sin(\angle BAA')}{\sin(\angle CAA')} \cdot \frac{\sin(\angle CBB')}{\sin(\angle ABB')} \cdot \frac{\sin(\angle ACC')}{\sin(\angle BCC')} = 1 \qquad (18)$$

holds and an odd number of the lines AA', BB', CC' pass through the interior of triangle ABC. (This result is called the *Trigonometric Form of Ceva's Theorem*. One possible approach to this exercise is to combine Ceva's Theorem with the Law of Sines.)

3.12. In the Euclidean plane, let ABC be a triangle. Let A' be a point on BC other than B and C, let B' be a point on CA other than C and A, and let C' be a point on AB other than A and B. Use Ceva's Theorem 3.4 and Exercise 2.10 to prove that the lines AA', BB', CC' either lie on a common point or are parallel if and only if the equation

$$\frac{\overline{A'B}}{\overline{A'C}} \cdot \frac{\overline{B'C}}{\overline{B'A}} \cdot \frac{\overline{C'A}}{\overline{C'B}} = -1$$

holds (Figures 3.7 and 2.12).

3.13. In the Euclidean plane, let ABC be a triangle. Let M be the midpoint of B and C, let N be the midpoint of C and A, and let O be the midpoint of A and B. Let A' and A'' be two points on BC other than B and C that have M as their midpoint, let B' and B'' be two points on CA other than C and A that have N as their midpoint, and let C' and C'' be two points on AB other than A and B that have O as their midpoint. Use Theorem 2.6 and Exercise 3.12 to prove that AA', BB', CC' either lie on a common point or are parallel if and only if AA'', BB'', CC'' either lie on a common point or are parallel. Illustrate this result with a figure in each of the following cases:

   (a) AA', BB', CC' lie on a common point, and AA'', BB'', CC'' lie on a common point;
   (b) AA', BB', CC' lie on a common point, and AA'', BB'', CC'' are parallel.

3.14. In the notation of Exercise 3.13, prove or disprove that it is possible for AA', BB', CC' to be parallel and for AA'', BB'', CC'' to be parallel at the same time.

3.15. In the Euclidean plane, let l and m be two parallel lines (Figure 3.14). Let A and B' be two points on l, and let B and A' be two points on m. Let C' be a point on AB other than A and B, and let n be the line on C' parallel to l and m. Choose corresponding positive ends for directed distances on the parallel lines l and m (as in Definition 1.15). Prove that AA', BB', n lie on a common point if and only if the equation

$$\frac{\overline{A'B}}{\overline{B'A}} \cdot \frac{\overline{C'A}}{\overline{C'B}} = -1$$

holds. (This result corresponds to Ceva's Theorem when C vanishes.)

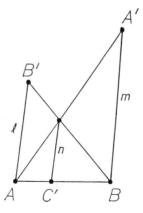

FIGURE 3.14.

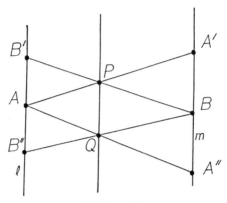

FIGURE 3.15.

3.16. In the Euclidean plane, let l and m be two parallel lines (Figure 3.15). Let B' and B'' be two points on l, and let A be their midpoint. Let A' and A'' be two points on m, and let B be their midpoint. Assume that AA' and BB' intersect at a point P and that AA'' and BB'' intersect at a point Q. Prove that line PQ exists and is parallel to l and m. (See Exercise 3.15.)

3.17. Consider the following result (Figure 3.7):

Theorem. In the Euclidean plane, let ABC be a triangle. Let A' be a point on BC other than B and C, let B' be a point on CA other than C and A, and let C' be a point on AB other than A and B. Assume that AA', BB', CC' lie on a point P. Then we have

$$\overline{A'P}/\overline{A'A} + \overline{B'P}/\overline{B'B} + \overline{C'P}/\overline{C'C} = 1. \qquad (19)$$

(a) Prove the theorem as follows. Use Theorem 3.5 and Theorem 1.10 to show that the left side of Equation 19 equals

$$\frac{1}{1 - c + ca} + \frac{1}{1 - a + ab} + \frac{1}{1 - b + bc}, \qquad (20)$$

where $a = \overline{A'B}/\overline{A'C}$, $b = \overline{B'C}/\overline{B'A}$, and $c = \overline{C'A}/\overline{C'B}$. Then use Ceva's Theorem to show that the sum in (20) equals 1; one possible approach uses direct computation, and another possible approach is based on showing that

$$1 - c + ca = -c(1 - a + ab),$$

$$1 - c + ca = ca(1 - b + bc),$$

and using these equations to rewrite the last two terms in (20) as fractions with denominators $1 - c + ca$.

(b) In the notation of part (a), prove that

$$\overline{AP}/\overline{AA'} + \overline{BP}/\overline{BB'} + \overline{CP}/\overline{CC'} = 2.$$

(*Hint:* One possible appproach uses Equation 19 and Theorem 1.10.)

3.18. In the Euclidean plane, let ABC be a triangle. Let A' be a point on BC other than B and C, and let B' be a point on CA other than C and A. Assume that AA' and BB' lie on a point P such that CP is parallel to AB (Figure 3.6). Prove that

$$\overline{A'P}/\overline{A'A} + \overline{B'P}/\overline{B'B} = 0. \tag{21}$$

(*Hint:* One possible approach is to use Theorem 1.11 to express $\overline{A'P}/\overline{A'A}$ and $\overline{B'P}/\overline{B'B}$ in terms of $\overline{A'B}, \overline{A'C}, \overline{B'C}, \overline{B'A}$ and then to use Theorem 3.3 to establish Equation 21.)

3.19. In the Euclidean plane, let ABC be a triangle. Let r be a number other than 0 and 1. Let A', B', C' be the points on BC, CA, AB, respectively, such that $r = \overline{A'B}/\overline{A'C} = \overline{B'C}/\overline{B'A} = \overline{C'A}/\overline{C'B}$. For what values of r is it true that AA' intersects BB' at a point D, AA' intersects CC' at a point E, and the distance from D to E equals the distance from A to A'? For any such value of r prove that BB' intersects CC' at a point F such that the distance from D to F equals the distance from B to B' and the distance from F to E equals the distance from C to C'. Illustrate each possible value of r with a figure.

3.20. Prove that both sides of Equation 1 have the same absolute value by applying the Law of Sines (Theorem 0.15) to each of the triangles $PA'B$, $PA'C$, $PB'C$, $PB'A$, $PC'A$, and $PC'B$ and combining the equations that result. Conclude that Theorem 3.2 holds by considering how the signs of the division ratios in Equation 1 depend on which of the seven regions in Figure 3.16 contain P.

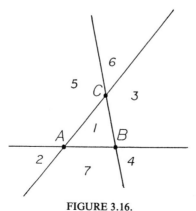

FIGURE 3.16.

3.21. Use the Law of Sines (Theorem 0.15) to prove that both sides of Equation 4 have the same absolute value. Deduce that Theorem 3.3 holds by considering how the signs of the division ratios in Equation 4 depend on which of the seven regions in Figure 3.16 contains P. (We can obtain another proof of Ceva's Theorem by substituting Exercise 3.20 for Theorem 3.2 and by substituting Exercise 3.21 for Theorem 3.3 in the proof of Theorem 3.4.)

3.22. For the situation in Exercise 3.1, use Exercises 1.6 (Stewart's Theorem) and 3.1 to compute the distances between the following pairs of points. The answers may look complicated.

(a) A and A'; P and A; P and A'.
(b) B and B'; P and B; P and B'.
(c) C and C'; P and C; P and C'.

3.23. Repeat Exercise 3.22 for the situation in Exercise 3.2.

3.24. Repeat Exercise 3.22 for the situation in Exercise 3.3.

3.25. Repeat Exercise 3.22 for the situation in Exercise 3.4.

Section 4.

The Euler Line

We saw in the last section that the medians of a triangle lie on a common point, the centroid. We now consider relationships among a number of other points and lines determined by a triangle.

DEFINITION 4.1. In the Euclidean plane, let A and B be two points. The *perpendicular bisector* of A and B is the line through the midpoint M of A and B that is perpendicular to line AB. □

For example, the line l in Figure 4.1 is the perpendicular bisector of A and B. The next result shows that the perpendicular bisector of A and B consists of exactly those points in the plane equidistant from A and B.

THEOREM 4.2. In the Euclidean plane, let A and B be two points. A point P in the plane lies on the perpendicular bisector of A and B if and only if P is equidistant from A and B (Figure 4.1).

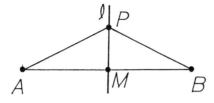

FIGURE 4.1.

Proof: Let M be the midpoint of A and B. Let l be the perpendicular bisector of A and B, the line through M perpendicular to AB.

First, assume that P lies on l and not on AB. Then AMP and BMP are triangles such that the corresponding sides AM and BM have equal length (since M is the midpoint of A and B), the corresponding sides MP and MP are equal, and the included angles $\angle AMP$ and $\angle BMP$ are equal (since these are right angles). Thus, triangles AMP and BMP are congruent (by Property 0.2), and so their corresponding sides AP and BP have equal length. Thus, P is equidistant from A and B.

Next, assume that P is equidistant from A and B and doesn't lie on AB. Then AMP and BMP are triangles whose corresponding sides have equal length: sides AM and BM have equal length (since M is the midpoint of A and B); sides MP and MP have equal length; and sides AP and BP have equal length (since P is equidistant from A and B). Thus, triangles AMP and BMP are congruent (by Property 0.1), and so their corresponding angles $\angle AMP$ and $\angle BMP$ are equal. These angles sum to 180° (since they combine to form a straight angle AMB), and so they are each 90°. Thus, PM is perpendicular to AB, and so it is the pependicular bisector l of A and B. Therefore, P lies on l, as desired.

The previous paragraphs prove the theorem when P doesn't lie on AB. The theorem also holds when P lies on AB, because M is the unique point on AB equidistant from A and B, and M is the unique point on AB that also lies on l. □

Let O be a point in the Euclidean plane, and let r be a positive real number. The *circle* of *radius* r with center O consists of the points in the plane at distance r from O. Thus, all points on a circle are equidistant from the center, and so Theorem 4.2 suggests that there may be a connection between perpendicular bisectors and circles. In fact, the next result shows that the perpendicular bisectors of the three sides of a triangle lie on a common point, the center of the circle circumscribed about the triangle (Figure 4.2).

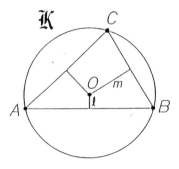

FIGURE 4.2.

Theorem 4.3. In the Euclidean plane, let ABC be a triangle (Figure 4.2).

(i) The perpendicular bisectors of the three sides of the triangle lie on a common point O. O is the unique point where any two of the perpendicular bisectors intersect. O is also the unique point equidistant from A, B, C.

(ii) There is a unique circle $\mathcal{K}$ that contains A, B, C. O is the center of $\mathcal{K}$.

Proof:

(i) Let l be the perpendicular bisector of A and B, and let m be the perpendicular bisector of B and C. Since AB and BC aren't parallel, neither are l and m. Thus, l and m intersect at a unique point O. Theorem 4.2 shows that $|\overline{OA}| = |\overline{OB}|$ and $|\overline{OB}| = |\overline{OC}|$, since O lies on the perpendicular bisectors l and m. Combining these equations shows that $|\overline{OA}| = |\overline{OC}|$, and so O also lies on the perpendicular bisector of A and C (by Theorem 4.2). Thus, O lies on all three perpendicular bisectors. No two of the perpendicular bisectors are parallel (since the sides of the triangle to which they are perpendicular aren't parallel), and so any two perpendicular bisectors intersect at the unique point O. Thus, O is the unique point equidistant from A, B, C (by Theorem 4.2).

(ii) By part (i), O is the unique point equidistant from A, B, C. Thus, A, B, C lie at the same distance r from O, and O is the only possible center from a circle containing A, B, C (since the points on a circle are equidistant from the center). A circle with center O contains A, B, C if and only if it has radius r. Combining the last two sentences shows that the circle with center O and radius r is the unique circle containing A, B, C. □

Theorem 4.3(ii) shows that *a unique circle can be circumscribed around any triangle. Thus, any three points not in a line lie on a unique circle*, since the points are the vertices of a triangle (Figure 4.2).

DEFINITION 4.4. In the notation of Theorem 4.3, the point O is called the *circumcenter* of triangle ABC, and the circle $\mathcal{K}$ is called the *circumcircle* of triangle ABC (Figure 4.2). □

We can use the terms in Definition 4.4 to restate Theorem 4.3 as follows: *The circumcenter of a triangle is the unique point where the three perpendicular bisectors meet. The circumcircle of a triangle is the unique circle through the vertices of the triangle. The circumcenter of a triangle is the center of the circumcircle* (Figure 4.2).

Theorem 4.3 shows that three points not in a line lie on a circle. Part (ii) of the next result shows that, conversely, any three points on a circle don't lie on a line.

THEOREM 4.5.
  (i) Three points on a line don't lie on a circle.
  (ii) Three points on a circle don't lie on a line.

Proof: (i) Let A, B, C be three points on a line l. By symmetry, we can assume that B lies between A and C (Figure 4.3). Let m be the perpendicular bisector of A and B, and let n be the perpendicular bisector of B and C. Then m and n are distinct parallel lines (since they are perpendicular to l and lie on opposite sides of B), and so they have no points in common. By Theorem 4.2, m consists of the points in the plane equidistant from A and B, and n consists of the points equidistant from B and C. Thus, since m and n have no common points, there is no point equidistant from A, B, C. Hence, no circle contains A, B, C (since the center of such a circle would be equidistant from A, B, C).

Part (ii) is a restatement of part (i), since both parts state that no three points lie on both a line and circle. □

The next result describes when the circumcenter of a triangle is the midpoint of a side.

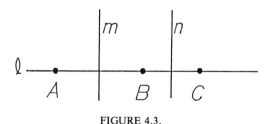

FIGURE 4.3.

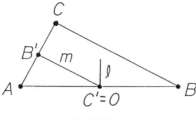

FIGURE 4.4.

THEOREM 4.6. In the Euclidean plane, let O be the circumcenter of triangle ABC. Then O is the midpoint of A and B if and only if $\angle ACB$ is a right angle (Figure 4.4).

Proof: Let B' be the midpoint of A and C, and let C' be the midpoint of A and B. Let l be the perpendicular bisector of A and B, and let m be the perpendicular bisector of A and C. O is the unique point on both l and m (by Theorem 4.3i). Thus, C' equals O if and only if C' lies on m (since C' lies on l, by Definition 4.1). C' lies on m if and only if $\angle AB'C'$ is a right angle (by Definition 4.1). This occurs if and only if $\angle ACB$ is a right angle (since BC and $B'C'$ are parallel, by Theorem 1.13). Combining the last three sentences shows that O is the midpoint C' of A and B if and only if $\angle ACB$ is a right angle. $\square$

   The circumcenter O of triangle ABC is equidistant from A, B, C (by Theorem 4.3i). Thus, O lies on AB if and only if O is the midpoint of A and B. Hence, Theorem 4.6 shows that O *lies on AB if and only if $\angle ACB$ is a right angle.* Accordingly, if a triangle doesn't have a right angle, then the circumcenter doesn't lie on any side of the triangle.
   Theorem 4.6 implies that right angles are exactly those angles that can be inscribed in semicircles (Figure 4.5). *Diameter AB* of a circle $\mathcal{K}$ is a line segment that has endpoints A and B on $\mathcal{K}$ and that contains the center of $\mathcal{K}$. Diameter AB and $\mathcal{K}$ intersect only at the points A and B (by Theorem 4.5).

THEOREM 4.7.
   (i) In the Euclidean plane, let A and B be two points. Then there is a unique circle $\mathcal{K}$ having diameter AB. The center of $\mathcal{K}$ is the midpoint C' of A and B (Figure 4.5).
   (ii) In the Euclidean plane, let A, B, C be three points. Then C lies on the circle having diameter AB if and only if $\angle ACB$ is a right angle (Figure 4.5).

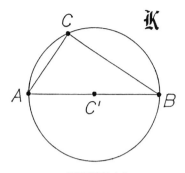

FIGURE 4.5.

Proof: (i) Since C' is the midpoint of A and B, we have $|\overline{AC'}| = |\overline{BC'}|$. If a circle $\mathcal{K}$ has diameter AB, the center of $\mathcal{K}$ is C' (since the center lies on AB equidistant from A and B), and the radius of $\mathcal{K}$ is $|\overline{AC'}| = |\overline{BC'}|$ (since the radius is the distance from the center to the points on $\mathcal{K}$). Conversely, the circle with center C' and radius $|\overline{AC'}| = |\overline{BC'}|$ has diameter AB (since $\mathcal{K}$ contains A and B, and its center C' lies on AB). Thus, there is a unique circle $\mathcal{K}$ with diameter AB, and the center of $\mathcal{K}$ is C'.

(ii) First assume that C lies on the circle $\mathcal{K}$ having diameter AB. The center of $\mathcal{K}$ is the midpoint C' of A and B (by part (i)). Since $\mathcal{K}$ contains A, B, C, these points are the vertices of a triangle (by Theorem 4.5ii), and the circumcenter O of triangle ABC is the center of $\mathcal{K}$ (by Theorem 4.3ii). The last two sentences show that the circumcenter O of triangle ABC is the midpoint C' of A and B. Thus, $\angle ACB$ is a right angle, by Theorem 4.6.

Conversely, assume that $\angle ACB$ is a right angle. Then A, B, C are the vertices of a triangle. The circumcenter O of triangle ABC is the midpoint C' of A and B (by Theorem 4.6). Thus, the circumcircle $\mathcal{K}$—which contains A, B, C—has diameter AB (since its center $O = C'$ lies on AB). $\mathcal{K}$ is the unique circle with diameter AB, by part (i). $\square$

The proof of Theorem 4.7 shows that, when a right triangle is inscribed in a circle, the same point is the center of the circle, the midpoint of the hypotenuse, and the circumcenter of the triangle (Figures 4.4 and 4.5).

So far we've considered two sets of three lines determined by a triangle, the medians and the perpendicular bisectors. We now consider a third set of three lines.

DEFINITION 4.8. In the Euclidean plane, the *altitude* on a vertex of a triangle is the line l through the vertex perpendicular to the line n through the other two vertices. The *foot* of the altitude is the point where l and n intersect. $\square$

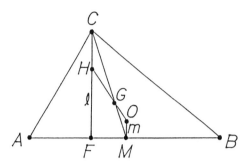

FIGURE 4.6.

For example, the line l in Figure 4.6 is the altitude on the vertex C of triangle ABC because l is the line on C perpendicular to the line AB through the other two vertices. The point F where l intersects line AB is the foot of the altitude. The next result discusses the relative positions of an altitude of a triangle, the centroid, and the circumcenter.

THEOREM 4.9. In the Euclidean plane, let ABC be a triangle such that sides AC and BC have unequal lengths (Figure 4.6).

(i) Then the centroid G doesn't equal the circumcenter O.
(ii) The altitude on C intersects GO at a unique point H.
(iii) H, G, O are distinct points such that $\overline{GH}/\overline{GO} = -2$.
(iv) H is the only point X on line GO such that X, G, O are distinct points satisfying $\overline{GX}/\overline{GO} = -2$.

Proof: Let l be the altitude on C, let M be the midpoint of A and B, and let m be the perpendicular bisector of A and B.

(i) The distances from C to A and B are unequal (by assumption), and so C doesn't lie on m (by Theorem 4.2). Thus, the median CM intersects the perpendicular bisector m at the unique point M. Then G doesn't lie on m, since G is a point on CM other than M (by Theorem 3.7). Hence, we have $G \neq O$ (since m contains O but not G), and so part (i) holds.

(ii) Line GO exists, since we have $G \neq O$ (by part (i)). GO and m contain O and aren't equal (since G doesn't lie on m, by the previous paragraph), and so they aren't parallel. Then GO and l aren't parallel, by Theorem 0.11 (since l and m are parallel because they're both perpendicular to AB). Hence, GO and l intersect at a unique point H, and so part (ii) holds.

(iii) First assume that $\angle ACB$ is not a right angle. The lines l and m are parallel (as in the previous paragraph) and distinct (since C lies on l but not on m, by the proof of part (i)). Thus, CM intersects l and m at the unique

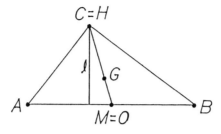

FIGURE 4.7.

points C and M, and so neither l nor m contains G (since G is a point on CM other than C and M, by Theorem 3.7). In short, l and m are distinct parallel lines that don't contain G, and so the points G, H, O are distinct, as well as the points G, C, M. The assumption that $\angle ACB$ isn't a right angle implies that $O \neq M$ (by Theorem 4.6), and so we have $m = OM$. G doesn't lie on $m = OM$ (by the proof of part (i)), and so GO and GM are distinct lines. In summary, G, H, O are distinct points, G, C, M are distinct points, GO and GM are distinct lines, and the lines $CH = l$ and $OM = m$ are parallel. Hence, Theorem 1.11 shows that

$$\overline{GH}/\overline{GO} = \overline{GC}/\overline{GM}. \tag{1}$$

The centroid G trisects the median CM, and so we have

$$\overline{GC}/\overline{GM} = -2 \tag{2}$$

(by Theorem 3.7iii). Substituting Equation 2 into Equation 1 shows that $\overline{GH}/\overline{GO} = -2$, as desired.

Next assume that $\angle ACB$ is a right angle (Figure 4.7). Then O equals M (by Theorem 4.6), and so the point H where l intersects $GO = GM$ is C. Thus, we have

$$\overline{GH}/\overline{GO} = \overline{GC}/\overline{GM} = -2$$

(by Theorem 3.7iii) in this case as well.

(iv) Choose a positive end for directed distances on GO. Let X be a point on GO such that X, G, O are distinct points satisfying $\overline{GX}/\overline{GO} = -2$. We also have $\overline{GH}/\overline{GO} = -2$ (by part (iii)), and so it follows that $\overline{GX} = \overline{GH}$. This implies that $X = H$, by Theorem 1.1(i). □

The key to proving part (iii) of Theorem 4.9 was to note that the altitude l on C is parallel to the perpendicular bisector m of A and B, and so triangles GHC and GOM are similar when $\angle ACB \neq 90°$ (Figure 4.6). Corresponding sides of similar triangles are proportional (by Property 0.4),

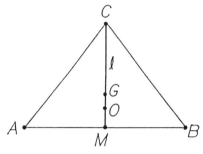

FIGURE 4.8.

and so we have

$$\overline{GH}/\overline{GO} = \overline{GC}/\overline{GM},$$

where Theorem 1.11 assures us that both sides of this equation have the same sign. Hence, the fact that $\overline{GC}/\overline{GM} = -2$ (by Theorem 3.7iii) implies that $\overline{GH}/\overline{GO} = -2$. We had to use a different argument when $\angle ACB = 90°$, since there are no triangles GHC and GOM in this case (Figure 4.7).

The next result is the analogue of Theorem 4.9 that holds when sides AC and BC of triangle ABC have the same length.

THEOREM 4.10. In the Euclidean plane, let ABC be a triangle whose sides AC and BC have equal length. Then the same line is the median on C, the altitude on C, and the perpendicular bisector of A and B. This line contains the centroid G and the circumcenter O of triangle ABC (Figure 4.8).

Proof: Let M be the midpoint of A and B. Since C is equidistant from A and B (by assumption), C lies on the perpendicular bisector l of A and B (by Theorem 4.2). Thus, we have $l = CM$ (since l contains M, by Definition 4.1). Hence, the perpendicular bisector l of A and B is also the median CM on C. In fact, l is also the altitude on C, since l contains C and is perpendicular to AB. The centroid G lies on each median, and the circumcenter O lies on each perpendicular bisector, and so both G and O lie on l. □

Theorem 4.9 shows that, when sides AC and BC of triangle ABC have different lengths, the altitude on C intersects GO at a point H uniquely determined by the condition $\overline{GH}/\overline{GO} = -2$. It follows that the three altitudes of any triangle lie on a common point H. The proof of this fact also uses Theorem 4.10 when at least two sides of the triangle have the same length. A triangle is called *equilateral* when all its sides have the same length.

THEOREM 4.11. In the Euclidean plane, let G be the centroid and let O be the circumcenter of triangle ABC.

  (i) The three altitudes of the triangle all lie on a point H, and H is the unique point where any two of the altitudes intersect.
 (ii) If triangle ABC isn't equilateral, the points G, H, O are distinct, lie on a line, and satisfy the relation $\overline{GH}/\overline{GO} = -2$.
(iii) If triangle ABC is equilateral, the points G, H, O are equal.

Proof: No two sides of a triangle are parallel, and the three altitudes are perpendicular to the three sides, and so no two altitudes are parallel. Thus, any two altitudes intersect at a unique point. There are three cases, depending on the number of sides of the triangle that have the same length.

Case 1: No two sides of triangle ABC have the same length (Figure 4.9). Then Theorem 4.9 applies with the altitude on C replaced by any of the three altitudes. Thus, G and O aren't equal, and each of the three altitudes intersects line GO at the unique point H such that G, H, O are distinct points satisfying $\overline{GH}/\overline{GO} = -2$. Hence, the three altitudes all lie on this one point H. H is the unique point where any two of the altitudes intersect, by the first paragraph of the proof.

Case 2: Exactly two sides of triangle ABC have equal lengths. By symmetry, we can assume that AC and BC are the two sides having equal lengths (Figure 4.10). Thus, we can apply Theorem 4.9 if we interchange C with either A or B. Hence, Theorem 4.9 shows that $O \neq G$ and the altitudes on A and B intersect line GO at the unique point H such that H, G, O are distinct and $\overline{GH}/\overline{GO} = -2$. Since sides AC and BC have equal lengths, the altitude on C contain G and O (by Theorem 4.10), and so it equals line GO and thus contains H. In short, all three altitudes contain H, and H, G, O are distinct points such that $\overline{GH}/\overline{GO} = -2$. Any two of the altitudes intersect at the unique point H, by the first paragraph of the proof.

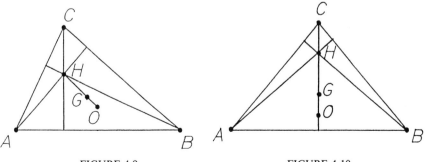

FIGURE 4.9.   FIGURE 4.10.

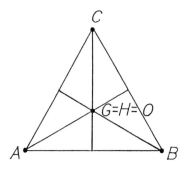

FIGURE 4.11.

Case 3: All three sides of triangle ABC have the same length; so ABC is an equilateral triangle (Figure 4.11). Then Theorem 4.10 applies with C replaced by any of the vertices, and so the three altitudes, the three medians, and the three perpendicular bisectors are all equal. The three medians intersect at the unique point G (by Theorem 3.9), and the three perpendicular bisectors intersect at the unique point O (by Theorem 4.3i), and so the three altitudes intersect at a unique point H, where $G = H = O$. ☐

Theorem 4.11(i) shows that the altitudes of a triangle meet at a unique point H (Figure 4.9). This point is called the *orthocenter* of the triangle. If the triangle isn't equilateral, Theorem 4.11(ii) shows that the centroid G, the orthocenter H, and the circumcenter O lie on a unique line and that the relation $\overline{GH}/\overline{GO} = -2$ holds. The line through G, H, O is called the *Euler line* of the triangle. The relation $\overline{GH}/\overline{GO} = -2$ shows that G lies between H and O (by Theorem 1.4), twice as far from H as from O. *Thus, G lies one-third of the way from O to H.*

Figures 4.12–4.15 show the relative positions of the medians, the altitudes, and the perpendicular bisectors in various triangles. Each figure

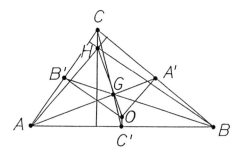

FIGURE 4.12.

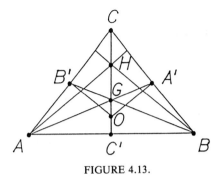

FIGURE 4.13.

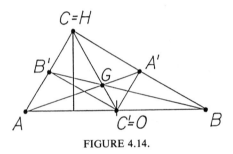

FIGURE 4.14.

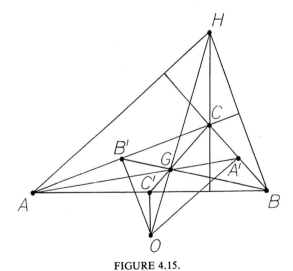

FIGURE 4.15.

shows the following: a triangle ABC that isn't equilateral, the midpoints A', B', C' of the three sides, the three medians AA', BB', CC' (which meet at the centroid G), the three altitudes (which meet at the orthocenter H, and which are the lines through the vertices perpendicular to the opposite sides), and the three perpendicular bisectors (which meet at the circumcenter O, and which are the lines through A', B', C' perpendicular to the sides). In each figure, the points G, H, O lie on a line, the Euler line, and G lies one-third of the way from O to H (by Theorem 4.11ii).

A triangle is called *isosceles* if it has at least two sides of the same length; otherwise, the triangle is called *scalene*. Thus, equilateral triangles are isosceles, and scalene triangles are those whose sides all have different lengths. An angle is called *acute* if it is less than 90°, and it is called *obtuse* if it is greater than 90°

Figures 4.12 and 4.13 show the same triangles as do Figures 4.9 and 4.10; thus, Figure 4.12 shows a scalene triangle whose angles are all acute, and Figure 4.13 shows an isosceles triangle that isn't equilateral and whose angles are all acute. Figure 4.14 shows a scalene triangle with a right angle at C; thus, O equals C' (by Theorem 4.6), and H equals C (either by the discussion accompanying Figure 4.7 or by the facts that AC and BC are the altitudes on A and B). Figure 4.15 shows a scalene triangle with an obtuse angle at C; we note that H and O lie outside triangle ABC here. We can think of the case of a right triangle (where H and O lie on the sides of the triangle, as in Figure 4.14) as a transition case separating the case where all angles in the triangle are acute (so H and O lie inside the triangle, as in Figure 4.12) from the case where an angle is obtuse (so H and O lie outside the triangle, as in Figure 4.15).

Figure 4.11 shows an equilateral triangle. Here the centroid, the orthocenter, and the circumcenter are equal (by Theorem 4.11iii). Thus, there is no Euler line in this case, and the medians, the altitudes, and the perpendicular bisectors are all the same three lines.

EXERCISES

4.1. In the Euclidean plane, let PQR be a triangle (Figure 4.16). Let S be the midpoint of P and R. Prove that triangle PQR has a right angle at Q if and only if S is equidistant from P, Q, and R.

4.2. In the Euclidean plane, let ABC be a triangle (Figure 4.17). Let A' be the midpoint of B and C, and let E and F be the feet of the altitudes on B and C.

(a) Use Exercise 4.1 to prove that E and F are equidistant from A'.

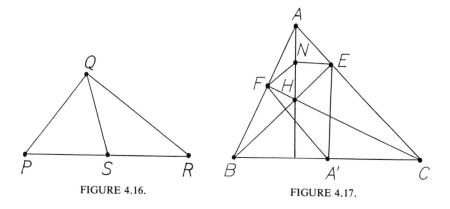

FIGURE 4.16.   FIGURE 4.17.

(b) Let H be the orthocenter of triangle ABC, and let N be the midpoint of A and H. Use Exercise 4.1 to prove that E and F are equidistant from N. (When A and H are the same point, we consider this point to be the midpoint of A and H.)

4.3. In the Euclidean plane, let $\mathcal{K}$ be a circle with center O (Figure 4.18). Let P be a point other than O. Prove that the circle $\mathcal{L}$ with diameter OP contains the midpoint M of every pair of points A and B on $\mathcal{K}$ such that AB contains P.

4.4. In the Euclidean plane, let ABC be a triangle. Let H be the orthocenter, let O be the circumcenter, and let M be the midpoint of A and B. Prove that the distance from C to H is twice the distance from O to M. (*Thus, the distance from the orthocenter to a vertex is twice the distance from the circumcenter to the opposite side.* One possible approach to this exercise is to combine Theorem 3.7(iii) with the discussions accompanying Figures 4.6, 4.7, 4.10, and 4.11.)

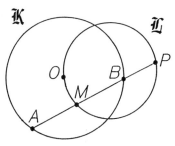

FIGURE 4.18.

4.5. In the Euclidean plane, let ABC be a triangle. Let H be the orthocenter, and let R be the radius of the circumcircle. Prove that $AB^2 + CH^2 = 4R^2$, where XY denotes the distance between points X and Y.

(*Hint:* One possible approach is as follows. Let O be the circumcenter, and let M be the midpoint of A and B. When $\angle ACB \neq 90°$, one can show that AOM is a right triangle, apply the Pythagorean Theorem 0.9, and combine the result with Exercise 4.4. The case where $\angle ACB = 90°$ needs separate consideration.)

4.6. In the Euclidean plane, let ABC be a triangle that doesn't have a right angle at A or C. Let H be the orthocenter. Let D be the other endpoint of the diameter of the circumcircle that has B as one endpoint. Prove that $AHCD$ is a parallelogram. In each of the following cases, illustrate this result with a figure where no two sides of triangle ABC have the same length.

  (a) All angles of triangle ABC are acute.
  (b) $\angle ABC$ is obtuse.
  (c) $\angle BAC$ is obtuse.
  (d) $\angle ABC$ is a right angle.

4.7. In the Euclidean plane, let ABC be a triangle without a right angle. Let H be the orthocenter of triangle ABC.

  (a) Prove that A, B, C, H are four points such that any three are the vertices of a triangle whose orthocenter is the fourth point.
  (b) Prove that none of the triangles whose vertices are three of the points A, B, C, H is a right triangle.
  (c) Prove that the same points are the feet of the altitudes of each of the four triangles whose vertices are three of the points A, B, C, H.
  (d) Prove that exactly three of the four triangles in part (c) have an obtuse angle and that the three obtuse angles occur at the same point.

4.8. In the Euclidean plane, let ABC be a triangle. Let H be the orthocenter, and let A', B', C' be the feet of the altitudes on A, B, C. Use Theorem 4.7 to prove that the points in each of the following sets lie on a circle:

$$\{A, B', C', H\}, \qquad \{B, C', A', H\}, \qquad \{C, A', B', H\}.$$

Illustrate this result with a figure in each of the following cases.

  (a) ABC is a scalene triangle whose angles are all acute.
  (b) ABC is a scalene triangle with an obtuse angle.
  (c) ABC is a scalene right triangle.

4.9. In the Euclidean plane, let ABC be a triangle.

(a) Prove that there is a triangle $A'B'C'$ such that $A'B'$ contains C and is parallel to AB, $B'C'$ contains A and is parallel to BC, and $C'A'$ contains B and is parallel to CA.

(b) Prove that the perpendicular bisectors of triangle $A'B'C'$ are the altitudes of triangle ABC. (Thus, Theorem 4.11i follows from Theorem 4.3i.)

(c) Illustrate parts (a) and (b) with a figure.

4.10. In the Euclidean plane, let ABC be a triangle that isn't equilateral. Prove that the Euler line contains C if and only if either $\angle ACB = 90°$ or sides AC and BC have equal length.

4.11. In the Euclidean plane, let ABC be a triangle that isn't equilateral. Prove that the Euler line contains the midpoint of A and B if and only if either $\angle ACB = 90°$ or sides AC and BC have equal length.

4.12. In the Euclidean plane, let ABC be a triangle without a right angle. Let D, E, F be the feet of the altitudes on A, B, C, respectively. Prove that the following conditions are equivalent.

(i) EF is parallel to BC.
(ii) D is the midpoint of B and C.
(iii) The altitude on A is the perpendicular bisector of B and C.
(iv) The sides AB and AC of triangle ABC have the same length.

(*Hint:* One possible approach is to use Theorems 4.2 and 4.11i and either Exercise 2.14 or 3.9. See the parenthetical note after Exercise 0.7.)

4.13. In the notation of Exercise 4.12, prove that the following conditions are equivalent to the conditions (i)–(iv) of Exercise 4.12.

(v) The altitude on A contains the midpoint of E and F.
(vi) The altitude on A is the perpendicular bisector of E and F.
(vii) A is equidistant from E and F.
(viii) The distance from E to C equals the distance from F to B.

4.14. Consider a triangle in the Euclidean plane. Prove that the following conditions are equivalent.

(i) The triangle is isosceles.
(ii) Two altitudes have the same length.
(iii) The orthocenter is equidistant from two vertices.
(iv) Two medians have the same length.
(v) The centroid is equidistant from two vertices.

(See the parenthetical note after Exercise 0.7.)

4.15. In the Euclidean plane, let ABC be a triangle without a right angle. Let D, E, F be the feet of the altitudes on A, B, C, respectively. Prove that D is equidistant from E and F if and only if one of the following conditions holds:

  (i) A is equidistant from B and C;

  (ii) BC is the perpendicular bisector of E and F.

Illustrate this result with two figures: one where condition (i) holds and triangle ABC isn't equilateral, and one where condition (ii) holds. (*Hint:* One possible approach is to use Exercises 4.2a and 4.12 and Theorem 4.2.)

4.16. In the Euclidean plane, let ABC be a scalene triangle without a right angle. Let A', B', C' be the feet of the altitudes on A, B, C, respectively. Prove that $B'C'$ intersects BC at a point A'', $C'A'$ intersects CA at a point B'', and $A'B'$ intersects AB at a point C'' such that A'', B'', C'' lie on a line. Illustrate this result with a figure in each of the following cases.

  (a) All angles of triangle ABC are acute.

  (b) Triangle ABC has an obtuse angle.

  (*Hint:* One possible approach is to combine Exercise 4.12, Theorem 4.11i, Ceva's Theorem, and repeated applications of Menelaus' Theorem.)

4.17. In the Euclidean plane, let ABC be a triangle without a right angle (Figure 4.19). Let H be the orthocenter. Let M be a point on BC, and let N be a point on AH such that $M \neq N$. Prove that MN is parallel to AB if and only if CN is perpendicular to MH. (As always, we consider a line to be parallel to itself. If C, M, N are noncollinear, one possible approach to this exercise is to prove that the condition that H is the orthocenter of triangle CMN is equivalent to each of the conditions that MN is parallel to AB and that CN is perpendicular to MH. Cases where C, M, N are collinear would then need separate consideration.)

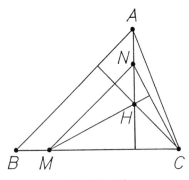

FIGURE 4.19.

4.18. Use Exercise 3.11 to prove Theorem 4.11(i). (*Hint:* One possible approach is to express the angles in Equation 18 of Section 3 in terms of the angles of triangle ABC when AA', BB', CC' are altitudes and the triangle doesn't have a right angle.)

Section 5.
The Nine-Point Circle and the Equicircles

We've already considered one way that a triangle determines a circle: the circumcircle of a triangle is the unique circle through the three vertices. We now study two other ways that triangles determine circles. First, we prove that the altitudes and medians of a triangle determine nine points that all lie on one circle, the nine-point circle of the triangle. Second, we prove that the sides of any triangle are tangent to four circles, the equicircles of the triangle; one of these circles lies inside the triangle, and the other three lie outside.

We start with a theorem that relates rectangles and circles. This result follows directly from Theorem 4.7(ii), which states that right angles are exactly the angles that can be inscribed in semicircles. We recall that two points X and Y determine a unique circle having diameter XY (by Theorem 4.7i).

THEOREM 5.1. In the Euclidean plane, let A B, C, D be four points. Then $ABCD$ is a rectangle if and only if AC and BD are diameters of the same circle (Figure 5.1).

Proof: First assume that AC and BD are diameters of the same circle. Then $\angle ABC$, $\angle BCD$, $\angle CDA$, $\angle DAB$ are right angles (by Theorem 4.7ii), and so $ABCD$ is a rectangle.

Conversely, assume that $ABCD$ is a rectangle. Then the circle with diameter AC contains B and D, and the circle with diameter BD contains

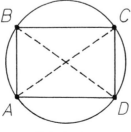

FIGURE 5.1.

A and C, by Theorem 4.7(ii). Moreover, the points A, B, C, D lie on at most one circle (by Theorem 4.3ii), and so AC and BD are diameters of the same circle. $\square$

We can now prove that the altitudes and medians of a triangle determine nine particular points that lie on a circle. We use the following convention in the statement and proof of the next theorem: When A and B are the same point, the statement that "M is the midpoint of A and B" means that $M = A = B$.

THEOREM 5.2. In the Euclidean plane, let $A_1 A_2 A_3$ be a triangle, and let H be its orthocenter. Let F_1, F_2, F_3 be the feet of the altitudes on A_1, A_2, A_3, let M_1, M_2, M_3 be the midpoints of the sides $A_2 A_3$, $A_3 A_1$, $A_1 A_2$, and let N_1, N_2, N_3 be the midpoints of H and A_1, A_2, A_3 (Figure 5.2).

(i) Then there is a unique circle $\mathcal{K}$ that contains

$$F_1, F_2, F_3, M_1, M_2, M_3, N_1, N_2, N_3. \tag{1}$$

(ii) The segments $M_1 N_1$, $M_2 N_2$, $M_3 N_3$ are diameters of $\mathcal{K}$. In particular, these segments all have the same midpoint T, the center of $\mathcal{K}$.

Proof: We first prove that there is a circle $\mathcal{K}$ having the three segments $M_1 N_1$, $M_2 N_2$, $M_3 N_3$ as diameters. There are two cases, depending on whether or not $A_1 A_2 A_3$ is a right triangle.

Case 1: $A_1 A_2 A_3$ is not a right triangle (Figure 5.3). Then the feet of the altitudes aren't vertices of the triangle. Thus, the orthocenter H, which lies on the three altitudes, doesn't lie on any side of the triangle (by Theorem 3.1). M_2 is the midpoint of A_1 and A_3, and M_3 is the midpoint of A_1 and A_2; so $M_2 M_3$ is parallel to $A_2 A_3$ (by Theorem 1.13). N_2 is the

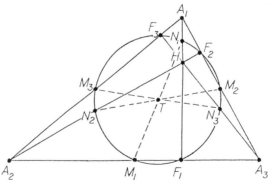

FIGURE 5.2.

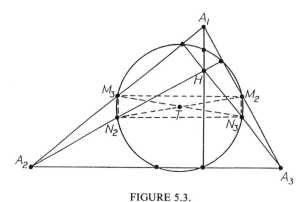

FIGURE 5.3.

midpoint of H and A_2, and N_3 is the midpoint of H and A_3; so $N_2 N_3$ is parallel to $A_2 A_3$ (by Theorem 1.13, which applies because H doesn't lie on any side of triangle $A_1 A_2 A_3$). M_3 is the midpoint of A_2 and A_1, and N_2 is the midpoint of A_2 and H; so $M_3 N_2$ is parallel to $A_1 H$ (by Theorem 1.13, which applies since H doesn't lie on any side of triangle $A_1 A_2 A_3$). By symmetry, $M_2 N_3$ is also parallel to $A_1 H$ (since M_2 is the midpoint of A_3 and A_1, while N_3 is the midpoint of A_3 and H). In short, $M_2 M_3$ and $N_2 N_3$ are parallel to $A_2 A_3$, while $M_2 N_3$ and $M_3 N_2$ are parallel to the altitude $A_1 H$ perpendicular to $A_2 A_3$. Hence, $M_2 M_3 N_2 N_3$ is a rectangle. Thus, $M_2 N_2$ and $M_3 N_3$ are diameters of the same circle (by Theorem 5.1). By symmetry, $M_1 N_1$ and $M_2 N_2$ are also diameters of the same circle, and so $M_1 N_1$, $M_2 N_2$, $M_3 N_3$ are all diameters of the same circle (by Theorem 4.7i).

Case 2: $A_1 A_2 A_3$ is a right triangle. By symmetry, we can assume that the right angle is at A_2 (Figure 5.4). Then $A_1 A_2$ is the altitude on A_1, and $A_2 A_3$ is the altitude on A_3. These two lines intersect at A_2, and so we have $H = A_2$ (by Theorem 4.11i). (Of course, the altitude on A_2 also contains $A_2 = H$, in agreement with Theorem 4.11i.) Thus, the midpoint N_2 of H and A_2 equals A_2, the midpoint N_1 of H and A_1 equals the midpoint M_3 of A_2 and A_1, and the midpoint N_3 of H and A_3 equals the midpoint M_1 of A_2 and A_3. M_2 and M_3 are the midpoints of the sides $A_1 A_3$ and $A_1 A_2$, and so $M_2 M_3$ is parallel to $A_2 A_3 = A_2 M_1$ (by Theorem 1.13). M_1 and M_2 are the midpoints of the sides $A_3 A_2$ and $A_3 A_1$, and so $M_1 M_2$ is parallel to $A_1 A_2 = M_3 A_2$ (by Theorem 1.13). Moreover, the sides $A_1 A_2$ and $A_2 A_3$ are perpendicular. Combining the last three sentences shows that $M_1 M_2 M_3 A_2$ is a rectangle. Thus, $M_1 M_3$ and $M_2 A_2$ are diameters of the same circle $\mathcal{K}$ (by Theorem 5.1). We've seen that $N_1 = M_3$, $N_2 = A_2$, and $N_3 = M_1$; so $\mathcal{K}$ has $M_1 N_1$, $M_2 N_2$, $M_3 N_3$ as diameters, completing this case.

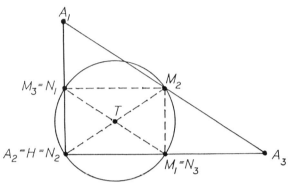

FIGURE 5.4.

In both cases, we've proved that M_1N_1, M_2N_2, M_3N_3 are diameters of the same circle $\mathcal{K}$ (Figures 5.2 and 5.4). The center T of $\mathcal{K}$ is the midpoint of each diameter (by Theorem 4.7i). In both cases, we've seen that the set $\{M_1, M_2, M_3, N_1, N_2, N_3\}$ contains the vertices of a rectangle, and so $\mathcal{K}$ is the unique circle containing these points (by Theorem 4.3ii). We must prove that $\mathcal{K}$ contains the feet F_1, F_2, F_3 of the altitudes. By symmetry, it's enough to prove that $\mathcal{K}$ contains F_1. If F_1 equals either M_1 or N_1, we're done, since these points lie on $\mathcal{K}$. On the other hand, suppose that neither M_1 nor N_1 equals F_1 (Figure 5.2). Then F_1M_1 equals A_2A_3, and F_1N_1 is the altitude on A_1; so F_1M_1 and F_1N_1 are perpendicular. Thus, F_1 lies on the circle $\mathcal{K}$ having diameter M_1N_1 (by Thoerem 4.7ii), as desired. $\square$

The circle $\mathcal{K}$ in Theorem 5.2 is called the *nine-point circle* of triangle $A_1A_2A_3$ because $\mathcal{K}$ contains the nine points in (1). These points are the feet of the three altitudes, the midpoints of the three sides, and the midpoints of the orthocenter and each of the three vertices. Three points on a circle don't lie on a line (by Theorem 4.5ii), and so the circle they lie on is unique (by Theorem 4.3ii). Thus, the nine-point circle $\mathcal{K}$, which contains all nine points in (1), is the unique circle through any three of these points that are distinct.

In fact, we should put quotation marks around the word "nine" when we talk about the nine points in (1), because these points are not always distinct. For example, these points aren't distinct when $A_1A_2A_3$ is a right triangle, as the discussion accompanying Figure 5.4 shows.

The proof of Case 1 of Theorem 5.2, when $A_1A_2A_3$ is not a right triangle, shows that $M_2M_3N_2N_3$ is a rectangle (Figure 5.3). Of course, $M_1M_2N_1N_2$ and $M_1M_3N_1N_3$ are also rectangles, by symmetry, when A_1A_2, A_3 is not a right triangle (Figures 5.5 and 5.6). The diagonals M_1N_1, M_2N_2, M_3N_3

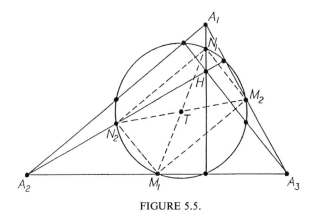

FIGURE 5.5.

of these rectangles are diameters of the nine-point circle, and their common midpoint T is the center of the nine-point circle (by Theorem 5.2).

Ceva's Theorem describes when three lines on the vertices of a triangle meet at a point. The medians and the altitudes of a triangle are two such triples of lines. We now consider another example, angle bisectors. We will use the results we obtain on angle bisectors to determine the four circles tangent to the three sides of a given triangle.

In the Euclidean plane, let l and m be two intersecting lines (Figure 5.7). The *angle bisectors* determined by l and m are the two lines p and q that divide the four angles formed by l and m into equal parts. Each pair of vertical angles formed by l and m is bisected by one of the lines p and q, since vertical angles are equal (by Theorem 0.6). As the next result shows, the angle bisectors p and q are perpendicular.

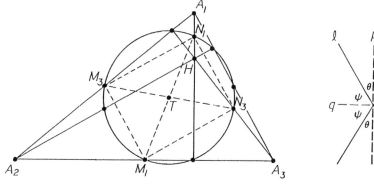

FIGURE 5.6.   FIGURE 5.7.

THEOREM 5.3. In the Euclidean plane, the angle bisectors determined by two lines on a point are perpendicular (Figure 5.7).

Proof: Let l and m be the two given lines, and let p and q be the angle bisectors. One pair of vertical angles is divided by p into angles of equal measure θ, and the other pair of vertical angles is divided by q into angles of equal measure ψ. The adjacent angles formed by l and m have measures 2θ and 2ψ, and so we have $2\theta + 2\psi = 180°$ (by Theorem 0.6). Dividing by 2 gives $\theta + \psi = 90°$. Since p and q form the angle $\theta + \psi$, p and q are perpendicular. $\square$

We want to prove that the angle bisectors at the vertices of a triangle lie by threes on common points. The proof is analogous to the proof in Section 4 that the perpendicular bisectors of the three sides of a triangle meet at a point. By Theorem 4.2, the perpendicular bisector of two points A and B consists of exactly those points equidistant from A and B. Similarly, we show that the angle bisectors of two lines l and m consist of exactly those points equidistant from l and m. We use the following definition.

DEFINITION 5.4. In the Euclidean plane, let P be a point and let l be a line. The *foot of the perpendicular* from P to l is the point F where l intersects the line on P perpendicular to l. The *distance from P to l* is the distance between the points P and F (Figure 5.8). $\square$

The next result is analogous to Theorem 4.2.

THEOREM 5.5. In the Euclidean plane, let l and m be two lines on a point O. Then the angle bisectors consist of exactly those points equidistant from l and m (Figure 5.8).

Proof: Let P be a point, and let F and G be the feet of the perpendiculars from P to l and m.

First, assume that P doesn't lie on l or m and that neither F nor G equals O (Figure 5.8). P is equidistant from l and m if and only if it is equidistant from F and G (by Definition 5.4). FOP and GOP are right triangles that have a common side OP. Thus, P is equidistant from F and G if and only if triangles FOP and GOP are congruent (by the Pythagorean Theorem 0.9 and the SSS Property 0.1). This occurs if and only if $\angle FOP$ equals $\angle GOP$ (by the ASA Property 0.3, since triangles FOP and GOP share side OP, have right angles at F and G, and have angles that sum to 180°). $\angle FOP$ equals $\angle GOP$ if and only if P lies on an angle bisector. In short, P is equidistant from l and m if and only if P lies on an angle bisector.

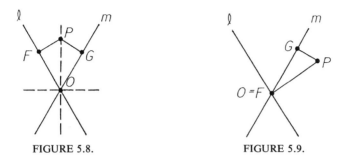

FIGURE 5.8.   FIGURE 5.9.

Next, assume that P doesn't lie on l or m and that either F or G equals O. By symmetry, we can assume that $F = O$ (Figure 5.9). Then $G \neq O$ (since l and m aren't parallel), and so FGP is a right triangle with hypotenuse FP. Thus, the distance from P to F is greater than the distance from P to G (by the Pythagorean Theorem 0.9), and so P is not equidistant from l and m. On the other hand, P doesn't lie on an angle bisector, since line PO is perpendicular to l but not m.

Finally, assume that P lies on l or m. Then P is equidistant from l and m if and only if P equals O, and this happens if and only if P lies on an angle bisector. □

Two sides of a triangle lie on a vertex O and determine two angle bisectors (Figure 5.10). These are called the angle bisectors of the triangle at O. One of the bisectors passes through the interior of the triangle and is called the *internal angle bisector* at O. The other bisector doesn't pass through the interior of the triangle and is called the *external angle bisector* at O. In Figure 5.10, p is the internal angle bisector at O, and q is the external angle bisector at O.

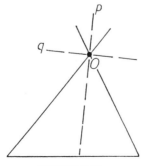

FIGURE 5.10.

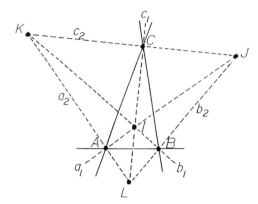

FIGURE 5.11.

A triangle determines six angle bisectors, an internal and an external angle bisector at each of the three vertices. It follows from Theorem 5.5 that the six angle bisectors lie by threes on four points. This is illustrated by Figure 5.11, where a_1, b_1, c_1 are the internal angle bisectors at the vertices A, B, C of triangle ABC, and where a_2, b_2, c_2 are the external angle bisectors at these vertices. As Figure 5.11 shows, the three internal angle bisectors meet at a point I, and there are three other points J, K, L where the internal bisector at one vertex intersects the external angle bisectors at the other two vertices.

THEOREM 5.6. In the Euclidean plane, the angle bisectors of a triangle lie by threes on four points. One of these points lies on the three inernal angle bisectors. The other three points each lie on the internal angle bisector at one vertex and on the external angle bisectors at the other two vertices (Figure 5.11).

Proof: Let A, B, C be the vertices of the triangle. We base the proof on three claims.

Claim 1: If a point P lies on angle bisectors at two vertices of a triangle, then it also lies on an angle bisector at the third vertex. To prove this, we can assume by symmetry that P lies on angle bisectors at A and B. Then the distance from P to AB equals the distance from P to AC and the distance from P to BC (by Theorem 5.5). Thus, P is equidistant from AC and BC, and so P also lies on an angle bisector at C (by Theorem 5.5).

Claim 2: The internal bisectors at two vertices intersect at a point inside the triangle (Figure 5.12). By symmetry, it's enough to consider the internal angle bisectors a_1 and b_1 at A and B. BC intersects a_1 at a point A' between

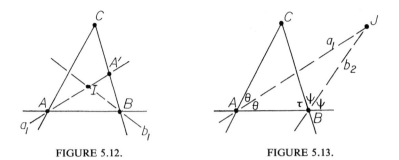

FIGURE 5.12.   FIGURE 5.13.

B and C. A and A' lie on opposite sides of b_1, and so $a_1 = AA'$ intersects b_1 at a point I between A and A'. I lies inside triangle ABC because it lies between A and A'.

Claim 3: The internal angle bisector and the external angle bisector at two vertices intersect at a point outside the triangle (Figure 5.13). By symmetry, it's enough to consider the internal angle bisector a_1 at A and the external angle bisector b_2 at B. Then a_1 divides $\angle BAC$ into two equal angles of measure θ, and b_2 divides the angles at B adjacent to $\angle ABC$ into equal angles of measure ψ. If we set $\tau = \angle ABC$, then the adjacent angles formed by AB and BC have measures 2ψ and τ, and so we have

$$\tau = 180° - 2\psi \tag{2}$$

(by Theorem 0.6). The angles of triangle ABC at A and B have measure 2θ and τ, and so we have

$$2\theta + \tau < 180° \tag{3}$$

(by Theorem 0.8). Substituting Equation 2 into Inequality 3 gives

$$2\theta + 180° - 2\psi < 180°,$$

which simplifies to

$$\theta < \psi. \tag{4}$$

If a_1 and b_2 were parallel, then θ and ψ would be corresponding angles formed by the parallel lines a_1 and b_2 with AB, and θ would equal ψ (by Property 0.5). Thus, the fact that $\theta < \psi$ (by (4)) implies that a_1 and b_2 aren't parallel. Hence, a_1 and b_2 intersect at a point J. J lies outside triangle ABC, since every point on b_2 except B lies outside the triangle.

We can now prove the theorem. The internal angle bisectors a_1 and b_1 at A and B intersect at a point I inside the triangle, by Claim 2 (Figure 5.12). Then I lies on an angle bisector at C, by Claim 1 (Figure 5.11). In fact,

I lies on the internal angle bisector at C, because I lies inside the triangle while a_1 intersects the external angle bisector at C outside the triangle (by Claim 3). Thus, I lies on all three internal angle bisectors.

The internal angle bisector a_1 at A intersects the external angle bisector b_2 at B in a point J outside the triangle, by Claim 3 (Figure 5.13). Then J also lies on angle bisector at C, by Claim 1 (Figure 5.11). In fact, J lies on the external angle bisector at C, because J lies outside the triangle while a_1 intersects the internal angle bisector at C inside the triangle (by Claim 2). Thus, J lies on the internal angle bisector at A and the external angle bisectors at B and C. By symmetry, there is a point K that lies on the internal angle bisector at B and on the external angle bisectors at A and C, and there is also a point L that lies on the internal angle bisector at C and on the external angle bisectors at A and B (Figure 5.11).

Neither I nor J equals either K or L, since I and J lie inside $\angle BAC$, while K and L lie outside. Thus, I, J, K, L are four distinct points, by symmetry. $\square$

The point I where the internal angle bisectors of a triangle meet is called the *incenter* of the triangle (Figure 5.11). The *excenters* of the triangle are the three points J, K, L that each lie on the internal angle bisector at one vertex and on the external angle bisectors at the other two vertices. The four points I, J, K, L are all called *equicenters*.

A line that intersects a circle in exactly one point is called a *tangent* of the circle. We end this section by using the previous results on angle bisectors to determine the circles that can be drawn tangent to the three sides of a given triangle. We start by proving that any point on a circle lies on a unique tangent.

THEOREM 5.7. In the Euclidean plane, let A be a point on a circle $\mathcal{K}$ with center O. Then there is exactly one line through A tangent to $\mathcal{K}$, the line l through A perpendicular to the radius OA. Every point on l other than A lies outside of $\mathcal{K}$. Any line m through A other than l intersects $\mathcal{K}$ in exactly two points, A and one other (Figures 5.14 and 5.15).

Proof: First, consider the line l through A perpendicular to OA (Figure 5.14). Let X be any point on l other than A. OAX is a right triangle with hypotenuse OX, and so the distance from O to X is greater than the distance from O to A (by the Pythagorean Theorem 0.9). Thus, since the distance from O to A is the radius of $\mathcal{K}$, X lies outside $\mathcal{K}$. Hence, A is the unique point of l that lies on $\mathcal{K}$, and so l is tangent to $\mathcal{K}$. We've also seen that every point X on l other than A lies outside $\mathcal{K}$.

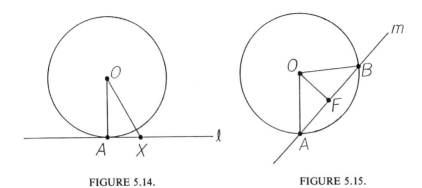

FIGURE 5.14.   FIGURE 5.15.

Next, consider a line m through A other than l (Figure 5.15). Let F be the point where m intersects the line through O perpendicular to m. F doesn't equal A, since m isn't perpendicular to OA. Let B be the point on m such that F is the midpoint of A and B. O lies on the perpendicular bisector of A and B, and so O is equidistant from A and B (by Theorem 4.2). Thus, B lies on $\mathcal{K}$, since the distance from O to A is the radius of $\mathcal{K}$. In short, m and $\mathcal{K}$ intersect at the two points A and B, and so m isn't tangent to $\mathcal{K}$. Moreover, m doesn't intersect $\mathcal{K}$ in any points other than A and B, by Theorem 4.5.

The two paragraphs preceding show that l is the unique line through A tangent to $\mathcal{K}$. □

Let A be any point on a circle $\mathcal{K}$. By Theorem 5.7, there is a unique line through A tangent to $\mathcal{K}$. We call this the *tangent at A*, and we write it as $\tan A$.

Theorem 4.3 shows that the circumcenter of a triangle—the point where the perpendicular bisectors meet—is the center of the unique circle containing the vertices of the triangle. Analogously, we show that the equicenters of a triangle—the points where the angle bisectors meet—are the centers of the four circles tangent to the sides of a triangle (Figure 5.16).

THEOREM 5.8. In the Euclidean plane, the three sides of a triangle are tangent to exactly four circles. The centers of the circles are the equicenters of the triangle (Figure 5.16).

Proof: First consider any point P and any line l (Figure 5.17). Let F be the foot of the perpendicular from P to l. Let $\mathcal{K}$ be a circle with center P and radius r. $\mathcal{K}$ is tangent to l if and only if $\mathcal{K}$ contains F (by Theorem 5.7). This occurs if and only if the radius r of $\mathcal{K}$ is the distance from P to F. Thus, $\mathcal{K}$ is tangent to l if and only if r is the distance from P to l (by Definition 5.4).

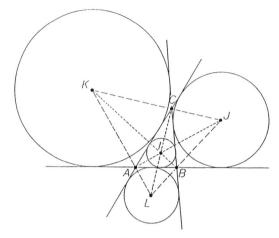

FIGURE 5.16.

Now let *ABC* be any triangle, and let *P* be any point. The previous paragraph shows that *P* is the center of a circle tangent to the sides of the triangle if and only if *P* is equidistant from the three sides. Moreover, if *P* is equidistant from the sides, the previous paragraph shows that the common distance from *P* to the sides is the radius of the unique circle with center *P* tangent to the sides. *P* is equidistant from the three sides if and only if *P* lies on an angle bisector at each of the three vertices (by Theorem 5.5). This occurs if and only if *P* is an equicenter, and so the theorem follows. □

By Theorem 5.8, there is a unique circle that is tangent to the sides of a triangle and that has the incenter of the triangle as its center (Figure 5.16). This circle is called the *incircle* of the triangle. By Theorem 5.8, there are also three circles tangent to the sides of a triangle that have the three excenters of the triangle as their respective centers. These three circles are called the *excircles* of the triangle. The incircle and the three excircles of a triangle are all called *equicircles*. By Theorem 5.8, the equicircles of a

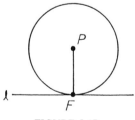

FIGURE 5.17.

triangle are the four circles tangent to the sides of a triangle. The incircle lies inside the triangle, and the three excircles lie outside the triangle.

Feuerbach's Theorem is the remarkable result that the nine-point circle of any triangle is tangent to the three excircles and is either tangent to or equal to the incircle. We use inversion in circles to outline a proof of Feuerbach's Theorem in Exercises 23.32–23.37.

EXERCISES

5.1. In the Euclidean plane, let A be a point on a line l, and let B be a point that doesn't lie on l. Prove that there is a unique circle that is tangent to l at A and contains B.

5.2. In the notation of Theorem 5.2, prove that $M_1 F_2$ and $N_1 F_2$ are perpendicular lines. Illustrate this result with a figure where $A_1 A_2 A_3$ is a scalene triangle with an obtuse angle.

5.3. In the notation of Theorem 5.2, assume that triangle $A_1 A_2 A_3$ doesn't have a right angle at A_1.

   (a) Use Exercise 4.2 to prove that $M_1 N_1$ is the perpendicular bisector of F_2 and F_3. Illustrate this result with a figure where $A_1 A_2 A_3$ is a scalene triangle with an obtuse angle at A_2.
   (b) Prove that the tangents to the nine-point circle at M_1 and N_1 are parallel to $F_2 F_3$. Illustrate this result with a figure where $A_1 A_2 A_3$ is a scalene triangle whose angles are all acute.
   (c) Prove that $M_1 F_2 T$ and $M_1 F_3 T$ are congruent triangles. Illustrate this result with a figure where $A_1 A_2 A_3$ has an obtuse angle at A_1.

5.4. In the notation of Theorem 5.2, let O be the circumcenter of triangle $A_1 A_2 A_3$. Set i equal to one of the values 1, 2, 3.

   (a) Assume that triangle $A_1 A_2 A_3$ doesn't have a right angle at A_i and that the two sides on A_i have different lengths. Prove that $N_i O M_i H$ and $A_i O M_i N_i$ are parallelograms. Illustrate this result with two figures, one where triangle $A_1 A_2 A_3$ doesn't have a right angle, and one where it does. (See Exercise 4.4 and its hint.)
   (b) If triangle $A_1 A_2 A_3$ has a right angle at A_i, prove that the points N_i, O, M_i, H, A_i lie on a line. Illustrate this result with a figure where $A_1 A_2 A_3$ is a scalene triangle.
   (c) If the two sides of triangle $A_1 A_2 A_3$ on A_i are equal, prove that the points N_i, O, M_i, H, A_i lie on a line. Illustrate this result with three figures: one where triangle $A_1 A_2 A_3$ isn't equilateral and doesn't have a right angle; one where $A_1 A_2 A_3$ is a right triangle; and one where $A_1 A_2 A_3$ is an equilateral triangle.

5.5. Let the notation be as in Theorm 5.2, and let O be the circumcenter of triangle $A_1A_2A_3$.

(a) If triangle $A_1A_2A_3$ isn't equilateral, use Exercise 5.4(a) to prove that N_iOM_iH and $A_iOM_iN_i$ are parallelograms for at least one of the values 1, 2, 3 of i.

(b) If triangle $A_1A_2A_3$ isn't equilateral, use part (a), Theorem 5.2(ii), and Exercise 0.2 to prove that the center T of the nine-point circle is the midpoint of the orthocenter H and the circumcenter O. Prove that $T = H = O$ if $A_1A_2A_3$ is an equilateral triangle.

(c) Prove that the radius of the nine-point circle is half the radius of the circumcircle. When triangle $A_1A_2A_3$ isn't equilateral, usc part (a), Theorem 5.2(ii), and Theorem 0.12. Consider separately the case when $A_1A_2A_3$ is an equilateral triangle.

5.6. Prove that a side of a triangle is tangent to the nine-point circle if and only if the other two sides are of equal length.

5.7. In the notation of Theorem 5.2, prove that the following conditions are equivalent.

(i) $F_1 = N_1$.
(ii) A_2 is equidistant from A_1 and H.
(iii) A_3 is equidistant from A_1 and H.
(iv) N_1 lies on A_2A_3.
(v) T lies on A_2A_3.
(vi) The nine-point circle is tangent to the altitude A_1H.

Illustrate this result with a figure where $F_1 = N_1$, $F_2 \neq N_2$, and $F_3 \neq N_3$.

5.8. In the notation of Theorem 5.2, prove that the following conditions are equivalent.

(i) $F_1 = N_1$, $F_2 = M_2$, and $F_3 = N_3$.
(ii) Two of the equations in (i) hold.
(iii) A_1HA_3 is an equilateral triangle.
(iv) $A_2 = T$.
(v) The nine-point circle is tangent to the altitudes on A_1 and A_3.

Illustrate this result with a figure.

5.9. In the Euclidean plane, let ABC be a triangle. Let I be the incenter, and let J, K, L be the excenters on the internal angle bisectors at A, B, C, respectively (Figure 5.11).

(a) Express each angle of triangle ABL in terms of the angles of triangle ABC.

(b) Prove that triangles JBC, AKC, ABL, and JKL are all similar.
(c) Express $\angle AIB$, $\angle BIC$, and $\angle AIC$ in terms of the angles of triangle ABC.

5.10. Use Exercise 3.11 to prove Theorem 5.6.

5.11. In the Euclidean plane, let ABC be a triangle. Let XY be the distance between points X and Y, except that "line XY" is the line through the two points.

(a) Let P be the point where the internal angle bisector at C intersects line AB. Use the Law of Sines (Theorem 0.15) to prove that

$$\overline{PA}/\overline{PB} = -AC/BC.$$

(Figures 5.18a, 5.18b, and 5.19).
(b) Prove that the external angle bisector at C is parallel to line AB if and only if $AC = BC$ (Figure 5.19).
(c) If $AC \neq BC$, the external angle bisector at C intersects line AB at a point Q (by part (b)). Use the Law of Sines (Theorem 0.15) to prove that

$$\overline{QA}/\overline{QB} = AC/BC$$

(Figures 5.18a and 5.18b).

(Thus, in general, *an angle bisector at a vertex of a triangle divides the opposite side into two segments proportional in length to the two adjacent sides*.)

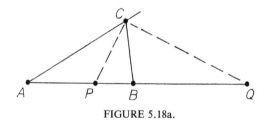

FIGURE 5.18a.

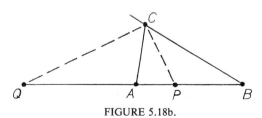

FIGURE 5.18b.

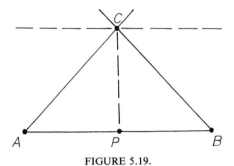

FIGURE 5.19.

5.12. In the Euclidean plane, let ABC be a scalene triangle. Prove that the external angle bisector at A intersects BC at a point A', the external angle bisector at B intersects CA at a point B', the external angle bisector at C intersects AB at a point C', and the points A', B', C' lie on a line. Illustrate this result with a figure. (See Exercise 5.11 and Menelaus' Theorem 2.4.)

5.13. In the Euclidean plane, let ABC be a triangle such that sides AB and AC have different lengths. Prove that the external angle bisector at A intersects BC at a point A', the internal angle bisector at B intersects CA at a point B', the internal angle bisector at C intersects AB at a point C', and the points A', B', C' lie on a line. Illustrate this result with a figure. (See Exercise 5.11 and Menelaus' Theorem 2.4.)

5.14. In the Euclidean plane, let A and B be two points (Figure 5.20). Let k be a positive number other than 1.

(a) Prove that there are exactly two points Y on line AB such that k is the ratio $|\overline{YA}/\overline{YB}|$ of the distances from Y to A and B.
(b) Let C and D be the two points on line AB determined in part (a). Let X be any point not on line AB such that k is the ratio of the distances

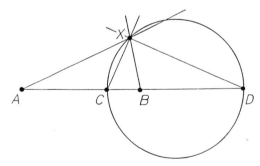

FIGURE 5.20.

from X to A and B. Use Exercise 5.11 to deduce that CX and DX bisect the two pairs of angles formed by AX and BX. Then use Theorem 4.7 to conclude that X lies on the circle with diameter CD.

(c) Let C and D be as in part (b). Let X be any point other than C and D on the circle with diameter CD. Prove that k is the ratio of the distances from X to A and B. (*Hint:* Show that we can assume that A doesn't lie between C and D. Prove that there is a point B' on AB such that CX and DX bisect the two pairs of angles formed by AX and $B'X$. Show that $\overline{DA}/\overline{DB} = -\overline{CA}/\overline{CB}$, and conclude from Exercise 5.11 that $\overline{DA}/\overline{DB'} = -\overline{CA}/\overline{CB'}$. Then prove that $B' = B$, and conclude that k is the ratio of the distances from X to A and B.)

(This exercise shows that there is a circle that consists of exactly the points X such that k is the ratio of the distances from X to A and B. This circle is called the *circle of Apollonius* for points A and B and ratio k.)

5.15. In the Euclidean plane, let ABC be a triangle. Prove that any three of the equicenters of triangle ABC are the vertices of a triangle $\mathfrak{I}$ that has the following properties (Figure 5.11):

   (i) The orthocenter of triangle $\mathfrak{I}$ is the fourth equicenter of triangle ABC.
   (ii) Triangle $\mathfrak{I}$ doesn't have a right angle.
   (iii) The feet of the altitudes of triangle $\mathfrak{I}$ are A, B, C.

5.16. Let triangles ABC and $\mathfrak{I}$ be as in Exercise 5.15. Prove that the nine-point circle of triangle $\mathfrak{I}$ is the circumcircle of triangle ABC.

5.17. In the Euclidean plane, let ABC be any triangle. Prove that the circumcircle of triangle ABC contains the midpoint of any two equicenters of triangle ABC. Illustrate this result with a figure where triangle ABC is scalene. (See Exercise 5.16.)

5.18. Consider the following result (Figure 5.21):

Theorem. In the Euclidean plane, let ABC be a triangle without right angles at A or B. Let F be the foot of the altitude on C, and let Q be a point on the altitude other than C and F. Assume that AQ intersects BC at a point A' and that BQ intersects AC at a point B'. Then $A'F$ and $B'F$ are distinct lines, and the angles they form are bisected by the side AB and the altitude CF of triangle ABC.

(a) Let l be the line though C parallel to AB (Figure 5.22). Choose corresponding positive ends for directed distances on l and AB. Let A'' and B'' be the points where l intersects $A'F$ and $B'F$, respectively.

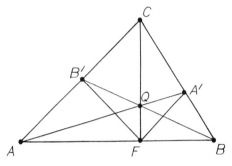

FIGURE 5.21.

Prove that the following equations hold:

$$\frac{\overline{A'B}}{\overline{A'C}} \cdot \frac{\overline{B'C}}{\overline{B'A}} \cdot \frac{\overline{FA}}{\overline{FB}} = -1, \qquad \frac{\overline{A'B}}{\overline{A'C}} = \frac{\overline{FB}}{\overline{A''C}}, \qquad \frac{\overline{B'C}}{\overline{B'A}} = \frac{\overline{B''C}}{\overline{FA}}.$$

(See Exercise 2.11.)
(b) Conclude from part (a) that $\overline{CB''}/\overline{CA''} = -1$.
(c) Use part (b), Theorem 1.5, and Property 0.2 to prove the stated theorem.

5.19. Prove the following result (Figure 5.23):

Theorem. In the Euclidean plane, let ABC be a triangle without a right angle. Let D, E, F be the feet of the altitudes on A, B, C, respectively. Then DEF is a triangle whose six angle bisectors are the three sides and the three altitudes of triangle ABC.

(Triangle DEF is called the *orthic triangle* of triangle ABC. One possible approach to this exercise uses Exercise 5.18.)

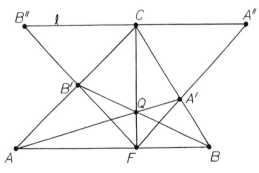

FIGURE 5.22.

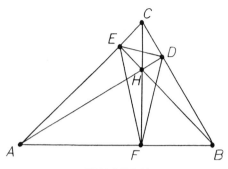

FIGURE 5.23.

5.20. In the Euclidean plane, let UVW be a triangle. Use Exercises 5.15 and 5.19 to prove that there are exactly four triangles that each have U, V, W as the feet of their altitudes.

5.21. In the Euclidean plane, let PQR be a triangle without a right angle. Use Exercises 5.15 and 5.19 to prove that there is a unique triangle $\mathfrak{I}'$ that has P, Q, R as three of its equicenters.

5.22. In the Euclidean plane, let ABC be a triangle. Let a, b, c be the respective lengths of sides BC, CA, AB, and set $s = \frac{1}{2}(a + b + c)$. Let r be the radius of the incircle, let r_1 be the radius of the excircle on the side of BC opposite A, let r_2 be the radius of the excircle on the side of CA opposite B, and let r_3 be the radius of the excircle on the side of AB opposite C (Figure 5.16).

    (a) Prove that area($\triangle ABC$) $= rs$. (*Hint:* One possible approach is as follows. Let I be the incenter, express the area of triangle ABC in terms of the areas of triangles IBC, AIC, and ABI, and use Theorem 0.13 to express the areas of the four triangles here in terms of r, a, b, c.)
    (b) Prove that

$$\text{area}(\triangle ABC) = (s - a)r_1 = (s - b)r_2 = (s - c)r_3.$$

(See the hint to part (a).)

5.23. In the notation of Exercise 5.22, use that exercise to prove that

$$\frac{1}{r_1} + \frac{1}{r_2} + \frac{1}{r_3} = \frac{1}{r}.$$

5.24. In the notation of Exercise 5.22, prove that

$$rr_1r_2r_3 = (\text{area } \triangle ABC)^2.$$

(*Hint:* One possible approach is to use Exercise 5.22 to express the left side of this equation in terms of a, b, c, s, and the area of triangle ABC. Then use Exercise 1.16b.)

5.25. In the notation of Exercise 5.22, let h_A, h_B, h_C be the respective lengths of the altitudes on A, B, C. Prove that

$$\frac{1}{h_A} + \frac{1}{h_B} + \frac{1}{h_C} = \frac{1}{r}.$$

(*Hint:* One possible approach is to use Theorem 0.13 to express the left side of this equation in terms of a, b, c, and the area of triangle ABC. Then apply Exercise 5.22a.)

Section 6.

Circles and Directed Distances

So far we've used directed distances primarily to study triangles. We now use directed distances to study circles. In particular, if a scalene triangle is inscribed in a circle, we prove that the tangents at the vertices intersect the opposite sides of the triangle in three collinear points. The results of this section are vital for studying triangles inscribed in and circumscribed about conic sections (in Section 18) and for studying inversion in circles (in Section 23).

The part of a circle that lies inside a given angle is called the arc that the angle *intercepts*. A *central angle* of a circle is one of the two angles formed by two rays that originate at the center of the circle; for example, the angles marked x and y in Figure 6.1 are both central angles. Central angles vary from 0° to 360°. We say that $\angle ACB$ is *inscribed* in a circle if the points A, B, C lie on the circle (Figure 6.2a). We recall from the discussion after Property 0.4 that we only use the notation $\angle ACB$ for an angle θ such that $0° \le \theta \le 180°$.

We begin with a fundamental result in Euclidean geometry: *An angle inscribed in a circle equals one-half of the central angle that intercepts the same arc.* (See Figures 6.2a and b, where $\angle ACB$ is the inscribed angle, and x is the corresponding central angle.)

THEOREM 6.1. In the Euclidean plane, let A, B, C be three points on a circle. Then $\angle ACB$ equals one-half the central angle that intercepts the same arc (Figures 6.2a and b).

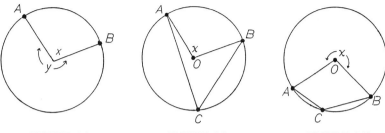

FIGURE 6.1.   FIGURE 6.2a.   FIGURE 6.2b.

Proof: Let x be the central angle that intercepts the same arc as $\angle ACB$. Let O be the center of the circle. There are three cases, depending on whether O lies on a side of $\angle ACB$, inside this angle, or outside this angle.

Case 1: O lies on a side of $\angle ACB$. By symmetry, we can assume that O lies on side AC (Figure 6.3). Set $y = \angle ACB$. The sides OB and OC of triangle BOC are equal, since these sides are radii of the circle. $\angle OBC$ and $\angle OCB$ lie opposite the equal sides OB and OC of triangle BOC, and so these angles are equal (by Theorem 0.10). Thus, we have $\angle OBC = \angle OCB = y$. $\angle BOC$ equals $180° - x$, since this angle combines with $\angle AOB = x$ to form a straight angle. The angles of triangle BOC total $180°$ (by Theorem 0.8), and so we have

$$(180° - x) + 2y = 180°,$$

which simplifies to $y = \frac{1}{2}x$, as desired.

Case 2: O lies inside $\angle ACB$ (Figure 6.4). Let D be the other endpoint of diameter OC. Case 1 shows that $\angle ACD = \frac{1}{2}\angle AOD$ and $\angle BCD = \frac{1}{2}\angle BOD$. Thus, we have

$$\angle ACB = \angle ACD + \angle BCD = \frac{1}{2}\angle AOD + \frac{1}{2}\angle BOD$$
$$= \frac{1}{2}(\angle AOD + \angle BOD) = \frac{1}{2}x.$$

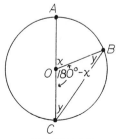

FIGURE 6.3.

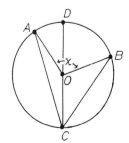

FIGURE 6.4.

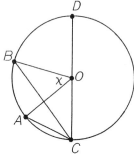

FIGURE 6.5.

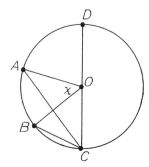

FIGURE 6.6.

Case 3: O lies outside $\angle ACB$. Either B lies inside $\angle ACO$ (Figure 6.5) or A lies inside $\angle BCO$ (Figure 6.6), depending on whether A or B lies closer to C. By symmetry, we can assume that B lies inside $\angle ACO$ (Figure 6.5). Let D be the other endpoint of diameter OC. Case 1 shows that $\angle ACD = \frac{1}{2}\angle AOD$ and $\angle BCD = \frac{1}{2}\angle BOD$. Thus, we have

$$\angle ACB = \angle ACD - \angle BCD = \frac{1}{2}\angle AOD - \frac{1}{2}\angle BOD$$

$$= \frac{1}{2}(\angle AOD - \angle BOD) = \frac{1}{2}\angle AOB = \frac{1}{2}x. \quad \square$$

The next result follows directly from the previous theorem. It states that, *if two angles inscribed in a circle have the same endpoints, then the angles are either equal or supplementary, depending on whether or not they intercept the same arc of the circle.*

THEOREM 6.2. In the Euclidean plane, let A, B, C, D be four points on a circle.

(i) If $\angle ACB$ and $\angle ADB$ intercept the same arc, then these angles are equal (Figure 6.7).

(ii) If $\angle ACB$ and $\angle ADB$ intercept different arcs, then these angles are supplementary (Figure 6.8).

Proof: Let x be the central angle that intercepts the same arc as $\angle ACB$, and let y be the central angle that intercepts the same arc as $\angle ADB$. Theorem 6.1 shows that

$$\angle ACB = \frac{1}{2}x \quad \text{and} \quad \angle ADB = \frac{1}{2}y. \tag{1}$$

(i) If $\angle ACB$ and $\angle ADB$ intercept the same arc, then x equals y (Figure 6.7). Thus, (1) shows that

$$\angle ACB = \frac{1}{2}x = \frac{1}{2}y = \angle ADB.$$

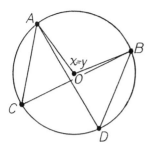

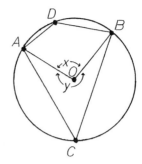

FIGURE 6.7.   FIGURE 6.8.

(ii) If $\angle ACB$ and $\angle ADB$ intercept different arcs, then x and y are the two angles formed by the rays $\overrightarrow{OA}$ and $\overrightarrow{OB}$ (Figure 6.8). Thus, we have $x + y = 360°$. Together with (1), this shows that

$$\angle ACB + \angle ADB = \tfrac{1}{2}x + \tfrac{1}{2}y = \tfrac{1}{2}(x + y) = \tfrac{1}{2}(360)° = 180°,$$

and so $\angle ACB$ and $\angle ADB$ are supplementary. □

In Theorem 6.1 and Figure 6.2a, imagine that the points O, A, B are fixed, while the point C moves around the circle toward A (Figure 6.9). As C approaches A, the line CA approaches tan A, and CB approaches AB. As C approaches A from one side, $\angle ACB$ maintains a constant value, one-half of the central angle that intercepts the same arc (by Theorem 6.1). The last two sentences suggest that *tan A and AB form an angle that is one-half of the central angle that intercepts the same arc*. The next theorem confirms this conjecture.

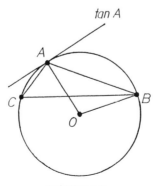

FIGURE 6.9.

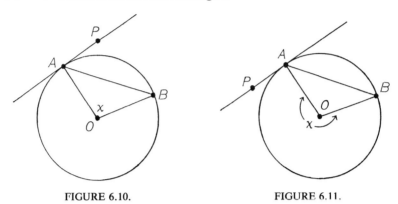

FIGURE 6.10.   FIGURE 6.11.

THEOREM 6.3. In the Euclidean plane, let A and B be two points on a circle. Let P be a point on tan A other than A. Then $\angle PAB$ equals one-half of the central angle that intercepts the same arc (Figures 6.10 and 6.11).

Proof: Let x be the central angle that intercepts the same arc as $\angle PAB$. We must prove that $\angle PAB = \frac{1}{2}x$.

Let O be the center of the circle. Since $AP = \tan A \neq AO$, P doesn't lie on AO. Thus, there are three cases, depending on whether B lies on AO, B and P lie on the same side of AO, or B and P lie on opposite sides of AO.

Case 1: B lies on AO (Figure 6.12). Then A and B are the endpoints of a diameter, and we have $x = \angle AOB = 180°$. By Theorem 5.7, we have

$$\angle PAB = 90° = \tfrac{1}{2}(180)° = \tfrac{1}{2}x.$$

Case 2: B and P lie on the same side of AO (Figure 6.13). Let D be the other endpoint of the diameter AO. We have

$$\begin{aligned}
\angle PAB &= \angle PAD - \angle BAD \\
&= 90° - \tfrac{1}{2}\angle BOD \quad \text{(by Theorems 5.7 and 6.1)} \\
&= \tfrac{1}{2}(180° - \angle BOD) \\
&= \tfrac{1}{2}x.
\end{aligned}$$

Case 3: B and P lie on opposite sides of AO (Figure 6.14). Let D be the other endpoint of the diameter AO. We have

$$\begin{aligned}
\angle PAB &= \angle PAD + \angle BAD \\
&= 90° + \tfrac{1}{2}\angle BOD \quad \text{(by Theorems 5.7 and 6.1)} \\
&= \tfrac{1}{2}(180° + \angle BOD) \\
&= \tfrac{1}{2}x. \quad \square
\end{aligned}$$

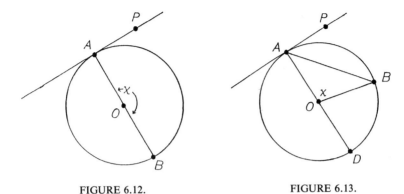

FIGURE 6.12.   FIGURE 6.13.

Combining Theorems 6.1–6.3 gives the following result.

THEOREM 6.4. In the Euclidean plane, let A, B, C be three points on a circle. Let P be a point on $\tan A$ other than A.

  (i) If $\angle PAB$ and $\angle ACB$ intercept the same arc, then these angles are equal (Figure 6.15).

  (ii) If $\angle PAB$ and $\angle ACB$ intercept different arcs, then these angles are supplementary (Figure 6.16).

Proof:
  (i) Let x be the central angle that intercepts the same arc as $\angle PAB$ (Figure 6.17). Since $\angle PAB$ and $\angle ACB$ intercept the same arc, so do x and $\angle ACB$. Thus, we have

$$\angle PAB = \tfrac{1}{2}x = \angle ACB$$

(by Theorems 6.3 and 6.1).

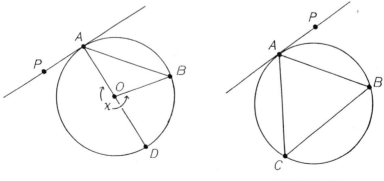

FIGURE 6.14.   FIGURE 6.15.

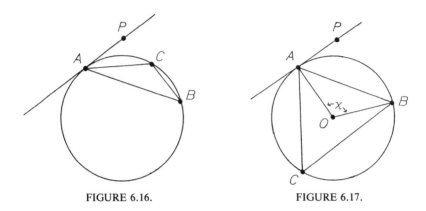

FIGURE 6.16.   FIGURE 6.17.

(ii) Let D be a point on the circle that lies inside $\angle ACB$ (Figure 6.18). $\angle ADB$ intercepts the same arc as $\angle PAB$ and a different arc from $\angle ACB$. Thus, we have

$$\angle PAB = \angle ADB = 180° - \angle ACB$$

(by part (i) and Theorem 6.2ii). □

The theorems so far in this chapter have concerned angles determined by points on circles. We now use these theorems to obtain results on directed distances determined by points on circles.

Let A, B, C be three points on a line l. The value of $(\overline{CB})^2$ doesn't depend on the choice of a positive end of l for directed distances; switching the positive end of l replaces $(\overline{CB})^2$ with $(-\overline{CB})^2 = (\overline{CB})^2$ (by Theorem 1.1ii), and so the value of $(\overline{CB})^2$ remains unchanged. Thus, since

$$\overline{CA} \cdot \overline{CB} = (\overline{CA}/\overline{CB}) \cdot (\overline{CB})^2, \tag{2}$$

the value of $\overline{CA} \cdot \overline{CB}$ doesn't depend on the choice of a positive end of l (by Theorem 1.2ii). Equation 2 also implies that $\overline{CA} \cdot \overline{CB}$ is negative if and only if C lies between A and B (by Theorem 1.4 and the fact that $(\overline{CB})^2$ is positive).

THEOREM 6.5. In the Euclidean plane, let A, B, C, D be four points on a circle $\mathcal{K}$ such that the lines AB and CD intersect at a point P. Then triangles APC and DPB are similar, and we have

$$\overline{PA} \cdot \overline{PB} = \overline{PC} \cdot \overline{PD}$$

(Figures 6.19–6.21).

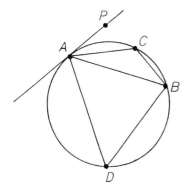

FIGURE 6.18.

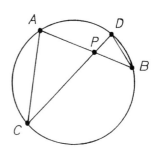

FIGURE 6.19.

Proof: We prove first that triangles APC and DPB are similar. P doesn't lie on $\mathcal{K}$ (by Theorem 4.5), and so P lies either inside or outside $\mathcal{K}$. If P lies outside $\mathcal{K}$, then either B lies between A and P or A lies between B and P; we can assume that B lies between A and P (by interchanging A with B and C with D, if necessary). Thus, it's enough to consider three cases: first, P lies inside $\mathcal{K}$ (Figure 6.19); second, P lies outside $\mathcal{K}$, B lies between A and P, and C lies between D and P (Figure 6.20); and third, P lies outside $\mathcal{K}$, B lies between A and P, and D lies between C and P (Figure 6.21).

Case 1: P lies inside $\mathcal{K}$ (Figure 6.19), $\angle BAC$ equals $\angle BDC$ (by Theorem 6.2i), and so triangles APC and DPB have equal angles at A and D. These triangles have equal angles at P (by Theorem 0.6), and so they also have equal angles at C and B (by Theorem 0.8). Thus, triangles APC and DPB are similar.

Case 2: P lies outside $\mathcal{K}$, B lies between A and P, and C lies between D and P (Figure 6.20). $\angle BAC$ equals $\angle BDC$ (by Theorem 6.2i), and so triangles APC and DPB have equal angles at A and D. These triangles

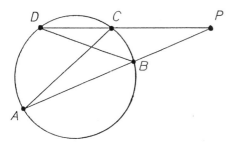

FIGURE 6.20.

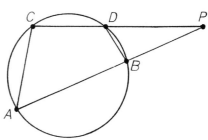

FIGURE 6.21.

have the same angle at P, and so they also have equal angles at C and B (by Theorem 0.8). Thus, triangles APC and DPB are similar.

Case 3: P lies outside $\mathcal{K}$, B lies between A and P, and D lies between C and P (Figure 6.21). We have

$$\angle BAC = 180° - \angle BDC = \angle BDP$$

(by Theorem 6.2ii), and so triangles APC and DPB have equal angles at A and D. These triangles have the same angle at P, and so they also have equal angles at C and B (by Theorem 0.8). Thus, triangles APC and DPB are similar.

Combining these three cases shows that triangles APC and DPB are always similar. Corresponding sides of similar triangles are proportional (by Property 0.4), and so we have

$$|\overline{PA}/\overline{PC}| = |\overline{PD}/\overline{PB}|.$$

We can rewrite this as

$$|\overline{PA} \cdot \overline{PB}| = |\overline{PC} \cdot \overline{PD}|.$$

It follows that

$$\overline{PA} \cdot \overline{PB} = \overline{PC} \cdot \overline{PD},$$

since the discussion before the theorem shows that both sides of this equation are negative when P lies inside $\mathcal{K}$ (Figure 6.19) and that both sides of this equation are positive when P lies outside $\mathcal{K}$ (Figures 6.20 and 6.21). □

In Figures 6.20 and 6.21, imagine that line CP revolves around P until it becomes tangent to the circle. As this happens, D approaches C, and CD approaches $\tan C$. Thus, we're led to the next result, which is obtained by replacing D with C in Theorem 6.5.

THEOREM 6.6. In the Euclidean plane, let A, B, C be three points on a circle $\mathcal{K}$ such that the lines AB and $\tan C$ intersect at a point P. Then triangles APC and CPB are similar, and we have

$$\overline{PA} \cdot \overline{PB} = (\overline{PC})^2$$

(Figure 6.22).

Proof: Since P lies on $\tan C$, it lies outside $\mathcal{K}$ (by Theorem 5.7), and so either A lies between B and P or B lies between A and P. By symmetry, we can assume that B lies between A and P.

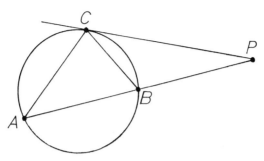

FIGURE 6.22.

∠*CAB* equals ∠*PCB* (by Theorem 6.4i), and so triangles *APC* and *CPB* have equal angles at *A* and *C*. These triangles have the same angle at *P*, and so they also have equal angles at *C* and *B* (by Theorem 0.8). Thus, triangles *APC* and *CPB* are similar.

Corresponding sides of triangles *APC* and *CPB* are proportional (by Property 0.4), and so we have

$$|\overline{PA}/\overline{PC}| = |\overline{PC}/\overline{PB}|.$$

We can rewrite this as

$$|\overline{PA} \cdot \overline{PB}| = |\overline{PC}|^2.$$

It follows that

$$\overline{PA} \cdot \overline{PB} = (\overline{PC})^2,$$

since *P* lies outside 𝒦, and so both sides of this equation are positive (by the discussion before Theorem 6.5). □

When a triangle is inscribed in a circle, the next result determines the ratio in which a side of the triangle is divided by the tangent through the opposite vertex. This result follows directly from Theorem 6.6. In the statement and proof of this result, we let *XY* be the distance between points *X* and *Y* except when we write "line *XY*."

THEOREM 6.7. In the Euclidean plane, let *ABC* be a triangle, and let 𝒦 be its circumcircle. Let tan *C* be the tangent to 𝒦 at *C*.

(i) If *AC* ≠ *BC*, then tan *C* intersects line *AB* at a point *P* such that

$$\overline{PA}/\overline{PB} = (AC/BC)^2$$

(Figure 6.22).

(ii) If *AC* = *BC*, then tan *C* is parallel to line *AB* (Figure 6.23).

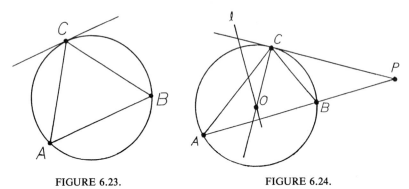

FIGURE 6.23.   FIGURE 6.24.

Proof: The circumcenter O of triangle ABC is the center of $\mathcal{K}$. (See Definition 4.4.)

(i) Since $AC \neq BC$ in this case, C doesn't lie on the perpendicular bisector l of A and B (by Theorem 4.2), which is the line through O perpendicular to line AB (Figure 6.24). Thus, the lines CO and AB aren't perpendicular. The lines CO and $\tan C$, however, are perpendicular (by Theorem 5.7). The last two sentences imply that the lines AB and $\tan C$ aren't parallel (by Property 0.5), and so they intersect at a point P.

Triangles APC and CPB are similar (by Theorem 6.6). Corresponding sides of similar triangles are proportional (by Property 0.4), and so we have

$$PA/AC = PC/CB \quad \text{and} \quad PC/AC = PB/CB.$$

It follows that

$$PA = \frac{PC \cdot AC}{CB} \quad \text{and} \quad \frac{1}{PB} = \frac{AC}{PC \cdot CB}.$$

Multiplying these equations together gives

$$PA/PB = (AC/CB)^2.$$

Thus, we have

$$\overline{PA}/\overline{PB} = (AC/BC)^2,$$

because P lies outside $\mathcal{K}$ (by Theorem 5.7, since P lies on $\tan C$), and so $\overline{PA}/\overline{PB}$ is positive (by Theorem 1.4).

(ii) Since $AC = BC$ in this case, C lies on the perpendicular bisector l of A and B (by Theorem 4.2), which is the line through O perpendicular to line AB (Figure 6.25). Thus, the lines CO and AB are perpendicular. The lines CO and $\tan C$ are also perpendicular (by Theorem 5.7). The last two sentences imply that the lines $\tan C$ and AB are parallel (by Property 0.5). □

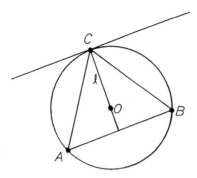

FIGURE 6.25.

Consider three points on a circle. These points are the vertices of a triangle inscribed in the circle (by Theorem 4.5ii). If the triangle is scalene, Theorem 6.7(i) shows that *each side of the triangle intersects the tangent through the opposite vertex at a point that divides the side in proportion to the squares of the lengths of the other two sides*. In this way, we determine three points, one on each side of the triangle. As Figure 6.26 shows, these three points lie on a line. We prove this statement here by combining the previous theorem with Menelaus' Theorem.

THEOREM 6.8. In the Euclidean plane, let A, B, C be three points on a circle. ABC is a triangle, by Theorem 4.5(ii).

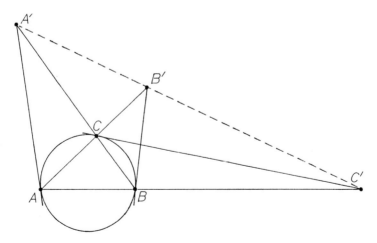

FIGURE 6.26.

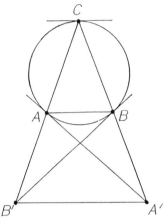

FIGURE 6.27.

(i) If triangle ABC is scalene, then tan A intersects side BC at a point A', tan B intersects side AC at a point B', tan C intersects side AB at a point C', and the points A', B', C' lie in a line (Figure 6.26).

(ii) If sides AC and BC have the same length, and if side AB has a different length, then tan A intersects side BC at a point A', tan B intersects side AC at a point B', and the points A' and B' lie on a line parallel to both tan C and AB (Figure 6.27).

(iii) If triangle ABC is equilateral, the tangent at each vertex is parallel to the opposite side of the triangle (Figure 6.28).

Proof: We let XY denote the distance between points X and Y except when we write "line XY."

(i) Theorem 6.7(i) shows that the points A', B', C' exist and that

$$\frac{\overline{A'B}}{\overline{A'C}} \cdot \frac{\overline{B'C}}{\overline{B'A}} \cdot \frac{\overline{C'A}}{\overline{C'B}} = \left(\frac{BA}{CA}\right)^2 \left(\frac{CB}{AB}\right)^2 \left(\frac{AC}{BC}\right)^2 = 1$$

(since $XY = YX$). Thus, A', B', C' lie on a line (by Menelaus' Theorem 2.4).

(ii) Theorem 6.7(i) shows that the points A' and B' exist and that

$$\frac{\overline{A'B}}{\overline{A'C}} \cdot \frac{\overline{B'C}}{\overline{B'A}} = \left(\frac{BA}{CA}\right)^2 \left(\frac{CB}{AB}\right)^2 = 1$$

(since $XY = YX$ and we have $AC = BC$ in this case). Thus, the lines $A'B'$ and AB are parallel (by Theorem 2.2). The lines AB and tan C are also parallel (by Theorem 6.7ii), and so the lines $A'B'$, AB, and tan C are all parallel (by Theorem 0.11).

Part (iii) follows from Theorem 6.7(ii). $\square$

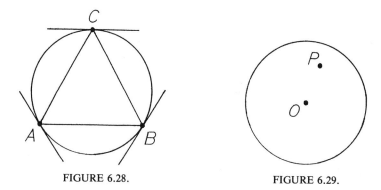

FIGURE 6.28.   FIGURE 6.29.

Theorem 6.6 provides information about the tangents through a point P not on a circle. We end this section with an easy result about the number of tangents through P.

THEOREM 6.9. In the Euclidean plane, let $\mathcal{K}$ be a circle with center O. Let P be a point that doesn't lie on $\mathcal{K}$.

   (i) If P lies inside $\mathcal{K}$, then P lies on no tangents to $\mathcal{K}$ (Figure 6.29).
   (ii) If P lies outside $\mathcal{K}$, then there are exactly two points on $\mathcal{K}$ whose tangents contain P. These points are equidistant from P; in fact, the square of the distance from either of these points to P is $d^2 - r^2$, where d is the distance from P to O and where r is the radius of $\mathcal{K}$ (Figure 6.30).

Proof: Part (i) follows from Theorem 5.7, which shows that every point on a line tangent to $\mathcal{K}$ lies either on or outside $\mathcal{K}$.

   (ii) Since P lies outside $\mathcal{K}$ in this case, the distance d from O to P is greater than the radius r of $\mathcal{K}$. Construct a right triangle EFG whose legs have lengths r and $\sqrt{d^2 - r^2}$ by marking off these distances on the sides

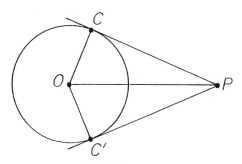

FIGURE 6.30.

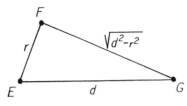

FIGURE 6.31.

of a right angle at F (Figure 6.31). The hypotenuse has length d (by the Pythagorean Theorem 0.9). Since the distance from P to O is also d, there are exactly two points C and C' such that triangles OCP and $OC'P$ are congruent to triangle EFG (Figure 6.30). Thus, C and C' are exactly the two points that lie r units from O and that determine perpendicular lines through O and P (by the Pythegorean Theorem 0.9 and the SSS Property 0.1). In other words, C and C' are exactly the two points on $\mathcal{K}$ at which the tangents contain P (by Theorem 5.7). □

EXERCISES

6.1. Prove the following result (Figure 6.32):

Theorem. In the Euclidean plane, let l and m be two lines on a point Z. Let $\mathcal{K}$ be a circle that intersects l at two points R and T other than Z and that intersects m at two points S and U. Let $\mathcal{K}'$ be a circle that intersects l at T and a second point V other than Z and that intersects m at U and a second point W. Then RS and VW are parallel lines.

(*Hint:* One possible approach is to combine Theorem 1.11 with two applications of Theorem 6.5.)

6.2. Prove the following result (Figure 6.33):

Theorem. In the Euclidean plane, let l and m be two lines on a point Z. Let $\mathcal{K}$ be a circle that intersects l at two points R and T other than Z and that is tangent to m at a point S. Let $\mathcal{K}'$ be a circle that intersects l at T and a second point V other than Z and that intersects m at S and a second point W. Then RS and VW are parallel lines.

6.3. Prove the following result (Figure 6.34):

Theorem. In the Euclidean plane, let l and m be two lines on a point Z. Let $\mathcal{K}$ be a circle that intersects l at two points R and T other than Z and that intersects m at two points S and U. Let $\mathcal{K}'$ be a circle through T, U, and Z. Then RS is parallel to the tangent n to $\mathcal{K}'$ at Z.

(*Hint:* One possible approach is to combine Theorems 6.2 and 6.4 with Property 0.5. Be sure to take into account the fact that Z may or may not lie inside $\mathcal{K}$.)

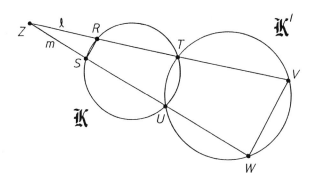

FIGURE 6.32.

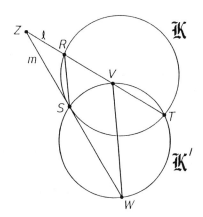

FIGURE 6.33.

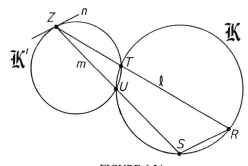

FIGURE 6.34.

6.4. In the Euclidean plane, let ABC be a triangle without a right angle. Let H be the orthocenter, and let A', B', C' be the feet of the altitudes on A, B, C. In each part of this exercise, two angles are given. What condition on the angles of triangle ABC is equivalent to the condition that the two given angles are equal? What condition on the angles of triangle ABC is equivalent to the condition that the two given angles are supplementary? Justify your answers.

   (a) $\angle B'AC'$ and $\angle B'HC'$
   (b) $\angle B'AH$ and $\angle B'C'H$
   (c) $\angle A'AB'$ and $\angle A'BB'$

6.5. In the Euclidean plane, let H be the orthocenter of triangle ABC. Let A', B', C' be the respective feet of the altitudes on A, B, C. Prove that

$$\overline{HA} \cdot \overline{HA'} = \overline{HB} \cdot \overline{HB'} = \overline{HC} \cdot \overline{HC'}, \tag{3}$$

where we set $\overline{XY} = 0$ when $X = Y$.

6.6.
   (a) In the Euclidean plane, let ABC be a triangle without a right angle. Let A', B', C' be the feet of the altitudes on A, B, C, respectively. Prove that the triangles $AB'C'$, $A'BC'$, $A'B'C$ are each similar to triangle ABC.
   (b) Use part (a) and Theorem 5.3 to prove the theorem in Exercise 5.19.

   (In each part of this exercise be sure to take into account that triangle ABC may or may not have an obtuse angle.)

6.7. In the Euclidean plane, let ABC be a triangle, and let A', B', C' be the feet of the altitudes on A, B, C.

   (a) If ABC isn't a right triangle, use Theorem 6.5 to prove that

$$\overline{AB} \cdot \overline{C'B} = \overline{CB} \cdot \overline{A'B}, \qquad \overline{BC} \cdot \overline{A'C} = \overline{AC} \cdot \overline{B'C},$$

$$\overline{CA} \cdot \overline{B'A} = \overline{BA} \cdot \overline{C'A}.$$

   (b) If ABC isn't a right triangle, use part (a), Ceva's Theorem, and Theorem 1.10 to prove that the three altitudes lie on a common point.
   (c) If ABC is a right triangle, conclude directly from the definition of altitudes that the three altitudes of triangle ABC lie on a common point.

   (Parts (b) and (c) provide another proof of Theorem 4.11i.)

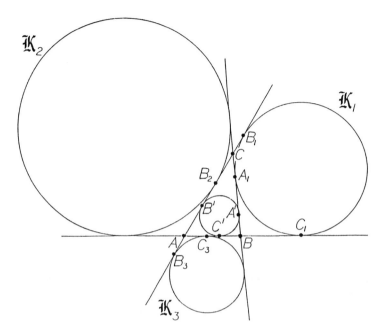

FIGURE 6.35.

Exercises 6.8–6.13 use the following notation (Figure 6.35). In the Euclidean plane, let ABC be a triangle, and let $\mathcal{K}$ be the incircle. Let $\mathcal{K}$ be tangent to BC at A', tangent to CA at B', and tangent to AB at C'. Let $\mathcal{K}_1$ be the excircle tangent to BC at a point A_1 between B and C, let $\mathcal{K}_2$ be the excircle tangent to CA at a point B_2 between C and A, and let $\mathcal{K}_3$ be the excircle tangent to AB at a point C_3 between A and B. Let B_1 and C_1 be the respective points where $\mathcal{K}_1$ is tangent to CA and AB, and let B_3 be the point where $\mathcal{K}_3$ is tangent to CA. Let a, b, c be the respective lengths of sides BC, CA, AB of triangle ABC, and set $s = \frac{1}{2}(a + b + c)$. (We call s the *semiperimeter* of triangle ABC.)

6.8. Prove that the distance from C' to B is $s - b$ and that the distance from C' to A is $s - a$.

(*Hint:* One possible approach is as follows. As in Figure 6.36, let u, v, w, x, y, z be the distances from A', B', C' to the adjacent vertices of triangle ABC. Use the fact that the sides of triangle ABC have lengths a, b, c to obtain three equations relating the unknowns $u, \ldots, z$ and the quantities a, b, c. Use Theorem 6.9(ii) to obtain three more equations relating $u \ldots, z$. Solve the resulting system of six equations in the six unknowns $u, \ldots, z$ for u and v in terms of a, b, c.)

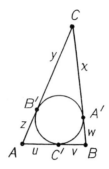

FIGURE 6.36.

6.9.

(a) Prove that the distance from C_3 to B is $s - a$ and that the distance from C_3 to A is $s - b$. (*Hint:* One possible approach is to adapt the hint to Exercise 6.8.)

(b) Prove that the distance from B_3 to A is $s - b$ and that the distance from B_3 to C is s.

6.10. Use Exercises 6.8 and 6.9 to prove the following statements.

(a) The distance between C' and C_3 is $|b - a|$.

(b) C' and C_3 have the same midpoint as A and B. (As usual, we take the midpoint of X and Y to be X when X equals Y.)

6.11. Prove that the lines AA', BB', CC' lie on a common point. Illustrate this result with a figure.

6.12. Prove that the lines AA_1, BB_1, CC_1 lie on a common point. Illustrate this result with a figure.

6.13. Prove that the lines AA_1, BB_2, CC_3 lie on a common point Q. Illustrate this result with a figure. (Q is called the *Nagel point* of triangle ABC.)

6.14. In the Euclidean plane, let ABC be a triangle. Let $\mathcal{K}$ be the circumcircle, and let F be the foot of the altitude on C. Let D be the other endpoint of the diameter that has C as one endpoint. In each of the following cases, prove that ACF and DCB are similar triangles, and illustrate this result with a figure.

(a) Triangle ABC has acute angles at both A and B.
(b) Triangle ABC has an obtuse angle at A.
(c) Triangle ABC has an obtuse angle at B.
(d) Triangle ABC has a right angle at B.

6.15. In the Euclidean plane, let ABC be a triangle. Let a, b, c be the respective lengths of sides BC, CA, AB. Let h be the length of the altitude on C, and let R be the radius of the circumcircle.

(a) Prove that $ab = 2hR$. (See Exercise 6.14.)

(b) Prove that

$$\text{area}(\triangle ABC) = \frac{abc}{4R}.$$

(c) Let r be the radius of the incircle, and set $s = \frac{1}{2}(a + b + c)$. Use part (b) and Exercise 5.22(a) to conclude that $abc = 4Rrs$.

6.16. In this exercise, let XY denote the distance between any points X and Y. Consider the following result (Figure 6.37):

Ptolemy's Theorem. In the Euclidean plane, let A, B, C, D be four points arranged consecutively around a circle. Then we have the relation

$$AB \cdot CD + BC \cdot DA = AC \cdot BD.$$

In other words, *if a quadrilateral is inscribed in a circle, the product of the diagonals is the sum of the products of the opposite sides.*

(a) Prove that there is a point E on the line segment with endpoints A and C such that $\angle ABE = \angle DBC$ and $\angle EBC = \angle ABD$ (Figure 6.38).

(b) Prove that triangles BAE and BDC are similar and that triangles BCE and BDA are similar.

(c) Use part (b) and Property 0.4 to conclude that

$$AB \cdot CD = AE \cdot BD \quad \text{and} \quad BC \cdot DA = CE \cdot BD.$$

(d) Deduce Ptolemy's Theorem.

Exercises 6.17–6.25 use the following definition. In the Euclidean plane, let A, B, C be three points that don't lie on a line (Figure 6.39). The *l-angle* $l\angle ACB$ is the angle θ that lies counterclockwise after line AC and before

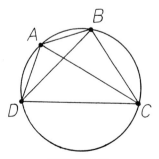

FIGURE 6.37.

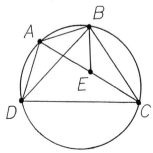

FIGURE 6.38.

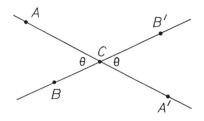

FIGURE 6.39.

line BC. That is, $l \angle ACB$ is the angle θ between $0°$ and $180°$ such that a counterclockwise rotation through angle θ about C maps line AC to line BC. Note that $l \angle ACB$ depends only on the lines AC and BC and not on the rays $\overrightarrow{CA}$ and $\overrightarrow{CB}$: If $\overrightarrow{CA}$ and $\overrightarrow{CA}'$ are the two rays on line AC originating at C, and if $\overrightarrow{CB}$ and $\overrightarrow{CB}'$ are the two rays on line BC originating at C, then we have

$$l \angle ACB = l \angle A'CB = l \angle ACB' = l \angle A'CB'.$$

Note also that

$$l \angle BCA = 180° - l \angle ACB,$$

and so the expression $l \angle ACB$ is not symmetric in A and B. The term l-angle is not standard, but we can avoid considering many separate cases in Exercises 6.21–6.25 by using this term. We use the letter l to stand for "line."

6.17. In the Euclidean plane, let A, B, C, D be four points on a circle. Prove that $l \angle ACB = l \angle ADB$ (Figures 6.7 and 6.8).

6.18. In the Euclidean plane, let A, B, C be three points on a circle, and let P be a point other than A on the tangent at A. Prove that $l \angle ACB = l \angle PAB$ (Figures 6.15 and 6.16).

6.19. In the Euclidean plane, let B be a point, and let m at a line that doesn't contain B. Let θ be an angle such that $0° < \theta < 180°$. Prove that there is a unique point D on m such that $l \angle EDB = \theta$ for a point E on m other than D.

6.20. In the Euclidean plane, let A, B, C be three points on a circle $\mathcal{K}$. Let D be a point that doesn't lie on line AB. Prove that D lies on $\mathcal{K}$ if and only if $l \angle ACB = l \angle ADB$.

(See Exercises 6.17–6.19.)

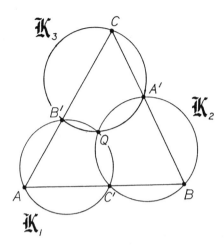

FIGURE 6.40.

6.21. In the Euclidean plane, let ABC be a triangle (Figure 6.40). Let A' be a point on line BC other than B and C, let B' be a point on line CA other than C and A, and let C' be a point on line AB other than A and B. A, B', C' lie on a circle $\mathcal{K}_1$; A', B, C' lie on a circle $\mathcal{K}_2$; and A', B', C lie on a circle $\mathcal{K}_3$ (by Theorem 4.3ii). If two of the circles $\mathcal{K}_1$, $\mathcal{K}_2$, $\mathcal{K}_3$ pass through a point Q that doesn't lie on any of the lines AB, BC, CA, prove that the third circle also passes through Q.

(*Q* is called the *Miquel point* determined by triangle ABC and the points A', B', C'. One possible approach to this exercise is to prove that $l\angle A'QB'$, $l\angle B'QC'$, $l\angle C'QA'$ sum to an integral multiple of 180° and that $l\angle ACB$, $l\angle BAC$, $l\angle CBA$ sum to an integral multiple of 180° and then to use Exercise 6.20 to compare the first three quantities with the second three.)

6.22. In the notation of Exercise 6.21, prove that Q lies on the circumcircle of triangle ABC if and only if A', B', C' lie on a line. Illustrate this result with a figure.

(*Hint:* One possible approach is to use Exercise 6.20 to prove that $l\angle QB'C' = l\angle QAB$ and $l\angle QB'A' = l\angle QCB$. Then prove that the four l-angles in these equations are all equal if and only if Q lies on the circumcircle of triangle ABC and if and only if A', B', C' lie on a line.)

6.23. In the Euclidean plane, let ABC be a triangle, and let Q be a point (Figure 6.41). Let A', B', C' be the feet of the perpendiculars from Q to BC, CA, AB, respectively. Prove that Q lies on the circumcircle of triangle ABC if and only if A', B', C' lie on a line.

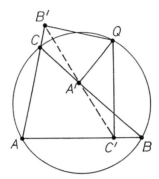

FIGURE 6.41.

(A', B', C' are not all equal, and so they lie on at most one line. When Q lies on the circumcircle of triangle ABC, the line through A', B', C' is called the *Simson line* determined by the triangle ABC and the point Q. One possible approach to this exercise is to use Theorem 4.7(ii) and Exercise 6.22 when no two of the points Q, A', B', C', A, B, C are equal. The remaining cases are straightforward.)

6.24. In the Euclidean plane, let Q be a point on the circumcircle $\mathcal{K}$ of triangle ABC. Assume that the perpendicular from Q to AB intersects $\mathcal{K}$ at a point R other than Q and C. In the terminology of Exercise 6.23, prove that line CR is parallel to the Simson line determined by triangle ABC and the point Q.

(*Hint:* In the notation of Exercise 6.23, if Q doesn't equal A, B, or C, and if B doesn't equal A' or C', one possible approach is to use Exercise 6.20 and Theorem 4.7(ii) to prove that $l \angle QRC = l \angle QC'A'$ by proving that each of these l-angles equals $l \angle QBC$. The remaining cases can be handled directly.)

6.25. In the Euclidean plane, let ABC be a triangle (Figure 6.40). Let A' be a point on line BC other than B and C, let B' be a point on line CA other than C and A, and let C' be a point on line AB other than A and B. A, B', C' lie on a circle $\mathcal{K}_1$; A', B, C' lie on a circle $\mathcal{K}_2$; and A', B', C lie on a circle $\mathcal{K}_3$ (by Theorem 4.3(ii)). Prove that the three circles $\mathcal{K}_1$, $\mathcal{K}_2$, $\mathcal{K}_3$ have a unique point Q in common.

(This exercise extends the construction of the Miquel point in Exercise 6.21 to include cases where two of the circles $\mathcal{K}_1$, $\mathcal{K}_2$, $\mathcal{K}_3$ are tangent. One possible approach in such cases is to use Exercise 6.18 to adapt the hint to Exercise 6.21.)

Chapter II

Transformation Geometry

INTRODUCTION AND HISTORY

Much of the beauty of geometric figures lies in their symmetry. Every culture has produced art based on intricate repeated designs. Notable examples include the tiled walls of Islamic religious buildings, lattice frameworks for paper windows in China, hand-stamped and printed cloth from Africa, and woven and embroidered fabrics from India.

Isometries are the key to studying symmetry. An isometry of the plane is a rigid motion, a map of the plane onto itself that doesn't change distances. In the sixth century B.C., the Greek geometer Thales apparently proved theorems by reflecting one part of a figure onto another. Euclid used motions of the plane to prove congruence theorems of triangles in the *Elements*. In the eighteenth century, the brilliant analyst Leonhard Euler classified all isometries in order to study the symmetries of curves. He determined that there are only five types of isometries, a fact we prove in Section 7.

The symmetries of a two-dimensional figure are the isometries that map the figure back onto itself. For example, a figure is symmetric across a line if it is unchanged by reflection in the line. A figure has n-fold symmetry about a point if it is unaffected by rotation through $(360/n)°$ about the point.

133

We note in Section 7 that the symmetries of a figure form a group. This means that the combined effect of two symmetries is a symmetry, the inverse of a symmetry is a symmetry, and the identity map (which fixes every point) is a symmetry. In the Renaissance, Leonardo da Vinci essentially determined all finite symmetry groups in the plane as part of his architectural studies. Exercises 11.6–11.8 outline a proof that every finite symmetry group is of type C_n or D_n for some positive integer n, where C_n consists of the identity map and the rotations through integral multiples of $(360/n)°$ about a point P, and D_n is formed from C_n by adding reflections in lines through P spaced by integral multiplies of $(180/n)°$. The three-dimensional finite symmetry groups were determined in the nineteenth century by Johann Friedrich Christian Hessel and Auguste Bravais. These groups arise either from spatial analogues of C_n and D_n or from the symmetry groups of regular polyhedra.

A wallpaper group is the symmetry group of a figure in the plane that repeats itself at regular intervals in more than one direction. In Section 8, we determine the translations in a wallpaper group, and we prove that they map any point to the vertices of a lattice of parallelograms. Such lattices and their higher-dimensional analogues are of central importance in sphere packing and approximating irrational numbers by rationals.

In Sections 9–12, we prove that there are seventeen types of wallpaper groups. This result was first published by Evgraf Stepanovič Fedorov in 1891 and later rediscovered several times. The graphic artist M. C. Escher based his popular prints of intricately interlocking figures on the analysis of wallpaper groups published by George Pólya in 1924. Classifying wallpaper groups is a key step in studying periodic tilings and patterns in the plane, subjects with easily accessible open questions.

Much of the interest in wallpaper groups arose from studies of their three-dimensional analogues, the space crystallographic groups. These are the symmetry groups of figures that repeat at regular intervals in three-dimensional space. As their name suggests, crystallographic groups are important to chemists studying the symmetries of crystals. The fact that there are 230 types of crystallographic groups was proved by Fedorov in 1885 (six years before he classified the wallpaper groups) and independently by Arthur Schönflies in 1891.

In 1900, the great mathematician David Hilbert proposed a series of questions to guide mathematical research in the twentieth century. One of the questions was whether there are finitely many types of crystallographic groups in Euclidean space of any dimension. This question was answered affirmatively in 1910 by Ludwig Bieberbach, using results on matrix groups. An algorithm developed by Hans Zassenhaus in 1948 was used in 1973 to prove that there are 4783 types of crystallographic groups in four-dimensional Euclidean space.

We study dilations in Section 13. These are generalizations of isometries that, instead of preserving distances, multiply them by a constant factor. Dilations were apparently first studied by Apollonius in the third century B.C., and we use them to obtain results on nine-point circles. Section 13 is independent of Sections 7–12 and can be read at any time.

Symmetry groups can be generalized to transformation groups. A nonempty set of maps from a space to itself is called a transformation group if it is closed under composition and taking inverses. In 1872, Felix Klein asserted that geometry is the study of the properties of a space that are preserved by a transformation group. For example, Euclidean geometry is the study of those properties such as length and angle that are preserved by isometries. Different transformation groups preserve different properties and determine different geometries. This point of view, called the *Erlanger Programm*, extended the idea of invariants from algebra to geometry and provided a theoretical framework for the special theory of relativity.

The following books and articles are very good sources of further reading. They provided much of the material in this chapter.

Danzer, Ludwig, Grünbaum, Branko, and Shephard, G. C., "Equitransitive Tilings, or How to Discover New Mathematics," *Mathematics Magazine*, Vol. 60, No. 2, April 1987, pp. 67–89.

Grünbaum, Branko, and Shephard, G. C., *Tilings and Patterns*, Freeman, New York, 1987.

Hilbert, D. and Cohn-Vossen, S., *Geometry and the Imagination* (translated by P. Nemenyi), Chelsea, New York, 1952.

Lockwood, E. H. and Macmillan, R. H., *Geometric Symmetry*, Cambridge University Press, Cambridge, 1978.

Martin, George E., *Transformation Geometry, An Introduction to Symmetry*, Springer-Verlag, New York, 1982.

Milnor, J., "Hilbert's Problem 18: On Crystallographic Groups, Fundamental Domains, and on Sphere Packing," *Proceedings of Symposia in Pure Mathematics*, Vol. 28, American Mathematical Society, Providence, 1976.

Wieting, Thomas W., *The Mathematical Theory of Chromatic Plane Ornaments*, Marcel Dekker, New York, 1982.

Section 7.

Isometries

Our goal in Sections 7–12 is to study the symmetries of figures in the Euclidean plane. We start by studying isometries, maps from the plane to itself that preserve distances.

DEFINITION 7.1. In the Euclidean plane, an *isometry* is a map φ that sends each point X in the plane to a point $\varphi(X)$ such that the distance between every pair of points A and B equals the distance between their images $\varphi(A)$ and $\varphi(B)$.   □

We can think of an isometry as a rigid motion of the plane, a movement of the plane that doesn't change the distances between points. We say that a map φ from the Euclidean plane to itself *fixes* a point A if $\varphi(A) = A$. Thus the fixed points of φ are the points that φ doesn't move.

The next five definitions give examples of isometries.

DEFINITION 7.2. In the Euclidean plane, let l be a line. The *reflection* σ_l in l is the map that fixes the points on l and maps any point A not on l to the point $\sigma_l(A)$ such that l is the perpendicular bisector of A and $\sigma_l(A)$.   □

Figure 7.1 shows the images of several points under the reflection σ_l. Intuitively, σ_l "flips" the plane across l. Points on l aren't moved, and points not on l map to their "mirror images" in l.

DEFINITION 7.3. In the Euclidean plane, let P be a point, and let θ be an angle such that $0° < \theta < 360°$. The *rotation* $\rho_{P,\theta}$ of angle θ about P is the map that revolves the plane counterclockwise through angle θ about P.   □

Figure 7.2 shows the images of various points under the rotation $\rho = \rho_{P,\theta}$.

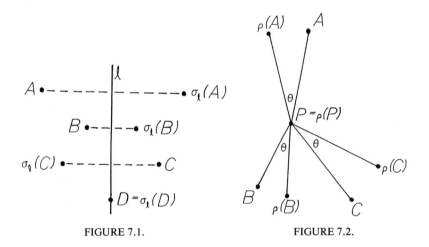

FIGURE 7.1.   FIGURE 7.2.

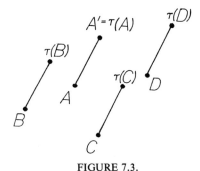

FIGURE 7.3.

DEFINITION 7.4. In the Euclidean plane, let A and A' be two points. The *translation* $\tau_{A,A'}$ is the map that sends A to A' and moves all points in the plane the same distance in the same direction. $\square$

Figure 7.3 shows the images of various points under $\tau = \tau_{A,A'}$.

DEFINITION 7.5. In the Euclidean plane, let A and A' be two points. The *glide reflection* $\gamma_{A,A'}$ is the map obtained by first applying the reflection $\sigma_{AA'}$ in line AA' and then applying the translation $\tau_{A,A'}$ that takes A to A'. $\square$

Figure 7.4 shows the images of several points under the glide reflection $\gamma = \gamma_{A,A'}$. γ first reflects points across line AA' and then translates them parallel to AA'. γ fixes no points. If D is any point on AA', and if we set $D' = \gamma(D)$, then we can also write γ as $\gamma_{D,D'}$, since we have $\sigma_{DD'} = \sigma_{AA'}$ and $\tau_{D,D'} = \tau_{A,A'}$.

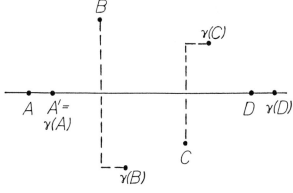

FIGURE 7.4.

DEFINITION 7.6. In the Euclidean plane, the *identity map I* is the map that fixes every point. □

The identity map is the "motion" of the plane that doesn't move any point. The maps in Definitions 7.2–7.6—reflections, rotations, translations, glide reflections, and the identity map—are isometries because they all preserve distances. That is, these maps are rigid motions, motions that don't involve any stretching or shrinking of the plane. The main result of this section is the Classification Theorem for Isometries, which shows that the maps in Definitions 7.2–7.6 are all the isometries there are. Thus, these maps are the only ways to move points in the plane without changing distances.

We start by showing that *isometries preserve lines*. Since isometries preserve distances, it's enough to show that we can use distances to determine when points lie on a line.

THEOREM 7.7. In the Euclidean plane, let A, B, C be three points. Let d_1, d_2, d_3 be the respective distances between the points in each of the pairs $\{A, B\}$, $\{B, C\}$, $\{A, C\}$.

   (i) B lies on line AC between A and C if and only if the relation $d_1 + d_2 = d_3$ holds (Figure 7.5).
   (ii) A, B, C lie on a line if and only if one of the distances d_1, d_2, d_3 is the sum of the other two.
   (iii) Let φ be an isometry. Then A, B, C lie on a line if and only if $\varphi(A)$, $\varphi(B)$, $\varphi(C)$ lie on a line.

Proof:
   (i) If B lies on line AC between A and C, it's clear that the relation $d_1 + d_2 = d_3$ holds (Figure 7.5). On the other hand, we have $d_1 + d_2 > d_3$ if B doesn't lie on line AC (Figure 7.6) or if B lies on line AC outside the segment with endpoints A and C (Figures 7.7a and b).
   (ii) When three points lie on a line, one of the points lies between the other two. Thus, part (ii) follows from part (i).
   (iii) Since φ is an isometry, d_1, d_2, d_3 are also the distances between the points in the pairs $\{\varphi(A), \varphi(B)\}$, $\{\varphi(B), \varphi(C)\}$, $\{\varphi(A), \varphi(C)\}$. Thus, part (iii) follows from part (ii). □

We consider two isometries to be equal when they have the same effect on every point in the plane. We use the previous result to prove that two isometries are equal if they map the vertices of a triangle to the same three points. Thus, an isometry is uniquely determined by the images of the vertices of a triangle.

FIGURE 7.5.   FIGURE 7.6.

THEOREM 7.8. In the Euclidean plane, let φ and ψ be isometries. If φ and ψ map three noncollinear points A, B, C to the same three points A', B', C', then φ and ψ are equal.

Proof: Let X be any point in the plane. The distances from $\varphi(X)$ and $\psi(X)$ to A' both equal the distance from X to A (since φ and ψ are isometries). Thus, A' is equidistant from $\varphi(X)$ and $\psi(X)$. By symmetry, B' and C' are also equidistant from $\varphi(X)$ and $\psi(X)$. If $\varphi(X)$ and $\psi(X)$ were distinct, the points equidistant from them would form a line (by Theorem 4.2). Thus, since A', B', C' are equidistant from $\varphi(X)$ and $\psi(X)$ and are noncollinear (by Theorem 7.7 and the assumption that A, B, C are noncollinear), $\varphi(X)$ and $\psi(X)$ must be equal. This holds for every point X in the plane, and so φ and ψ are equal. $\square$

We've seen that an isometry is uniquely determined by the image of three points that don't lie on a line. As the next result shows, it follows that an isometry is limited to two possibilities by the images of just two points.

THEOREM 7.9. In the Euclidean plane, let A and B be two points, and let A' and B' be two points. Then there are at most two isometries that map A to A' and B to B'.

Proof: Let C be a point such that ABC is an equilateral triangle (Figure 7.8). There are exactly two points D and E that form equilateral triangles with A' and B'. An isometry that maps A to A' and B to B' must map C to either D or E. Moreover, an isometry is uniquely determined by the images of the vertices of a triangle (by Theorem 7.8). Thus, there are at most two isometries mapping A to A' and B to B'. $\square$

FIGURE 7.7a.   FIGURE 7.7b.

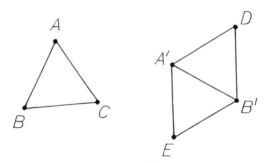

FIGURE 7.8.

We can now determine all isometries.

Theorem 7.10 (Classification Theorm for Isometries). Every isometry of the Euclidean plane is either a reflection, a rotation, a translation, a glide reflection, or the identity map.

Proof: Let φ be an isometry; we must prove that φ is one of the five types listed. If φ fixes every point, then φ is the identity map. Thus, we can assume there is a point A such that $\varphi(A) \neq A$. We set $B = \varphi(A)$ and $C = \varphi(B)$. Thus, φ maps $A \to B$ and $B \to C$, where $A \neq B$. If XY denotes the distance between points X and Y, we have

$$BC = \varphi(A)\varphi(B) = AB \tag{1}$$

(by Definition 7.1). There are three cases, depending on the relative positions of A, B, C.

Case 1: C equals A. Then φ maps $A \to B$ and $B \to A$. Let σ_l be the reflection in the perpendicular bisector l of A and B, and let $\rho_{M, 180°}$ be the rotation through $180°$ around the midpoint M of A and B (Figure 7.9). σ_l and $\rho_{M, 180°}$ map $A \to B$ and $B \to A$, and they are the only isometries to do so (by Theorem 7.9). Thus, φ is either the reflection σ_l or the rotation $\rho_{M, 180°}$.

Case 2: C lies on line AB and doesn't equal A. Equation 1 and the assumption that $C \neq A$ imply that B is the midpoint of A and C (Figure 7.10). Thus, the translation $\tau_{A, B}$ and the glide reflection $\gamma_{A, B}$ map $A \to B$ and $B \to C$. They are the only isometries to do so (by Thoerem 7.9), and so one of them equals φ.

Case 3: C doesn't lie on line AB. Then A, B, C lie on a circle $\mathcal{K}$ (by Theorem 4.3ii). Let O be the center of $\mathcal{K}$, and let ρ be the rotation around O that maps A to B. Since B is equidistant from A and C (by Equation 1), AB and BC are not diameters of $\mathcal{K}$, and so AOB and BOC are triangles.

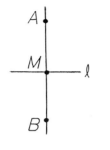

FIGURE 7.9.   FIGURE 7.10.

We set $\theta = \angle AOB$, and so ρ is either $\rho_{O,\theta}$ (Figure 7.11) or $\rho_{O,360°-\theta}$ (Figure 7.12). Corresponding sides of triangles AOB and BOC are equal (by Equation 1 and the fact that O is the center of $\mathcal{K}$). Thus, triangles AOB and BOC are congruent (by the SSS Property 0.1), and so we have $\angle BOC = \angle AOB = \theta$. Thus, since $A \neq C$, the rotation ρ about O that maps A to B also maps B to C.

Let M be the midpoint of A and B, and let N be the midpoint of B and C (Figure 7.13). Since $A \neq C$, we have $M \neq N$. Since A, B, C don't all lie on one line, none of these points lies on line MN. Let A', B', C' be the feet of the perpendiculars dropped from A, B, C to MN. The lines MN and AC are parallel (by Theorem 1.13), and so the lines BB' and AC are perpendicular and intersect at a point F. The sides AB and BC of triangle ABC are of equal length (by Equation 1), and F is the foot of the altitude on B of triangle ABC, and so F is the midpoint of A and C (by Theorem 4.10). Thus, B' is the midpoint of A' and C', and so $\gamma_{A',B'}$, which maps $A' \to B'$, also maps $B' \to C'$. Triangles $AA'M$ and $BB'M$ have equal angles at M (by Theorem 0.6), equals angles at A and B (by Theorem 0.7), and sides AM and BM of equal length, and so these triangles are congruent (by Property 0.3). Thus, their sides AA' and BB' are of equal length, and

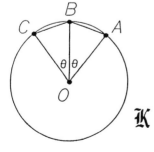

FIGURE 7.11.

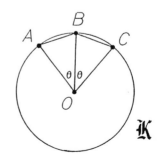

FIGURE 7.12.

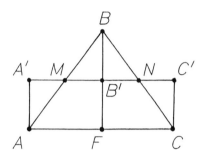

FIGURE 7.13.

so A and B are equidistant from line MN on opposite sides of that line. By symmetry, A, B, C are equidistant from line MN, and B lies on the opposite side of MN from A and C. Hence, since $\gamma_{A',B'}$ maps $A' \to B'$ and $B' \to C'$, it also maps $A \to B$ and $B \to C$.

The two previous paragraphs show that the rotation ρ and the glide reflection $\gamma_{A',B'}$ map $A \to B$ and $B \to C$. These isometries are distinct (since ρ fixes O, while $\gamma_{A',B'}$ has no fixed points). Thus, φ is one of these two isometries, by Theorem 7.9. □

We are interested in isometries primarily in order to study patterns in the plane. We now consider the relationship between patterns and isometries.

DEFINITION 7.11. In the Euclidean plane, a *figure* $\mathfrak{F}$ is a set of points. The *image* $\varphi(\mathfrak{F})$ of a figure $\mathfrak{F}$ under an isometry φ is the set of all points $\varphi(A)$ where A is a point of $\mathfrak{F}$. A *symmetry* of a figure $\mathfrak{F}$ is an isometry φ such that $\varphi(\mathfrak{F}) = \mathfrak{F}$. □

Thus, a symmetry of a figure $\mathfrak{F}$ is an isometry that maps $\mathfrak{F}$ onto itself. Speaking roughly, the number of symmetries of a figure $\mathfrak{F}$ measures how symmetrical $\mathfrak{F}$ is. The Classification Theorem 7.10 makes it practical to determine the symmetries of figures, as the next two examples illustrate.

EXAMPLE 7.12. Determine the symmetries of an equilateral triangle (Figure 7.14).

Solution: Let A, B, C be the vertices of the triangle, and let G be the centroid (Figure 7.15). Any symmetry of the triangle maps the vertices among themselves and the midpoints of the sides among themselves, and so it fixes the centroid G. Thus, any symmetry is either a reflection in a line through G, a rotation about G, or the identity map (by the Classification

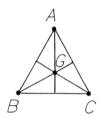

FIGURE 7.14.   FIGURE 7.15.

Theorem 7.10 and the facts that a reflection σ_l fixes only the points on l, a rotation $\rho_{P,\theta}$ fixes only P, the identity map fixes every point, and translations and glide reflections don't fix any points). The reflections in lines through G that map the triangle onto itself are σ_{AG}, σ_{BG}, σ_{CG}. The rotations about G that map the triangle onto itself are $\rho_{G,120°}$ and $\rho_{G,240°}$. Of course, the identity map I maps the triangle onto itself. Thus, the triangle has exactly six symmetries:

$$\sigma_{AG}, \qquad \sigma_{BG}, \qquad \sigma_{CG}, \qquad \rho_{G,120°}, \qquad \rho_{G,240°}, \qquad I. \quad \square \qquad (2)$$

EXAMPLE 7.13. Let $\mathfrak{F}$ be the figure obtained by continuing the pattern in Figure 7.16 indefinitely to the left and the right. Determine the symmetries of $\mathfrak{F}$.

Solution: Let m be the line that bisects $\mathfrak{F}$ horizontally (Figure 7.17). Let the points where m intersects $\mathfrak{F}$ be labeled consecutively as A_i for all integers i. Let l_i be the perpendicular bisector of A_i and A_{i+1} for each integer i. The reflections that map $\mathfrak{F}$ onto itself are σ_{l_i} for all integers i. The rotations that map $\mathfrak{F}$ onto itself are $\rho_{A_i,180°}$ for all integers i. The translations that map $\mathfrak{F}$ onto itself are $\tau_{A_0,A_{2i}}$ for all nonzero integers i (since $2i$ varies over all nonzero even integers as i varies over all nonzero integers). The glide reflections that map $\mathfrak{F}$ onto itself are $\gamma_{A_0,A_{2i+1}}$ for all integers i (since $2i+1$ varies over all odd integers as i varies over all integers). Of course, the identity map I maps $\mathfrak{F}$ onto itself. The preceding isometries are all the symmetries of $\mathfrak{F}$ (by the Classification Theorem 7.10). $\square$

FIGURE 7.16.

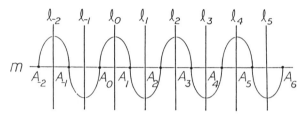

FIGURE 7.17.

In the previous example, there was no need to list such isometries as τ_{A_1,A_3} and γ_{A_1,A_2} among the symmetries of $\mathfrak{F}$, since they equal the isometries τ_{A_0,A_2} and γ_{A_0,A_1} already listed.

In the Euclidean plane, let φ and ψ be isometries. Their *product* $\varphi\psi$ is the map of the plane that first applies ψ and then φ. Thus, $\varphi\psi$ takes any point X in the plane to $\varphi(\psi(X))$, and so we have

$$(\varphi\psi)(X) = \varphi(\psi(X)). \tag{3}$$

Eliminating extra parentheses, we let $\varphi\psi(X)$ denote the quantity on either side of Equation 3.

The product $\varphi\psi$ of isometries φ and ψ is the map that follows ψ with φ. For example, the glide reflection $\gamma_{A,A'}$ consists of the reflection $\sigma_{AA'}$ followed by the translation $\tau_{A,A'}$ (by Definition 7.5), and so we have

$$\gamma_{A,A'} = \tau_{A,A'}\sigma_{AA'}. \tag{4}$$

Another way besides products to form new maps from given isometries is to take the inverses of the isometries. Specifically, let φ be an isometry of the Euclidean plane. For any point Y in the plane, there is a unique point X such that $\varphi(X) = Y$; this follows from the Classification Theorem 7.10, since it obviously holds for each type of isometry listed there. We define φ^{-1}, the *inverse* of φ, to be the map of the plane that takes any point Y to the unique point X such that $\varphi(X) = Y$. In other words, the equations

$$\varphi(X) = Y \quad \text{and} \quad \varphi^{-1}(Y) = X$$

are equivalent for all points X and Y. Substituting each of these equations into the other shows that

$$X = \varphi^{-1}(\varphi(X)) \quad \text{and} \quad \varphi(\varphi^{-1}(Y)) = Y$$

for all points X and Y, and so we have

$$\varphi^{-1}\varphi = I = \varphi\varphi^{-1}. \tag{5}$$

We might say that φ^{-1} "undoes" or "reverses" the isometry φ. For example, $\rho_{P,\theta}$ is the counterclockwise rotation through angle θ about P, and we reverse it by rotating the plane clockwise through θ about P, which has the same effect as rotating the plane counterclockwise through $360° - \theta$ about P. Thus, we have

$$\rho_{P,\theta}^{-1} = \rho_{P,\,360°-\theta}. \tag{6}$$

For any two points A and B, the translations $\tau_{A,B}$ and $\tau_{B,A}$ "undo" each other (since they move points the same distance in opposite directions), and so we have

$$\tau_{A,B}^{-1} = \tau_{B,A}. \tag{7}$$

The next result shows that we can take products and inverses of isometries or symmetries of a figure and obtain more such maps.

THEOREM 7.14.

(i) In the Euclidean plane, if φ and ψ are isometries, then so are $\varphi\psi$ and φ^{-1}.

(ii) In the Euclidean plane, if φ and ψ are symmetries of a figure $\mathfrak{F}$, then so are $\varphi\psi$ and φ^{-1}.

Proof: (i) Let A and B be any two points in the plane.

The distance between $\varphi\psi(A)$ and $\varphi\psi(B)$ equals the distance between $\psi(A)$ and $\psi(B)$ (since φ is an isometry), and this equals the distance between A and B (since ψ is an isometry). Thus, $\varphi\psi$ preserves distances between points, and so it is an isometry.

φ maps $\varphi^{-1}(A)$ to A, and it maps $\varphi^{-1}(B)$ to B. Thus, the distance between $\varphi^{-1}(A)$ and $\varphi^{-1}(B)$ equals the distance between A and B (since φ is an isometry). Hence, φ^{-1} preserves distances between points, and so it is an isometry.

(ii) Since φ and ψ are symmetries of $\mathfrak{F}$, they are isometries such that

$$\varphi(\mathfrak{F}) = \mathfrak{F} \quad \text{and} \quad \psi(\mathfrak{F}) = \mathfrak{F}. \tag{8}$$

$\varphi\psi$ and φ^{-1} are isometries (by part (i)), and we have

$$\varphi\psi(\mathfrak{F}) = \varphi(\mathfrak{F}) = \mathfrak{F}$$

(by the equations in (8)) and

$$\varphi^{-1}(\mathfrak{F}) = \varphi^{-1}(\varphi(\mathfrak{F})) = \mathfrak{F}$$

(by the equations in (8) and (5)). Hence, $\varphi\psi$ and φ^{-1} are symmetries of $\mathfrak{F}$. $\square$

Stated roughly, part (i) of the previous theorem shows that the combined effect of two rigid motions is a rigid motion and that reversing a rigid motion gives a rigid motion. Equations 4, 6, and 7 illustrate part (i) by showing in particular cases that the product of two isometries is an isometry and that the inverse of an isometry is an isometry. An important part of the work in Sections 8–12 is to determine exactly which isometry we obtain when we multiply two given isometries.

The isometries in (2) are the symmetries of an equilateral triangle. Thus, part (ii) of the previous theorem shows that the product of any two of these symmetries is again one of these symmetries and that the inverse of any of these symmetries is also one of these symmetries. For example, $\rho_{G, 120°}$ appears in (2) and its inverse is $\rho_{G, 240°}$ (by Equation 6), which also appears in (2). The reflections σ_{AG} and σ_{BG} appear in (2), and it's clear from Figure 7.15 that

$$\sigma_{AG}\sigma_{BG}(A) = \sigma_{AG}(C) = B,$$

$$\sigma_{AG}\sigma_{BG}(B) = \sigma_{AG}(B) = C,$$

$$\sigma_{AG}\sigma_{BG}(C) = \sigma_{AG}(A) = A.$$

Since $\rho_{G, 120°}$ also maps $A \to B$, $B \to C$, and $C \to A$, it follows from Theorems 7.8 and 7.14(i) that

$$\sigma_{AG}\sigma_{BG} = \rho_{G, 120°}, \tag{9}$$

illustrating the fact that the product of two elements of (2) is again an element of (2). Similarly, we can see from Figure 7.15 that

$$\sigma_{BG}\sigma_{AG}(A) = \sigma_{BG}(A) = C,$$

$$\sigma_{BG}\sigma_{AG}(B) = \sigma_{BG}(C) = A,$$

$$\sigma_{BG}\sigma_{AG}(C) = \sigma_{BG}(B) = B.$$

Since $\rho_{G, 240°}$ also maps $A \to C$, $B \to A$, and $C \to B$, it follows that

$$\sigma_{BG}\sigma_{AG} = \rho_{G, 240°}. \tag{10}$$

Equation 10 illustrates once again the fact that the product of two elements of (2) is an element of (2), in agreement with Theorem 7.14(ii).

Comparing Equations 9 and 10 shows that

$$\sigma_{AG}\sigma_{BG} \neq \sigma_{BG}\sigma_{AG}, \tag{11}$$

so the commutative law of multiplication does *not* generally hold for isometries. *The order of the isometries in a product matters:* We can *not* generally change the order of the isometries in a product without affecting the value of the product.

In the Euclidean plane, let $\mathcal{F}$ be any figure. Let $\mathcal{G}$ be the set of symmetries of $\mathcal{F}$. Theorem 7.14(ii) shows that the product of two elements of $\mathcal{G}$ belongs to $\mathcal{G}$ and that the inverse of an element of $\mathcal{G}$ belongs to $\mathcal{G}$. The identity map I is an isometry that maps $\mathcal{F}$ onto itself (since I fixes every point), and so I belongs to $\mathcal{G}$. We introduce a term for any set of isometries that has these properties.

DEFINITION 7.15. In the Euclidean plane, a *group* of isometries is a set $\mathcal{G}$ of isometries that has the following properties.

 (i) If φ and ψ are elements of $\mathcal{G}$, then so is $\varphi\psi$.
 (ii) If φ is an element of $\mathcal{G}$, then so is φ^{-1}.
 (iii) $\mathcal{G}$ contains the identity map I. □

Let $\mathcal{G}$ be the set of symmetries of a figure $\mathcal{F}$ in the Euclidean plane. $\mathcal{G}$ satisfies conditions (i)–(iii) of Definition 7.15 (by the discussion before the definition), and so $\mathcal{G}$ is a group. We call $\mathcal{G}$ the *symmetry group* of $\mathcal{F}$. We've proved the following result.

THEOREM 7.16. In the Euclidean plane, the set of symmetries of a figure is a group of isometries. □

In the Euclidean plane, let φ_1, φ_2, φ_3 be isometries. For any point A, repeated application of Equation 3 shows that

$$[(\varphi_1\varphi_2)\varphi_3](A) = (\varphi_1\varphi_2)(\varphi_3(A)) = \varphi_1(\varphi_2(\varphi_3(A)))$$

and

$$[\varphi_1(\varphi_2\varphi_3)](A) = \varphi_1((\varphi_2\varphi_3)(A)) = \varphi_1(\varphi_2(\varphi_3(A))).$$

Thus, the associative law

$$(\varphi_1\varphi_2)\varphi_3 = \varphi_1(\varphi_2\varphi_3).$$

holds.

If $\varphi_1, \ldots, \varphi_n$ are isometries, it makes sense to write the product $\varphi_1\cdots\varphi_n$ without parentheses, since the associative law shows that we can insert parentheses in any way without changing the value of the product. In short, we can use parentheses to group the factors in a product of isometries in any way. On the other hand, Inequality 11 shows that the commutative law doesn't hold, and so we cannot generally change the order of the factors in a product of isometries.

Since the identity map I fixes every point, we have

$$\varphi I(A) = \varphi(A) = I\varphi(A)$$

for any isometry φ and every point A, and so we have

$$\varphi I = \varphi = I\varphi. \tag{12}$$

In Sections 8–12 we determine all isometry groups of a particular type. We use conditions (i)–(iii) of Definition 7.15 to deduce that a group containing certain isometries must also contain others. The next result makes it convenient to do so.

THEOREM 7.17. In the Euclidean plane, let φ, ψ, ξ be isometries such that $\varphi\psi = \xi$. If two of the isometries φ, ψ, ξ belong to a group $\mathcal{G}$, then so does the third.

Proof: If $\mathcal{G}$ contains φ and ψ, then it also contains $\varphi\psi = \xi$ (by Definition 7.15i). Combining the equation $\varphi\psi = \xi$ with Equations 5 and 12 shows that

$$\xi\psi^{-1} = \varphi\psi\psi^{-1} = \varphi I = \varphi;$$

thus, if $\mathcal{G}$ contains ψ and ξ, then it also contains φ (by Definition 7.15i and ii). Finally, the equation $\varphi\psi = \xi$ and Equations 5 and 12 show that

$$\varphi^{-1}\xi = \varphi^{-1}\varphi\psi = I\psi = \psi;$$

thus, if $\mathcal{G}$ contains φ and ξ, then it also contains ψ (by Definition 7.15i and ii). $\square$

EXERCISES

7.1. Determine all symmetries of the following figures.

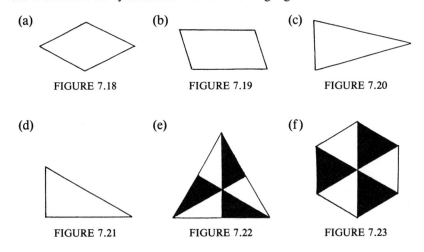

(a)

FIGURE 7.18

(b)

FIGURE 7.19

(c)

FIGURE 7.20

(d)

FIGURE 7.21

(e)

FIGURE 7.22

(f)

FIGURE 7.23

(g)

(h)

(i)

(j)

(k)

(l)

(m)

(n)

(o)

(p)

(q)

(r)

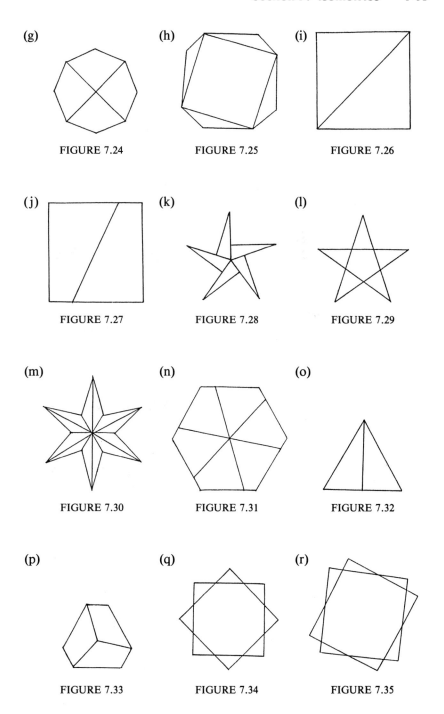

FIGURE 7.24

FIGURE 7.25

FIGURE 7.26

FIGURE 7.27

FIGURE 7.28

FIGURE 7.29

FIGURE 7.30

FIGURE 7.31

FIGURE 7.32

FIGURE 7.33

FIGURE 7.34

FIGURE 7.35

(s) (t)

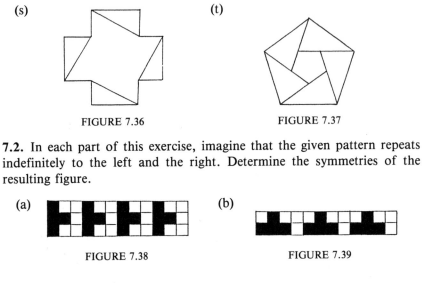

FIGURE 7.36   FIGURE 7.37

7.2. In each part of this exercise, imagine that the given pattern repeats indefinitely to the left and the right. Determine the symmetries of the resulting figure.

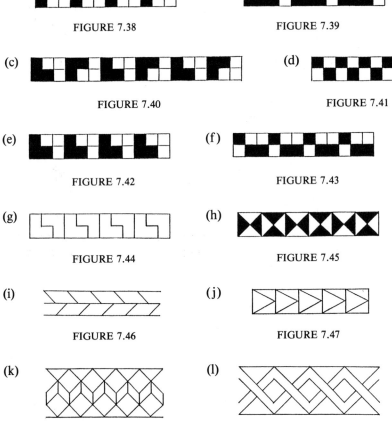

(a)

FIGURE 7.38

(b)

FIGURE 7.39

(c)

FIGURE 7.40

(d)

FIGURE 7.41

(e)

FIGURE 7.42

(f)

FIGURE 7.43

(g)

FIGURE 7.44

(h)

FIGURE 7.45

(i)

FIGURE 7.46

(j)

FIGURE 7.47

(k)

FIGURE 7.48

(l)

FIGURE 7.49

(m) 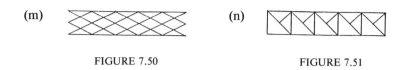 (n)

FIGURE 7.50   FIGURE 7.51

Exercises 7.3 and 7.4 use the following terminology. Let $\mathcal{G}$ be a finite group of isometries; say $\mathcal{G}$ has d elements. A *multiplication table* for $\mathcal{G}$ is a d-by-d grid of boxes such that the rows are labeled with the elements of $\mathcal{G}$, the columns are labeled with the elements of $\mathcal{G}$, and the box in row φ and column ψ contains the value of the product $\varphi\psi$ for any elements φ and ψ of $\mathcal{G}$.

7.3. Let $\mathcal{G}$ be the symmetry group of an equilateral triangle (Figure 7.14). Let the vertices of the triangle be labeled counterclockwise A, B, C, and let G be the centroid (Figure 7.15). By Example 7.12, we can label the rows and columns of a multiplication table for $\mathcal{G}$ as in Figure 7.52. We have used Equations 9 and 10 to fill in two entries of the table. For each part (a)–(hh) of this exercise, determine the entry in the corresponding box of the table.

	I	$P_{G,120°}$	$P_{G,240°}$	σ_{AG}	σ_{BG}	σ_{CG}
I	a	b	c	d	e	f
$P_{G,120°}$	g	h	i	j	k	l
$P_{G,240°}$	m	n	o	p	q	r
σ_{AG}	s	t	u	v	$P_{G,120°}$	w
σ_{BG}	x	y	z	$P_{G,240°}$	aa	bb
σ_{CG}	cc	dd	ee	ff	gg	hh

FIGURE 7.52.

7.4. Let $\mathcal{G}$ be the symmetry group of a square (Figure 7.53). Label the corners of the square counterclockwise A–D, and let G be the center of the square (Figure 7.54). Let l be the perpendicular bisector of A and B,

FIGURE 7.53.

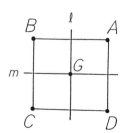

FIGURE 7.54.

	I	$P_{G,90°}$	$P_{G,180°}$	$P_{G,270°}$	σ_{AC}	σ_{BD}	σ_l	σ_m
I	a	b	c	d	e	f	g	h
$P_{G,90°}$	i	j	k	l	m	n	o	p
$P_{G,180°}$	q	r	s	t	u	v	w	x
$P_{G,270°}$	y	z	aa	bb	cc	dd	ee	ff
σ_{AC}	gg	hh	ii	jj	kk	ll	mm	nn
σ_{BD}	oo	pp	qq	rr	ss	tt	uu	vv
σ_l	ww	xx	yy	zz	aaa	bbb	ccc	ddd
σ_m	eee	fff	ggg	hhh	iii	jjj	kkk	lll

FIGURE 7.55.

and let m be the perpendicular bisector of B and C. Figure 7.55 shows a multiplication table of $\mathcal{G}$, and the elements of $\mathcal{G}$ are listed along the top and the side of this table. For each part (a)–(lll) of this exercise, determine the entry in the corresponding box of the table.

7.5. Equations 6 and 7 evaluate each of the isometries $\rho_{P,\theta}^{-1}$ and $\tau_{A,B}^{-1}$ as one of the maps in Definitions 7.2–7.6. Evaluate each of the following isometries as one of the maps in these definitions. (Theorems 7.10 and 7.14i ensure that this is possible.)

(a) σ_l^{-1}, where l is any line in the Euclidean plane
(b) $\gamma_{A,B}^{-1}$, where A and B are any two points in the Euclidean plane
(c) I^{-1}

7.6. In the Euclidean plane, prove that $\tau\rho \neq \rho\tau$ for any translation τ and rotation ρ. (*Hint:* One possible approach is to prove that $\tau\rho$ and $\rho\tau$ map the center of ρ to different points.)

7.7. In the Euclidean plane, let A and B be two points.

(a) For what pairs of points A' and B' does $\tau_{A,B} = \tau_{A',B'}$?
(b) For what pairs of points A' and B' does $\gamma_{A,B} = \gamma_{A',B'}$?

7.8. In the Euclidean plane, let ABC and $A'B'C'$ be congruent triangles.

(a) Prove that there is a reflection σ_l that maps A to A'.
(b) Prove that there is a reflection σ_m that fixes A' and maps $\sigma_l(B)$ to B'.
(c) If we set $C_1 = \sigma_m\sigma_l(C)$, prove that C' equals either C_1 or $\sigma_{A'B'}(C_1)$.

7.9. In the Euclidean plane, let A and B be two points, and let A' and B' be two points.

(a) Prove that there is an isometry that maps A to A' and B to B' if and only if the distance between A and B equals the distance between A' and B'.

(b) If the distance between A and B equals the distance between A' and B', prove that there are exactly two isometries that map A to A' and B to B'.

(Parts (a) and (b) of Exercise 7.8 may be helpful in doing this exercise, which sharpens Theorem 7.9.)

7.10. In the Euclidean plane, let ABC and $A'B'C'$ be triangles. Prove that these triangles are congruent if and only if there is an isometry mapping $A \to A'$, $B \to B'$, and $C \to C'$. (If such an isometry exists, it is unique, by Theorem 7.8. See Exercise 7.8.)

7.11. In the Euclidean plane, let φ be an isometry.

(a) Use Exercise 7.8 and Theorem 7.8 to prove that φ can be written as a product of at most three reflections.
(b) If φ has a fixed point, prove that φ can be written as a product of at most two reflections.

7.12. In the Euclidean plane, let φ, ψ, ξ be isometries. Prove that the following conditions are equivalent.

(i) $\varphi\psi = \xi$
(ii) $\varphi = \xi\psi^{-1}$
(iii) $\psi = \varphi^{-1}\xi$

7.13. In the Euclidean plane, let φ and ψ be isometries. Prove that the following conditions are equivalent.

(i) $\varphi\psi = \psi\varphi$
(ii) $\varphi^{-1}\psi = \psi\varphi^{-1}$
(iii) $\varphi\psi^{-1} = \psi^{-1}\varphi$
(iv) $\varphi^{-1}\psi^{-1} = \psi^{-1}\varphi^{-1}$

Section 8.

Wallpaper Groups and Translations

We now consider figures that repeat at regular, discrete intervals in non-parallel directions. Remarkably, the symmetry groups of such figures fall into 17 types. Thus, although the figures vary infinitely in design, the patterns of symmetry they display are sharply limited. We consider these patterns of symmetry in Sections 8–12.

We start with a formal definition of the groups of isometries that we're considering. We characterize these groups by the translations they contain. The *length of a translation* $\tau_{A,B}$ is the distance from A to B, the distance that the translation moves every point. We call translations $\tau_{A,B}$ and $\tau_{C,D}$ *parallel* (or *perpendicular*) if the lines AB and CD are parallel (or perpendicular). Likewise, we say that a translation $\tau_{A,B}$ is *parallel* (or *perpendicular*) to a line l if the lines AB and l are parallel (or perpendicular).

DEFINITION 8.1. In the Euclidean plane, a *wallpaper group* is a group $\mathcal{G}$ of isometries that has the following two properties:

(i) $\mathcal{G}$ contains translations that aren't parallel.

(ii) There is a positive number b such that $\mathcal{G}$ contains no translations of length less than b.   □

A symmetry group is a wallpaper group when its translations are not all parallel and are not arbitrarily short. Accordingly, the symmetry group of a figure is a wallpaper group when the figure repeats in nonparallel directions and at regular intervals that aren't arbitrarily small. Thus, if Figure 8.1 and 8.2 are extended to cover the plane, the symmetry groups of the resulting figures are wallpaper groups. When the patterns in Figures 7.38–7.51 are repeated horizontally to form figures of infinite length, the symmetry groups of the resulting figures are not wallpaper groups; all translations in these symmetry groups are horizontal, and so condition (i) of Definition 8.1 doesn't hold. When Figure 8.3 is extended to cover the plane, the symmetry group of the resulting figure is not a wallpaper group; the symmetry group contains vertical translations of every positive length, and so condition (ii) of Definition 8.1 doesn't hold. *Whenever we consider symmetry groups of figures in the rest of this section and in Sections 9–12, we imagine that the figures are extended to cover the plane by repeating the pattern shown.*

As Definition 8.1 suggests, translations play a key role in wallpaper groups. The next result shows that the product of two translations is another translation or the identity map. This result is intuitively clear if we think of a translation as a rigid motion of the plane that doesn't involve

FIGURE 8.1.

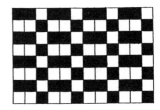

FIGURE 8.2.

FIGURE 8.3.

turning or reflecting. *Here and in the future we use an arrow to show the length and direction of a translation.*

THEOREM 8.2. In the Euclidean plane, let τ_1 and τ_2 be translations. Let A be any point.

(i) If $\tau_2\tau_1(A) \neq A$, then $\tau_2\tau_1$ is the translation $\tau_{A,\,\tau_2\tau_1(A)}$ that maps A to $\tau_2\tau_1(A)$ (Figure 8.4).
(ii) If $\tau_2\tau_1(A) = A$, then $\tau_2\tau_1$ is the identity map I.

Proof: Let X be any point in the plane (Figure 8.4). Since τ_1 is a translation, $\tau_1(X)$ lies the same distance in the same direction from X as $\tau_1(A)$ lies from A. Likewise, since τ_2 is a translation, $\tau_2\tau_1(X)$ lies the same distance in the same direction from $\tau_1(X)$ as $\tau_2\tau_1(A)$ lies from $\tau_1(A)$. Combining the last two sentences shows that $\tau_2\tau_1(X)$ lies the same distance in the same direction from X as $\tau_2\tau_1(A)$ lies from A. Thus, if $\tau_2\tau_1(A) \neq A$, the translation $\tau_{A,\,\tau_2\tau_1(A)}$ maps X to $\tau_2\tau_1(X)$; this holds for every point X, and so $\tau_{A,\,\tau_2\tau_1(A)}$ equals $\tau_2\tau_1$, and we've proved part (i). If $\tau_2\tau_1(A) = A$, then the second-to-last sentence shows that $\tau_2\tau_1(X) = X$ for every point X, and we've proved part (ii). $\square$

In the Euclidean plane, let τ_1 and τ_2 be distinct translations in a group $\mathcal{G}$ of isometries. For any point A in the plane, we have $\tau_1(A) \neq \tau_2(A)$ (Figure 8.5). $\tau_2\tau_1^{-1}$ is an element of $\mathcal{G}$ (by Definition 7.15), and it maps $\tau_1(A)$ to

$$\tau_2\tau_1^{-1}\tau_1(A) = \tau_2 I(A) = \tau_2(A)$$

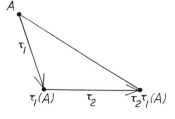

FIGURE 8.4.

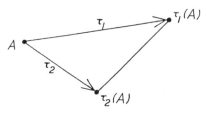

FIGURE 8.5.

(by Equations 5 and 12 of Section 7). Thus, $\tau_2 \tau_1^{-1}$ is a translation (by Equation 7 of Section 7, Theorem 8.2i, and the last two sentences). In short, $\tau_2 \tau_1^{-1}$ is a translation in $\mathcal{G}$ that maps $\tau_1(A)$ to $\tau_2(A)$.

As we noted after Thoerem 7.16, the associative law implies that products of isometries can be written without parentheses. Thus, we can define the powers of an isometry φ as follows: We set

$$\varphi^p = \underbrace{\varphi \cdot \varphi \cdots \varphi}_{p \text{ factors}} \quad \text{and} \quad \varphi^{-p} = \underbrace{\varphi^{-1} \cdot \varphi^{-1} \cdots \varphi^{-1}}_{p \text{ factors}}$$

for every positive integer p, and we let φ^0 be the identity map I. Thus, φ^n is an isometry for every integer n (by Theorem 7.14i).

In the next three theorems we analyze the translations in a wallpaper group. We say that a translation τ in a set $\mathcal{S}$ of isometries has *shortest length* if the length of τ is less than or equal to the length of every other translation in $\mathcal{S}$. The symmetry group of Figure 8.3 has no translation of shortest length because the symmetry group contains vertical translations of every positive length. On the other hand, because the symmetry groups of Figures 8.1 and 8.2 are wallpaper groups, each of these symmetry groups has translations of shortest length, as the next result shows. For example, the symmetry group of Figure 8.2 has two translations of shortest length: one moves points two squares up, and the other moves points two squares down.

THEOREM 8.3. In the Euclidean plane, let $\mathcal{G}$ be a wallpaper group. Let $\mathcal{S}$ be a subset of $\mathcal{G}$ that contains at least one translation. Then $\mathcal{S}$ has at least one translation of shortest length.

Proof: Choose any point A in the plane. For any two translations τ_1 and τ_2 in $\mathcal{G}$, $\tau_2 \tau_1^{-1}$ is a translation in $\mathcal{G}$ that maps $\tau_1(A)$ to $\tau_2(A)$ (by the discussion accompanying Figure 8.5), and so the distance from $\tau_1(A)$ to $\tau_2(A)$ is at least the positive number b in Definition 8.1(ii).

$\mathcal{S}$ contains a translation τ, by assumption, let d be the length of τ, and let $\mathcal{K}$ be the circle of radius $d + b/2$ with center A. Consider the circles of radius $b/2$ whose centers are the points $\tau'(A)$ as τ' varies over all translations in S of length at most d. The interiors of these circles lie entirely inside $\mathcal{K}$ and don't overlap (since the centers of the circles lie at least b units apart, by the previous paragraph). Thus, the sum of the areas of these circles is at most the area of $\mathcal{K}$. Hence, there are only finitely many such circles (since they all have the same area). Accordingly, there are only finitely many translations in S whose length is less than or equal to the length of τ. Therefore, S has a translation of shortest length. $\square$

We can now determine the translations in a wallpaper group that are parallel to a given translation.

THEOREM 8.4. In the Euclidean plane, let τ_1 be a translation in a wallpaper group $\mathcal{G}$. Then there is a translation τ of shortest length among the translations in $\mathcal{G}$ parallel to τ_1. The translations in $\mathcal{G}$ parallel to τ_1 are exactly τ^i for all nonzero integers i.

Proof: Consider the set of translations in $\mathcal{G}$ parallel to τ_1. This set contains τ_1, and so it has an element τ of shortest length d (by Theorem 8.3). Choose a point A, and let l be the line through A parallel to τ (Figure 8.6). The points $\tau^i(A)$ for all integers i divide l into intervals of length d (since τ moves every point in the plane d units in one direction parallel to l, and τ^{-1} moves every point d units in the opposite direction, by Equation 7 of Section 7). Thus, the maps τ^i are distinct translations in $\mathcal{G}$ parallel to τ_1 for all nonzero integers i (by Theorem 8.2i and Equation 7 of Section 7).

Conversely, let τ' be a translation in $\mathcal{G}$ parallel to τ_1. Let t be a nonzero integer such that the distance from $\tau'(A)$ to $\tau^t(A)$ is as small as possible. This distance is less than d (since $\tau'(A)$ lies on l, and the points $\tau^i(A)$ lie d units apart). Thus, if τ' didn't equal τ^t, $\tau^t\tau'^{-1}$ would be a translation in $\mathcal{G}$ that maps $\tau'(A)$ to $\tau^t(A)$ (by the discussion accompanying Figure 8.5), and so it would be parallel to τ_1 and have length less than d. This would contradict the definition of d, and so τ' equals τ^t. Together with the previous paragraph, this shows that the translations in $\mathcal{G}$ parallel to τ_1 are exactly the maps τ^i for all nonzero integers i. $\square$

To illustrate the previous theorem, let $\mathcal{G}$ be the symmetry group of Figure 8.2. Let τ_1 be the translation that shifts the figure six squares to the right. Then the translation τ that shifts the figure three squares to the right

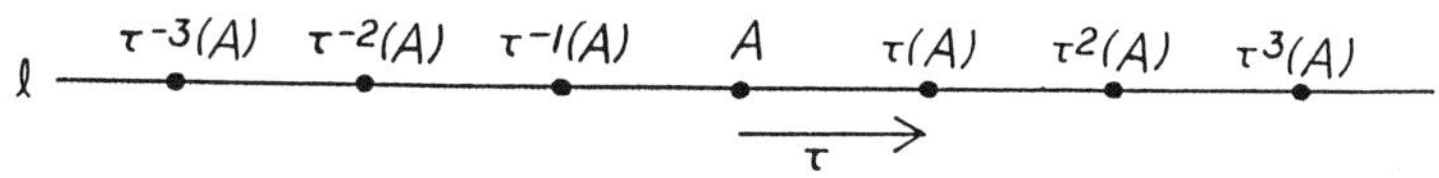

FIGURE 8.6.

is of shortest length among the translations in G parallel to τ_1. For any positive integer p, τ^p translates the figure $3p$ squares to the right, and τ^{-p} translates the figure $3p$ squares to the left. These are exactly the translations in G parallel to τ_1, as Theorem 8.4 states.

We can now determine all translations in a wallpaper group.

THEOREM 8.5. In the Euclidean plane, let G be a wallpaper group.

(i) G has a translation τ of shortest length, and it has a translation τ' of shortest length among those not parallel to τ.

(ii) The translations in G and the identity map are exactly the maps $\tau^i \tau'^j$ for all integers i and j.

Proof: Part (i) follows from Theorem 8.3 and Definition 8.1(i).

(ii) Choose a point A in the plane. The points $\tau^i \tau'^j(A)$ for all integers i and j are the vertices of a lattice such that each side of the lattice is parallel to and has the same length as τ or τ'. (See Figure 8.7, where we don't write τ^0 or τ'^0 because each of these is the identity map.) Thus, the maps $\tau^i \tau'^j$ are distinct translations in G for all integers i and j not both zero (by Equation 7 of Section 7, Definition 7.15, and Theorem 8.2i). G also

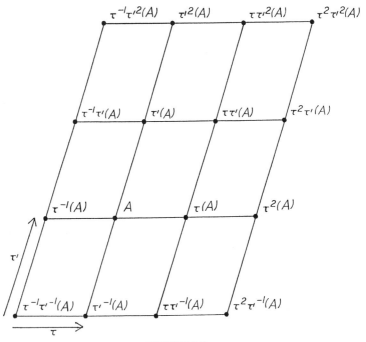

FIGURE 8.7.

contains the identity map $I = \tau^0\tau'^0$ (by Definition 7.15iii). We must prove that every translation in $\mathcal{G}$ equals $\tau^i\tau'^j$ for some integers i and j.

Let τ_1 be any translation in $\mathcal{G}$. Let s and t be integers such that $\tau^s\tau''(A)$ is as close as possible to $\tau_1(A)$. We claim that $\tau_1(A)$ equals $\tau^s\tau''(A)$. If so, then τ_1 equals $\tau^s\tau''$, and every translation in $\mathcal{G}$ has the form $\tau^i\tau'^j$ for some integers i and j, as desired. Accordingly, we assume that $\tau_1(A) \neq \tau^s\tau''(A)$ and seek a contradiction. The assumption that $\tau_1(A) \neq \tau^s\tau''(A)$ implies that $\tau^s\tau''\tau_1^{-1}$ is a translation in $\mathcal{G}$ mapping $\tau_1(A)$ to $\tau^s\tau''(A)$; this follows from the discussion accompanying Figure 8.5 if s and t are not both zero (since $\tau^s\tau''$ is a translation when s and t are not both zero, by the previous paragraph), and it follows from Equation 12 of Section 7 if $s = 0 = t$ (since $\tau^0\tau'^0$ is the identity map I). Let l be the line through $\tau^s\tau''(A)$ parallel to τ. We consider two cases, depending on whether or not $\tau_1(A)$ lies on l.

First, suppose that $\tau_1(A)$ doesn't lie on l. $\tau_1(A)$ lies on or within a parallelogram of lattice points (since the parallelograms in Figure 8.7 cover the plane). Label the vertices of the parallelogram $PQRS$ so that $Q = \tau(P)$, $R = \tau\tau'(P)$, and $S = \tau'(P)$ (Figure 8.8). Let d and e be the respective lengths of τ and τ'. Let $\mathcal{K}_1$ and $\mathcal{K}_2$ be circles of radius e with respective centers P and R. $\mathcal{K}_1$ contains S, and $\mathcal{K}_2$ contains Q. Since $d \leq e$ (because τ is a translation of shortest length in $\mathcal{G}$), Q lies on or inside $\mathcal{K}_1$, and S lies on or inside $\mathcal{K}_2$. Thus, the interior of $\mathcal{K}_1$ contains every point on or inside triangle PQS except for S and possibly Q, and the interior of $\mathcal{K}_2$ contains every point on or inside triangle RQS except for Q and possibly S. Hence, every point on or inside parallelogram $PQRS$ lies less than e units from one of the lattice points P, Q, R, S. In particular, $\tau_1(A)$ lies less than e units from a lattice point, and so the distance from $\tau_1(A)$ to the nearest lattice point $\tau^s\tau''(A)$ is less than e. Then the previous paragraph shows that $\tau^s\tau''\tau_1^{-1}$ is a translation in $\mathcal{G}$ that has length less than e and that isn't parallel to τ (since it maps $\tau_1(A)$ to $\tau^s\tau''(A)$, and since $\tau_1(A)$ doesn't lie on the line l through

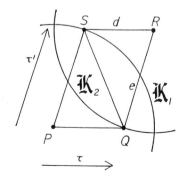

FIGURE 8.8.

$\tau^s\tau''(A)$ parallel to τ). This contradicts the definition of e as the length of the translation τ' of shortest length among those in $\mathcal{G}$ not parallel to τ.

Next, suppose that $\tau_1(A)$ lies on l. If d is the length of τ, the lattice points on l lie d units apart (as in Figure 8.7, where l is one of the horizontal lines shown). Thus, the distance from $\tau_1(A)$ to the nearest lattice point $\tau^s\tau''(A)$ is less than d. Hence, $\tau^s\tau''\tau_1^{-1}$ is a translation in $\mathcal{G}$ of length less than d (by the second-to-last paragraph), which contradicts the fact that τ is a translation of shortest length in $\mathcal{G}$.

The last three paragraphs show that any translation τ_1 in $\mathcal{G}$ equals $\tau^s\tau''$ for some integers s and t, which completes the proof. $\square$

Consider a figure whose symmetry group is a wallpaper group. Let τ and τ' be as in Theorem 8.5(i). For any point A, the points $\tau^i\tau'^j(A)$ are the vertices of a lattice of parallelograms, as in Figure 8.7. Figures 8.9 and 8.10 illustrate this for Figures 8.1 and 8.2; the heavy dots in Figures 8.9 and 8.10 are the vertices $\tau^i\tau'^j(A)$ of the lattice, and the heavy lines are the sides of the lattice. The point A is chosen arbitrarily; taking another point A translates the lattice with respect to the figure. For example, Figures 8.10 and 8.11 both show lattices for Figure 8.2.

The fact that the symmetry group $\mathcal{G}$ of a figure $\mathcal{F}$ is a wallpaper group says that $\mathcal{F}$ repeats at regular, discrete intervals in nonparallel directions. For any point A, the repetitions of $\mathcal{F}$ are shown by the parallelograms with vertices $\tau^i\tau'^j (A)$ (Figures 8.9–8.11). The entire figure is obtained by covering the plane with copies of the pattern inside any one of the parallelograms. The parallelograms correspond to the translations in $\mathcal{G}$ and the identity map: If we pick a parallelogram $\mathcal{P}$, we associate the map $\tau^i\tau'^j$ with the parallelogram $\tau^i\tau'^j(\mathcal{P})$.

The converse of Theorem 8.5 also holds: A group $\mathcal{G}$ of isometries is a wallpaper group if it contains nonparallel translations τ and τ' as in Theorem 8.5(ii). This result gives a way besides Definition 8.1 to characterize wallpaper groups.

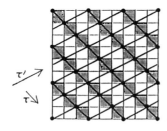

FIGURE 8.9.

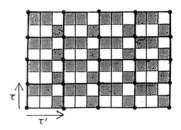

FIGURE 8.10.

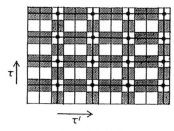

FIGURE 8.11.

THEOREM 8.6. In the Euclidean plane, let G be a group of isometries. Then G is a wallpaper group if and only if it satisfies the following condition:

> There are nonparallel translations τ and τ' in G such that the translations in G and the identity map are exactly the maps $\tau^i\tau'^j$ for all integers i and j. (1)

Proof: If G is a wallpaper group, then G satisfies the condition in (1), by Theorem 8.5. Conversely, let G be a group of isometries that satisfies the condition in (1). Then G contains nonparallel translations, and so condition (i) of Definition 8.1 holds. Choose a circle $\mathcal{K}$, and let A be its center. Consider the lattice of points $\tau^i\tau'^j(A)$ for all integers i and j. (See Figure 8.12, where the points $\tau^i\tau'^j(A)$ are shown by dots.) For any integer s, let l_s be the line parallel to τ' that contains the points $\tau^s\tau'^j(A)$ for all integers j. For any integer t, let m_t be the line parallel to τ that contains the points $\tau^i\tau'^t(A)$ for all integers i.

The l_s are evenly spaced parallel lines, and so are the m_t. Thus, only finitely many of these lines intersect the circle $\mathcal{K}$. If the lattice point $\tau^s\tau'^t(A)$ lies inside $\mathcal{K}$, then $\mathcal{K}$ intersects the lines l_s and m_t that contain the point. Combining the last two sentences shows that only finitely many of the lattice points $\tau^s\tau'^t(A)$ lie inside $\mathcal{K}$. Thus, there is a positive number b that

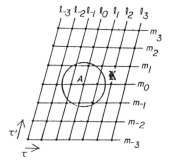

FIGURE 8.12.

is less than the distance from A to $\tau_1(A)$ for all translations τ_1 in $\mathcal{G}$ (by the condition in (1)). Then b is less than the length of every translation in $\mathcal{G}$, and so condition (ii) of Definition 8.1 holds. Hence, $\mathcal{G}$ is a wallpaper group, by Definition 8.1. $\square$

We define the *axis of a reflection* σ_l to be the line l, and we define the *axis of a glide reflection* $\gamma_{A, A'}$ to be the line AA'. In the rest of this section, we use our knowledge of the translations in a wallpaper group $\mathcal{G}$ to analyze the glide reflections in $\mathcal{G}$ that have a particular line as their axis. We start by considering how to use another isometry to change the direction of a translation.

THEOREM 8.7. In the Euclidean plane, let φ be an isometry, and let $\tau_{A, B}$ be a translation. Then we have

$$\varphi\tau_{A, B}\varphi^{-1} = \tau_{\varphi(A), \varphi(B)}.$$

Proof: Let X be any point that doesn't lie on line AB, and set $Y = \tau_{A, B}(X)$. Then $ABYX$ is a parallelogram (Figure 8.13). It follows that $\varphi(A)\varphi(B)\varphi(Y)\varphi(X)$ is also a parallelogram (by the Classification Theorem 7.10, for example), and so we have

$$\varphi(Y) = \tau_{\varphi(A), \varphi(B)}(\varphi(X)).$$

Then Equations 5 and 12 of Section 7 imply that

$$\varphi\tau_{A, B}\varphi^{-1}[\varphi(X)] = \varphi\tau_{A, B}(X)$$
$$= \varphi(Y)$$
$$= \tau_{\varphi(A), \varphi(B)}(\varphi(X)).$$

This holds for every point X not on line AB, and so the theorem follows (by Theorem 7.8). $\square$

We can now determine the glide reflections in a wallpaper group $\mathcal{G}$ that have the same axis as a reflection in $\mathcal{G}$.

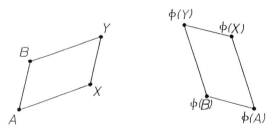

FIGURE 8.13.

THEOREM 8.8. In the Euclidean plane, let σ_l be a reflection in a wallpaper group $\mathcal{G}$. Then there is a translation τ in $\mathcal{G}$ parallel to l such that the translations in $\mathcal{G}$ parallel to l are exactly the maps τ^i for all nonzero integers i. The glide reflections in $\mathcal{G}$ with axis l are the maps $\tau^i\sigma_l$ for all nonzero integers i. For any point A on l, these glide reflections are the maps $\gamma_{A, \tau^i(A)}$ for all nonzero integers i (Figure 8.6).

To illustrate Theorem 8.8, let $\mathcal{G}$ be the symmetry group of Figure 8.2. The line l in Figure 8.14 is the axis of a reflection in $\mathcal{G}$; here and in the future, *we use double lines to mark axes of reflections.* By Theorem 8.8, the glide reflections in $\mathcal{G}$ with axis l are the maps $\gamma_{A, \tau^i(A)}$ for all nonzero integers i, where τ translates the figures three squares to the right, and where A is any point on l. When i is positive, $\gamma_{A, \tau^i(A)}$ reflects the figure across l and translates it $3i$ squares to the right. When i is negative, $\gamma_{A, \tau^i(A)}$ reflects the figure across l and translates it $3|i|$ squares to the left.

Proof: $\mathcal{G}$ contains a translation that isn't perpendicular to l (by Definition 8.1i). We can write this translation as $\tau_{A, B}$, where A is any point of l, and B is the image of A under the translation (Figure 8.15). If we set $B' = \sigma_l(B)$, we have

$$\sigma_l \tau_{A, B} \sigma_l^{-1} = \tau_{A, B'} \qquad (2)$$

(by Theorem 8.7, since σ_l fixes A). $\tau_{A, B'}$ maps B to a point C of l, and C doesn't equal A (since $\tau_{A, B}$ isn't perpendicular to l). It follows that

$$\tau_{A, B'} \tau_{A, B} = \tau_{A, C} \qquad (3)$$

(by Theorem 8.2i, since $\tau_{A, B}$ maps A to B, and $\tau_{A, B'}$ maps B to C). In short, $\tau_{A, C}$ is a translation parallel to l that belongs to $\mathcal{G}$ (by Definition 7.15 and Equations 2 and 3). Accordingly, there is a translation τ such that the translations in $\mathcal{G}$ parallel to l are the maps τ^i for all nonzero integers i (by Theorem 8.4).

The glide reflections γ' with axis l are the maps $\gamma' = \tau'\sigma_l$ as τ' varies over all translations parallel to l (by Equation 4 of Section 7). Since $\mathcal{G}$ contains σ_l, it contains γ' if and only if it contains τ' (by Theorem 7.17). Thus, the glide reflections in $\mathcal{G}$ with axis l are the maps $\tau'\sigma_l$ for all translations τ' in

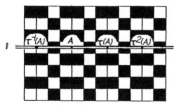

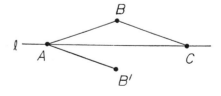

FIGURE 8.14.   FIGURE 8.15.

$\mathcal{G}$ parallel to l. Then these glide reflections are the maps $\tau^i \sigma_l$ for all nonzero integers i (by the previous paragraph). The glide reflection $\tau^i \sigma_l$ maps any point A on its axis l to $\tau^i \sigma_l(A) = \tau^i(A)$, and so $\tau^i \sigma_l$ is the glide reflection $\gamma_{A, \tau^i(A)}$. $\quad\square$

In the Euclidean plane, let $\mathcal{G}$ be a wallpaper group. If γ is a glide reflection in $\mathcal{G}$, we call γ *trivial* if $\mathcal{G}$ contains a reflection that has the same axis as γ. The previous theorem shows why this terminology is natural: Every trivial glide reflection is the product of a reflection and a translation that each belong to $\mathcal{G}$.

The next theorem describes the nontrivial glide reflections in a wallpaper group $\mathcal{G}$ that have a particular line l as their axis. The condition that the glide reflections are nontrivial means that l is not the axis of a reflection in $\mathcal{G}$.

Two general observations about products of isometries will be useful. First, we have

$$\sigma_l^2 = I \tag{4}$$

for any line l, since reflecting twice across l sends each point back to its original position. Second, if τ_1 is a translation parallel to a line l, then we have

$$\sigma_l \tau_1 \sigma_l^{-1} = \tau_1$$

(by Theorem 8.7, since τ_1 equals $\tau_{A, B}$ for two points A and B on l, and σ_l fixes these points); it follows that

$$\tau_1 \sigma_l = \sigma_l \tau_1 \sigma_l^{-1} \sigma_l = \sigma_l \tau_1 \tag{5}$$

(by Equations 5 and 12 of Section 7). In other words, although the commutative law does not generally hold for products of isometries, it does hold for the product of a reflection and a translation parallel to the axis of the reflection.

THEOREM 8.9. In the Euclidean plane, let l be the axis of a nontrivial glide reflection in a wallpaper group $\mathcal{G}$. Then there is a translation τ in $\mathcal{G}$ such that the translations in $\mathcal{G}$ parallel to l are exactly the maps τ^i for all nonzero integers i. For any point A on l, the glide reflections in $\mathcal{G}$ with axis l are the maps γ_{A, A_i} for all integers i, where A_i is the midpoint of $\tau^i(A)$ and $\tau^{i+1}(A)$ (Figure 8.16).

To illustrate this theorem, let $\mathcal{G}$ be the symmetry group of Figure 8.17. The line l in this figure is the axis of glide reflections in $\mathcal{G}$, and these glide reflections are nontrivial because $\mathcal{G}$ doesn't contain σ_l; here and in the future, *we use dotted lines to mark axes of nontrivial glide reflections.* By

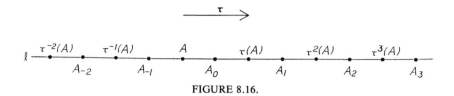

FIGURE 8.16.

Theorem 8.9, the glide reflections in G with axis l are the maps γ_{A,A_i} for all integers i, where A is any point on l, τ is the translation that moves the figure four squares to the right, and A_i is the midpoint of $\tau^i(A)$ and $\tau^{i+1}(A)$. If $i \geq 0$, γ_{A,A_i} reflects the figure across l and translates it $4i + 2$ squares to the right. If $i < 0$, γ_{A,A_i} reflects the figures across l and translates it $4|i| - 2$ squares to the left.

Proof: By assumption, G contains a nontrivial glide reflection γ with axis l. We can write

$$\gamma = \tau_1 \sigma_l \tag{6}$$

for a translation τ_1 parallel to l (by Equation 4 of Section 7). Note that σ_l doesn't belong to G (by assumption), and so neither does τ_1 (by Equation 6, Theorem 7.17, and the fact that G contains γ but not σ_l).

Combining Equations 4–6 with Equation 12 of Section 7 shows that

$$\gamma^2 = \tau_1 \sigma_l \sigma_l \tau_1 = \tau_1 I \tau_1 = \tau_1^2. \tag{7}$$

γ^2 belongs to G (by Definition 7.15), and τ_1^2 is a translation parallel to l (by Theorem 8.2i). Thus, Equation 7 shows that G contains a translation parallel to l. Hence, by Theorem 8.4, there is a translation τ such that the translations in G parallel to l are the maps τ^i for all nonzero integers i.

γ^2 is a translation in G parallel to l, and so γ^2 equals τ^s for some nonzero integer s (by the previous paragraph). If s were an even integer $2k$ for some integer k, the relations $\tau_1^2 = \gamma^2 = \tau^{2k}$ (by Equation 7) would imply that $\tau_1 = \tau^k$ (since τ_1 and τ are translations); then τ_1 would belong to G, which

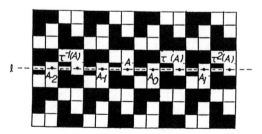

FIGURE 8.17.

would contradict the first paragraph of the proof. Thus, s is an odd integer $2k + 1$ for some integer k. Since γ acts on l as a translation, $\gamma(A)$ is the midpoint of A and $\gamma^2(A) = \tau^{2k+1}(A)$, and so $\gamma(A)$ is the midpoint A_k of $\tau^k(A)$ and $\tau^{k+1}(A)$. Thus, γ is the glide reflection γ_{A,A_k}.

As j varies over all integers, $\tau^j(A_k)$ varies over the points A_i for all integers i. Thus, since $\tau_1(A) = \gamma(A) = A_k$ (by Equation 6 and the previous paragraph), $\tau^j\tau_1(A)$ varies over all the points A_i, and $\tau^j\tau_1$ varies over all the translations τ_{A,A_i} (by Theorem 8.2i). Consequently, as j varies, $\tau^j\gamma = \tau^j\tau_1\sigma_l$ varies over all the glide reflections γ_{A,A_i}. Since $\mathcal{G}$ contains τ and γ, it contains all these glide reflections (by Definition 7.15). $\mathcal{G}$ doesn't contain any other glide reflections with axis l (by the previous paragraph), and so the maps γ_{A,A_i} are exactly the glide reflections in $\mathcal{G}$ with axis l. $\square$

EXERCISES

8.1. For each of the following figures, let $\mathcal{G}$ be the symmetry group of the figure. Let τ be a translation of shortest length in $\mathcal{G}$, and let τ' be a translation of shortest length among those in $\mathcal{G}$ not parallel to τ (as in Theorem 8.5). On a copy of the figure, draw τ, τ', and a lattice of parallelograms whose vertices are the points $\tau^i\tau'^j(A)$ for some point A (as in Figures 8.9–8.11).

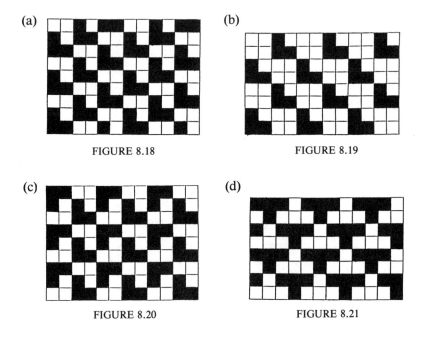

(a)

FIGURE 8.18

(b)

FIGURE 8.19

(c)

FIGURE 8.20

(d)

FIGURE 8.21

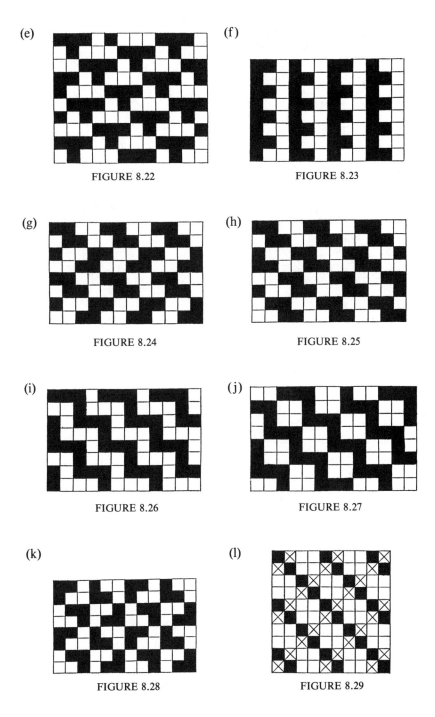

(e)

FIGURE 8.22

(f)

FIGURE 8.23

(g)

FIGURE 8.24

(h)

FIGURE 8.25

(i)

FIGURE 8.26

(j)

FIGURE 8.27

(k)

FIGURE 8.28

(l)

FIGURE 8.29

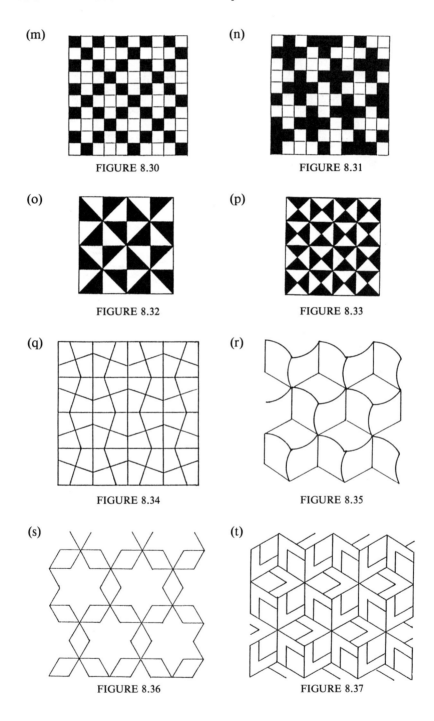

(m)

FIGURE 8.30

(n)

FIGURE 8.31

(o)

FIGURE 8.32

(p)

FIGURE 8.33

(q)

FIGURE 8.34

(r)

FIGURE 8.35

(s)

FIGURE 8.36

(t)

FIGURE 8.37

(u)

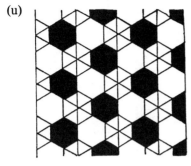

FIGURE 8.38

(v)

FIGURE 8.39

8.2. In each part of Exercise 8.1, let G be the symmetry group of the given figure. On a copy of the figure, use double lines to mark as many lines l_j as possible such that the l_j are axes of reflections in G and no two of the lines l_j are parallel. For each line l_j, choose a point A and a translation τ_j and draw the points $\tau_j^i(A)$ such that the glide reflections in G with axis l_j are the maps $\gamma_{A,\tau_j^i(A)}$ for all nonzero integers i. (See Theorem 8.8.)

8.3. In each part of Exercise 8.1, let τ and τ' be as in Exercise 8.1, and let τ_j be as in Exercise 8.2. For each τ_j, find integers s and t such that $\tau_j = \tau^s \tau'^t$.

8.4. In each part of Exercise 8.1, let G be the symmetry group of the given figure. On a copy of the figure, use dotted lines to mark as many lines l_j as possible such that the l_j are axes of nontrivial glide reflections in G and no two of the lines l_j are parallel. For each line l_j, draw points $\tau_j^i(A)$ and A_i such that A is a point of l_j, τ_j is a translation parallel to l_j, A_i is the midpoint of $\tau_j^i(A)$ and $\tau_j^{i+1}(A)$ for each integer i, and the glide reflections in G with axis l_j are the maps γ_{A,A_i} for all integers i. (See Theorem 8.9.)

8.5. In each part of Exercise 8.1, let τ and τ' be as in Exercise 8.1, and let τ_j be as in Exercise 8.4. For each τ_j, find integers s and t such that $\tau_j = \tau^s \tau'^t$.

8.6. In the Euclidean plane, let A and B be two points.

(a) Determine all isometries φ such that $\varphi(A) = B$ and $\varphi(B) = A$.
(b) Determine all isometries φ such that $\varphi(\{A, B\}) = \{A, B\}$.

8.7. In the Euclidean plane, let P be a point, and let l be a line. In each of the following cases, determine all isometries φ such that $\varphi(P) = P$ and $\varphi(l) = l$.

(a) P lies on l.
(b) P doesn't lie on l.

8.8. In the Euclidean plane, let l be a line.

(a) Determine all isometries that map l to itself.
(b) Determine all isometries that map l to a parallel line.

8.9. In the Euclidean plane, let l and m be two parallel lines.

(a) Determine all isometries φ such that $\varphi(l) = l$ and $\varphi(m) = m$.
(b) Determine all isometries φ such that $\varphi(l) = m$ and $\varphi(m) = l$.

8.10. In the Euclidean plane, let l and m be perpendicular lines.

(a) Determine all isometries φ such that $\varphi(l) = l$ and $\varphi(m) = m$.
(b) Determine all isometries φ such that $\varphi(l) = m$ and $\varphi(m) = l$.

8.11. In the Euclidean plane, let l and m be lines that are neither parallel nor perpendicular.

(a) Determine all isometries φ such that $\varphi(l) = l$ and $\varphi(m) = m$.
(b) Determine all isometries φ such that $\varphi(l) = m$ and $\varphi(m) = l$.

8.12. In the Euclidean plane, let φ and ψ be isometries. Prove that the equations $\varphi\psi\varphi^{-1} = \psi$ and $\varphi\psi = \psi\varphi$ are equivalent.

8.13. In the Euclidean plane, prove that $\tau\tau' = \tau'\tau$ for any translations τ and τ'.

8.14. In the Euclidean plane, let $\mathcal{G}$ be a wallpaper group. Let τ and τ' be translations such that the translations in $\mathcal{G}$ and the identity map are exactly the maps $\tau^i\tau'^j$ for all integers i and j. Set $\tau'' = \tau\tau'$. Prove that the translations in $\mathcal{G}$ and the identity map are exactly the maps $\tau''^i\tau'^j$ for all integers i and j.

8.15. For each part of Exercise 8.1, let $\mathcal{G}$ be the symmetry group of the figure shown. Let τ be a translation of shortest length in $\mathcal{G}$, and let τ' be a translation of shortest length among those in $\mathcal{G}$ not parallel to τ. Set $\tau'' = \tau\tau'$. On a copy of the given figure, show τ, τ', τ'' and a lattice of parallelograms whose vertices are the points $\tau''^i\tau'^j(A)$ for some point A and all integers i and j. (See Theorem 8.5 and Exercise 8.14.)

8.16.

(a) In the Euclidean plane, let $\mathcal{G}$ be a wallpaper group. Assume that $\mathcal{G}$ contains nonparallel translations τ and τ' that are each of shortest length in $\mathcal{G}$. Let A be any point, and let θ be the angle at A in the triangle with vertices A, $\tau(A)$, and $\tau'(A)$. Prove that $\theta \geq 60°$ by using the discussion accompanying Figure 8.5.

(b) In the Euclidean plane, let $\mathcal{G}$ be a wallpaper group. Use part (a) to prove that $\mathcal{G}$ has either exactly· 2, exactly 4, or exactly 6 translations of shortest length.

(c) For each of the numbers 2, 4, 6 draw a figure whose symmetry group is a wallpaper group that has exactly the specified number of translations of shortest length.

8.17. In the Euclidean plane, let $\mathcal{G}$ be a wallpaper group. Let τ be a translation of shortest length in $\mathcal{G}$, and let τ' be a translation of shortest length among those in $\mathcal{G}$ not parallel to τ (as in Theorem 8.5). Let θ be the angle at A in the triangle with A, $\tau(A)$, $\tau'(A)$ as vertices, and let F be the foot of the altitude on $\tau'(A)$ (Figure 8.40).

(a) Prove that the distance from A to F is less than or equal to half the length of τ. (*Hint:* See the discussion accompanying Figure 8.5.)

(b) Prove that $\theta \geq 60°$.

(c) Assume that $\mathcal{G}$ contains a translation τ'' that has the same length as τ' and that isn't parallel to τ. Assume further that $\tau'(A)$ and $\tau''(A)$ lie on the same side of the line through A and $\tau(A)$. Prove that the line through $\tau'(A)$ and $\tau''(A)$ is parallel to τ. (*Hint:* One possible approach is to use part (b) to prove that the angle at A is the triangle with vertices A, $\tau'(A)$, $\tau''(A)$ is at most 60°. Then consider the distance from $\tau'(A)$ to $\tau''(A)$.)

(d) Prove that there are either exactly 2 or exactly 4 translations of shortest length among those in $\mathcal{G}$ not parallel to τ. For each of these two possibilities, draw a figure whose symmetry group is a wallpaper group $\mathcal{G}$ that illustrates the possibility and is such that τ and τ' have different lengths.

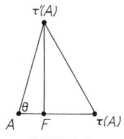

FIGURE 8.40.

(e) Prove that there are exactly 4 translations of shortest length among those in G not parallel to τ if and only if the distance from A to F is half the length of τ.

Section 9.

Axes and Centers of Wallpaper Groups

We shall see in Sections 10–12 that there are seventeen types of wallpaper groups. In preparation for this, we continue to study the general properties of a wallpaper group G in this section. We've already considered the translations in G and the glide reflections in G that have a common axis. We now consider the spacing of parallel lines that are axes of reflections or nontrivial glide reflections in G. We also determine the possible angles of rotations in G.

To start, we consider a reflection σ_l and a translation τ that isn't parallel to l, and we prove that the product $\tau\sigma_l$ is a reflection or a glide reflection, depending on whether or not τ is perpendicular to l.

THEOREM 9.1. In the Euclidean plane, let A be a point on a line l. Let τ be a translation that isn't parallel to l. Let m be the line parallel to l that lies midway between l and $\tau(l)$.

  (i) If τ is perpendicular to l, then we have

$$\tau\sigma_l = \sigma_m$$

    (Figure 9.1).

 (ii) If τ is not perpendicular to l, then we have

$$\tau\sigma_l = \gamma_{B,C},$$

where B and C are the feet of the perpendiculars drawn to m from A and $\tau(A)$, respectively (Figure 9.2).

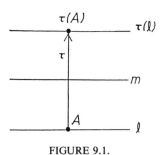
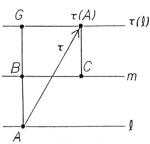

FIGURE 9.1.   FIGURE 9.2.

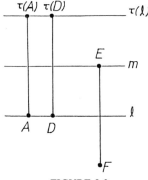

FIGURE 9.3.

Proof: (i) Let D be a point on l other than A (Figure 9.3). Then σ_l fixes A and D (since these points lie on l), and σ_m maps A to $\tau(A)$ and D to $\tau(D)$ (since m is the perpendicular bisector of A and $\tau(A)$ and of D and $\tau(D)$). Thus, we have

$$\tau\sigma_l(A) = \tau(A) = \sigma_m(A) \quad \text{and} \quad \tau\sigma_l(D) = \tau(D) = \sigma_m(D).$$

Let E be any point on m. Let F be the point such that l is the perpendicular bisector of E and F. Then we have

$$\tau\sigma_l(E) = \tau(F) = E = \sigma_m(E).$$

The two previous paragraphs show that $\tau\sigma_l$ and σ_m agree on the three points A, D, E. Since these points are the vertices of a triangle, $\tau\sigma_l$ and σ_m are equal (by Theorems 7.8 and 7.14i).

(ii) Let G be the foot of the perpendicular dropped from A to $\tau(l)$ (Figure 9.2). Since

$$\tau_{G,\tau(A)}\tau_{A,G}(A) = \tau_{G,\tau(A)}(G) = \tau(A),$$

Theorem 8.2(i) shows that

$$\tau_{G,\tau(A)}\tau_{A,G} = \tau.$$

Thus, we have

$$\tau\sigma_l = \tau_{G,\tau(A)}\tau_{A,G}\sigma_l$$

$$= \tau_{G,\tau(A)}\sigma_m$$

(by part (i), since $\tau_{A,G}$ is perpendicular to l)

$$= \tau_{B,C}\sigma_m$$

(since we travel the same distance in the same direction in going from G to $\tau(A)$ and from B to C)

$$= \gamma_{B,C}$$

(by Equation 4 of Section 7). $\square$

We use part (i) of the previous theorem to determine the spacing of parallel axes of reflections in a wallpaper group. We say that lines l_i are *spaced by* $\frac{1}{2}\tau'$ if τ' is a translation perpendicular to the lines and if we obtain l_{i+1} by translating l_i half the length of τ' in the direction of τ' for every integer i (Figure 9.4). Lines l_i are spaced by $\frac{1}{2}\tau'$ if and only if l_0 is perpendicular to τ' and l_i is the perpendicular bisector of A and $\tau'^i(A)$ for any point A on l_0 and each nonzero integer i.

THEOREM 9.2. In the Euclidean plane, let $\mathcal{G}$ be a wallpaper group. Let l be the axis of a reflection in $\mathcal{G}$. Then there is a translation τ' in $\mathcal{G}$ such that the translations in $\mathcal{G}$ perpendicular to l are the maps τ'^i for all nonzero integers i. The lines parallel to l that are axes of reflections in $\mathcal{G}$ are spaced by $\frac{1}{2}\tau'$.

To illustrate this theorem, let $\mathcal{G}$ be the symmetry group of Figure 8.2. The vertical translations that lie in $\mathcal{G}$ are those that move the figure an even number of squares. These are the maps τ'^i for all nonzero integers i, where τ' is the translation shifting the figure two squares upwards. As Figure 9.5 shows, the lines perpendicular to τ' that are axes of reflections in $\mathcal{G}$ are spaced by $\frac{1}{2}\tau'$.

Proof: Let A be any point of l. $\mathcal{G}$ contains a translation that isn't parallel to l (by Definition 8.1i). We can write this translation as $\tau_{A,B}$, where B is the image of A under the translation (Figure 9.6). We have

$$\sigma_l \tau_{A,B} \sigma_l^{-1} = \tau_{A,B'} \tag{1}$$

for $B' = \sigma_l(B)$ (by Theorem 8.7, since σ_l fixes A). We have

$$\tau_{A,B'}^{-1} = \tau_{B',A} \tag{2}$$

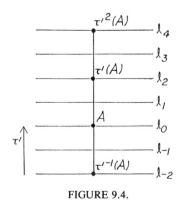

FIGURE 9.4.

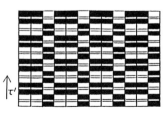

FIGURE 9.5.

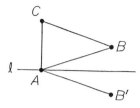

FIGURE 9.6.

(by Equation 7 of Section 7), and this translation maps B to a point C on the line through A perpendicular to l. C doesn't equal A (since $\tau_{A,B}$ isn't parallel to l), and so we have

$$\tau_{B',A}\tau_{A,B} = \tau_{A,C} \tag{3}$$

(by Theorem 8.2i, since $\tau_{A,B}$ maps A to B, and $\tau_{B',A}$ maps B to C). Thus, $\tau_{A,C}$ is a translation perpendicular to l that belongs to $\mathcal{G}$ (by Definition 7.15 and Equations 1–3). Accordingly, there is a translation τ' such that the translations in $\mathcal{G}$ perpendicular to l are the maps τ'^i for all nonzero integers i (by Theorem 8.4). Let l_i be the perpendicular bisector of A and $\tau'^i(A)$ for each nonzero integer i (as in Figure 9.4, if we take l_0 to be l).

Let τ be a translation perpendicular to l (Figure 9.1). We have $\tau\sigma_l = \sigma_m$, where m is the perpendicular bisector of A and $\tau(A)$ (by Theorem 9.1i). Thus, $\mathcal{G}$ contains τ if and only if it contains σ_m (by Theorem 7.17 and the fact that $\mathcal{G}$ contains σ_l). Hence, the lines parallel to and distinct from l that are axes of reflections in $\mathcal{G}$ are the lines l_i for all nonzero integers i (by the last two sentences of the previous paragraph). Thus, the lines parallel to l that are axes of reflections in $\mathcal{G}$ are spaced by $\frac{1}{2}\tau'$. $\square$

We want the analogue of the previous theorem obtained by replacing reflections with nontrivial glide reflections. We start with the analogue of Theorem 9.1: We determine the product of a translation τ and a glide reflection whose axis isn't parallel to τ.

THEOREM 9.3. In the Euclidean plane, let A and A' be two points. Let τ be a translation that isn't parallel to the line AA'. Let m be the line parallel to AA' that lies midway between AA' and $\tau(AA')$.

(i) If the lines $A\tau(A')$ and AA' are perpendicular, we have

$$\tau\gamma_{A,A'} = \sigma_m.$$

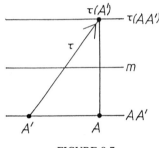

FIGURE 9.7.

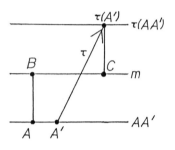

FIGURE 9.8.

(ii) If the lines $A\tau(A')$ and AA' are not perpendicular, we have

$$\tau\gamma_{A,A'} = \gamma_{B,C},$$

where B and C are the feet of the perpendiculars dropped to m from A and $\tau(A')$, respectively (Figure 9.8).

Proof: We have

$$\tau\tau_{A,A'}(A) = \tau(A') = \tau_{A,\tau(A')}(A) \tag{4}$$

(Figures 9.7 and 9.8). It follows that

$$\tau\gamma_{A,A'} = \tau\tau_{A,A'}\sigma_{AA'}$$

(by Equation 4 of Section 7)

$$= \tau_{A,\tau(A')}\sigma_{AA'}$$

(by Theorem 8.2i and Equation 4). Thus, the conclusions of parts (i) and (ii) follow by applying Theorem 9.1, where we take the line l in Theorem 9.1 to be AA', and where we take the translation in Theorem 9.1 to be the translation $\tau_{A,\tau(A')}$ that maps A to $\tau(A')$. $\square$

We can now determine the spacing of parallel axes of nontrivial glide reflections in a wallpaper group.

THEOREM 9.4. In the Euclidean plane, let $\mathcal{G}$ be a wallpaper group. Let l be the axis of a nontrivial glide reflection in $\mathcal{G}$. Then there is a translation τ' in $\mathcal{G}$ such that the translations in $\mathcal{G}$ perpendicular to l are the maps τ'^i for all nonzero integers i. The lines parallel to l that are axes of nontrivial glide reflections in $\mathcal{G}$ are spaced by $\frac{1}{2}\tau'$.

To illustrate this theorem, let $\mathcal{G}$ be the symmetry group of Figure 8.17. The vertical translations that lie in $\mathcal{G}$ are those that move the figure a multiple of three squares. These are the translations τ'^i for all nonzero integers i,

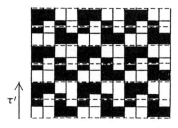

τ′

FIGURE 9.9.

where τ' is the translation shifting the figure three squares upward. As Figure 9.9 shows, the lines perpendicular to τ' that are axes of nontrivial glide reflections in $\mathcal{G}$ are spaced by $\frac{1}{2}\tau'$.

Proof: $\mathcal{G}$ contains a nontrivial glide reflection γ with axis l. Let A be any point of l. $\mathcal{G}$ contains a translation that isn't parallel to l (by Definition 8.1i). We can write this translation as $\tau_{A,B}$, where B is the image of A under the translation (Figure 9.10). We have

$$\gamma\tau_{A,B}\gamma^{-1} = \tau_{\gamma(A),\gamma(B)} \tag{5}$$

(by Theorem 8.7). We have

$$\tau_{\gamma(A),\gamma(B)}^{-1} = \tau_{\gamma(B),\gamma(A)} \tag{6}$$

(by Equation 7 of Section 7), and this translation maps B to a point C on the line through A perpendicular to l. C doesn't equal A (since $\tau_{A,B}$ isn't parallel to l), and so we have

$$\tau_{\gamma(B),\gamma(A)}\tau_{A,B} = \tau_{A,C} \tag{7}$$

(by Theorem 8.2i, since $\tau_{A,B}$ maps A to B and $\tau_{\gamma(B),\gamma(A)}$ maps B to C). In short, $\tau_{A,C}$ is a translation perpendicular to l that belongs to $\mathcal{G}$ (by Definition 7.15 and Equations 5–7). Thus, there is a translation τ' such that the translations in $\mathcal{G}$ perpendicular to l are the maps τ'^i for all nonzero integers i (by Theorem 8.4). Let l_i be the perpendicular bisector of A and $\tau'^i(A)$ for each nonzero integer i (as in Figure 9.4, if we take l_0 to be l).

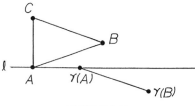

FIGURE 9.10.

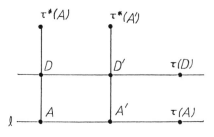

FIGURE 9.11.

There is a translation τ of shortest length among those in G parallel to l (by Theorem 8.9). Let A' be the midpoint of A and $\tau(A)$ (Figure 9.11). Let τ^* be a translation perpendicular to l, let D be the midpoint of A and $\tau^*(A)$, and let D' be the midpoint of A' and $\tau^*(A')$. Theorem 9.3(ii) shows that

$$\tau^* \gamma_{A,A'} = \gamma_{D,D'} . \tag{8}$$

Since $\gamma_{A,A'}$ belongs to G (by Theorem 8.9), G contains τ^* if and only if it contains $\gamma_{D,D'}$ (by Equation 8 and Theorem 7.17). Thus, G contains τ^* if and only if it contains a nontrivial glide reflection with axis DD' (by Theorems 8.8 and 8.9, since D' is the midpoint of D and $\tau(D)$). Hence, the lines parallel to and distinct from l that are axes of nontrivial glide reflections in G are the lines l_i for all nonzero integers i (by the last two sentences of the previous paragraph). Thus, the lines parallel to l that are axes of nontrivial glide reflections in G are spaced by $\frac{1}{2}\tau'$. $\square$

When we multiply two rotations about the same point, it's clear that the angles of rotation add, provided that we discard suitable multiplies of $360°$. Thus, we have the following result.

THEOREM 9.5. In the Euclidean plane, let $\rho_{P,\theta}$ and $\rho_{P,\psi}$ be rotations about the same point P.

(i) If $\theta + \psi < 360°$, then we have

$$\rho_{P,\psi}\rho_{P,\theta} = \rho_{P,\theta+\psi} .$$

(ii) If $\theta + \psi = 360°$, then $\rho_{P,\psi}\rho_{P,\theta}$ is the identity map I.

(iii) If $\theta + \psi > 360°$, then we have

$$\rho_{P,\psi}\rho_{P,\theta} = \rho_{P,\theta+\psi-360°} . \quad \square$$

We now determine the possible angles of rotations in a wallpaper group. We say that a rotation $\rho_{P,\theta}$ has *center* P. We say that a wallpaper group G has *n-center* P if n is an integer greater than 1 such that the rotations about P in $\mathcal{G}$ are those through the angles

$$(360/n)°, (2 \cdot 360/n)°, \ldots, ((n-1) \cdot 360/n)°. \qquad (9)$$

These angles are the $n-1$ positive multiplies of $(360/n)°$ that are less than $360°$. For example, the rotations in $\mathcal{G}$ about an n-center P are as follows for certain values of n:

$$\rho_{P,180°}, \quad \text{if } n = 2;$$

$$\rho_{P,120°}, \rho_{P,240°}, \quad \text{if } n = 3;$$

$$\rho_{P,90°}, \rho_{P,180°}, \rho_{P,270°} \quad \text{if } n = 4; \qquad (10)$$

$$\rho_{P,60°}, \rho_{P,120°}, \rho_{P,180°}, \rho_{P,240°}, \rho_{P,300°}, \quad \text{if } n = 6.$$

The symmetry groups of Figures 9.12 and 9.14 are wallpaper groups (by the discussion after Definition 8.1). Figures 9.13 and 9.15 show the centers of the rotations in the symmetry groups. Here and in the future, *dots show 2-centers, triangles show 3-centers, squares show 4-centers, and hexagons show 6-centers.* These figures show that centers of rotations in wallpaper

FIGURE 9.12.

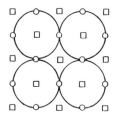

FIGURE 9.13.

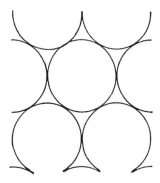

FIGURE 9.14.

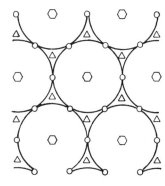

FIGURE 9.15.

groups can be 2-centers, 3-centers, 4-centers, and 6-centers. The next result shows that there are no other possibilities: *The center of any rotation in a wallpaper group is either a 2-center, a 3-center, a 4-center, or a 6-center.* Accordingly, the only rotations that can occur in a wallpaper group are those in (10).

THEOREM 9.6. In the Euclidean plane, let $\mathcal{G}$ be a wallpaper group, and let P be the center of a rotation in $\mathcal{G}$. Then P is an n-center, where n is one of the integers 2, 3, 4, 6.

Proof: First, assume that $\mathcal{G}$ contains a rotation ρ through an angle θ about P, where $0° < \theta < 90°$. We claim that $\theta = 60°$. To see this, let τ be a translation in $\mathcal{G}$ of shortest length (by Theorem 8.5i). If ρ' is the rotation through $360° - \theta$ about P, we have

$$\rho' = \rho^{-1} \tag{11}$$

(by Equation 6 of Section 7). We set

$$\tau_1 = \rho\tau\rho^{-1} \quad \text{and} \quad \tau_2 = \rho'\tau\rho'^{-1}. \tag{12}$$

We also set $Q = \tau(P)$ and $R = \tau_1(P)$, and we let S be the point that τ_2 maps to Q. (See Figures 9.16a–c, which illustrate the three cases $0° < \theta < 60°$, $\theta = 60°$, $60° < \theta < 90°$.) Theorem 8.7 and (12) imply that

$$\angle RPQ = \theta = \angle PQS \tag{13}$$

and that R and S lie on the same side of line PQ. $\mathcal{G}$ contains $\tau_1\tau^{-1}\tau_2$ (by Equations 11 and 12 and Definition 7.15), and we have

$$\tau_1\tau^{-1}\tau_2(S) = \tau_1\tau^{-1}(Q) = \tau_1(P) = R. \tag{14}$$

If R equals S, then PQR is an equilateral triangle (since τ, τ_1, and τ_2 have equal length), and we have $\theta = 60°$ (by (13)) (Figure 9.16b). Thus, if θ

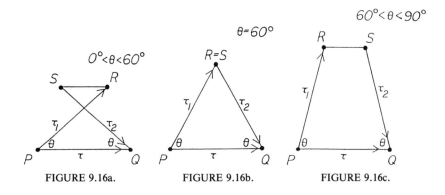

FIGURE 9.16a.   FIGURE 9.16b.   FIGURE 9.16c.

didn't equal $60°$ (as in Figures 9.16a and c), we would have $R \neq S$, and $\tau_1 \tau^{-1} \tau_2$ would be a translation in $\mathcal{G}$ mapping S to R (by (14), Equation 7 of Section 7, Definition 7.15, and Theorem 8.2i); since the distance from S to R is less than the length of τ (by (13) and the facts that $0° < \theta < 90°$ and τ, τ_1, τ_2 are of equal length), we've obtained a contradiction to the fact that τ is a translation in $\mathcal{G}$ of shortest length. This contradiction shows that $\theta = 60°$, as claimed.

Next, assume that $\mathcal{G}$ contains rotations through two angles θ and ψ about P. By relabeling these angles if necessary, we can assume that $\theta > \psi$. Then $\mathcal{G}$ contains

$$\rho_{P,\theta}\rho_{P,\psi}^{-1} = \rho_{P,\theta}\rho_{P,360°-\psi} = \rho_{P,\theta-\psi}$$

(by Equation 6 of Section 7, Definition 7.15, and Theorem 9.5iii). Thus, $\theta - \psi$ is at least $60°$ (by the previous paragraph). In short, any two rotations about P in $\mathcal{G}$ cover angles that differ by at least $60°$. Accordingly, since these angles lie between $0°$ and $360°$, there are only finitely many of them. Hence, there is a smallest positive number d such that $\mathcal{G}$ contains the rotation through $d°$ about P (since $\mathcal{G}$ contains at least one rotation about P, by assumption).

Let n be the least positive integer such that

$$nd - 360 \geq 0. \tag{15}$$

We have

$$n > 1 \tag{16}$$

(since $d < 360$). The relation

$$nd - 360 \geq d$$

would imply that

$$(n - 1)d - 360 \geq 0;$$

by (16), this would contradict the fact that n is the least positive integer such that (15) holds. Thus, we have

$$nd - 360 < d. \tag{17}$$

If $nd - 360$ were positive, then we would have

$$\rho_{P,d°}^n = \rho_{P,(nd-360)°}$$

(by Theorem 9.5 and (17)); this would be a rotation about P in $\mathcal{G}$ through an angle less than $d°$ (by Definition 7.15 and (17)), which would contradict the choice of d. Thus, we have $nd - 360 = 0$ (by (15)), which shows that $d = 360/n$.

Since $\mathcal{G}$ contains a rotation through $d°$ about P, we have $d = 360/n \geq 60$ (by the first paragraph of the proof). This shows that $n \leq 6$. If we had $n = 5$, then $\mathcal{G}$ would contain a rotation through $d° = (360/5)° = 72°$ about P, which would contradict the first paragraph of the proof. Combining the last two sentences with (16) shows that n is one of the integers 2, 3, 4, 6.

Let ρ'' be the rotation about P through $d°$. The maps ρ'', ρ''^2, ..., ρ''^{n-1} are the rotations about P through the angles in (9) (by Theorem 9.5i), and these rotations belong to $\mathcal{G}$ (by Definition 7.15). To prove that P is an n-center, we must show that these are the only rotations about P that belong to $\mathcal{G}$. Accordingly, let $\rho_{P,\,t°}$ be any rotation about P that belongs to $\mathcal{G}$. We claim that $t°$ is one of the angles in (9). Let s be the least positive integer such that

$$sd - t \geq 0. \tag{18}$$

The relation $sd - t \geq d$ would imply that $(s - 1)d - t \geq 0$; this would show that $s - 1 > 0$ and contradict the fact that s is the least positive integer such that (18) holds. Thus, we have

$$sd - t < d. \tag{19}$$

If $sd - t$ were positive, then we would have

$$\rho_{P,\,d°}^{s}\rho_{P,\,t°}^{-1} = \rho_{P,\,d°}^{s}\rho_{P,\,(360-t)°} = \rho_{P,\,(sd-t)°}$$

(by Equation 6 of Section 7, Theorem 9.5, (18), and (19)); this would be a rotation about P in $\mathcal{G}$ through an angle less than $d°$ (by Definition 7.15 and (19)), which would contradict the choice of d. Thus, we have $sd - t = 0$ (by (18)), which gives $t = sd = 360s/n$. This shows that $s < n$ (since $t < 360$) and that $t°$ is one of the angles in (9), as claimed.

The last two paragraphs show that P is an n-center, where n is one of the integers 2, 3, 4, 6. □

We now prove that *the product of two reflections with intersecting axes is a rotation*. We start with a preliminary observation. In the Euclidean plane, let P and X be two points (Figure 9.17). Let θ be an angle such that $0° < \theta < 180°$, and set $Y = \rho_{P,\,\theta}(X)$ and $Z = \rho_{P,\,\theta}(Y)$. The points X, Y, Z are equidistant from P, and the fact that triangles XYP and YZP are congruent (by the SAS Property 0.2) implies that X and Z are equidistant from Y. Thus, the perpendicular bisector of X and Z contains both P and Y (by Theorem 4.2), and so it is the line PY.

THEOREM 9.7. In the Euclidean plane, let l and m be two lines on a point P. Then we have

$$\sigma_m \sigma_l = \rho_{P,\,2\psi},$$

where ψ is the angle at P that lies counterclockwise after l and before m (Figure 9.18).

FIGURE 9.17.   FIGURE 9.18.

Proof: Let A be a point on l other than P, and set $B = \rho_{P,2\psi}(A)$ (Figure 9.19). Since m is the line through P that lies at angle ψ counterclockwise after PA and before PB, m is the perpendicular bisector of A and B (by the discussion accompanying Figure 9.17). Thus, we have

$$\sigma_m \sigma_l(A) = \sigma_m(A) = B = \rho_{P,2\psi}(A).$$

Let D be a point other than P on the line through P that lies at angle $\psi/2$ counterclockwise after l and before m (Figure 9.20). Let C be the point such that $\rho_{P,\psi}(C) = D$, and set $E = \rho_{P,\psi}(D)$. Since l is the line through P that lies at angle $\psi/2$ counterclockwise after PC and before PD, l is the perpendicular bisector of C and D (by the discussion accompanying Figure 9.17). Likewise, since m lies at angle $\psi/2$ counterclockwise after PD and before PE, m is the perpendicular bisector of D and E. Thus, we have

$$\sigma_m \sigma_l(C) = \sigma_m(D) = E = \rho_{P,2\psi}(C).$$

Finally, since the maps σ_m, σ_l, and $\rho_{P,2\psi}$ all fix P, we have

$$\sigma_m \sigma_l(P) = \sigma_m(P) = P = \rho_{P,2\psi}(P).$$

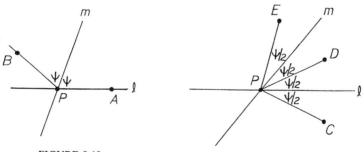

FIGURE 9.19.   FIGURE 9.20.

Together with the two previous paragraphs, this shows that $\sigma_m \sigma_l$ and $\rho_{P, 2\psi}$ agree on the three points A, C, P. These points are the vertices of a triangle, and so $\sigma_m \sigma_l$ and $\rho_{P, 2\psi}$ are equal (by Theorems 7.8 and 7.14i). □

We used Theorem 9.1(i) in Theorem 9.2 to relate the reflections and the translations in a wallpaper group by considering reflections with parallel axes. To end this section, we use the previous theorem to relate the reflections and the rotations in a wallpaper group by considering reflections with intersecting axes.

Let P be an n-center of a wallpaper group $\mathcal{G}$, and assume that P lies on the axis l of a reflection in $\mathcal{G}$ (Figure 9.18). Let m be another line on P, and let ψ be the angle that lies counterclockwise after l and before m. $\mathcal{G}$ contains σ_m if and only if it contains $\rho_{P, 2\psi}$ (by Theorems 9.7 and 7.17.) Moreover, $\mathcal{G}$ contains $\rho_{P, 2\psi}$ if and only if 2ψ is one of the angles in (9), and so $\mathcal{G}$ contains σ_m if and only if the angle ψ between l and m is one of the angles

$$(180/n)°, (2 \cdot 180/n)°, \ldots, ((n-1) \cdot 180/n)°.$$

Thus, P lies on n axes of reflections in $\mathcal{G}$ spaced by angles of $(180/n)°$.

The possible values of n for an n-center are 2, 3, 4, 6 (by Theorem 9.6). If we substitute these values for n in the last sentence of the previous paragraph, we obtain the following results. If P is a 2-center, it lies on 2 axes of reflections in $\mathcal{G}$ spaced by 90° angles (Figure 9.21). If P is a 3-center, it lies on 3 axes of reflections in $\mathcal{G}$ spaced by 60° angles (Figure 9.22). If P is a 4-center, it lies on 4 axes of reflections in $\mathcal{G}$ spaced by 45° angles (Figure 9.23). If P is a 6-center, it lies on 6 axes of reflections in $\mathcal{G}$ spaced by 30° angles (Figure 9.24).

The symmetry groups of Figures 9.12 and 9.14 illustrate the previous paragraph. Figures 9.25 and 9.26 show the n-centers and axes of reflections in the symmetry groups, and the axes of reflections on each n-center are spaced as in Figures 9.21–9.24.

Now consider any two nonparallel axes of reflections in a wallpaper group $\mathcal{G}$. The axes intersect at a point P. $\mathcal{G}$ contains the product of the reflections in the axes (by Definition 7.15i), and this product is a rotation

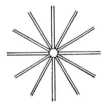

FIGURE 9.21.   FIGURE 9.22.   FIGURE 9.23.   FIGURE 9.24.

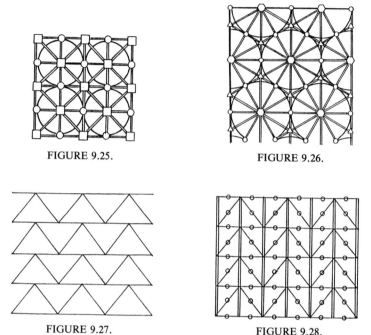

FIGURE 9.25.

FIGURE 9.26.

FIGURE 9.27.

FIGURE 9.28.

about P (by Theorem 9.7). Then P is an n-center, and so the lines through P that are axes of reflections in $\mathcal{G}$ are spaced as in the discussion accompanying Figures 9.21–9.24. Thus Figures 9.21–9.24 depict the intersection of any two nonparallel axes of reflections in a wallpaper group.

We emphasize that Figures 9.21–9.24 show how n-centers lie on axes of reflections in a wallpaper group $\mathcal{G}$. Of course, it can happen that an n-center doesn't lie on the axis of any reflection in $\mathcal{G}$, and it can also happen that the axis of a reflection in $\mathcal{G}$ doesn't contain any n-centers. The symmetry group of Figure 9.27 illustrates both of these possibilities, since Figure 9.28 shows the n-centers and the axes of reflections in the symmetry group.

EXERCISES

9.1. In each part of Exercise 8.1, let $\mathcal{G}$ be the symmetry group of the given figure. For each family $\mathcal{F}_j$ of parallel axes of reflections in $\mathcal{G}$, let τ_j' be a translation in $\mathcal{G}$ such that the axes in $\mathcal{F}_j$ are spaced by $\frac{1}{2}\tau_j'$ (as in Theorem 9.2). Draw a separate copy of the given figure for each family $\mathcal{F}_j$, use double lines to show the axes in $\mathcal{F}_j$, and use an arrow to show that translation τ_j' (as in Figure 9.5).

9.2. In each part of Exercise 8.1, let τ and τ' be as in Exercise 8.1, and let the translations τ'_j be as in Exercise 9.1. For each τ'_j, find integers s and t such that $\tau'_j = \tau^s \tau'^t$.

9.3. In each part of Exercise 8.1, let $\mathcal{G}$ be the symmetry group of the given figure. For each family $\mathcal{F}_j$ of parallel axes of nontrivial glide reflections in $\mathcal{G}$, let τ'_j be a translation in $\mathcal{G}$ such that the axes in $\mathcal{F}_j$ are spaced by $\frac{1}{2}\tau'_j$ (as in Theorem 9.4). Draw a separate copy of the given figure for each family $\mathcal{F}_j$, use dotted lines to show the axes in $\mathcal{F}_j$, and use an arrow to show τ'_j (as in Figure 9.9).

9.4. In each part of Exercise 8.1, let τ and τ' be as in Exercise 8.1, and let the translations τ'_j be as in Exercise 9.3. For each τ'_j, find integers s and t such that $\tau'_j = \tau^s \tau'^t$.

9.5. For each part of Exercise 8.1, draw a copy of the given figure, and mark the n-centers of the symmetry group by using dots, triangles, squares, and hexagons (as in Figures 9.13 and 9.15).

9.6. For each part of Exercise 8.1, let $\mathcal{G}$ be the symmetry group of the given figure. As in Figures 9.25, 9.26, and 9.28, mark the n-centers of $\mathcal{G}$ on a copy of the given figure by using dots, triangles, squares, and hexagons, and mark the axes of the reflections in $\mathcal{G}$ by using double lines. (For each n-center that lies on axes of reflections in $\mathcal{G}$, observe that the axes are spaced as in one of the Figures 9.21–9.24.)

9.7. In the Euclidean plane, what triangles have at least one reflection as a symmetry? Justify your answer completely.

9.8. In the Euclidean plane, what quadrilaterals have the following as symmetries? Justify your answers completely.

(a) At least one reflection.
(b) At least two reflections.
(c) More than two reflections.

9.9. In the Euclidean plane, let $\mathcal{K}$ and $\mathcal{K}'$ be two circles. Prove that the following conditions are equivalent.

(i) $\mathcal{K}$ and $\mathcal{K}'$ intersect at exactly one point.
(ii) $\mathcal{K}$ and $\mathcal{K}'$ have distinct centers that lie on a line through a point of intersection of the circles.
(iii) $\mathcal{K}$ and $\mathcal{K}'$ are tangent to the same line at the same point.

($\mathcal{K}$ and $\mathcal{K}'$ are called *tangent* when these conditions hold.)

9.10. In the Euclidean plane, let $A_1 A_2 A_3$ be a triangle. Let M_1, M_2, M_3 be the midpoints of the sides $A_2 A_3$, $A_3 A_1$, $A_1 A_2$. Let $\mathcal{K}$, $\mathcal{K}_1$, $\mathcal{K}_2$, $\mathcal{K}_3$ be the nine-point circles of triangles $M_1 M_2 M_3$, $A_1 M_2 M_3$, $M_1 A_2 M_3$,

$M_1 M_2 A_3$. In the terminology of Exercise 9.9, prove that $\mathcal{K}$ is tangent to $\mathcal{K}_1$, $\mathcal{K}_2$, $\mathcal{K}_3$.

(*Hint:* If P is the midpoint of M_2 and M_3, one possible way to prove that $\mathcal{K}$ is tangent to $\mathcal{K}_1$ is to prove that $\rho_{P,180°}$ interchanges the triangles $M_1 M_2 M_3$ and $A_1 M_2 M_3$.)

9.11. In the Euclidean plane, let $\mathcal{C}$ and $\mathcal{D}$ be two circles that intersect at two points A and B. Prove that there is a reflection that interchanges A and B and maps $\mathcal{C}$ and $\mathcal{D}$ to themselves. Conclude that the tangents to $\mathcal{C}$ and $\mathcal{D}$ at A form the same angles as the tangents to $\mathcal{C}$ and $\mathcal{D}$ at B.

9.12. In the Euclidean plane, let $\mathcal{C}$ be a circle that intersects a line l at two points A and B. Prove that there is a reflection that interchanges A and B and maps $\mathcal{C}$ and l to themselves. Conclude that the tangents to $\mathcal{C}$ at A and B form equal angles with l.

9.13. In the Euclidean plane, let l and m be parallel lines. Evaluate the product $\sigma_m \sigma_l$.

(*Hint:* One possible approach is to use Theorem 9.1i.)

9.14. In the Euclidean plane, let ϕ be an isometry. Prove that ϕ is a glide reflection if an only if it does not equal a product of fewer than three reflections. (See Exercises 7.11 and 9.13.)

9.15. In the Euclidean plane, let l be a line, and let τ be a translation that isn't parallel to l. Evaluate the product $\sigma_l \tau$ in the following cases.

   (a) τ is perpendicular to l.
   (b) τ is not perpendicular to l.

9.16. In the Euclidean plane, let A and A' be two points, and let τ be a translation that isn't parallel to the line AA'. Evaluate the product $\gamma_{A,A'} \tau$. (As in Theorem 9.3, it may be necessary to divide the problem into separate cases.)

9.17. In the Euclidean plane, let P be a point, and let l be a line that doesn't contain P (Figure 9.29). Prove that

$$\sigma_l \rho_{P,180°} = \gamma_{P,Q}, \tag{20}$$

where Q is the reflection of P across l.

(*Hint:* One possible approach is to prove that each side of Equation 20 equals

$$\tau_{P,Q} \sigma_n \sigma_n \sigma_{PQ},$$

where n is the line through P parallel to l.)

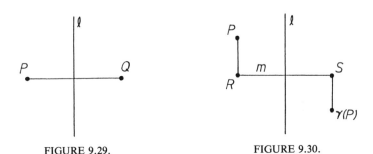

FIGURE 9.29.   FIGURE 9.30.

9.18. In the Euclidean plane, let P be a point, and let γ be a glide reflection whose axis l doesn't contain P (Figure 9.30). Let m be the line perpendicular to l that is equidistant from P and $\gamma(P)$, and let R and S be the feet of the perpendiculars drawn to m from P and $\gamma(P)$, respectively. Prove that

$$\gamma\rho_{P,180°} = \gamma_{R,S}.$$

(*Hint:* One possible approach is to use Equation 4 of Section 7, Exercise 9.17, and Theorem 9.3ii.)

9.19.
(a) In the notation of Exercise 9.17, evaluate $\rho_{P,180°}\sigma_l$.
(b) In the notation of Exercise 9.18, evaluate $\rho_{P,180°}\gamma$.

9.20. In the Euclidean plane, let P be a point on a line l. Let γ be a glide reflection with axis l.

(a) Evaluate $\gamma\rho_{P,180°}$.
(b) Evaluate $\rho_{P,180°}\gamma$.

9.21. In the Euclidean plane, let $\rho_{P,\theta}$ be a rotation, and let σ_l be a reflection.

(a) Evaluate $\rho_{P,\theta}\sigma_l$ if l contains P.
(b) Evaluate $\sigma_l\rho_{P,\theta}$ if l contains P.
(c) Evaluate $\sigma_l\rho_{P,\theta}$ if l doesn't contain P.
(d) Evaluate $\rho_{P,\theta}\sigma_l$ if l doesn't contain P.

9.22.
(a) In the Euclidean plane, determine all isometries φ such that $\varphi^4 = I$, the identity map.
(b) In the Euclidean plane, determine all isometries φ such that $\varphi^4 = \varphi$.
(c) In the Euclidean plane, determine all isometries φ such that $\varphi^4 = \varphi^2$.

9.23. Let $n \geq 3$ be an integer. A *regular n-gon* is a polygon that has n sides of equal length and n equal angles inside the polygon. Let $\mathfrak{I}$ be a collection of regular n-gons in the Euclidean plane that are all of the same size. We call

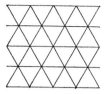

FIGURE 9.31.   FIGURE 9.32.   FIGURE 9.33.

$\mathfrak{I}$ an *edge-to-edge tiling* of the plane if every point in the plane lies on or inside one of the n-gons in $\mathfrak{I}$ and if the intersection of any two n-gons in $\mathfrak{I}$ is either the empty set, a vertex of both n-gons, or an entire edge of both n-gons. Conclude from Theorem 9.6 that the plane can only be tiled edge-to-edge by triangles, squares, or hexagons, as in Figures 9.31–9.33.

Section 10.
Wallpaper Groups without Rotations

We've studied general properties of wallpaper groups in the last two sections. We use these properties in this section to prove that there are just four types of wallpaper groups without rotations. This is the first step in classifying all wallpaper groups into seventeen types in this and the next two sections.

We start with three general results about products of isometries. We can write a rotation as a product of two reflections (by Theorem 9.7), and we can evaluate the product of a translation and a reflection (by Theorem 9.1). Combining these two theorems lets us show that the product of a translation and a rotation is a rotation.

THEOREM 10.1. In the Euclidean plane, let τ be a translation, and let $\rho_{P,\theta}$ be a rotation. Let m be the perpendicular bisector of P and $\tau(P)$, let l be the line through P parallel to m, and let n be the line obtained by rotating l *clockwise* about P through an angle of $\theta/2$. Then m and n intersect at a point Q, and we have

$$\tau\rho_{P,\theta} = \rho_{Q,\theta}$$

(Figure 10.1).

Proof: Since $0° < \theta < 360°$, we have $0° < \theta/2 < 180°$. Thus, n isn't parallel to l, and so it isn't parallel to m either (by Theorem 0.11), and it intersects m at a point Q. We have

$$\tau\rho_{P,\theta} = \tau\sigma_l\sigma_n$$

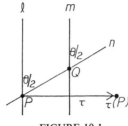

FIGURE 10.1.

(by Theorem 9.7, since an angle of $\theta/2$ lies counterclockwise after n and before l at P)

$$= \sigma_m \sigma_n$$

(by Theorem 9.1i, since τ and l are perpendicular)

$$= \rho_{Q,\theta}$$

(by Theorem 9.7, since Property 0.5 shows that an angle of $\theta/2$ lies counterclockwise after n and before m at Q). $\quad\square$

The product of two reflections with intersecting axes is a rotation (by Theorem 9.7). Combining this result with the previous theorem implies that we obtain rotations when we multiply a glide reflection and a reflection with intersecting axes and when we multiply two glide reflections with intersecting axes.

THEOREM 10.2.
(i) In the Euclidean plane, let γ be a glide reflection, and let σ_l be a reflection. If the axes of γ and σ_l aren't parallel, the product $\gamma\sigma_l$ is a rotation.
(ii) In the Euclidean plane, let γ and γ' be glide reflections. If the axes of γ and γ' aren't parallel, the product $\gamma\gamma'$ is a rotation.

Proof:
(i) If γ has axis m, there is a translation τ such that

$$\gamma\sigma_l = \tau\sigma_m\sigma_l$$

(by Equation 4 of Section 7)

$$= \tau\rho$$

for some rotation ρ (by Theorem 9.7, since the axes l and m of σ_l and γ aren't parallel). The product $\tau\rho$ is a rotation (by Theorem 10.1).

(ii) If γ and γ' have respective axes m and l, there are translations τ and τ' such that

$$\gamma\gamma' = \tau\sigma_m\tau'\sigma_l$$

(by Equation 4 of Section 7)

$$= \tau\sigma_m\tau'\sigma_m^{-1}\sigma_m\sigma_l$$

(by Equations 5 and 12 of Section 7). Moreover, $\sigma_m\tau'\sigma_m^{-1}$ is a translation τ'' (by Theorem 8.7), and $\sigma_m\sigma_l$ is a rotation ρ (by Theorem 9.7, since the axes l and m of γ and γ' aren't parallel), and so the previous sentence shows that

$$\gamma\gamma' = \tau\tau''\rho.$$

The product $\tau\tau''$ is either a translation or the identity map (by Theorem 8.2), and so $\tau\tau''\rho$ is a rotation (by Theorem 10.1 and Equation 12 of Section 7), as desired. $\square$

We saw in Theorem 8.7 how to use multiplication to transform one translation into another of equal length. Similarly, the next result shows how to use multiplication to transform one reflection into another.

THEOREM 10.3. In the Euclidean plane, let φ be an isometry, and let σ_l be a reflection. Then we have

$$\varphi\sigma_l\varphi^{-1} = \sigma_{\varphi(l)}.$$

Proof: Let X be any point that doesn't lie on l (Figure 10.2). Set $Y = \sigma_l(X)$, and so l is the perpendicular bisector of X and Y. Then $\varphi(l)$ is the perpendicular bisector of $\varphi(X)$ and $\varphi(Y)$, and so we have

$$\varphi(Y) = \sigma_{\varphi(l)}(\varphi(X)).$$

It follows that

$$\varphi\sigma_l\varphi^{-1}[\varphi(X)] = \varphi\sigma_l(X)$$

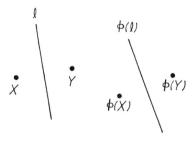

FIGURE 10.2.

(by Equations 5 and 12 of Section 7)

$$= \varphi(Y)$$

$$= \sigma_{\varphi(l)}(\varphi(X)).$$

This holds for every point X not on l, and so the theorem follows (by Theorems 7.8 and 7.14i). $\square$

We can now prove that wallpaper groups without rotations fall into four types, named p1, pm, pg, cm. Here and in the next two sections, the letter "m" (for mirror) in the name of a group shows that the group contains a family of reflections with parallel axes. The letter "g" (for glide reflection) shows that the group contains a family of nontrivial glide reflections with parallel axes. The letter "c" (for "centered cell") is used to distinguish certain types of wallpaper groups where the translations τ and τ' in Theorem 8.5 are not necessarily perpendicular. Each case requires two steps. First, we determine the form of the wallpaper groups in that case. Then, we prove that these wallpaper groups exist by observing that they are the symmetry groups of figures.

In the rest of this section, we let $\mathcal{G}$ be a wallpaper group that doesn't contain any rotations. In order to classify $\mathcal{G}$, we consider four cases, depending on whether $\mathcal{G}$ contains reflections or nontrivial glide reflections.

Case 1: $\mathcal{G}$ doesn't contain any reflections or nontrivial glide reflections. $\mathcal{G}$ doesn't contain any trivial glide reflections either (since it doesn't contain any reflections). Thus, $\mathcal{G}$ doesn't contain any rotations, reflections, or glide reflections, and so every element of $\mathcal{G}$ is either a translation or the identity map (by the Classification Theorem 7.10). There are nonparallel translations τ and τ' such that the translations in $\mathcal{G}$ and the identity map are the maps $\tau^i\tau'^j$ for all integers i and j (by Theorem 8.5). Thus, the elements of $\mathcal{G}$ are exactly the maps $\tau^i\tau'^j$ for all integers i and j. This type of group is named p1.

Conversely, if τ and τ' are any two nonparallel translations, the symmetries of Figure 10.3 are exactly the maps $\tau^i\tau'^j$ for all integers i and j. Thus, these isometries form a wallpaper group of type p1 (by Theorems 7.16 and 8.6).

Case 2: $\mathcal{G}$ contains reflections but no nontrivial glide reflections. Let l_0 be the axis of a reflection in $\mathcal{G}$. There is a translation τ such that the translations in $\mathcal{G}$ parallel to l_0 are the maps τ^i for all nonzero integers i (by Theorem 8.8) (Figure 10.4). There is a translation τ' such that the translations in $\mathcal{G}$ perpendicular to l_0 are the maps τ'^j for all nonzero integers

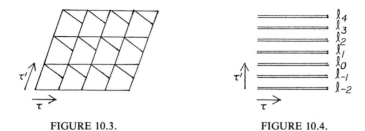

FIGURE 10.3.   FIGURE 10.4.

j and such that the lines parallel to l_0 that are axes of reflections in $\mathcal{G}$ are spaced by $\frac{1}{2}\tau'$ (by Theorem 9.2). We label the axes consecutively as l_j for all integers j.

The product of two reflections in $\mathcal{G}$ with intersecting axes would be a rotation in $\mathcal{G}$ (by Theorem 9.7 and Definition 7.15). Since $\mathcal{G}$ has no rotations, the reflections in $\mathcal{G}$ have parallel axes, and so they are exactly the maps σ_{l_j} for all integers j. $\mathcal{G}$ has no nontrivial glide reflections (by assumption), and so Theorem 8.8 determines all glide reflections in $\mathcal{G}$: They are the maps $\tau^i \sigma_{l_j}$ for all nonzero integers i and all integers j.

We claim that the translations in $\mathcal{G}$ are exactly the maps $\tau^i \tau'^j$ for all integers i and j not both zero. These maps are translations in $\mathcal{G}$ (by Theorem 8.2i, Equations 7 and 12 of Section 7, and Definition 7.15). Conversely, let τ^* be any translation in $\mathcal{G}$; we must prove that τ^* equals $\tau^s \tau'^t$ for some integers s and t not both zero. We can assume that τ^* is not parallel or perpendicular to l_0 (since the translations in $\mathcal{G}$ parallel or perpendicular to l_0 are the maps $\tau^i = \tau^i \tau'^0$ and $\tau'^j = \tau^0 \tau'^j$ for nonzero integers i and j). Let A be any point on l_0, and let F be the foot of the perpendicular drawn from A to $\tau^*(l_0)$ (Figure 10.5). $\mathcal{G}$ contains $\tau^* \sigma_{l_0}$ (by Definition 7.15), and this product is a glide reflection whose axis is the perpendicular bisector of A and F (by Theorem 9.1ii, since τ^* isn't parallel or perpendicular to l_0). This axis is l_t for some nonzero t (by the previous paragraph). Thus, we have $F = \tau'^t(A)$ (by the discussion accompanying Figure 9.4). Let τ_1 be the translation that maps F to $\tau^*(A)$. We have

$$\tau^* = \tau_1 \tau'^t \tag{1}$$

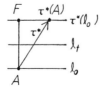

FIGURE 10.5.

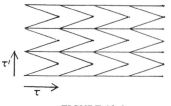

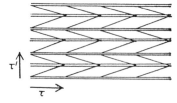

FIGURE 10.6.   FIGURE 10.7.

(by Theorem 8.2i, since $\tau_1 \tau''^t(A) = \tau_1(F) = \tau^*(A)$). $\mathcal{G}$ contains τ^* and τ''^t, and so it also contains τ_1 (by Theorem 7.17 and Equation 1). Thus, since τ_1 is parallel to l_0, it equals τ^s for some nonzero integer s. Then Equation 1 shows that $\tau^* = \tau^s \tau''^t$, as desired.

In summary, there are perpendicular translations τ and τ' such that the translations in $\mathcal{G}$ and the identity map are the maps $\tau^i \tau'^j$ for all integers i and j (Figure 10.4). The axes of reflections in $\mathcal{G}$ are parallel to τ and spaced by $\frac{1}{2}\tau'$. The glide reflections in $\mathcal{G}$ are determined by these axes and Theorem 8.8. Since $\mathcal{G}$ doesn't contain any rotations (by assumption), we've determined all elements of $\mathcal{G}$ (by the Classification Theorem 7.10). This type of group is named pm.

Conversely, if τ and τ' are any two perpendicular translations, the symmetries of Figure 10.6 are exactly the isometries determined in the last paragraph, as Figure 10.7 shows. Thus, these isometries form a wallpaper group of type pm (by Theorems 7.16 and 8.6).

Case 3: $\mathcal{G}$ contains nontrivial glide reflections but no reflections. Let l_0 be the axis of a nontrivial glide reflection in $\mathcal{G}$. There is a translation τ such that the translations in $\mathcal{G}$ parallel to l_0 are the maps τ^i for all nonzero integers i (by Theorem 8.9) (Figure 10.8). There is a translation τ' such that the translations in $\mathcal{G}$ perpendicular to l_0 are the maps τ'^j for all nonzero integers j and such that the lines parallel to l_0 that are axes of nontrivial glide reflections in $\mathcal{G}$ are spaced by $\frac{1}{2}\tau'$ (by Theorem 9.4). We label these axes consecutively as l_j for all integers j.

The product of two glide reflections in $\mathcal{G}$ with intersecting axes would be a rotation in $\mathcal{G}$ (by Theorem 10.2ii and Definition 7.15). Thus, since $\mathcal{G}$ doesn't contain any rotations, all glide reflections in $\mathcal{G}$ have parallel axes. Moreover, all glide reflections in $\mathcal{G}$ are nontrivial (since $\mathcal{G}$ has no reflections). Hence, the glide reflections in $\mathcal{G}$ are exactly those having the lines l_j as their axes, as determined by Theorem 8.9.

We claim that the translations in $\mathcal{G}$ are exactly the maps $\tau^i \tau'^j$ for all integers i and j not both zero. These maps are translations in $\mathcal{G}$ (by Theorem 8.2i, Equations 7 and 12 of Section 7, and Definition 7.15). Conversely, let τ^* be any translation in $\mathcal{G}$; we must prove that τ^* equals

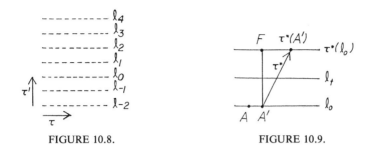

FIGURE 10.8.   FIGURE 10.9.

$\tau^s \tau''$ for some integers s and t not both zero. We can assume that τ^* is not parallel to l_0 (since the translations in G parallel to l_0 are the maps $\tau^i = \tau^i \tau'^0$ for nonzero integers i). Let A be any point on l_0 (Figure 10.9). Let γ be a glide reflection in G that has axis l_0, and set $A' = \gamma(A)$. Let F be the foot of the perpendicular drawn from A' to $\tau^*(l_0)$. G contains $\tau^*\gamma$ (by Definition 7.15), and this product is a glide reflection whose axis is the perpendicular bisector of A' and F (by Theorem 9.3ii, since the fact that G has no reflections shows that the product cannot be evaluated by Theorem 9.3i). This axis is l_t for some nonzero integer t (by the previous paragraph). Thus, we have $F = \tau''(A')$ (by the discussion accompanying Figure 9.4). Let τ_1 be the translation that maps F to $\tau^*(A')$. We have

$$\tau^* = \tau_1 \tau'' \tag{2}$$

(by Theorem 8.2i, since $\tau_1 \tau''(A') = \tau_1(F) = \tau^*(A')$). G contains τ^* and τ'', and so it also contains τ_1 (by Theorem 7.17 and Equation 2). Thus, since τ_1 is parallel to l_0, it equals τ^s for some nonzero integer s. Then Equation 2 shows that $\tau^* = \tau^s \tau''$, as desired.

In summary, there are perpendicular translations τ and τ' such that the translations in G and the identity map are the maps $\tau^i \tau'^j$ for all integers i and j (Figure 10.8). The glide reflections in G are nontrivial and determined by Theorem 8.9 along axes parallel to τ and spaced by $\frac{1}{2}\tau'$. Since G doesn't contain any rotations or reflections (by assumption), we've determined all elements of G (by the Classification Theorem 7.10). This type of group is named pg.

Conversely, if τ and τ' are any two perpendicular translations, the symmetries of Figure 10.10 are exactly the isometries determined in the last paragraph, as Figure 10.11 shows. Thus, these isometries form a wallpaper group of type pg (by Theorems 7.16 and 8.6).

Case 4: G contains reflections and nontrivial glide reflections. Let l_0 be the axis of a reflection in G. There is a translation τ such that the translations in G parallel to l_0 are the maps τ^i for all nonzero integers i (by Theorem 8.8)

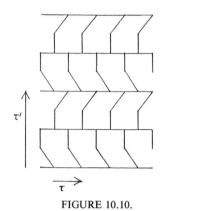

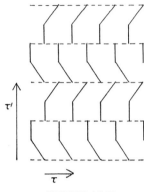

FIGURE 10.10.   FIGURE 10.11.

(Figure 10.12). There is a translation τ_1 such that the translations in $\mathcal{G}$ perpendicular to l_0 are the maps τ_1^j for all nonzero integers j and such that the lines parallel to l_0 that are axes of reflections in $\mathcal{G}$ are spaced by $\frac{1}{2}\tau_1$ (by Theorem 9.2). We label the axes consecutively as l_j for all integers j.

The product of two reflections in $\mathcal{G}$ with intersecting axes would be a rotation in $\mathcal{G}$ (by Theorem 9.7 and Definition 7.15). Since $\mathcal{G}$ has no rotations, the reflections in $\mathcal{G}$ have parallel axes, and they are exactly the maps σ_{l_j} for all integers j. The trivial glide reflections in $\mathcal{G}$ are the maps $\tau^i \sigma_{l_j}$ for all nonzero integers i and all integers j (by Theorem 8.8).

For each integer j, let m_j be the line parallel to l_0 that lies midway between l_j and l_{j+1}. Let m be the axis of a nontrivial glide reflection γ in $\mathcal{G}$. If m weren't parallel to l_0, $\gamma\sigma_{l_0}$ would be a rotation in $\mathcal{G}$ (by Theorem 10.2i and Definition 7.15). Since $\mathcal{G}$ has no rotations, m is parallel to l_0. $\mathcal{G}$ contains $\gamma\sigma_{l_0}\gamma^{-1} = \sigma_{\gamma(l_0)}$ (by Definition 7.15 and Theorem 10.3), and so γ maps l_0 to l_k for some integer k. Thus, since m is parallel to l_0 and doesn't equal any of the lines l_j (because γ is nontrivial), m is one of the lines m_j. Hence, the lines m_j are all axes of nontrivial glide reflections in $\mathcal{G}$, since the axes are spaced by $\frac{1}{2}\tau_1$ (by Theorem 9.4). Conversely, as we've seen, the axis m of any nontrivial glide reflection in $\mathcal{G}$ is one of the lines m_j. In short, the nontrivial glide reflections in $\mathcal{G}$ are exactly those having the lines m_j as their axes, as determined by Theorem 8.9.

Let B be any point on m_0, and let C be the midpoint of B and $\tau(B)$ (Figure 10.13). Let A be the foot of the perpendicular from B to l_0, and let D be the foot of the perpendicular from C to l_1. Let τ' be the translation that maps A to D. We have

$$\tau'\sigma_{l_0} = \gamma_{B,C}$$

(by Theorem 9.1ii). Thus, since $\mathcal{G}$ contains σ_{l_0} and $\gamma_{B,C}$ (by the two previous paragraphs and Theorem 8.9), it also contains τ' (by Theorem 7.17).

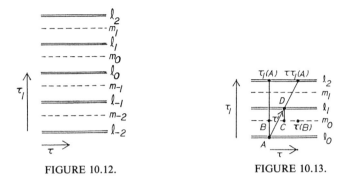

FIGURE 10.12.   FIGURE 10.13.

Since $\tau'(A) = D$ is the midpoint of A and $\tau\tau_1(A)$, the translation τ' maps any point X to the midpoint of X and $\tau\tau_1(X)$.

We claim that the translations in $\mathcal{G}$ are exactly the maps $\tau^i\tau'^j$ for all integers i and j not both zero. These maps are translations in $\mathcal{G}$ (by combining the fact that τ and τ' are translations in $\mathcal{G}$ with Theorem 8.2i, Equations 7 and 12 of Section 7, and Definition 7.15). Conversely, let τ^* be any translation in $\mathcal{G}$; we must prove that τ^* equals $\tau^s\tau''$ for some integers s and t not both zero. $\mathcal{G}$ contains

$$\tau^*\sigma_{l_0}\tau^{*-1} = \sigma_{\tau^*(l_0)}$$

(by Definition 7.15 and Theorem 10.3), and so $\tau^*(l_0)$ equals l_t for some integer t. Since τ' maps l_j to l_{j+1} for each integer j, τ'' maps l_0 to l_t, and we have

$$\tau^*\tau'^{-t}(l_t) = \tau^*(l_0) = l_t.$$

Thus, $\tau^*\tau'^{-t}$ is either the identity map or a translation in $\mathcal{G}$ parallel to l_0 (by Theorem 8.2, Equations 7 and 12 of Section 7, and Definition 7.15), and so it equals τ^s for some integer s. Thus, we have

$$\tau^* = \tau^*\tau'^{-t}\tau'^t = \tau^s\tau'^t$$

(by Equations 5 and 12 of Section 7). Since τ^* is not the identity map, s and t are not both zero, and so τ^* has the desired form.

In summary, there are perpendicular translations τ and τ_1 such that the translations in $\mathcal{G}$ and the identity map are the maps $\tau^i\tau'^j$ for all integers i and j, where τ' is the translation that maps each point X in the plane to the midpoint of X and $\tau\tau_1(X)$ (Figures 10.12 and 10.13). The axes of reflections in $\mathcal{G}$ are parallel to τ and spaced by $\frac{1}{2}\tau_1$. The trivial glide reflections in $\mathcal{G}$ are determined by these axes and Theorem 8.8. The nontrivial glide reflections in $\mathcal{G}$ are determined by Theorem 8.9 and the fact that their axes are parallel to τ, spaced by $\frac{1}{2}\tau_1$, and midway between consecutive axes of reflections

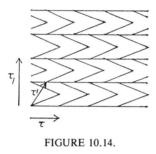

FIGURE 10.14.

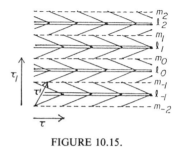

FIGURE 10.15.

in $\mathcal{G}$. Since $\mathcal{G}$ doesn't contain any rotations (by assumption), we've determined all elements of $\mathcal{G}$ (by the Classification Theorem 7.10). This type of group is named cm.

Conversely, if τ and τ_1 are any two perpendicular translations, the symmetries of Figure 10.14 are exactly the isometries described in the previous paragraph, as Figure 10.15 shows. Thus, these isometries form a wallpaper group of type cm (by Theorems 7.16 and 8.6).

The four cases we've considered cover all possibilities for a wallpaper group without rotations. Thus, we've proved the following result.

THEOREM 10.4. In the Euclidean plane, a wallpaper group without rotations is of one of the four types pl, pm, pg, cm. □

Let $\mathcal{G}$ be a wallpaper group that doesn't contain rotations. The descriptions of the four cases we've considered show how to determine the form of $\mathcal{G}$. If $\mathcal{G}$ doesn't contain any reflections or nontrivial glide reflections, it has form pl. If $\mathcal{G}$ contains reflections but no nontrivial glide reflections, it has form pm. If $\mathcal{G}$ contains nontrivial glide reflections but no reflections, it has form pg. If $\mathcal{G}$ contains both reflections and nontrivial glide reflections, it has form cm. Thus, we've classified all wallpaper groups without rotations. We classify wallpaper groups with rotations in the next two sections.

EXERCISES

10.1. A figure is given in each part of this exercise. Determine whether the symmetry group of the figure is of type pl, pm, pg, or cm. On a copy of the figure, label translations τ and τ' (and τ_1 for groups of type cm) as described in Cases 1–4 of this section, use double lines to mark axes of reflections, and use dotted lines to mark axes of nontrivial glide reflections.

(a)

(b)

(c)

(d)

(e)

(f)

(g)

(h)

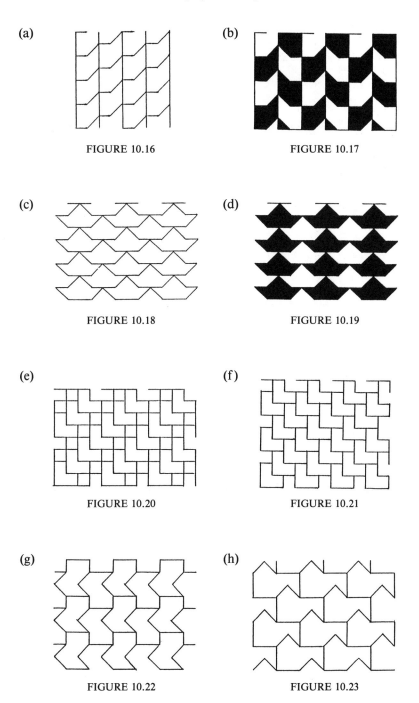

FIGURE 10.16

FIGURE 10.17

FIGURE 10.18

FIGURE 10.19

FIGURE 10.20

FIGURE 10.21

FIGURE 10.22

FIGURE 10.23

(i)

(j)

(k)

(l)

(m)

(n)

(o)

(p)

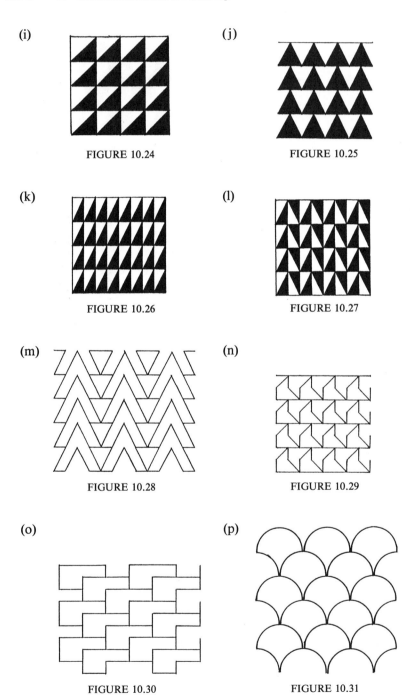

FIGURE 10.24

FIGURE 10.25

FIGURE 10.26

FIGURE 10.27

FIGURE 10.28

FIGURE 10.29

FIGURE 10.30

FIGURE 10.31

(q)

(r)

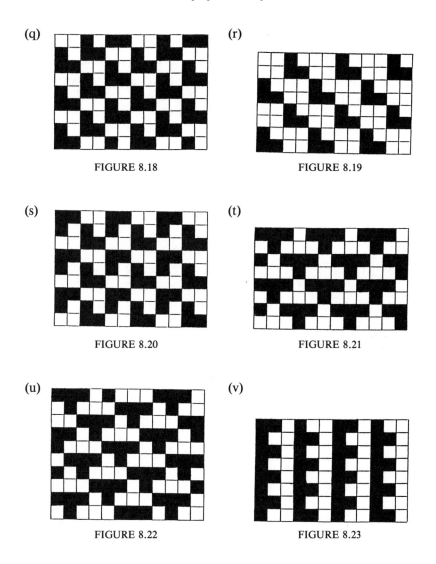

FIGURE 8.18

FIGURE 8.19

(s)

(t)

FIGURE 8.20

FIGURE 8.21

(u)

(v)

FIGURE 8.22

FIGURE 8.23

10.2. In each part of Exercise 10.1, let τ and τ' be as described in that exercise. On a copy of the given figure, draw τ, τ', and a lattice of parallelograms whose vertices are the points $\tau^i \tau'^j(A)$ for some point A (as in Figures 8.9–8.11).

10.3. Each part of this exercise specifies a letter of the alphabet, a wallpaper group, and a number. Use the given letter to create a figure $\mathcal{F}$ whose symmetry group is the given wallpaper group. The letter may be rotated or reflected in any way. On a copy of $\mathcal{F}$, show translations τ and τ' as described in Cases 1–4 of this section, use double lines to show axes of reflections, and use dotted lines to show axes of nontrivial glide reflections. On a second copy of $\mathcal{F}$, show τ and τ' and the lattice of parallelograms with vertices $\tau^i\tau'^j(A)$ for some point A and all integers i and j. The number specified is to be the number of times (including any fractions) that the given letter appears in each parallelogram. For example, if "T, pg, 2" were specified, Figures 10.32–10.34 would show one possible solution. Note that each rectangle in Figure 10.34 contains two copies of the letter "T," as specified.

(a) F, p1, 1	(b) F, pm, 2
(c) F, pg, 2	(d) F, cm, 2
(e) E, p1, 1	(f) E, pm, 1
(g) E, pg, 2	(h) E, cm, 1
(i) N, p1, 2	(j) N, pm, 2
(k) N, pg, 2	(l) N, cm, 2
(m) M, p1, 1	(n) M, pm, 1
(o) M, pg, 2	(p) M, cm, 1
(q) H, p1, 2	(r) H, pm, 2
(s) H, pg, 2	(t) H, cm, 2

FIGURE 10.32.

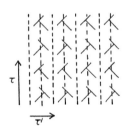

FIGURE 10.33.

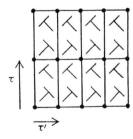

FIGURE 10.34.

10.4. In the Euclidean plane, let $\mathcal{G}$ be a wallpaper group of type cm. Prove that there are translations τ_2 and τ_3 of equal length such that the translations in $\mathcal{G}$ and the identity map are exactly the maps $\tau_2^i \tau_3^j$ for all integers i and j. (Thus, the lattice with vertices $\tau_2^i \tau_3^j(A)$ for some point A consists of rhombi, parallelograms whose sides are all of equal length.)

10.5. In the Euclidean plane, let $\mathcal{G}$ be a wallpaper group of type cm. Let τ, τ', and τ_1 be as in Case 4 of this section (Figure 10.13). Find integers s and t such that $\tau_1 = \tau^s \tau'^t$. Justify your answer.

10.6. In the Euclidean plane, let $\rho_{P,\theta}$ be a rotation, and let τ be a translation. Evaluate the product $\rho_{P,\theta} \tau$.

10.7.

(a) In the Euclidean plane, let γ be a glide reflection, and let σ_l be a reflection. If the axes of γ and σ_l are distinct parallel lines, evaluate the product $\gamma\sigma_l$.

(b) In the Euclidean plane, let γ and γ' be glide reflections whose axes are distinct parallel lines. Evaluate the product $\gamma\gamma'$.

10.8. In the Euclidean plane, let P and Q be two points, and let l be a line. In each part of this exercise, do the given isometries form a group? Justify your answers completely.

(a) The isometries that fix P

(b) The isometries that interchange P and Q

(c) The isometries that either fix or interchange P and Q

(d) The isometries that fix l

(e) The isometries that map l to a parallel line

(f) The isometries that map l to a perpendicular line

(g) The isometries that map l to either a parallel or a perpendicular line

10.9. In the Euclidean plane, let φ, ψ, and ξ be isometries. Prove that the equations $\varphi\psi\varphi^{-1} = \xi$ and $\varphi\psi = \xi\varphi$ are equivalent.

10.10. In the Euclidean plane, let τ be a translation.

(a) Determine all isometries φ such that $\varphi\tau = \tau\varphi$. (See Exercise 8.12 or Exercise 10.9.)

(b) Determine all isometries φ such that $\varphi\tau\varphi^{-1} = \tau^{-1}$.

10.11. In the Euclidean plane, let σ_l be a reflection. Determine all isometries φ such that $\varphi\sigma_l = \sigma_l\varphi$. (See Exercise 8.12 or Exercise 10.9.)

10.12. In the Euclidean plane, let φ be an isometry, and let $\gamma_{A,B}$ be a glide reflection. Evaluate $\varphi\gamma_{A,B}\varphi^{-1}$ and justify your answer.

10.13. In the Euclidean plane, let $\gamma_{A,B}$ be a glide reflection.

(a) Determine all isometries φ such that $\varphi\gamma_{A,B} = \gamma_{A,B}\varphi$.
(b) Determine all isometries φ such that $\varphi\gamma_{A,B}\varphi^{-1} = \gamma_{A,B}^{-1}$.

10.14. In the Euclidean plane, let φ and ψ be isometries. Prove that the following conditions are equivalent.

(i) $\varphi\psi = I$
(ii) $\psi = \varphi^{-1}$
(iii) $\psi\varphi = I$
(iv) $\varphi = \psi^{-1}$

10.15. In the Euclidean plane, let φ and ψ be isometries.

(a) Prove that $(\varphi^{-1})^{-1} = \varphi$.
(b) Prove that $(\varphi\psi)^{-1} = \psi^{-1}\varphi^{-1}$.
(c) Prove that the equation $(\varphi\psi)^{-1} = \varphi^{-1}\psi^{-1}$ holds if and only if $\varphi\psi = \psi\varphi$.

10.16. In the Euclidean plane, let ψ be an isometry such that $\psi^{-1} \neq \psi$. In each part of this exercise, do the given isometries form a group? Justify your answers completely.

(a) The isometries φ such that $\varphi\psi = \psi\varphi$
(b) The isometries φ such that $\varphi\psi = \psi^{-1}\varphi$
(c) The isometries φ such that either $\varphi\psi = \psi\varphi$ or $\varphi\psi = \psi^{-1}\varphi$

Section 11.

Wallpaper Groups with 90° Rotations

We've proved that there are four types of wallpaper groups without rotations. We complete the classification of wallpaper groups in this section and the next. In this section, we determine the wallpaper groups with 90° rotations. In the next section, we determine the wallpaper groups that have rotations but not through 90° angles.

The center of every rotation in a wallpaper group is an n-center, where n is either 2, 3, 4, or 6 (by Theorem 9.6). Our first goal in this section is to determine which values of n can occur in the same wallpaper group. We start by proving that the product of two rotations about different points is a rotation through the sum of the angles of the two given rotations when this sum is less than 360°.

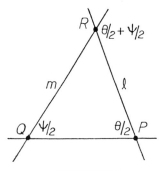

FIGURE 11.1.

THEOREM 11.1. In the Euclidean plane, let $\rho_{P,\theta}$ and $\rho_{Q,\psi}$ be rotations about distinct points P and Q. Assume that $\theta + \psi < 360°$. Let l be the line obtained by rotating line PQ through angle $\theta/2$ *clockwise* about P, and let m be the line obtained by rotating PQ through angle $\psi/2$ counterclockwise about Q. Then l and m intersect at a point R, and we have

$$\rho_{Q,\psi}\rho_{P,\theta} = \rho_{R,\theta+\psi}$$

(Figure 11.1).

Proof: We have

$$\rho_{Q,\psi} = \sigma_m \sigma_{PQ} \quad \text{and} \quad \rho_{P,\theta} = \sigma_{PQ} \sigma_l$$

(by Theorem 9.7). It follows that

$$\rho_{Q,\psi}\rho_{P,\theta} = \sigma_m \sigma_{PQ} \sigma_{PQ} \sigma_l$$

$$= \sigma_m \sigma_l \qquad (1)$$

(by Equation 4 of Section 8).

Since $\theta + \psi < 360°$, we have $\theta/2 + \psi/2 < 180°$. It follows that l and m intersect at a point R such that PQR is a triangle having angles $\theta/2$ and $\psi/2$ at P and Q. Thus, we have

$$\angle PRQ = 180° - \theta/2 - \psi/2$$

(by Theorem 0.8). Then the angles formed by l and m at R that are adjacent to $\angle PRQ$ have measure $\theta/2 + \psi/2$ (by Theorem 0.6), and these are the angles at R that lie counterclockwise after l and before m. Thus, Theorem 9.7 shows that

$$\sigma_m \sigma_l = \rho_{R,\theta+\psi}.$$

Together with Equation 1, this proves the theorem.   □

Every rotation in a wallpaper group is one of those listed in (10) of Section 9. Accordingly, we can use the previous theorem to restrict the values of n for which n-centers can belong to the same wallpaper group.

THEOREM 11.2. In the Euclidean plane, let $\mathcal{G}$ be a wallpaper group.

(i) If $\mathcal{G}$ has a 3-center or a 6-center, then it has no 4-centers.
(ii) If $\mathcal{G}$ has both a 2-center and a 3-center, then it has a 6-center.

Proof:

(i) Suppose that $\mathcal{G}$ had a 4-center P and a point Q that is either a 3-center or a 6-center. Then $\mathcal{G}$ would contain $\rho_{P, 90°}$ and $\rho_{Q, 120°}$ (by (10) of Section 9), and so it would also contain

$$\rho_{Q, 120°} \rho_{P, 90°} = \rho_{R, 210°}$$

for some point R (by Definition 7.15i and Theorem 11.1). This would contradict Theorem 9.6 and (10) of Section 9.

(ii) If $\mathcal{G}$ has both a 2-center P and a 3-center Q, then $\mathcal{G}$ contains

$$\rho_{Q, 120°} \rho_{P, 180°} = \rho_{R, 300°}$$

for some point R (by (10) of Section 9, Definition 7.15i, and Theorem 11.1). Thus, R is a 6-center (by Theorem 9.6 and (10) of Section 9). $\square$

Combining Theorems 9.6 and 11.2 gives the following result.

THEOREM 11.3. In the Euclidean plane, every wallpaper group $\mathcal{G}$ satisfies one of the following conditions.

(i) $\mathcal{G}$ has no n-centers.
(ii) $\mathcal{G}$ has a 2-center but no 3-centers, 4-centers, or 6-centers.
(iii) $\mathcal{G}$ has a 4-center. In this case, $\mathcal{G}$ has no 3-centers or 6-centers.
(iv) $\mathcal{G}$ has a 3-center but no 6-centers. In this case, $\mathcal{G}$ has no 2-centers or 4-centers.
(v) $\mathcal{G}$ has a 6-center. In this case, $\mathcal{G}$ has no 4-centers. $\square$

Of the five cases in Theorem 11.3, we considered case (i)—where $\mathcal{G}$ has no rotations—in Section 10. We study case (iii)—where $\mathcal{G}$ has 90° rotations—in the rest of this section. We consider case (ii)—where $\mathcal{G}$ has only 180° rotations—and cases (iv) and (v)—where $\mathcal{G}$ has 120° rotations—in the next section. Theorem 11.3 shows that we determine all wallpaper groups by studying these five cases.

In order to classify wallpaper groups with 90° rotations, we start by determining how to use multiplication to transform one rotation into another. This result is analogous to Theorems 8.7 and 10.3 about transforming translations and reflections.

THEOREM 11.4. In the Euclidean plane, let φ be an isometry, and let $\rho_{P,\theta}$ be a rotation.

(i) If φ is a translation, a rotation, or the identity map, then we have

$$\varphi\rho_{P,\theta}\varphi^{-1} = \rho_{\varphi(P),\theta}.$$

(ii) If φ is a reflection or a glide reflection, then we have

$$\varphi\rho_{P,\theta}\varphi^{-1} = \rho_{\varphi(P),360°-\theta}.$$

Proof: Let X be any point other than P, and set $Y = \rho_{P,\theta}(X)$. Equations 5 and 12 of Section 7 imply that

$$\varphi\rho_{P,\theta}\varphi^{-1}[\varphi(X)] = \varphi\rho_{P,\theta}(X) = \varphi(Y). \tag{2}$$

If $\theta \neq 180°$, then XPY and $\varphi(X)\varphi(P)\varphi(Y)$ are congruent triangles (by Definition 7.1 and Property 0.1). (Figure 11.2 illustrates part (i), and Figure 11.3 illustrates part (ii).) Since $Y = \rho_{P,\theta}(X)$, P is equidistant from X and Y, and we have $\angle XPY = \theta$. The last two sentences imply that $\varphi(P)$ is equidistant from $\varphi(X)$ and $\varphi(Y)$ and that $\angle \varphi(X)\varphi(P)\varphi(Y) = \theta$. Thus, we obtain $\varphi(Y)$ by rotating $\varphi(X)$ through angle θ either counterclockwise or clockwise about P. Moreover, directions of rotation are preserved by translations, rotations, and the identity map, and they are reversed by reflections and glide reflections. Thus, we have

$$\varphi(Y) = \rho_{\varphi(P),\theta}(\varphi(X)) \tag{3}$$

in part (i) and

$$\varphi(Y) = \rho_{\varphi(P),360°-\theta}(\theta(X)) \tag{4}$$

in part (ii). Equations 3 and 4 still hold when $\theta = 180°$, since P is then the midpoint of X and Y, and so $\varphi(P)$ is the midpoint of $\varphi(X)$ and $\varphi(Y)$ (by Theorem 7.7 and Definition 7.1) (Figure 11.4). In short, Equations 3 and 4 hold for every angle θ in parts (i) and (ii), respectively.

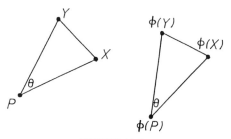

FIGURE 11.2.

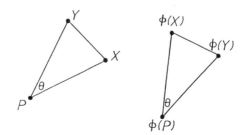

FIGURE 11.3.

In part (i), Equations 2 and 3 show that $\varphi \rho_{P,\theta} \varphi^{-1}$ and $\rho_{\varphi(P)}$ agree on $\varphi(X)$ for every point X other than P. Thus, part (i) holds (by Theorems 7.8 and 7.14i). Likewise, in part (ii), Equations 2 and 4 show that $\varphi \rho_{P,\theta} \varphi^{-1}$ and $\rho_{\varphi(P),360°-\theta}$ agree on $\varphi(X)$ for every point X other than P. Thus, part (ii) holds (by Theorems 7.8 and 7.14i). $\square$

Theorem 10.1 evaluates the product of a translation and a rotation. To classify wallpaper groups with 4-centers, we consider Theorem 10.1 in the specific cases of rotations through 180° and 90°. Accordingly, let τ be a translation, and let P be a point in the Euclidean plane (Figure 11.5). If Q is the midpoint of P and $\tau(P)$, setting $\theta = 180°$ in Theorem 10.1 shows that

$$\tau \rho_{P,180°} = \rho_{Q,180°} \tag{5}$$

(since $\theta/2 = 90°$, and so the line n in Theorem 10.1 is parallel to τ in this case). Let R be the point where the perpendicular bisector of P and $\tau(P)$ intersects the line obtained by rotating line PQ through 45° counter-clockwise about P. Setting $\theta = 90°$ in Theorem 10.1 shows that

$$\tau \rho_{P,90°} = \rho_{R,90°} \tag{6}$$

(since $\theta/2 = 45°$, and so the line n in Theorem 10.1 forms a 45° angle with the perpendicular to line PQ and thus with PQ itself).

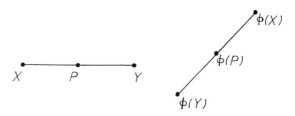

FIGURE 11.4.

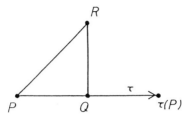

FIGURE 11.5.

In the rest of this section, we let $\mathcal{G}$ *be a wallpaper group that has a
4-center* P. $\mathcal{G}$ has a translation τ of shortest length (by Theorem 8.5i).
If ρ is the counterclockwise rotation through 90° about P, $\mathcal{G}$ contains $\rho\tau\rho^{-1}$
(by Definition 7.15), which is a translation τ' that also has shortest length
in $\mathcal{G}$ and is perpendicular to τ (by Theorem 8.7) (Figure 11.6). Thus, the
translations in $\mathcal{G}$ and the identity map are the maps $\tau^{i}\tau'^{j}$ for all integers i
and j (by Theorem 8.5ii).

Next, we determine the rotations in $\mathcal{G}$. Since $\mathcal{G}$ contains $\rho_{P, 90°}$, it contains

$$\tau^{i}\tau'^{j}\rho_{P, 90°}(\tau^{i}\tau'^{j})^{-1} = \rho_{\tau^{i}\tau'^{j}(P), 90°}$$

(by Definition 7.15 and Theorem 11.4i). Thus, the points $\tau^{i}\tau'^{j}(P)$ are
4-centers for all integers i and j (by Theorem 9.6 and (10) of Section 9).
These points are shown by white squares in Figure 11.6.

Let R be the midpoint of P and $\tau\tau'(P)$. $\mathcal{G}$ contains the 90° rotation about
R (by Equation 6 and Definition 7.15i, since $\mathcal{G}$ contains τ and $\rho_{P,90°}$).
Thus, the points $\tau^{i}\tau'^{j}(R)$ are 4-centers for all integers i and j (by replacing
P with R in the previous paragraph). These points are shown by black
squares in Figure 11.6.

Let d be the distance from P to R. Every point in the plane lies within
distance d of one of the 4-centers $\tau^{i}\tau'^{j}(P)$ and $\tau^{i}\tau'^{j}(R)$. Thus, if $\mathcal{G}$ had any
4-centers other than these, it would have two 4-centers less than d units

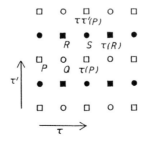

FIGURE 11.6.

apart; then G would contain a translation shorter than τ (by Equation 6 and Theorem 7.17), which would contradict the choice of τ. Hence, the 4-centers of G are exactly the points $\tau^i\tau'^j(P)$ and $\tau^i\tau'^j(R)$ for all integers i and j (as shown by the white and black squares in Figure 11.6).

The 2-centers are the only centers of rotations in G other than 4-centers (by Theorems 9.6 and 11.3iii). Let Q be the midpoint of P and $\tau(P)$, and let S be the midpoint of R and $\tau(R)$ (Figure 11.6). Since G contains the 180° rotation about P (by (10) of Section 9), it also contains the 180° rotation about Q (by Equation 5 and Definition 7.15i). Thus, G contains

$$\tau^i\tau'^j\rho_{Q,180°}(\tau^i\tau'^j)^{-1} = \rho_{\tau^i\tau'^j(Q),180°}$$

(by Definition 7.15 and Theorem 11.4i), and so the points $\tau^i\tau'^j(Q)$ are 2-centers for all integers i and j (by the first sentence of this paragraph). These points are shown by white circles in Figure 11.6. By the symmetry of the 4-centers P and R, the points $\tau^i\tau'^j(S)$ are also 2-centers for all integers i and j. These points are shown by black circles in Figure 11.6.

Let e be the distance from P to Q. Every point in the plane lies within distance e of one of the 4-centers or 2-centers in Figure 11.6. Thus, if G had any other 2-centers, it would have two centers of 180° rotations less than e units apart; then G would contain a translation shorter than τ (by Equation 5 and Theorem 7.17), which would contradict the choice of τ. Hence, the 2-centers of G are exactly the points $\tau^i\tau'^j(Q)$ and $\tau^i\tau'^j(S)$ for all integers i and j (as shown by the white and black dots in Figure 11.6).

In short, the centers of the rotations in G form the checkerboard pattern of 4-centers and 2-centers in Figure 11.7. We no longer color n-centers as in Figure 11.6 because we no longer distinguish among 4-centers or among 2-centers.

So far we've determined the translations and the rotations in G. We now consider the possible reflections and glide reflections in G.

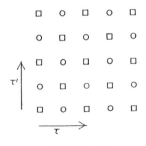

FIGURE 11.7.

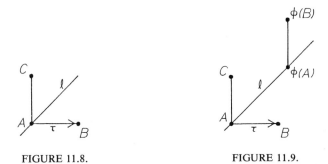

FIGURE 11.8.   FIGURE 11.9.

Let φ be a reflection or a glide reflection in $\mathcal{G}$, and let l be its axis. Let A be a point on l, and set

$$B = \tau(A) \quad \text{and} \quad C = \sigma_l(B). \tag{7}$$

(Figures 11.8 and 11.9 illustrate the cases where φ is a reflection and a glide reflection, respectively.) By Definition 7.15, $\mathcal{G}$ contains

$$\varphi\tau\varphi^{-1} = \varphi\tau_{A,B}\varphi^{-1}$$

(since τ maps A to B)

$$= \tau_{\varphi(A),\varphi(B)}$$

(by Theorem 8.7). If φ is the reflection σ_l, we have

$$\tau_{\varphi(A),\varphi(B)} = \tau_{A,C}, \tag{8}$$

since $\varphi(A) = A$ and $\varphi(B) = C$ (Figure 11.8). If φ is a glide reflection with axis l, Equation 8 still holds, since the motions from $\varphi(A)$ to $\varphi(B)$ and from A to C cover the same distance in the same direction (Figure 11.9). In short, the last three sentences show that $\mathcal{G}$ contains $\tau_{A,C}$ regardless of whether φ is a reflection or a glide reflection.

The translations in $\mathcal{G}$ are the maps $\tau^i\tau'^j$ for all integers i and j not both zero. Those of shortest length are τ, τ', τ^{-1}, and τ'^{-1} (since τ and τ' are perpendicular and of equal length) (Figure 11.10). $\tau_{A,C}$ is one of these

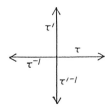

FIGURE 11.10.

FIGURE 11.11.   FIGURE 11.12.

translations, by the previous paragraph. If we combine this fact with the equations in (7) and with the fact that A lies on l, we obtain the following possibilities. If $\tau_{A,C} = \tau$, then l is parallel to τ (Figure 11.11). If $\tau_{A,C} = \tau^{-1}$, then l is perpendicular to τ (Figure 11.12). If $\tau_{A,C}$ is τ' or τ'^{-1}, then l forms a 45° angle with τ (Figures 11.13 and 11.14). In short, the axis l of any reflection or glide reflection in $\mathcal{G}$ is either parallel to τ, perpendicular to τ, or forms a 45° angle with τ (Figures 11.11–11.14).

If X is any 4-center of $\mathcal{G}$, and if φ is any reflection or glide reflection in $\mathcal{G}$, then $\mathcal{G}$ contains

$$\varphi\rho_{X,\,90°}\varphi^{-1} = \rho_{\varphi(X),\,270°}$$

(by Definition 7.15 and Theorem 11.4ii). Thus, $\varphi(X)$ is a 4-center (by (10) of Section 9).

In summary, if φ is any reflection or glide reflection in $\mathcal{G}$, the last two paragraphs show that the axis of φ is parallel to one of the double lines or dotted lines in Figures 11.15 and 11.16 and that φ maps the 4-centers among themselves. It follows that the double lines in Figures 11.15 and 11.16 are the only possible axes of reflections in $\mathcal{G}$ and that the dotted lines in these figures are the only possible axes of nontrivial glide reflections in $\mathcal{G}$.

Any two double lines in Figure 11.15 that aren't parallel either intersect at a 2-center and are perpendicular, or intersect at a 4-center and form angles that are multiples of 45°. Each of these two lines is the axis of a reflection in $\mathcal{G}$ if and only if the other is (by the discussion accompanying Figures 9.21 and 9.23). Applying this fact repeatedly shows that, if one of the double lines in Figure 11.15 is the axis of a reflection in $\mathcal{G}$, then they all are.

Let A be a 4-center of $\mathcal{G}$, and let l be the line through A parallel to $\tau\tau'$ (Figure 11.17). Let m be the line parallel to l that lies midway between l and $\tau(l)$. Let B and C be the feet of the perpendiculars drawn to m from

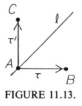

FIGURE 11.13.

FIGURE 11.14.

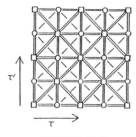

τ'

τ

FIGURE 11.15.

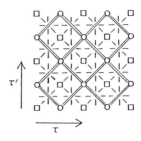

τ'

τ

FIGURE 11.16.

A and $\tau(A)$, respectively. We have

$$\tau\sigma_l = \gamma_{B,C}$$

(by Theorem 9.1ii). Thus, since $\mathcal{G}$ contains τ, it contains σ_l if and only if it contains $\gamma_{B,C}$ (by Theorem 7.17). Moreover, C is the midpoint of B and $D = \tau\tau'(B)$, where $\tau\tau'$ is a translation of shortest length in $\mathcal{G}$ parallel to m (since the translations in $\mathcal{G}$ are the maps $\tau^i\tau'^j$ for all integers i and j not both zero); thus, $\mathcal{G}$ contains $\gamma_{B,C}$ if and only if m is the axis of a nontrivial glide reflection in $\mathcal{G}$ (by Theorems 8.8 and 8.9). The last two sentences show that l is the axis of a reflection in $\mathcal{G}$ if and only if m is the axis of a nontrivial glide reflection in $\mathcal{G}$. By symmetry, for each dotted line in Figure 11.15 there is a double line in the figure such that the dotted line is the axis of a nontrivial glide reflection in $\mathcal{G}$ if and only if the double line is the axis of a reflection in $\mathcal{G}$.

The last two paragraphs show that, if any double line in Figure 11.15 is the axis of a reflection in $\mathcal{G}$, or if any dotted line in this figure is the axis of a nontrivial glide reflection in $\mathcal{G}$, then every double line in the figure is the axis of a reflection in $\mathcal{G}$ and every dotted line in the figure is the axis of a nontrivial glide reflection in $\mathcal{G}$.

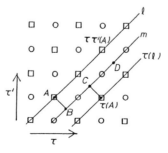

FIGURE 11.17.

Any two perpendicular double lines in Figure 11.16 intersect at a 2-center, and so one of these lines is the axis of a reflection in G if and only if the other is (by the discussion accompanying Figure 9.21). Applying this fact repeatedly shows that, if one of the double lines in Figure 11.16 is the axis of a reflection in G, then they all are.

Let A be a 2-center of G, and let l be the line through A parallel to $\tau\tau'$ (Figure 11.18). Let m be the line parallel to l that lies midway between l and $\tau(l)$. Let B and C be the feet of the perpendiculars drawn to m from A and $\tau(A)$, respectively. We have

$$\tau\sigma_l = \gamma_{B,C}$$

(by Theorem 9.1ii). Thus, since G contains τ, it contains σ_l if and only if it contains $\gamma_{B,C}$ (by Theorem 7.17). Moreover, C is the midpoint of B and $D = \tau\tau'(B)$, where $\tau\tau'$ is a translation of shortest length in G parallel to m (since the translations in G are $\tau^i\tau'^j$ for all integers i and j not both zero); thus, G contains $\gamma_{B,C}$ if and only if m is the axis of a nontrivial glide reflection in G (by Theorems 8.8 and 8.9). The last two sentences show that l is the axis of a reflection in G if and only if m is the axis of a nontrivial glide reflection in G. By symmetry, for each dotted line in Figure 11.16 that isn't parallel to τ or τ' there is a double line in this figure such that the dotted line is the axis of a nontrivial glide reflection in G if and only if the double line is the axis of a reflection in G.

Now let A be a 4-center of G, and let G be the midpoint of A and $\tau\tau'(A)$ (Figure 11.19). Let m be the perpendicular bisector of A and G, and let l be the line through A parallel to m. Let E be the midpoint of A and $\tau(A)$. Let p be the perpendicular bisector of A and E, and let n be the line through A parallel to p. Let B and C be the feet of the perpendiculars drawn to p from A and G, respectivley. Then we have

$$\sigma_m\rho_{A,90°} = \sigma_m\sigma_l\sigma_n$$

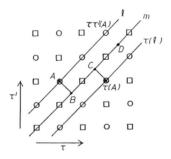

FIGURE 11.18.

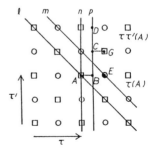

FIGURE 11.19.

(by Theorem 9.7 with $\psi = 45°$)

$$= \tau_{A,G} \sigma_l \sigma_l \sigma_n$$

(by Theorem 9.1i)

$$= \tau_{A,G} \sigma_n$$

(by Equation 4 of Section 8)

$$= \gamma_{B,C}$$

(by Theorem 9.1ii). Thus, since G contains $\rho_{A,90°}$, it contains σ_m if and only if it contains $\gamma_{B,C}$ (by Theorem 7.17). Since C is the midpoint of B and $D = \tau'(B)$, G contains $\gamma_{B,C}$ if and only if p is the axis of a nontrivial glide reflection in G (by Theorems 8.8 and 8.9). The last two sentences show that m is the axis of a reflection in G if and only if p is the axis of a nontrivial glide reflection in G. By symmetry, for each dotted line in Figure 11.16 parallel to τ or τ' there is a double line in the figure such that the dotted line is the axis of a nontrivial glide reflection in G if and only if the double line is the axis of a reflection in G.

The last three paragraphs show that, if any double line in Figure 11.16 is the axis of a reflection in G, or if any dotted line in this figure is the axis of a nontrivial glide reflection in G, then every double line in the figure is the axis of a reflection in G and every dotted line in the figure is the axis of a nontrivial glide reflection in G.

Finally, we note that the dotted lines in Figure 11.15 cannot be axes of nontrivial glide reflections in G at the same time as the double lines in Figure 11.16 are axes of reflections in G, because a line cannot simultaneously be the axis of a nontrivial glide reflection in G and a reflection in G.

In summary, the discussion accompanying Figures 11.15–11.19 shows that there are three possibilities for G. In the first case, G doesn't contain any reflections or nontrivial glide reflections. In the second case, the axes of the reflections and the nontrivial glide reflections in G are as shown in Figure 11.15. In the third case, the axes of the reflections and the nontrivial glide reflections in G are as shown in Figure 11.16.

We can distinguish among these three cases as follows. The first case is the only one where G doesn't contain any reflections. The second case is the only one where there is a reflection in G whose axis contains a 4-center. The third case is the only one where there is a reflection in G whose axis doesn't contain a 4-center. Thus, the three possibilities for G are as follows.

Case 1: G doesn't contain any reflections. Then G doesn't contain any nontrivial glide reflections (by the two previous paragraphs) or any trivial glide reflections (since there are no reflections in G). There are perpendicular translations τ and τ' of equal length such that the translations in G

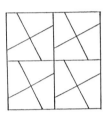

FIGURE 11.20.

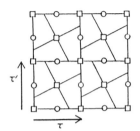

FIGURE 11.21.

and the identity map are the maps $\tau^i \tau'^j$ for all integers i and j. The 4-centers and 2-centers of $\mathcal{G}$ form the checkerboard pattern in Figure 11.7. Thus, we've determined the translations, the identity map, and the rotations in $\mathcal{G}$, and these are all the isometries in $\mathcal{G}$ (by the Classification Theorem 7.10, since $\mathcal{G}$ doesn't contain any reflections or glide reflections). This type of group is named p4.

Conversely, the symmetries of Figure 11.20 are exactly the isometries in the last paragraph, as Figure 11.21 shows. Thus, these isometries form a wallpaper group of type p4 (by Theorems 7.16 and 8.6).

Case 2: There is a reflection in $\mathcal{G}$ whose axis contains a 4-center. The translations, the identity map, and the rotations in $\mathcal{G}$ are as in Case 1. The double lines in Figure 11.15 are the axes of the reflections in $\mathcal{G}$. The trivial glide reflections in $\mathcal{G}$ are determined by these lines and Theorem 8.8. The nontrivial glide reflections in $\mathcal{G}$ have the dotted lines in Figure 11.15 as their axes and are determined by Theorem 8.9. $\mathcal{G}$ contains no other isometries (by the Classification Theorem 7.10). This type of group is named p4m.

Conversely, the symmetries of Figure 11.22 are exactly the isometries in the last paragraph, as Figure 11.23 shows. Thus, these isometries form a wallpaper group of type p4m (by Theorems 7.16 and 8.6).

FIGURE 11.22.

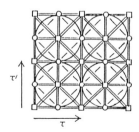

FIGURE 11.23.

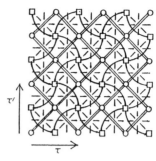

FIGURE 11.24.   FIGURE 11.25.

Case 3: There is a reflection in G whose axis doesn't contain any 4-centers. The translations, the identity map, and the rotations in G are as in Case 1. The double lines in Figure 11.16 are the axes of the reflections in G. The trivial glide reflections in G are determined by these lines and Theorem 8.8. The nontrivial glide reflections in G have the dotted lines in Figure 11.16 as their axes and are determined by Theorem 8.9. G contains no other isometries (by the Classification Theorem 7.10). This type of group is named p4g.

Conversely, the symmetries of Figure 11.24 are exactly the isometries in the last paragraph, as Figure 11.25 shows. Thus, these isometries form a wallpaper group of type p4g (by Theorems 7.16 and 8.6).

We've proved the following result.

THEOREM 11.5. In the Euclidean plane, a wallpaper group that contains a rotation through 90° is of one of the three types p4, p4m, p4g. □

Let G be a wallpaper group with 4-centers. The descriptions of the three cases before Theorem 11.5 make it easy to determine whether G is of type p4, p4m, or p4g. One simply determines whether G contains a reflection and, if so, whether the axis of the reflection contains a 4-center.

EXERCISES

11.1. A figure is given in each part of this exercise. Determine whether the symmetry group of the figure is of type p4, p4m, or p4g. On a copy of the figure, label translations τ and τ' as desribed in Cases 1–3 of this section, use squares to mark 4-centers, circles to mark 2-centers, double lines to mark axes of reflections, and dotted lines to mark axes of nontrivial glide reflections.

(a)

(b)

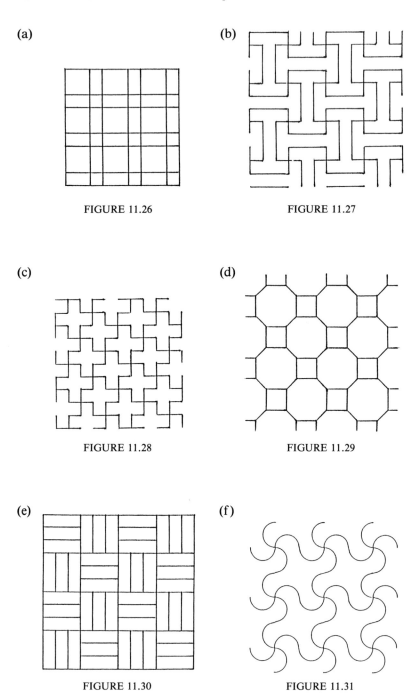

FIGURE 11.26

FIGURE 11.27

(c)

(d)

FIGURE 11.28

FIGURE 11.29

(e)

(f)

FIGURE 11.30

FIGURE 11.31

(g)

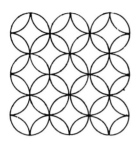

FIGURE 11.32

(h)

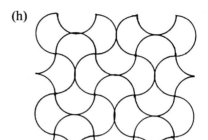

FIGURE 11.33

(i)

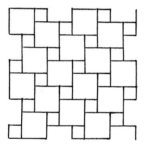

FIGURE 11.34

(j)

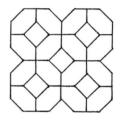

FIGURE 11.35

(k)

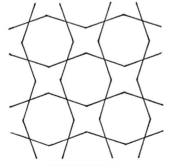

FIGURE 11.36

(l)

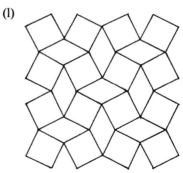

FIGURE 11.37

(m)

(n)

(o)

(p)

(q)

(r)

(s)

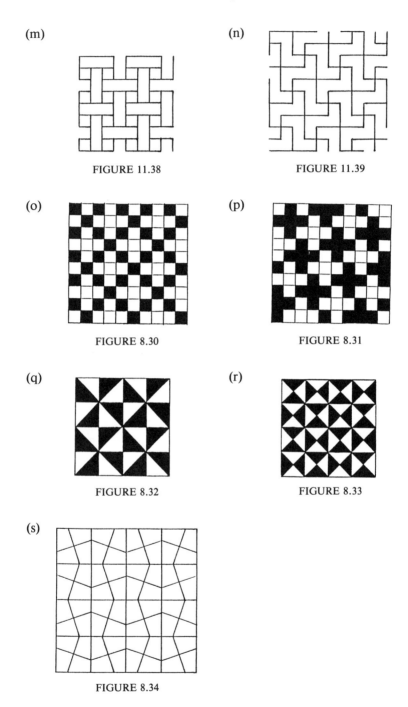

FIGURE 11.38

FIGURE 11.39

FIGURE 8.30

FIGURE 8.31

FIGURE 8.32

FIGURE 8.33

FIGURE 8.34

11.2. Each part of this exercise specifies a symbol, a wallpaper group and a number. The symbol is either a letter of the alphabet or a mathematical symbol. Use the given symbol to create a figure $\mathfrak{F}$ whose symmetry group is the given wallpaper group. The symbol may be rotated or reflected in any way. On a copy of $\mathfrak{F}$, show translations τ and τ' as described in Cases 1–3 of this section, use squares to show 4-centers, use circles to show 2-centers, use double lines to show axes of reflections, and use dotted lines to show axes of nontrivial glide reflections. On a second copy of $\mathfrak{F}$, show τ, τ', and the lattice of parallelograms with vertices $\tau^i \tau'^j(A)$ for some point A and all integers i and j. The number specified is to be the number of times (including any fractions) that the given symbol appears in each parallelogram. See the discussion accompanying Figures 10.32–10.34 for an example.

(a) J, p4, 4   (b) J, p4m, 8
(c) J, p4g, 8   (d) T, p4, 4
(e) T, p4m, 4   (f) T, p4g, 4
(g) ÷, p4, 2   (h) ÷, p4m, 2
(i) ÷, p4g, 2   (j) +, p4, 1
(k) +, p4m, 1   (l) +, p4g, 2
(m) Z, p4, 2   (n) Z, p4m, 8
(o) Z, p4g, 8

11.3. In the Euclidean plane, let $\rho_{P,\theta}$ and $\rho_{Q,\psi}$ be rotations about distinct points P and Q. Evaluate the product $\rho_{Q,\psi}\rho_{P,\theta}$ in each of the following cases.

(a) $\theta + \psi = 360°$
(b) $\theta + \psi > 360°$

(*Hint:* One possible approach is to adapt the proof of Theorem 11.1.)

11.4. In the Euclidean plane, let $\rho_{P,\theta}$ be a rotation. In each of the following cases, determine all isometries φ such that $\varphi\rho_{P,\theta} = \rho_{P,\theta}\varphi$. (See Exercise 8.12 or Exercise 10.9.)

(a) $\theta \neq 180°$
(b) $\theta = 180°$

11.5. In the Euclidean plane, let $\mathcal{G}$ be an isometry group that contains a rotation $\rho_{P,\theta}$ and an isometry that doesn't fix P.

(a) Use Theorem 11.4 and Equation 6 of Section 7 to prove that there is a point $Q \neq P$ such that $\mathcal{G}$ contains $\rho_{Q,\theta}$.
(b) Use part (a), Theorem 10.1, and Theorem 7.17 to prove that $\mathcal{G}$ contains a translation.

Exercises 11.6–11.8 use the following notation. In the Euclidean plane, a *rosette group* is an isometry group that contains only finitely many isometries. Let $\mathcal{G}$ be a rosette group, and let n be an integer greater than 1. We say that $\mathcal{G}$ is of type C_n if there is a point P such that $\mathcal{G}$ consists of the following n elements: the identity map, and the rotations about P through the angles $(360i/n)°$ for $i = 1, 2, \ldots, n - 1$. $\mathcal{G}$ is of type D_n if there is a point P and a line l through P such that $\mathcal{G}$ consists of the following $2n$ elements: the identity map, the rotations about P through the angles $(360i/n)°$ for $i = 1, \ldots, n - 1$, and the reflections across the lines

$$\rho^j_{P,(180/n)°}(l)$$

for $j = 0, 1, \ldots, n - 1$. $\mathcal{G}$ is of type C_1 if it consists only of the identity map. $\mathcal{G}$ is of type D_1 if it consists of two elements: the identity map and the reflection across a line.

For any integer $n \geq 2$, Figure 11.40 shows how to draw a figure whose symmetry group is a rosette group of type C_n: Draw a regular n-gon, and draw lines spiraling from the center of the n-gon to the vertices. For any integer $n \geq 2$, the symmetry group of a regular n-gon (Figure 11.41) is a rosette group of type D_n: Figure 11.42 shows the axes of reflections. The symmetry groups of Figures 11.43 and 11.44 are rosette groups of types C_1 and D_1, respectively: The double line in Figure 11.45 is the axis of the reflection in the symmetry group of Figure 11.44. Exercises 11.7 and 11.8 show that *every rosette group is of type C_n or D_n for some integer $n \geq 1$.*

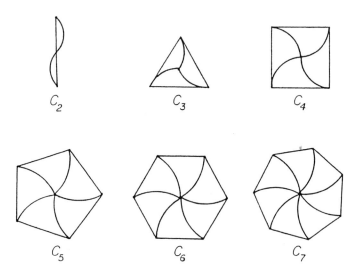

C_2 C_3 C_4

C_5 C_6 C_7

FIGURE 11.40.

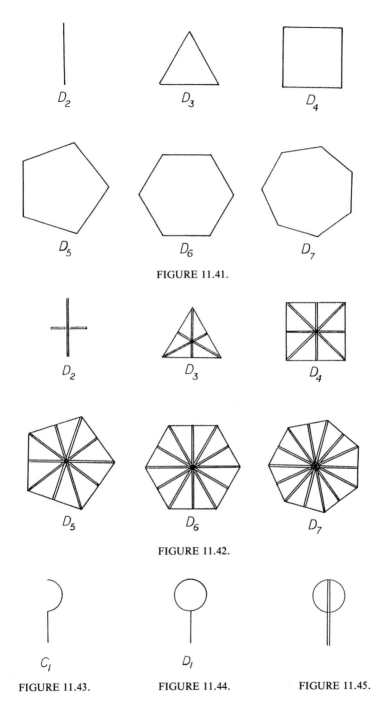

D_2 D_3 D_4

D_5 D_6 D_7

FIGURE 11.41.

D_2 D_3 D_4

D_5 D_6 D_7

FIGURE 11.42.

C_l D_l

FIGURE 11.43.   FIGURE 11.44.   FIGURE 11.45.

11.6. For each part of Exercise 7.1, classify the symmetry group of the given figure as one of the groups $C_1, C_2, \ldots, D_1, D_2, \ldots$. On a copy of the given figure, use double lines to mark the axes of reflections in the symmetry group, as in Figures 11.42 and 11.45.

11.7. In the Euclidean plane, let $\mathcal{G}$ be a rosette group that contains a rotation about a point P.

(a) Prove that every element of $\mathcal{G}$ is either a rotation about P, a reflection across a line that contains P, or the identity map. (*Hint:* One possible approach is to use Exercise 11.5 to prove that every element of $\mathcal{G}$ fixes P.)

(b) Prove that $\mathcal{G}$ is of type C_n or D_n for some integer $n \geq 2$.

11.8. In the Euclidean plane, let $\mathcal{G}$ be a rosette group that doesn't contain any rotations.

(a) Prove that $\mathcal{G}$ contains at most one reflection. (See Theorems 9.1, 9.7, and 7.17.)

(b) Prove that $\mathcal{G}$ is of type C_1 or D_1.

Section 12.

Wallpaper Groups with only 180° or with 120° Rotations

We've considered two categories of wallpaper groups so far: those without rotations, and those with 90° rotations. We complete the classification of wallpaper groups in this section by determining the wallpaper groups that contain only 180° rotations and those that contain 120° rotations. In terms of the categories of wallpaper groups in Theorem 11.3, we've classified the groups in categories (i) and (iii), and must determine those in categories (ii), (iv), and (v).

We don't prove the classifications described in this section, because the proofs are time consuming and similar to those in the last two sections. We outline the proofs in the exercises, however. No results from this section are used elsewhere in the book.

We start by classifying the wallpaper groups in part (ii) of Theorem 11.3. Accordingly, *let $\mathcal{G}$ be a wallpaper group that has a 2-center but no 3-centers, 4-centers, or 6-centers.* $\mathcal{G}$ contains a rotation through 180° but does not contain rotations through any other angles. We consider five cases determined by the reflections and the nontrivial glide reflections in $\mathcal{G}$.

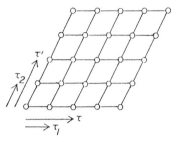

FIGURE 12.1.

Case 1: $\mathcal{G}$ doesn't contain any reflections or nontrivial glide reflections. There are nonparallel translations τ and τ' such that the translations in $\mathcal{G}$ and the identity map are the maps $\tau^i \tau'^j$ for all integers i and j (Figure 12.1). Let τ_1 be the translation that moves points in the same direction as τ but only half the distance, and let τ_2 be the translation that moves points in the same direction as τ' but only half the distance. The 2-centers of $\mathcal{G}$ are the points $\tau_1^i \tau_2^j(P)$ for all integers i and j, where P is any 2-center of $\mathcal{G}$. (These points are the vertices of the lattice in Figure 12.1. The 2-centers are spaced by half the length of τ along lines parallel to τ, and they are spaced by half the length of τ' along lines parallel to τ'.) $\mathcal{G}$ contains no elements besides the maps $\tau^i \tau'^j$ and the 180° rotations about the points P_{ij}. This type of group is named p2.

Conversely, if τ and τ' are two nonparallel translations, the symmetries of Figure 12.2 are the isometries in the last paragraph, as Figure 12.3 shows. Thus, these isometries form a wallpaper group of type p2 (by Theorems 7.16 and 8.6).

Case 2: $\mathcal{G}$ contains reflections but no nontrivial glide reflections. There are perpendicular translations τ and τ' such that the translations in $\mathcal{G}$ and the identity map are the maps $\tau^i \tau'^j$ for all integers i and j (Figure 12.4). The axes of the reflections in $\mathcal{G}$ are lines l_s and l_t' for all integers s and t, where the lines l_s are spaced by $\frac{1}{2}\tau'$ and the lines l_t' are spaced by $\frac{1}{2}\tau$.

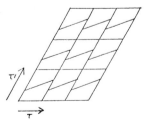

FIGURE 12.2.   FIGURE 12.3.

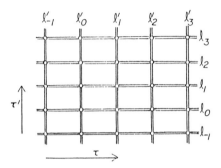

FIGURE 12.4.

The glide reflections in G are trivial, have the lines l_s and l'_t as their axes, and are determined by Theorem 8.8. The 2-centers of G are the points of intersection of the lines l_s and l'_t as s and t vary over all integers. This type of group is named pmm.

Conversely, for any perpendicular translations τ and τ', the symmetries of Figure 12.5 are the isometries in the last paragraph, as Figure 12.6 shows. Thus, these isometries form a wallpaper group of type pmm (by Theorems 7.16 and 8.6).

Case 3: G contains nontrivial glide reflections but no reflections. There are perpendicular translations τ and τ' such that the translations in G and the identity map are the maps $\tau^i \tau'^j$ for all integers i and j (Figure 12.7). The axes of the glide reflections in G are lines m_s and m'_t for all integers s and t, where the lines m_s are spaced by $\frac{1}{2}\tau'$ and the lines m'_t are spaced by $\frac{1}{2}\tau$.

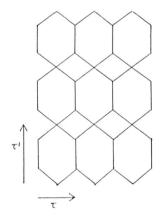

FIGURE 12.5.

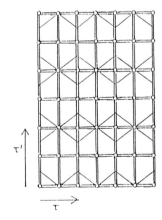

FIGURE 12.6.

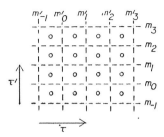

FIGURE 12.7.

All glide reflections in $\mathcal{G}$ are nontrivial, and they are determined by the lines m_s and m'_t and Theorem 8.9. The 2-centers of $\mathcal{G}$ are the centers of the rectangles formed by the lines m_s, m_{s+1}, m'_t, m'_{t+1} for all pairs of integers s and t. This type of group is named pgg.

Conversely, for any perpendicular translations τ and τ', the symmetries of Figure 12.8 are the isometries in the last paragraph, as Figure 12.9 shows. Thus, these isometries form a wallpaper group of type pgg (by Theorems 7.16 and 8.6).

Case 4: $\mathcal{G}$ contains a reflection and a nontrivial glide reflection that have parallel axes. There are perpendicular translations τ and τ_1 such that the translations in $\mathcal{G}$ and the identity map are the maps $\tau^i \tau'^j$ for all integers i and j, where τ' is the translation that maps any point X in the plane to the midpoint of X and $\tau\tau_1(X)$ (Figure 12.10). The axes of the reflections in $\mathcal{G}$ are lines l_s and l'_t for all integers s and t, where the lines l_s are spaced by $\frac{1}{2}\tau_1$ and the lines l'_t are spaced by $\frac{1}{2}\tau$. The trivial glide reflections in $\mathcal{G}$ are determined by the lines l_s and l'_t and Theorem 8.8. The axes of the nontrivial glide reflections in $\mathcal{G}$ are lines m_s and m'_t for all integers s and t, where each line m_s is parallel to τ and lies midway between l_s and l_{s+1}, and each line m'_t is parallel to τ_1 and lies midway between l'_t and l'_{t+1}. The nontrivial glide

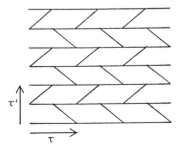

FIGURE 12.8.

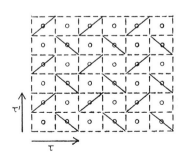

FIGURE 12.9.

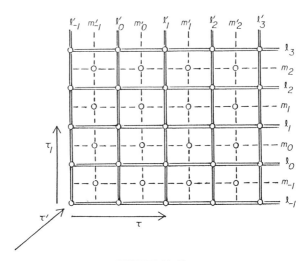

FIGURE 12.10.

reflections in G are determined by the lines m_s and m_t' and Theorem 8.9. The 2-centers of G are the points where the lines l_s and l_t' intersect and the points where the lines m_s and m_t' intersect for all pairs of integers s and t. This type of group is named cmm.

Conversely, for any perpendicular translations τ and τ_1, the symmetries of Figure 12.11 are the isometries in the last paragraph, as Figure 12.12 shows. Thus, these isometries form a symmetry group of type cmm (by Theorems 7.16 and 8.6).

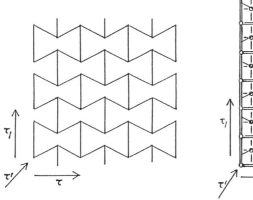

FIGURE 12.11.

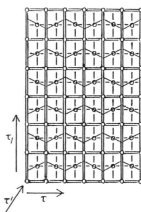

FIGURE 12.12.

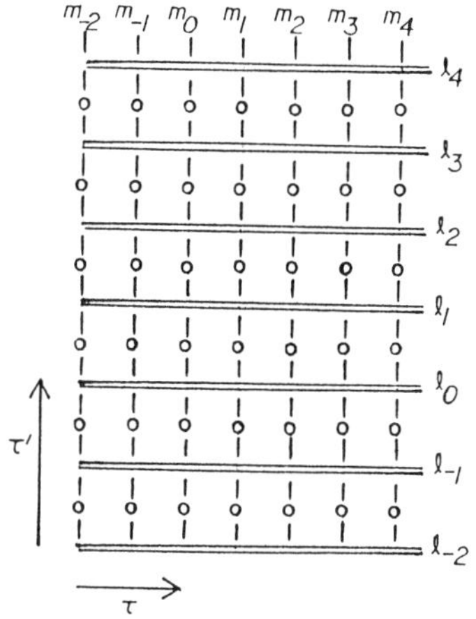

FIGURE 12.13.

Case 5: $\mathcal{G}$ contains both reflections and nontrivial glide reflections, but it does not contain a reflection and a nontrivial glide reflection that have parallel axes. There are perpendicular translations τ and τ' such that the translations in $\mathcal{G}$ and the identity map are the maps $\tau^i \tau'^j$ for all integers i and j (Figure 12.13). The axes of the reflections in $\mathcal{G}$ are lines l_s spaced by $\frac{1}{2}\tau'$ for all integers s. The trivial glide reflections in $\mathcal{G}$ are determined by the lines l_s and Theorem 8.8. The axes of the nontrivial glide reflections in $\mathcal{G}$ are lines m_t spaced by $\frac{1}{2}\tau$ for all integers t. The nontrivial glide reflections in $\mathcal{G}$ are determined by the lines m_t and Theorem 8.9. The 2-centers of $\mathcal{G}$ are the points that each lie midway between l_s and l_{s+1} on a line m_t as s and t vary over all pairs of integers. This type of group is named pmg.

Conversely, for any perpendicular translations τ and τ', the symmetries of Figure 12.14 are the isometries in the last paragraph, as Figure 12.15 shows. Thus, these isometries form a wallpaper group of type pmg (by Theorems 7.16 and 8.6).

The five cases we've considered cover all possibilities for a wallpaper group that has only 180° rotations. Thus, we have the following result.

THEOREM 12.1. In the Euclidean plane, a wallpaper group that contains a rotation through 180° but does not contain rotations through any other angles is of one of the five types p2, pmm, pgg, cmm, pmg. □

Next we classify the wallpaper groups in part (iv) of Theorem 11.3. Accordingly, *let $\mathcal{G}$ be a wallpaper group that has a 3-center but no 6-centers.* This means that $\mathcal{G}$ contains a 120° rotation but no 60° rotations (by (10)

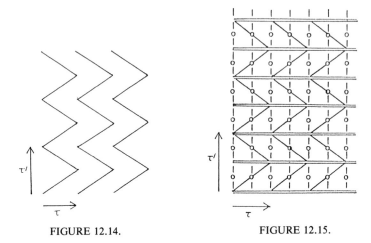

FIGURE 12.14.   FIGURE 12.15.

of Section 9). $\mathcal{G}$ doesn't have 2-centers or 4-centers (by Theorem 11.3iv), and so the center of every rotation in $\mathcal{G}$ is a 3-center (by Theorem 9.6). There are three cases, determined by the reflections in $\mathcal{G}$.

Case 1: $\mathcal{G}$ doesn't contain reflections. It can be shown that $\mathcal{G}$ doesn't contain glide reflections either, and so the only elements of $\mathcal{G}$ are translations, rotations, and the identity map. There are translations τ and τ' of equal length such that lines parallel to τ and τ' form $120°$ angles and such that the translations in $\mathcal{G}$ and the identity map are the maps $\tau^i \tau'^j$ for all integers i and j (Figure 12.16). If P is a 3-center of $\mathcal{G}$, the points $\tau^i \tau'^j(P)$ for all integers

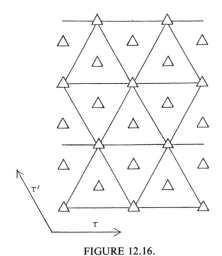

FIGURE 12.16.

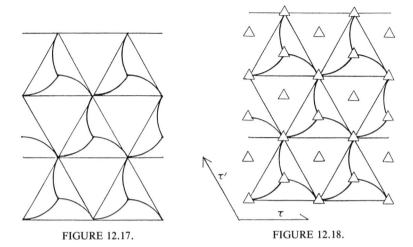

FIGURE 12.17.   FIGURE 12.18.

i and *j* are the vertices of a lattice of equilateral triangles. (The sides of these triangles are shown by lines in Figure 12.16.) The vertices and centroids of these triangles are the 3-centers of $\mathcal{G}$. This type of group is named p3.

Conversely, the symmetries of Figure 12.17 are the isometries in the last paragraph, as Figure 12.18 shows. Thus, these isometries form a wallpaper group of type p3 (by Theorems 7.16 and 8.6).

Case 2: Some but not all of the 3-centers lie on axes of reflections. The translations, the identity map, and the rotations in $\mathcal{G}$ are as described in Case 1. The double lines in Figure 12.19 show the axes of the reflections

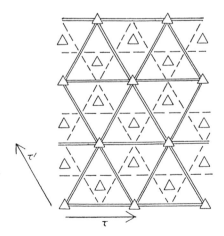

FIGURE 12.19.

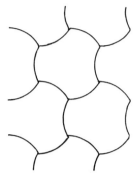

FIGURE 12.20.

in G, and the dotted lines show the axes of the nontrivial glide reflections in G. The trivial glide reflections in G are determined by the double lines in Figure 12.19 and Theorem 8.8, and the nontrivial glide reflections are determined by the dotted lines in the figure and Theorem 8.9. This type of group is named p31m.

Conversely, the symmetries of Figure 12.20 are the isometries in the last paragraph, as Figure 12.21 shows. Thus, these isometries form a wallpaper group of type p31m (by Theorems 7.16 and 8.6).

Case 3: All of the 3-centers lie on axes of reflections. The translations, the identity map, and the rotations in G are as described in Case 1. (See Figure 12.22, where the solid lines show the same lattice of equilateral triangles as in Figure 12.16.) The double lines in Figure 12.22 show the axes of the reflections in G, and the dotted lines show the axes of the nontrivial

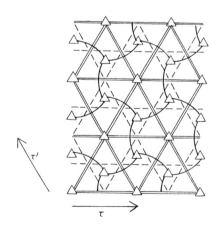

FIGURE 12.21.

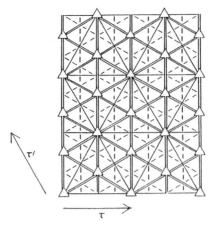

FIGURE 12.22.

glide reflections in $\mathcal{G}$. The trivial glide reflections in $\mathcal{G}$ are determined by the double lines in Figure 12.22 and Theorem 8.8, and the nontrivial glide reflections are determined by the dotted lines in the figure and Theorem 8.9. This type of group is named p3m1.

Conversely, the symmetries of Figure 12.23 are the isometries in the last paragraph, as Figure 12.24 shows. Thus, these isometries form a wallpaper group of type p3m1 (by Theorems 7.16 and 8.6).

The three cases we've considered cover all possibilities for a wallpaper group that has 3-centers but no 6-centers. Thus, we have the following result.

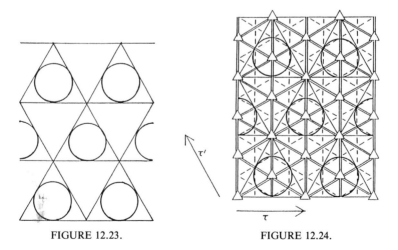

FIGURE 12.23.   FIGURE 12.24.

THEOREM 12.2. In the Euclidean plane, every wallpaper group that contains a 120° rotation but does not contain any 60° rotations is of one of the three types p3, p31m, p3m1. □

We complete the classification of wallpaper groups by determining those in part (v) of Theorem 11.3. Accordingly, *let G be a wallpaper group that has a 6-center*, which means that G is a wallpaper group that contains a 60° rotation. We consider two cases determined by the reflections in G.

Case 1: G doesn't contain reflections. It can be shown that G doesn't contain glide reflections either, and so the only elements of G are translations, rotations, and the identity map. There are translations τ and τ' of equal length such that lines parallel to τ and τ' form 120° angles and such that the translations in G and the identity map are the maps $\tau^i\tau'^j$ for all integers i and j (Figure 12.25). The 6-centers of G are the points $\tau^i\tau'^j(P)$ for all integers i and j, where P is any 6-center of G. The 6-centers are the vertices of a lattice of equilateral triangles whose sides are shown by solid lines in Figure 12.25. The 3-centers of G are the centroids of these triangles, and the 2-centers of G are the midpoints of the sides of the triangles. Since G doesn't have 4-centers (by Theorem 11.3v), we've determined all the rotations in G (by Theorem 9.6). This type of group is named p6.

Conversely, the symmetries of Figure 12.26 are the isometries in the previous paragraph, as Figure 12.27 shows. Thus, these isometries form a wallpaper group of type p6 (by Theorems 7.16 and 8.6).

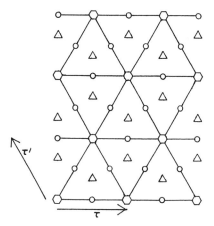

FIGURE 12.25.

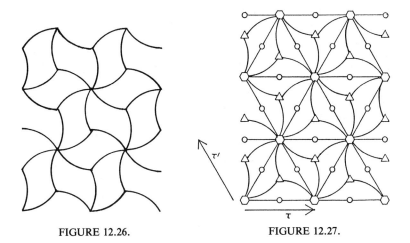

FIGURE 12.26.   FIGURE 12.27.

Case 2: $\mathcal{G}$ contains reflections. The translations, the identity map, and the rotations in $\mathcal{G}$ are as described in Case 1. The double lines in Figure 12.28 show the axes of the reflections in $\mathcal{G}$, and the dotted lines show the axes of the nontrivial glide reflections in $\mathcal{G}$. The trivial glide reflections in $\mathcal{G}$ are determined by the double lines in Figure 12.28 and Theorem 8.8, and the nontrivial glide reflections are determined by the dotted lines in the figure and Theorem 8.9. This type of group is named p6m.

Conversely, the symmetries of Figure 12.29 are the isometries in the last paragraph, as Figure 12.30 shows. Thus, these isometries form a wallpaper group of type p6m (by Theorems 7.16 and 8.6).

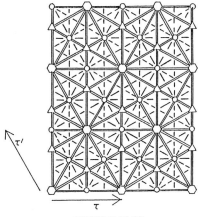

FIGURE 12.28.

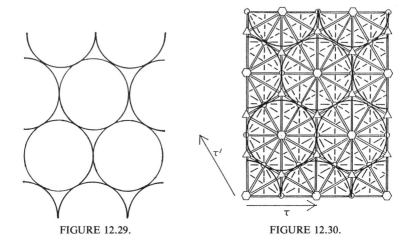

FIGURE 12.29.   FIGURE 12.30.

The two cases we've considered cover all possibilities for a wallpaper group that has a 6-center. Thus, we have the following result.

THEOREM 12.3. In the Euclidean plane, every wallpaper group that contains a rotation through 60° is of one of the two types p6 or p6m. □

We have now classified all wallpaper groups by considering the five cases in Theorem 11.3. Theorems 10.4, 11.5, 12.1, 12.2, and 12.3 show that there are four types of wallpaper groups without rotations, three types with 90° rotations, five types with only 180° rotations, three types with 120° but not 60° rotations, and two types with 60° rotations. *Thus, there are a total of seventeen types of wallpaper groups*, by Theorem 11.3. In short, we have the following classification of wallpaper groups.

THEOREM 12.4. In the Euclidean plane, every wallpaper group belongs to one of the seventeen types listed in Theorems 10.4, 11.5, 12.1, 12.2, and 12.3. □

We summarize the classification of wallpaper groups by listing the seventeen types of wallpaper groups and criteria for distinguishing them.

Groups without rotations
    p1—no reflections or nontrivial glide reflections
    pm—reflections but no nontrivial glide reflections
    pg—nontrivial glide reflections but not reflections
    cm—reflections and nontrivial glide reflections

Groups with 90° rotations
  p4—no reflections
  p4m—reflections whose axes contain 4-centers
  p4g—reflections whose axes do not contain 4-centers
Groups with only 180° rotations
  p2—no reflections or nontrivial glide reflections
  pmm—reflections but no nontrivial glide reflections
  pgg—nontrivial glide reflections but no reflections
  cmm—reflections with perpendicular axes and nontrivial glide
    reflections
  pmg—reflections whose axes are all parallel and nontrivial glide
    reflections
Groups with 120° but not 60° rotations
  p3—no reflections
  p31m—reflections whose axes do not contain all 3-centers
  p3m1—relections whose axes contain all 3-centers
Groups with 60° rotations
  p6—no reflections
  p6m—reflections

EXERCISES

12.1. A figure is given in each part of this exercise. Determine which type
of group listed in Theorems 12.1–12.3 describes the symmetry group of the
figure. On a copy of the figure, label translations τ and τ' (and τ_1 for
a group of type cmm) as described in the section, use circles to mark
2-centers, triangles to mark 3-centers, hexagons to mark 6-centers, double
lines to mark axes of reflections, and dotted lines to mark axes of nontrivial
glide reflections.

(a)

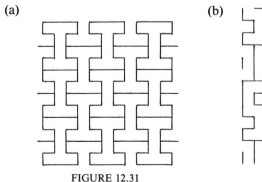

FIGURE 12.31

(b)

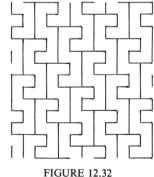

FIGURE 12.32

(c)

(d)

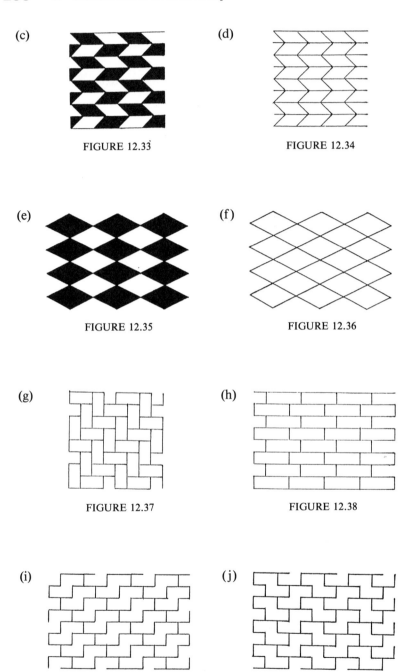

FIGURE 12.33

FIGURE 12.34

(e)

(f)

FIGURE 12.35

FIGURE 12.36

(g)

(h)

FIGURE 12.37

FIGURE 12.38

(i)

(j)

FIGURE 12.39

FIGURE 12.40

(k)

(l)

FIGURE 12.41

FIGURE 12.42

(m)

(n)

FIGURE 12.43

FIGURE 12.44

(o)

(p)

FIGURE 12.45

FIGURE 12.46

(q)

(r)

FIGURE 12.47

FIGURE 12.48

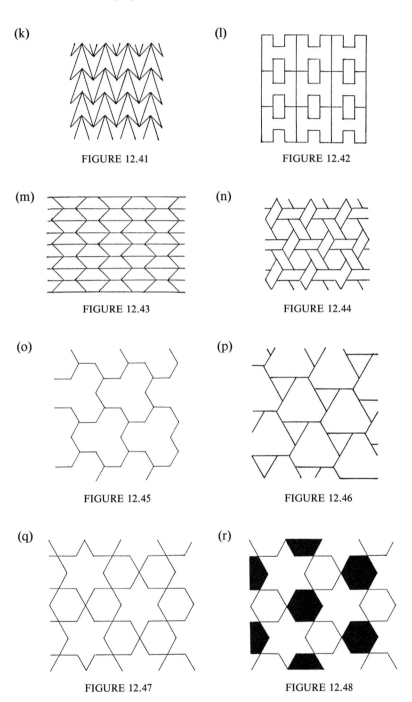

(s)

(t)

FIGURE 12.49

FIGURE 12.50

(u)

(v)

FIGURE 12.51

FIGURE 12.52

(w)

(x)

FIGURE 12.53

FIGURE 12.54

(y)

(z)

FIGURE 8.24

FIGURE 8.25

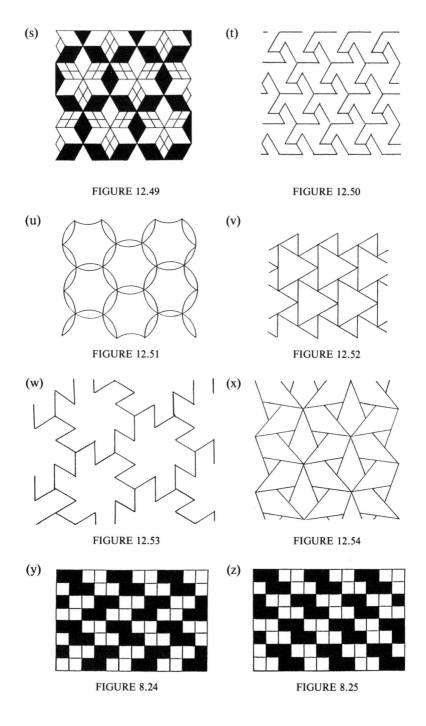

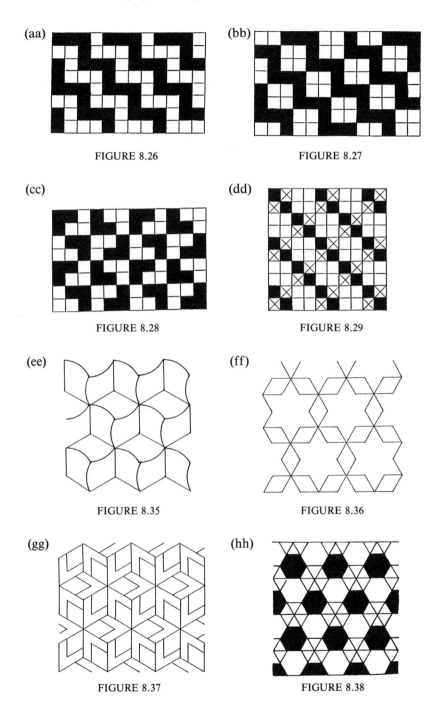

(aa)

FIGURE 8.26

(bb)

FIGURE 8.27

(cc)

FIGURE 8.28

(dd)

FIGURE 8.29

(ee)

FIGURE 8.35

(ff)

FIGURE 8.36

(gg)

FIGURE 8.37

(hh)

FIGURE 8.38

(ii)

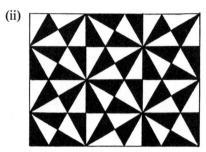

FIGURE 8.39

12.2. Each part of this exercise specifies a letter of the alphabet, a wallpaper group, and a number. Use the given letter to create a figure $\mathfrak{F}$ whose symmetry group is the given wallpaper group. The letter may be rotated or reflected in any way. On a copy of $\mathfrak{F}$, draw arrows to show translations τ and τ' as described in this section for the given symmetry group, and use circles to show 2-centers, triangles to show 3-centers, hexagons to show 6-centers, double lines to show axes of reflections, and dotted lines to show axes of nontrivial glide reflections. On a second copy of $\mathfrak{F}$, show τ, τ', and a lattice of parallelograms with vertices $\tau^i \tau'^j(A)$ for some point A and all points i and j. The specified number is to be the number of times (including any fractions) that the given letter appears inside each parallelogram. See the discussion accompanying Figures 10.32–10.34 for an example.

(a) P, p2, 2
(b) P, pmm, 4
(c) P, pgg, 4
(d) P, cmm, 4
(e) P, pmg, 4
(f) P, p3, 3
(g) P, p31m, 6
(h) P, p3m1, 6
(i) P, p6, 6
(j) P, p6m, 12
(k) V, p2, 2
(l) V, pmm, 2
(m) V, pgg, 4
(n) V, cmm, 2
(o) V, pmg, 2
(p) V, p31m, 3
(q) V, p3m1, 3
(r) V, p6m, 6
(s) Z, p2, 1
(t) Z, pmm, 4
(u) Z, pgg, 2
(v) Z, cmm, 2
(w) Z, pmg, 2
(x) I, p2, 1
(y) I, pmm, 1
(z) I, pgg, 2
(aa) I, cmm, 1
(bb) I, pmg, 2
(cc) I, p6m, 3

In Exercises 12.3–12.8, let G be a wallpaper group that contains rotations through 180° and no other angles. These exercises show that G belongs to one of the five types of groups in Theorem 12.1.

12.3. Assume that G doesn't contain any reflections or nontrivial glide reflections. Prove that G is of type p2. (See Equation 5 of Section 11.)

12.4. If two lines are each the axis of a reflection or a glide reflection in G, prove that the lines are parallel or perpendicular. (*Hint:* See the proof of Theorem 10.2.)

12.5. Assume that G contains reflections but not nontrivial glide reflections. Prove as follows that G is of type pmm (Figure 12.4).

(a) Prove that G contains reflections whose axes are perpendicular. (See Exercise 9.17.)

(b) Prove that there are perpendicular translations τ and τ', lines l_s spaced by $\frac{1}{2}\tau'$, and lines l_t' spaced by $\frac{1}{2}\tau$ such that the axes of reflections in G are exactly the lines l_s and l_t' for all integers s and t.

(c) Prove that the translations in G and the identity map are exactly the maps $\tau^i \tau'^j$ for all integers i and j. (See Theorem 9.1.)

(d) Prove that the centers of the 180° rotations in G are exactly the intersections of the lines l_s and l_t' for all pairs of integers s and t. (See Equation 5 of Section 11.)

12.6. Assume that G contains nontrivial glide reflections but not reflections. Prove as follows that G is of type pgg (Figure 12.7).

(a) Prove that G contains nontrivial glide reflections whose axes are perpendicular. (See Exercise 9.18.)

(b) Prove that there are perpendicular translations τ and τ', lines m_s spaced by $\frac{1}{2}\tau'$ and lines m_t' spaced by $\frac{1}{2}\tau$ such that the axes of the nontrivial glide reflections in G are exactly the lines m_s and m_t' for all integers s and t. (See Exercise 12.4.)

(c) Prove that the translations in G and the identity map are exactly the maps $\tau^i \tau'^j$ for all integers i and j. (See Theorem 9.3.)

(d) Prove that the centers of the 180° rotations in G are exactly the centers of the rectangles formed by the lines m_s, m_{s+1}, m_t', and m_{t+1}' for all pairs of integers s and t. (See Exercise 9.18 and Equation 5 of Section 11.)

12.7. Assume that G contains a reflection and a nontrivial glide reflection that have parallel axes. Prove as follows that G is of type cmm (Figure 12.10).

(a) Prove that there is a translation τ_1 in G that is perpendicular to axes of nontrivial glide reflections in G and is such that the lines perpendicular to τ_1 that are axes of reflections in G are lines l_s spaced by $\frac{1}{2}\tau_1$. Prove that the lines perpendicular to τ_1 that are axes of nontrivial glide reflections in $\mathcal{G}$ are lines m_s for all integers s, where m_s lies midway between l_s and l_{s+1} for each integer s. (See Theorem 10.3.)

(b) Prove that there is a translation τ of shortest length among the translations in $\mathcal{G}$ perpendicular to τ_1. Let τ' be the translation that maps any point X in the plane to the midpoint of X and $\tau\tau_1(X)$. Prove that the translations in $\mathcal{G}$ and the identity map are exactly the maps $\tau^i\tau'^j$ for all integers i and j. (See Theorem 9.1.)

(c) Use Theorem 10.3 to prove that any 2-center of $\mathcal{G}$ lies on one of the lines l_s or m_s. Then use Exercise 9.17 and Theorems 9.7, 7.17, and 9.2 to prove that there are lines l_t' spaced by $\frac{1}{2}\tau$ such that the lines perpendicular to τ that are axes of reflections in $\mathcal{G}$ are exactly the lines l_t' for all integers t.

(d) Use parts (b) and (c) and Theorem 9.1 to prove that the lines perpendicular to τ that are axes of nontrivial glide reflections in $\mathcal{G}$ are lines m_t' for all integers t, where m_t' lies midway between l_t' and l_{t+1}' for each integer t.

(e) Prove that the 2-centers of $\mathcal{G}$ are exactly the points where the lines l_s and l_t' intersect and the points where the lines m_s and m_t' intersect for all pairs of integers s and t. (See Theorem 9.7 and Equation 5 of Section 11.)

(f) Prove that the axes of the reflections and the nontrivial glide reflections in $\mathcal{G}$ are exactly the lines l_s, m_s, l_t', m_t' for all integers s and t. (See Exercise 12.4.)

12.8. Assume that $\mathcal{G}$ contains both reflections and nontrivial glide reflections but does not contain a reflection and a nontrivial glide reflection that have parallel axes. Prove as follows that $\mathcal{G}$ is of type pmg (Figure 12.13).

(a) Prove that the axes of all reflections in $\mathcal{G}$ are perpendicular to the axes of all nontrivial glide reflections in $\mathcal{G}$. (See Exercise 12.4.)

(b) Prove that there are perpendicular translations τ and τ' in $\mathcal{G}$ such that the axes of the reflections in $\mathcal{G}$ are lines l_s spaced by $\frac{1}{2}\tau'$ and such that the axes of the nontrivial glide reflections in $\mathcal{G}$ are lines m_t spaced by $\frac{1}{2}\tau$.

(c) Prove that the translations in $\mathcal{G}$ and the identity map are the maps $\tau^i\tau'^j$ for all integers i and j. (See Theorem 9.1.)

(d) Prove that the 2-centers of $\mathcal{G}$ are the points that each lie midway between l_s and l_{s+1} on a line m_t for integers s and t. (See Exercise 9.17 and Equation 5 of Section 11.)

In Exercises 12.9–12.11, let $\mathcal{G}$ be a wallpaper group that contains rotations through 120° but not 60°. These exercises show that $\mathcal{G}$ is of type p3, p31m, or p3m1, as stated in Theorem 12.2.

12.9. Prove that the translations and rotations in $\mathcal{G}$ are as described in the discussion accompanying Figure 12.16. (See Theorems 8.5, 8.7, 10.1, and 7.17.)

12.10.
(a) Prove that the axis of any reflection or nontrivial glide reflection in $\mathcal{G}$ is parallel to one of the double or dotted lines in Figures 12.19 and 12.22. (*Hint:* One possible approach is to use Exercise 12.9 to adapt the discussion accompanying Figures 11.8–11.14.)
(b) Use part (a) and Theorem 11.4 to prove that the axis of any reflection in $\mathcal{G}$ is in the position of one of the double lines in Figure 12.19 or Figure 12.22 and that the axis of any nontrivial glide reflection in $\mathcal{G}$ is in the position of one of the dotted lines in these figures.
(c) Use part (b), Theorems 9.1 and 7.17, and the discussion accompanying Figure 9.22 to prove that, if any double line in Figure 12.19 is the axis of a reflection in $\mathcal{G}$, or if any dotted line in this figure is the axis of a nontrivial glide reflection in $\mathcal{G}$, then all double lines in this figure are axes of reflections in $\mathcal{G}$ and all dotted lines in this figure are axes of nontrivial glide reflections in $\mathcal{G}$. Prove the corresponding result for Figure 12.22.

12.11. Use Exercises 12.9 and 12.10, the discussion accompanying Figure 9.22, and Theorem 9.2 to prove that $\mathcal{G}$ is of type p3, p31m, or p3m1.

In Exercises 12.12–12.14, let $\mathcal{G}$ be a wallpaper group that contains 60° rotations. These exercises show that $\mathcal{G}$ is of type p6 or p6m, as stated in Theorem 12.3.

12.12. Prove that the translations and rotations in $\mathcal{G}$ are as described in the discussion accompanying Figure 12.25. (See Theorems 8.5, 8.7, 10.1, and 7.17.)

12.13.
(a) Prove that the axis of any reflection or nontrivial glide reflection in $\mathcal{G}$ is parallel to one of the double or dotted lines in Figure 12.28. (One possible approach is to use Exercise 12.12 to adapt the discussion accompanying Figures 11.8–11.14.)
(b) Use part (a) and Theorem 11.4 to prove that the axis of any reflection in $\mathcal{G}$ is one of the double lines in Figure 12.30 and that the axis of any nontrivial glide reflection in $\mathcal{G}$ is one of the dotted lines in this figure.

12.14. Use Exercises 12.12 and 12.13, the discussion accompanying Figures 9.21–9.24, and Theorems 9.1 and 7.17 to prove that $\mathcal{G}$ is of type p6 or p6m.

Section 13.

Dilations

In Sections 7–12 we've studied isometries, maps of the plane that preserve distances. We now broaden our attention to include maps of the plane that multiply distances by a constant factor. We concentrate on particular examples of such maps, namely, dilations. After we introduce dilations, we illustrate their applicability by using them to determine where the center of the nine-point circle of a triangle lies in relation to the triangle.

In the Euclidean plane, let P be a point, and let r be a real number other than 0 and 1. Let X be a point in the plane other than P (Figure 13.1). Choose a coordinate system on line PX (as in Definition 1.6) so that P has coordinate 0. Let X' be any point on line PX other than P and X, and let x and x' be the respective coordinates of X and X'. We have

$$\frac{\overline{PX'}}{\overline{PX}} = \frac{x' - 0}{x - 0} = \frac{x'}{x}$$

(by Theorem 1.8), and so $\overline{PX'}/\overline{PX} = r$ if and only if $x' = rx$. Moreover, the facts that $r \neq 0$, $r \neq 1$, and $x \neq 0$ imply that $rx \neq 0$ and $rx \neq x$. Thus, the point X' on line PX with coordinate rx doesn't equal P or X and is the unique point on this line such that $\overline{PX'}/\overline{PX} = r$. This justifies the following definition.

DEFINITION 13.1. In the Euclidean plane, let P be a point, and let r be a real number other than 0 and 1. The *dilation* $\delta_{P,r}$ with *center* P and *ratio* r is the map of the plane that takes P to itself and sends any point X other than P to the unique point X' on line PX such that $\overline{PX'}/\overline{PX} = r$. $\square$

For example, if we take $r = 2$, the equation $\overline{PX'}/\overline{PX} = 2$ implies that X' is the point on ray $\overrightarrow{PX}$ that lies twice as far from P as X does. Figure 13.2 shows various points X and their images X' under $\delta_{P,2}$. We might imagine that this dilation stretches the plane about P by a factor of 2.

If we take $r = \frac{1}{2}$, the equation $\overline{PX'}/\overline{PX} = \frac{1}{2}$ implies that X' is the point on ray $\overrightarrow{PX}$ that lies half as far from P as X does. Figure 13.3 shows various points X and their images X' under $\delta_{P,1/2}$. We might imagine that this dilation shrinks the plane about P by a factor of $\frac{1}{2}$.

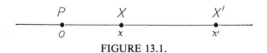

FIGURE 13.1.

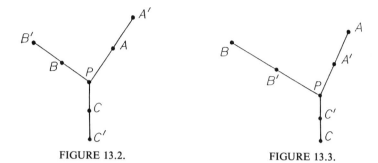

FIGURE 13.2.   FIGURE 13.3.

If we take $r = -2$, the equation $\overline{PX'}/\overline{PX} = -2$ implies that X' is the point on line PX that lies on the opposite side of P as X and that lies twice as far from P as X does. Figure 13.4 shows various points X and their images X' under $\delta_{P,\,-2}$. This dilation reflects points across P and multiplies their distances from P by a factor of 2.

Generalizing the preceding examples shows that, if $r > 0$, the dilation $\delta_{P,r}$ multiplies the distance of each point from P by r and maps the points on each ray originating at P among themselves. If $r < 0$, $\delta_{P,r}$ reflects each point across P and multiplies its distance from P by $|r|$.

The next result shows that *a dilation of ratio r multiplies all distances by* $|r|$, and so a dilation multiplies all distances by the same factor. Accordingly, since isometries preserve distances, dilations are examples of generalizations of isometries.

THEOREM 13.2. In the Euclidean plane, let P be a point, and let r be a real number other than 0 and 1. let X and Y be two points, and let X' and Y' be their respective images under the dilation $\delta_{P,r}$.

(i) Then the distance from X' to Y' is $|r|$ times the distance from X to Y.
(ii) The lines $X'Y'$ and XY are parallel.

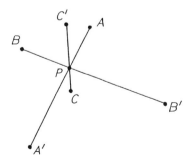

FIGURE 13.4.

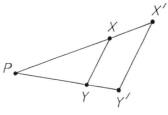

FIGURE 13.5.

Proof: First, assume that line XY doesn't contain P (Figure 13.5). We have

$$\overline{PX'}/\overline{PX} = r = \overline{PY'}/\overline{PY}$$

(by Definition 13.1). Thus, the lines $X'Y'$ and XY are parallel (by Theorem 1.11), and so part (ii) holds in this case. The proof of Theorem 1.11 shows that triangles XYP and $X'Y'P$ are similar, and so we have

$$|\overline{X'Y'}/\overline{XY}| = |\overline{PX'}/\overline{PX}| = |r|$$

(by Property 0.4). Accordingly, we have

$$|\overline{X'Y'}| = |r||\overline{XY}|,$$

and so part (i) holds in this case.

Next, assume that line XY contains P (Figure 13.6). X' and Y' lie on this line as well (by Definition 13.1). Choose a coordinate system on this line so that P has coordinate 0. Let X, Y, X', Y' have respective coordinates x, y, x', y'. If neither X nor Y equals P, we have

$$x' = rx \quad \text{and} \quad y' = ry \tag{1}$$

(by the discussion before Definition 13.1). These equations also hold when X and Y equals P, since the fact that the dilation maps P to itself implies that $x' = 0 = x$ if $X = P$ and that $y' = 0 = y$ if $Y = P$. In short, the equations in (1) hold whenever line XY contains P. It follows that

$$\frac{\overline{X'Y'}}{\overline{XY}} = \frac{y' - x'}{y - x} = \frac{ry - rx}{y - x} = r$$

(by Theorem 1.8). Thus, we have

$$|\overline{X'Y'}| = |r||\overline{XY}|, \tag{2}$$

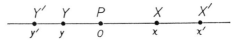

FIGURE 13.6.

and so part (i) holds in this case. Since $X \neq Y$ and $r \neq 0$, Equation 2 shows that $X' \neq Y'$, and so line $X'Y'$ exists. Thus, since X' and Y' lie on line XY, the lines $X'Y'$ and XY are equal and therefore parallel, and so part (ii) holds in this case as well. $\square$

Part (ii) of the previous theorem implies another property of dilations, that a dilation maps each line to a parallel line.

THEOREM 13.3. In the Euclidean plane, let P be a point, and let r be a number other than 0 and 1. For any line l, the dilation $\delta_{P,r}$ maps the points on l to the points on a line parallel to l.

Proof: Let A and B be two points on l, and let A' and B' be their images under $\delta_{P,r}$. $A'B'$ is a line parallel to line $AB = l$ (by Theorem 13.2ii). If X is any point on l other than A, and if X' is the image of X under $\delta_{P,r}$, then $A'X'$ is a line parallel to line $AX = l$ (by Theorem 13.2ii). Accordingly, the lines $A'X'$ and $A'B'$ are both parallel to l and both contain A', and so they are equal. Thus, any point X on l other than A maps to a point X' on line $A'B'$. Moreover, A maps to A', and so every point on l maps to a point on the parallel line $A'B'$. $\square$

Let P be any point in the plane, and let r be any real number other than 0 and 1. Let X' be any point in the plane other than P (Figure 13.1). Choose a coordinate system on line PX' so that P has coordinate 0, and let x' be the coordinate of X'. If we set $x = x'/r$, the point X on line PX' with coordinate x is the unique point that maps to X' under $\delta_{P,r}$ (by the discussion before Figure 13.1). Thus, since $\delta_{P,r}$ also maps P to itself, every point in the plane is the image of exactly one point under $\delta_{P,r}$.

Let T be the center of the nine-point circle of triangle $A_1A_2A_3$. We use dilations to prove that T is the midpoint of the orthocenter H and the circumcenter O of the triangle. Figure 13.7 illustrates this result, and we review the information in this figure. The points F_1, F_2, F_3, M_1, M_2, M_3, N_1, N_2, N_3 in Theorem 5.2 lie on the nine-point circle. These points are equidistant from the center T of the circle, and T is the midpoint of the diameters M_1N_1, M_2N_2, M_3N_3 (by Theorem 5.2, as Figure 5.2 shows). The altitudes A_1F_1, A_2F_2, A_3F_3 in Figure 13.7 intersect at the orthocenter H. The medians A_1M_1, A_2M_2, A_3M_3 intersect at the centroid G. The perpendicular bisectors of the three sides intersect at the circumcenter O, the center of the circle circumscribed about triangle $A_1A_2A_3$. If triangle $A_1A_2A_3$ isn't equilateral (as in Figure 13.7), then O doesn't equal H, and G lies one-third

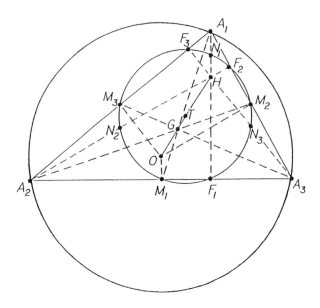

FIGURE 13.7.

of the way from O to H on the Euler line (by Theorem 4.11ii). The next result shows that T is the midpoint of H and O, and so it also lies on the Euler line. When triangle $A_1A_2A_3$ is equilateral, we have $O = H = G = T$, by Theorem 4.11(iii) and the next result (Figure 13.8). In this case, we have $M_1 = F_1$, $M_2 = F_2$, and $M_3 = F_3$ (since the median on each vertex equals the altitude, by Theorem 4.10). In the next theorem, we use the convention that the midpoint of a point X and itself is also X. Thus, we can say that T is the midpoint of H and O whether or not triangle $A_1A_2A_3$ is equilateral (Figures 13.7 and 13.8).

THEOREM 13.4. Consider a triangle in the Euclidean plane (Figures 13.7 and 13.8).

   (i) The center T of the nine-point circle is the midpoint of the orthocenter H and the circumcenter O.
  (ii) The radius of the nine-point circle is half the radius of the circumcircle.
 (iii) A point X lies on the circumcircle if and only if the midpoint X' of H and X lies on the nine-point circle.

In the notation of Theorem 5.2, part (ii) of Theorem 13.4 implies that the distance from T to the points M_i, N_i, F_i on the nine-point circle is half the distance from O to the points A_i on the circumcircle. Part (iii) is

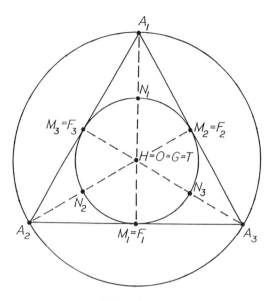

FIGURE 13.8.

illustrated by Figure 13.9, which shows the same triangle as Figure 13.7. The nine-point circle contains the foot F_i of each altitude, and so the ray from H through any F_i intersects the circumcircle at a point R_i such that F_i is the midpoint of H and R_i (by part (iii)). Similarly, the nine-point circle contains the midpoint M_i of each side, and so the ray from H through any M_i intersects the circumcircle at a point S_i such that M_i is the midpoint of H and S_i (by part (iii)).

Proof: We use the notation of Theorem 5.2. We consider the dilation $\delta_{H,1/2}$ of ratio $\frac{1}{2}$ whose center is the orthocenter H of triangle $A_1 A_2 A_3$. Let X be any point in the plane, and let X' be the image of X under $\delta_{H,1/2}$. We claim that X' is the midpoint of H and X. If $X \neq H$ (as in Figure 13.10), this claim follows from the fact that $\overline{HX'}/\overline{HX} = \frac{1}{2}$ (by Definition 13.1), and so H lies on line XX', half as far from X' as from X, and doesn't lie between X' and X (by Theorem 1.4). If $X = H$, we also have $X' = H$ (by Definition 13.1), and so it is still true that X' is the midpoint of H and X, as claimed.

For any point Y in the plane, if Y' is the image of Y under $\delta_{H,1/2}$, the dilation $\delta_{H,1/2}$ maps the points that lie at a given distance from Y onto the points that lie at half that distance from Y' (by Theorem 13.2i and the first paragraph after the proof of Theorem 13.3). Thus, $\delta_{H,1/2}$ maps any circle with center Y onto the circle of half the radius with center Y'.

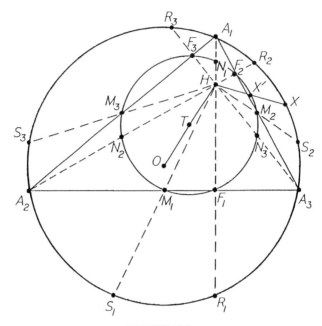

FIGURE 13.9.

The circumcircle $\mathcal{K}$ of triangle $A_1A_2A_3$ contains $A_1A_2A_3$ (Figure 13.7). The dilation $\delta_{H,1/2}$ maps A_1, A_2, A_3 to N_1, N_2, N_3 (by the first paragraph of the proof, since N_i is the midpoint of H and A_i for each i). Thus, $\delta_{H,1/2}$ maps $\mathcal{K}$ onto a circle $\mathcal{K}'$ containing N_1, N_2, N_3 (by the previous paragraph). The points N_1, N_2, N_3 are distinct (by Theorem 13.2i and the fact that A_1, A_2, A_3 are distinct). Thus, the fact that N_1, N_2, N_3 lie on $\mathcal{K}'$ implies that N_1, N_2, N_3 are noncollinear and that $\mathcal{K}'$ is the unique circle through N_1, N_2, N_3 (by Theorems 4.5ii and 4.3ii). Hence, $\mathcal{K}'$ is the nine-point circle of triangle $A_1A_2A_3$ (by Theorem 5.2).

We've proved that the dilation $\delta_{H,1/2}$ maps the circumcircle $\mathcal{K}$ of triangle $A_1A_2A_3$ onto the nine-point circle $\mathcal{K}'$. Then $\delta_{H,1/2}$ maps the center O of $\mathcal{K}$ to the center T of $\mathcal{K}'$ (by the second paragraph of the proof). Thus, T is the midpoint of H and O (by the first paragraph of the proof), and so part (i) holds. The radius of $\mathcal{K}'$ is half the radius of $\mathcal{K}$ (by the second paragraph of

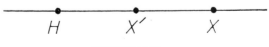

FIGURE 13.10.

the proof and the fact that $\delta_{H,1/2}$ maps $\mathcal{K}$ onto $\mathcal{K}'$), and so part (ii) holds. The dilation $\delta_{H,1/2}$ maps any point X in the plane to the midpoint X' of H and X (by the first paragraph of the proof), and X is the only point that maps to X' (by the first paragraph after the proof of Theorem 13.3). Thus, since $\delta_{H,1/2}$ maps $\mathcal{K}$ onto $\mathcal{K}'$, it follows that X lies on $\mathcal{K}$ if and only if X' lies on $\mathcal{K}'$, and so part (iii) holds. $\square$

EXERCISES

13.1. Let the notation be as in Theorem 5.2. In addition, let B_1, B_2, B_3 be the points such that T is the midpoint of A_i and B_i for each i.

(a) Use the map $\rho_{T,180°} = \delta_{T,-1}$ to prove that N_1 is the midpoint of B_2 and B_3, N_2 is the midpoint of B_1 and B_3, and that N_3 is the midpoint of B_1 and B_2. Illustrate this result with a figure.
(b) Prove that the orthocenter of each of the triangles $A_1A_2A_3$ and $B_1B_2B_3$ is the circumcenter of the other. Illustrate this result with a figure.
(c) Prove that T is the midpoint of the centroids of triangles $A_1A_2A_3$ and $B_1B_2B_3$. Illustrate this result with a figure.

13.2. Let the notation be as in Theorem 5.2, and let G be the centroid of triangle $A_1A_2A_3$ (Figure 13.7).

(a) Prove that the dilation $\delta_{G,-1/2}$ maps the circumcircle of triangle $A_1A_2A_3$ to the nine-point circle.
(b) For $i = 1, 2, 3$, prove that there is a point U_i on the circumcircle such that G lies between U_i and F_i twice as far from U_i as from F_i. Illustrate this result with a figure where $A_1A_2A_3$ is a scalene triangle.
(c) For $i = 1, 2, 3$, prove that there is a point V_i on the circumcircle such that G lies between V_i and N_i twice as far from V_i as from N_i. Prove that the circumcenter O is the midpoint of A_i and V_i and that A_iV_i is a diameter of the circumcircle. Illustrate these results with a figure where $A_1A_2A_3$ is a scalene triangle.

13.3. In the Euclidean plane, let $A_1A_2A_3$ be a triangle that isn't equilateral, let G be the centroid, let O be the circumcenter, and let T be the center of the nine-point circle (Figure 13.7). Prove that G lines on line OT between O and T, twice as far from O as from T:

(a) by using Exercise 13.2(a);
(b) by using Theorems 4.11(ii) and 13.4(i).

13.4. Let the notation be as in Theorem 5.2, and let S_1, S_2, S_3 be as in the discussion accompanying Figure 13.9.

(a) Use a dilation with center H to prove that $A_1 S_1$, $A_2 S_2$, $A_3 S_3$ are diameters of the circumcircle of triangle $A_1 A_2 A_3$. Illustrate this result with a figure where $A_1 A_2 A_3$ is a scalene triangle whose angles are all acute.
(b) If triangle $A_1 A_2 A_3$ doesn't have a right angle, prove that $A_1 A_2 S_1 S_2$, $A_1 A_3 S_1 S_3$, and $A_2 A_3 S_2 S_3$ are rectangles. Illustrate this result with a figure.

13.5. Let the notation be as in Theorem 5.2, and let the points R_i and S_i be as in the discussion accompanying Figure 13.9.

(a) If the sides $A_1 A_2$ and $A_1 A_3$ of triangle $A_1 A_2 A_3$ have different lengths, prove that $R_1 S_1$ and $A_2 A_3$ are parallel lines. Illustrate this result with two figures, one where triangle $A_1 A_2 A_3$ doesn't have a right angle, and one where $\angle A_1 A_2 A_3 = 90°$.
(b) If triangle $A_1 A_2 A_3$ doesn't have a right angle, prove that $R_1 R_2 R_3$ and $F_1 F_2 F_3$ are similar triangles such that corresponding sides lie on parallel lines. Illustrate this result with a figure where $A_1 A_2 A_3$ is a scalene triangle.
(c) If triangle $A_1 A_2 A_3$ doesn't have a right angle, prove that $A_1 S_1$ is the perpendicular bisector of R_2 and R_3, $A_2 S_2$ is the perpendicular bisector of R_1 and R_3, and $A_3 S_3$ is the perpendicular bisector of R_1 and R_2. Illustrate this result with a figure where $A_1 A_2 A_3$ is a scalene triangle. (See Exercise 5.3a.)

13.6. In the notation of Theorem 5.2, assume that triangle $A_1 A_2 A_3$ doesn't have a right angle.

(a) Prove that $H A_2 A_3$, $A_1 H A_3$, $A_1 A_2 H$ are triangles that have the same nine-point circle as triangle $A_1 A_2 A_3$. (See Exercise 4.7.)
(b) Prove that there are points U_1, U_2, U_3 on the circumcircle of triangle $H A_2 A_3$ such that F_1 is the midpoint of A_1 and U_1, F_2 is the midpoint of A_1 and U_2, and F_3 is the midpoint of A_1 and U_3. Illustrate this result with a figure where $A_1 A_2 A_3$ is a scalene triangle. (Of course, analogous results hold for triangles $A_1 H A_3$ and $A_1 A_2 H$.)

13.7. In the Euclidean plane, let $A_1 A_2 A_3$ be a triangle without a right angle. Let H be the orthocenter, and let O be the circumcenter. Let O_1, O_2, O_3 be the respective circumcenters of the triangles $H A_2 A_3$, $A_1 H A_3$, $A_1 A_2 H$. (See Exercise 4.7.)

(a) Prove that $A_2 A_3$ is the perpendicular bisector of O and O_1, $A_3 A_1$ is the perpendicular bisector of O and O_2, and $A_1 A_2$ is the perpendicular bisector of O and O_3. Illustrate this result with a figure where $A_1 A_2 A_3$ is a scalene triangle. (*Hint:* In the notation of the discussion accompanying Figure 13.9, one possible approach is to prove that the reflection $\sigma_{A_2A_3}$ interchanges the triangles A_2A_3H and $A_2A_3R_1$, and so it also interchanges their circumcircles.)

(b) Prove that A_1H is the perpendicular bisector of O_2 and O_3, A_2H is the perpendicular bisector of O_1 and O_3, and A_3H is the perpendicular bisector of O_1 and O_2. Illustrate this result with a figure where $A_1A_2A_3$ is a scalene triangle. (*Hint:* See part (a) and Exercise 4.7(a).)

13.8. In the notation of Exercise 13.7, let T be the center of the nine-point circle of triangle $A_1A_2A_3$.

(a) Prove that the map $\rho_{T,180°} = \delta_{T,-1}$ interchanges H with O, A_1 with O_1, A_2 with O_2, and A_3 with O_3. (See Exercise 13.6a.)

(b) Prove that the following pairs of triangles are congruent: $A_1A_2A_3$ and $O_1O_2O_3$; HA_2A_3 and OO_2O_3; A_1HA_3 and O_1OO_3; A_1A_2H and O_1O_2O. Illustrate this result with a figure where $A_1A_2A_3$ is a scalene triangle.

(c) If triangle $A_1A_2A_3$ is scalene, prove that A_1HO_1O, A_2HO_2O, A_3HO_3O are parallelograms. Illustrate this result with a figure.

(d) Prove that the points in each of the sets

$$\{A_1, A_2, O_1, O_2\}, \quad \{A_2, A_3, O_2, O_3\}, \quad \{A_1, A_3, O_1, O_3\} \quad (3)$$

are either the vertices of a parallelogram or lie on a line. Illustrate this result with two figures where $A_1A_2A_3$ is a scalene triangle: one figure where each set in (3) contains the vertices of a parallelogram, and one figure where the points in exactly one of the sets in (3) lie on a line. (See Exercise 5.7.)

13.9. Let the notation be as in Exercise 13.7. Use Exercise 13.8(a) to do each part of this exercise.

(a) Prove that any three of the points O, O_1, O_2, O_3 are the vertices of a triangle that doesn't have a right angle and whose orthocenter is the fourth point. Illustrate this result with a figure where $A_1A_2A_3$ is a scalene triangle.

(b) Prove that any three of the points O_1, O_2, O_3, O are the vertices of a triangle that has the same nine-point circle as triangle $A_1A_2A_3$.

(c) Let F_1, F_2, F_3 be the feet of the altitudes of triangle $A_1A_2A_3$ that lie on A_1, A_2, A_3, respectively. Let W_1, W_2, W_3 be the feet of the altitudes of triangle $O_1O_2O_3$ that lie on O_1, O_2, O_3, respectively. Prove that F_1W_1, F_2W_2, F_3W_3 are diameters of the nine-point circle of triangle $A_1A_2A_3$. Illustrate this result with a figure where $A_1A_2A_3$ is a scalene triangle.

13.10. In the Euclidean plane, let $A_1A_2A_3$ be a triangle without a right angle. Let H be the orthocenter, let G be the centroid, and let T be the center of the nine-point circle. Let G_1, G_2, G_3 be the respective centroids of the triangles HA_2A_3, A_1HA_3, A_1A_2H. (See Exercise 4.7.)

(a) Prove that the dilation $\delta_{T,-1/3}$ maps $H \to G$, $A_1 \to G_1$, $A_2 \to G_2$, and $A_3 \to G_3$.
(b) Prove that any three of the points G, G_1, G_2, G_3 are the vertices of a triangle without a right angle whose orthocenter is the fourth point. Illustrate this result with a figure where $A_1A_2A_3$ is a scalene triangle.

13.11. In the notation of Exercises 13.7 and 13.10, prove that the dilation $\delta_{T,3}$ maps $G \to O$, $G_1 \to O_1$, $G_2 \to O_2$, and $G_3 \to O_3$.

13.12. Let triangles ABC and $\mathfrak{I}$ be as in Exercise 5.15. Prove that the radius of the circumcircle of triangle $\mathfrak{I}$ is twice the radius of the circumcircle of triangle ABC. Illustrate this result with a figure in each of the following cases.

(a) Triangle ABC is scalene, and the three excenters of triangle ABC are the vertices of triangle $\mathfrak{I}$.
(b) Triangle ABC is scalene, and the incenter of triangle ABC is one of the vertices of triangle $\mathfrak{I}$.

(*Hint:* See Exercise 5.16.)

13.13. Let triangles ABC and $\mathfrak{I}$ be as in Exercise 5.15. Let O be the circumcenter of triangle ABC, let O' be the circumcenter of triangle $\mathfrak{I}$, and let E be the equicenter of triangle ABC that isn't a vertex of $\mathfrak{I}$. Prove that O is the midpoint of O' and E. Illustrate this result with a figure in each of the cases of Exercise 13.12. (*Hint:* One possible approach is to use Exercise 5.16 to apply Theorem 13.4i to the triangle $\mathfrak{I}$.)

Exercises 13.14–13.19 use the following notation. In the Euclidean plane, a *dilatation* is a map that sends the points on each line to points on a parallel line and that maps distinct points to distinct points. Note the difference between the words "dil-a-ta-tion" and "di-la-tion."

13.14. (a) In the Euclidean plane, prove that translations, dilations, and the identity map are all dilatations.
(b) In the Euclidean plane, let A and B be two points, and let A' and B' be two points. Prove that there is at most one dilatation mapping $A \to A'$ and $B \to B'$.

13.15. In the Euclidean plane, let φ be a dilatation other than the identity map. Let A be a point such that $\varphi(A) \neq A$. Let B be a point that doesn't lie on the line $A\varphi(A)$.

(a) Prove that $\varphi(B) \neq B$.

(b) If the lines $A\varphi(A)$ and $B\varphi(B)$ aren't parallel, use Theorem 1.11 to prove that there is a dilation that maps $A \to \varphi(A)$ and $B \to \varphi(B)$.

(c) If the lines $A\varphi(A)$ and $B\varphi(B)$ are parallel, prove that there is a translation that maps $A \to \varphi(A)$ and $B \to \varphi(B)$.

13.16. In the Euclidean plane, use Exercises 13.14 and 13.15 to prove that every dilatation is either a dilation, a translation, or the identity map. (This is the converse of Exercise 13.14a.)

13.17. In the Euclidean plane, let A and B be two points, and let A' and B' be two points. Prove that there is a dilatation mapping $A \to A'$ and $B \to B'$ if and only if the lines AB and $A'B'$ are parallel. (If such a dilatation exists, then it is unique, by Exercise 13.14b. One possible approach to this exercise is to adapt Exercise 13.15 when A' doesn't lie on line AB and to use coordinates on line AB when A' lies on AB.)

13.18. Identify points in the Euclidean plane by their Cartesian coordinates (x, y).

(a) Let r, s, t be real numbers with $r \neq 0$. Prove that the map $(x, y) \to (rx + s, ry + t)$ is a dilatation.

(b) Conversely, prove that every dilatation of the Euclidean plane has the form $(x, y) \to (rx + s, ry + t)$ for some real numbers r, s, t with $r \neq 0$. (See either Exercise 13.14b or 13.16.)

13.19. In the notation of Exercise 13.18, for what numbers r, s, t is the map $(x, y) \to (rx + s, ry + t)$ a dilation? For such numbers, express the coordinates of the center of the dilation in terms of r, s and t.

Exercises 13.20–13.26 use the following notation. Let r be a positive number. A *similarity* of ratio r is a map φ from the Euclidean plane to itself such that, for any two points X and Y in the plane, the distance from $\varphi(X)$ to $\varphi(Y)$ is r times the distance from X to Y. (In short, a similarity of ratio r multiplies all distances in the plane by r.) If φ and ψ are similarities, let $\varphi\psi$ be the map from the Euclidean plane to itself defined by $(\varphi\psi)(X) = \varphi(\psi(X))$ for all points X in the plane. Set $\varphi^2 = \varphi\varphi$ for any similarity φ. For any line l in the plane, let $\varphi(l)$ be the set of points $\varphi(X)$ for all points X on l.

13.20. In the Euclidean plane, let φ be a similarity.

(a) For any line l, prove that $\varphi(l)$ is a line.

(b) For any two lines l and m, prove that l and m are parallel if and only if $\varphi(l)$ and $\varphi(m)$ are parallel lines.

(c) Let A, B, C be three points on a line. Prove that $\varphi(A)$, $\varphi(B)$, $\varphi(C)$ are three points on a line such that

$$\overline{\varphi(C)\varphi(A)}/\overline{\varphi(C)\varphi(B)} = \overline{CA}/\overline{CB}.$$

13.21. In the Euclidean plane, let φ be a similarity. Assume that there are points A and B such that $A\varphi(A)$ and $B\varphi(B)$ are distinct parallel lines.

(a) If AB and $\varphi(A)\varphi(B)$ are parallel lines, prove that φ is an isometry.
(b) If AB and $\varphi(A)\varphi(B)$ are not parallel lines, prove that φ fixes the point of intersection of the lines AB and $\varphi(A)\varphi(B)$.

(*Hint:* One possible approach is to use Exercise 13.20c and Theorem 1.11.)

13.22. In the Euclidean plane, let φ be a similarity that doesn't fix any point and that has the property that X, $\varphi(X)$, $\varphi^2(X)$ lie on a line for any point X in the plane. For any point A in the plane and for any point B in the plane not on the line $A\varphi(A)$, prove that $A\varphi(A)$ and $B\varphi(B)$ are distinct parallel lines.

(*Hint:* One possible approach is to show that, if the lines $A\varphi(A)$ and $B\varphi(B)$ intersect at a point C, then C is fixed by φ.)

13.23. In the Euclidean plane, let φ be a similarity that doesn't fix any point. Assume that there is a point P such that P, $\varphi(P)$, $\varphi^2(P)$ don't lie on a line. Let l be the line through $\varphi^2(P)$ parallel to the line $P\varphi(P)$.

(a) Prove that $\varphi(l)$ and $\varphi(P)\varphi^2(P)$ are distinct parallel lines. (See Exercise 13.20.)
(b) Prove that l and $\varphi(l)$ intersect at a point Q such that $\varphi(P)\varphi^2(P)$ and $Q\varphi(Q)$ are distinct parallel lines.

13.24. In the Euclidean plane, prove that every similarity either fixes a point or is an isometry. (See Exercises 13.21–13.23.)

13.25. In the Euclidean plane, let φ be a similarity that has ratio $r \neq 1$ and that fixes a point P.

(a) Prove that $\delta_{P,1/r}\varphi$ is an isometry that fixes P.
(b) Prove that φ is one of the maps $\delta_{P,r}$, $\delta_{P,r}\rho_{P,\theta}$ for $0° < \theta < 360°$, or $\delta_{P,r}\sigma_l$ for a line l containing P.

13.26. In the Euclidean plane, a *stretch* is a dilation $\delta_{P,r}$, where P is a point and r is a positive number other than 1. A *stretch rotation* is a map of the form $\delta_{P,r}\rho_{P,\theta}$, where P is a point, r is a positive number other than 1, and $0° < \theta < 360°$. A *stretch reflection* is a map of the form $\delta_{P,r}\sigma_l$, where P is a point, r is a positive number other than 1, and l is a line containing P.

(a) Prove that *the similarities of the Euclidean plane are exactly the isometries, the stretches, the stretch reflections, and the stretch rotations.* (Note that one must prove both that all similarities are of these four types and that all maps of these four types are similarities.)

(b) Prove that every similarity other than an isometry can be written in exactly one way as one of the maps $\delta_{P,r}$, $\delta_{P,r}\rho_{P,\theta}$, $\delta_{P,r}\sigma_l$, where P is a point, r is positive number other than 1, $0° < \theta < 360°$, and l is a line containing P.

(c) Where do the dilations $\delta_{P,r}$ for negative numbers r fit into the classification of similarities in part (a)?

Exercises 13.27–13.31 use the following terminology. In the Euclidean plane, let $\mathcal{K}_1$ and $\mathcal{K}_2$ be circles that have distinct centers A and B. A *center of similitude* of $\mathcal{K}_1$ and $\mathcal{K}_2$ is a point P such that there is a dilation $\delta_{P,r}$ that maps the points of $\mathcal{K}_1$ to the points of $\mathcal{K}_2$, where r is a real number other than 0 and 1. A center of similitude is called *internal* if it lies on line AB between A and B, and otherwise it is called *external*.

13.27. In the Euclidean plane, let $\mathcal{K}_1$ and $\mathcal{K}_2$ be circles that have distinct centers A and B. Let s and t be the respective radii of the circles.

(a) Prove that the centers of similitude of $\mathcal{K}_1$ and $\mathcal{K}_2$ are exactly the points P on line AB such that $|\overline{PA}/\overline{PB}| = s/t$.

(b) Prove that $\mathcal{K}_1$ and $\mathcal{K}_2$ have a unique internal center of similitude.

(c) If $s \neq t$, prove that $\mathcal{K}_1$ and $\mathcal{K}_2$ have a unique external center of similitude. If $s = t$, prove that $\mathcal{K}_1$ and $\mathcal{K}_2$ have no external center of similitude.

(d) Prove that a point P is a center of similitude of $\mathcal{K}_1$ and $\mathcal{K}_2$ if and only if P is a center of similitude of $\mathcal{K}_2$ and $\mathcal{K}_1$.

13.28. In the Euclidean plane, let $\mathcal{K}_1$ and $\mathcal{K}_2$ be circles that have distinct centers A and B. Let P be a point. Prove that the following conditions are equivalent.

(i) P is a center of similitude of $\mathcal{K}_1$ and $\mathcal{K}_2$, and P lies outside $\mathcal{K}_1$.

(ii) P is the point where line AB intersects a line tangent to both $\mathcal{K}_1$ and $\mathcal{K}_2$.

(iii) P is the point of intersection of two lines that are each tangent to both $\mathcal{K}_1$ and $\mathcal{K}_2$.

(iv) P is a center of similitude of $\mathcal{K}_1$ and $\mathcal{K}_2$, and P lies outside both $\mathcal{K}_1$ and $\mathcal{K}_2$.

13.29. In the Euclidean plane, let $\mathcal{K}_1$, $\mathcal{K}_2$, $\mathcal{K}_3$ be circles whose centers don't lie on a line and whose radii have distinct lengths. Prove that the external centers of similitude of the three pairs of circles $\{\mathcal{K}_1, \mathcal{K}_2\}$, $\{\mathcal{K}_1, \mathcal{K}_3\}$,

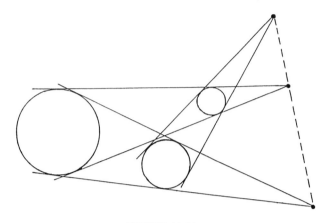

FIGURE 13.11.

$\{\mathcal{K}_2, \mathcal{K}_3\}$ lie on a line. (See Exercise 13.27 and Menelaus' Theorem. By Exercise 13.28, Figure 13.11 illustrates this result when the centers of similitude lie outside the circles.)

13.30. In the Euclidean plane, let $\mathcal{K}_1$, $\mathcal{K}_2$, $\mathcal{K}_3$ be three circles whose centers don't lie in a line. Assume that radii of $\mathcal{K}_2$ and $\mathcal{K}_3$ have the same length and that a radius of $\mathcal{K}_1$ has a different length. Prove that the line through the external centers of similitude of the two pairs of circles $\{\mathcal{K}_1, \mathcal{K}_2\}$ and $\{\mathcal{K}_1, \mathcal{K}_3\}$ is parallel to the line through the centers of $\mathcal{K}_2$ and $\mathcal{K}_3$. Illustrate this result with a figure analogous to Figure 13.11 where the centers of the circles are vertices of a scalene triangle and where the centers of similitude are determined as intersections of tangents (by Exercise 13.28). (See Exercise 13.27 and Theorem 2.2.)

13.31. In the Euclidean plane, let $\mathcal{K}_1$, $\mathcal{K}_2$, $\mathcal{K}_3$ be three circles whose centers don't lie on a line.

(a) If the radii of $\mathcal{K}_2$ and $\mathcal{K}_3$ have different lengths, prove that there is a line that contains the internal center of similitude of $\mathcal{K}_1$ and $\mathcal{K}_2$, the internal center of similitude of $\mathcal{K}_1$ and $\mathcal{K}_3$, and the external center of similitude of $\mathcal{K}_2$ and $\mathcal{K}_3$. (By Exercise 13.28, Figure 13.12 illustrates this result when the centers of similitude lie outside the circles.)

(b) Assume that $\mathcal{K}_1$, $\mathcal{K}_2$, $\mathcal{K}_3$ all have radii of different lengths. Part (a) and Exercise 13.29 imply that the six centers of similitude lie by threes on four lines. Illustrate this result with a figure where the centers of similitude are constructed as intersections of tangent lines (by Exercise 13.28).

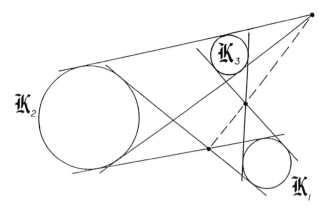

FIGURE 13.12.

(c) If the radii of $\mathcal{K}_2$ and $\mathcal{K}_3$ have equal lengths, prove that the line through the internal centers of similitude of the two pairs of circles $\{\mathcal{K}_1, \mathcal{K}_2\}$ and $\{\mathcal{K}_1, \mathcal{K}_3\}$ is parallel to the line through the centers of the circles $\mathcal{K}_2$ and $\mathcal{K}_3$. Illustrate this result with a figure where the centers of the circles are the vertices of a scalene triangle and where the centers of similitude are constructed as intersections of tangent lines (by Exercise 13.28).

Chapter III
Projective Geometry

INTRODUCTION AND HISTORY

Projective geometry grew out of the study of drawing in perspective. Painters in the Renaissance sought to portray the world around them as faithfully as possible. In the fifteenth and sixteenth centuries, such great artists as Piero della Francesca, Leonardo da Vinci, and Albrecht Dürer developed rules of perspective, the realistic depiction of three-dimensional objects in two-dimensional drawings.

Perspective drawing is based on the following idea (Figure III.1). The canvas is placed between the artist's eye and the subject. Each feature of the subject is drawn where the line from the feature to the artist's eye crosses the canvas. Then light rays travel to the artist's eye from the drawing along the same lines as light rays traveling from the subject. Thus, the drawing and the subject create the same image in the eye.

Instead of carrying out this procedure physically, artists developed mathematical principles, the laws of perspective, that determine the image on the canvas. To derive these principles, one considers two planes $\mathcal{P}$ and $\mathcal{Q}$ and a point O that doesn't lie on either plane (Figure III.2). A point X of $\mathcal{P}$ corresponds to the point X' where the line through X and O intersects $\mathcal{Q}$. The map $X \to X'$ from the points of $\mathcal{P}$ to the points of $\mathcal{Q}$ is called the

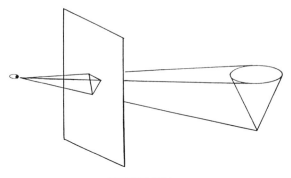

FIGURE III.1.

projection from $\mathcal{P}$ to $\mathcal{Q}$ through O. If X is a point of the subject, if the canvas lies in the plane $\mathcal{Q}$, and if the artist's eye is at O, then X' is the point on the canvas where the point X of the subject is drawn. Projective geometry is the study of the properties of a plane that are preserved by projection, that is, the properties shared by a planar figure and its image under a projection.

In the first half of the seventeenth century, Girard Desargues developed projective geometry as a branch of mathematics. He extended the Euclidean plane by adding new points, called *points at infinity*, so that parallel lines in the Euclidean plane intersect in a point at infinity. There is one point at infinity for each family of parallel lines. Desargues also added a new line, called the *line at infinity*, to contain the points at infinity. The *projective plane* is the result of adding the points at infinity and the line at infinity to the Euclidean plane.

The idea of the projective plane arises naturally from studying projections. For example, in a perspective drawing of railroad tracks that run straight across a plain, the rails of the tracks are drawn to intersect at a point on the

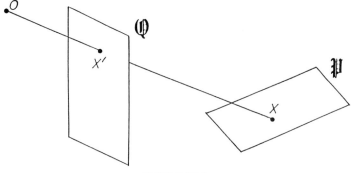

FIGURE III.2.

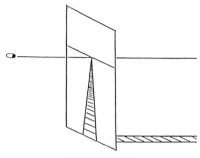

FIGURE III.3.

horizon (Figure III.3). The projection through the artist's eye maps the point at infinity on the parallel rails of the tracks to the point of the picture where the rails are drawn to meet. If the tracks ran in a different direction, the rails would be drawn to meet at another point on the horizon. Thus, the points at infinity of the plain project through the artist's eye to points of the horizon drawn on the canvas, and the line at infinity of the plain projects to the horizon line of the picture. In effect, the drawing on the canvas is a picture of parallel lines meeting in a point at infinity. We introduce the projective plane in Section 14, and we study projections between planes in Section 15.

To prove the main results of Sections 15 through 18, we project a given arrangement of points and lines to an arrangement easier to analyze. We use this technique in Section 15 to prove Pappus' Theorem on hexagons inscribed in two lines. Pappus worked in Alexandria at the end of the third century A.D., five to six centuries after Euclid, Archimedes, and Apollonius. He stated and proved his theorem in Euclidean terms, but it generalizes naturally to the projective plane, and it is easily proved by projecting between planes. Although Pappus used no projective techniques, his work foreshadows Desargues' results on projective geometry to a remarkable extent.

We use projections in Section 16 to prove Desargues' most famous result, his theorem on perspective triangles. In Section 17, we study harmonic sets of points, which were introduced by Pappus in Euclidean terms and generalized by Desargues to the projective plane.

Section 16 also contains the Duality Principle, which states that points and lines have symmetric properties in the projective plane. This principle was developed in the first half of the nineteenth century by Charles Julien Brianchon, Jean Victor Poncelet, and Joseph Diez Gergonne. The first derivations of the Duality Principle used the polar properties of conic sections, which are outlined in Exercises 20.10 and 20.16.

We end this chapter by introducing the subject of the next chapter, the projective study of conic sections. Section 18 contains theorems about triangles inscribed in and circumscribed about a conic section $\mathcal{K}$. Theorem 6.8 contains a corresponding result in the Euclidean plane when $\mathcal{K}$ is a circle, and this result extends to any conic section $\mathcal{K}$ in the projective plane by projecting between planes. We use Desargues' Theorem from Section 16 and properties of harmonic sets from Section 17 to deduce corollaries.

The following books are highly recommended for additional reading. They were the source of much of the material in this chapter.

Cremona, Luigi, *Elements of Projective Geometry*, translated by Charles Leudesdorf, Clarendon Press, Oxford, 1893.

Hughes, Daniel R., and Piper, Fred C., *Projective Planes*, Springer-Verlag, New York, 1973.

Seidenberg, A., *Lectures In Projective Geometry*, Van Nostrand, Princeton, 1962.

Veblen, Oswald, and Young, John Wesley, *Projective Geometry*, Vols. I and II, Blaisdell, New York, 1938.

Section 14.

The Extended Plane

As the discussion accompanying Figures 2.5 and 2.6 shows, it's natural to imagine that two lines become parallel when their point of intersection vanishes by moving off to infinity. Rather than saying that the point of intersection "vanishes," we say that it "becomes ideal." In other words, we imagine that parallel lines intersect at an "ideal point," a point that isn't part of the Euclidean plane but that we add to the plane.

As the discussions of Menelaus' and Ceva's Theorems in Sections 2 and 3 show, parallel lines in the Euclidean plane introduce special cases that make results awkward to state and prove. We eliminate these special cases when we imagine that parallel lines intersect at ideal points added to the Euclidean plane.

Thus, we're led to the following definitions, which are illustrated by Figure 14.1. The square in Figure 14.1 represents the Euclidean plane, and the dotted lines connect each ideal point to the parallel lines that contain it.

We begin with the Euclidean plane ε. We call its points and lines *ordinary points* and *ordinary lines*.

For each family of parallel lines in ε, we create a new point that lies on the lines of this family and on no other ordinary lines. We call such a point an *ideal point*. For example, there is one ideal point on all vertical lines in

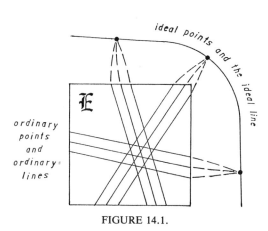

FIGURE 14.1.

&, a second ideal point on all lines of slope 5, a third on all lines of slope $-\frac{1}{2}$, and so on. We create a new line called the *ideal line*. The ideal line contains all the ideal points and no ordinary points.

The *extended plane* consists of the ordinary points, the ordinary lines, the ideal points, and the ideal line.

Two nonparallel ordinary lines intersect as usual in the Euclidean plane, and we don't want them to intersect again at an ideal point; that's why we add a different ideal point to each family of parallel lines. Any two points in the Euclidean plane lie on a line, and we want this property to continue to hold in the extended plane; that's why we add the ideal line through all the ideal points.

We don't claim that ideal points and the ideal line exist as physical realities. They are imaginary objects that we add to the Euclidean plane to form the extended plane. The first two paragraphs of this section show why it's natural to consider the extended plane, and other reasons appear in the next section.

In the Euclidean plane, any two lines intersect at a unique point, except when the lines are parallel. Thus, parallel lines create exceptions that must be considered in Euclidean geometry. Adding ideal points to the Euclidean plane eliminates these exceptions. As the next result shows, any two lines in the extended plane intersect at a unique point: two parallel ordinary lines intersect at an ideal point (Figure 14.2); two nonparallel ordinary lines intersect as usual at an ordinary point (Figure 14.3); and an ordinary line intersects the ideal line at the ideal point on the ordinary line (Figure 14.4).

THEOREM 14.1. In the extended plane, any two lines intersect at a unique point.

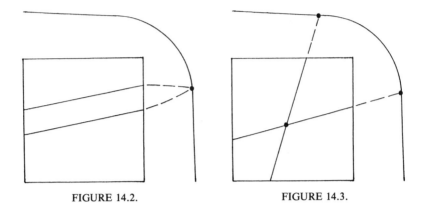

FIGURE 14.2.   FIGURE 14.3.

Proof: We consider the ways that two lines in the extended plane can lie with respect to the ideal line.

Case 1: The lines are ordinary, and they are parallel in the Euclidean plane (Figure 14.2). Since the lines are parallel in the Euclidean plane, they have no ordinary points in common, and they contain the same ideal point. This ideal point is the unique point that lies on both lines.

Case 2: The lines are ordinary, and they are not parallel in the Euclidean plane (Figure 14.3). Since the lines aren't parallel in the Euclidean plane, they intersect at a unique ordinary point, and they contain different ideal points. Thus, the lines have a unique point in common.

Case 3: One of the lines is ordinary, and the other is ideal (Figure 14.4). The ideal line contains all the ideal points and no ordinary points. The ordinary line contains exactly one ideal point. This is the unique point that lies on both lines.

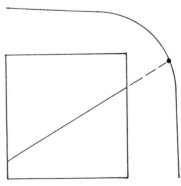

FIGURE 14.4.

The three cases include all possibilities, and we've shown in each case that the lines intersect at a unique point. □

A result analogous to Theorem 14.1 states that any two points in the extended plane lie on a unique line. Unlike Theorem 14.1, this result also holds in the Euclidean plane, and so we need only show that it remains true in the extended plane.

THEOREM 14.2. In the extended plane, any two points lie on a unique line.

Proof: We consider the ways that two points A and B in the extended plane can lie with respect to the ideal line.

Case 1: A and B are both ordinary points (Figure 14.5). Since two points determine a line in the Euclidean plane, A and B lie on exactly one ordinary line. Moreover, the ideal line doesn't contain any ordinary points. Thus, A and B lie on exactly one line.

Case 2: One of the points is ordinary, and one is ideal. By symmetry, we can assume that A is ordinary and B is ideal (Figure 14.6). The ideal line doesn't contain A because the ideal line contains only ideal points. The ordinary lines through B form a family of parallel lines in the Euclidean plane. This family consists of all lines in the Euclidean plane that have a particular slope. There is a unique line through A having this slope, and this is the unique line through A and B.

Case 3: A and B are both ideal (Figure 14.7). No ordinary line contains both points, since an ordinary line contains only one ideal point. The ideal

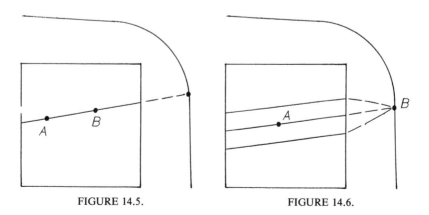

FIGURE 14.5.   FIGURE 14.6.

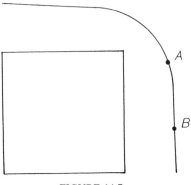

FIGURE 14.7.

line contains all the ideal points. Thus, the ideal line is the unique line containing both A and B.

The three cases include all possibilities, and we've shown in each case that A and B lie on a unique line. $\square$

If l and m are two lines in the extended plane, we let $l \cap m$ be their point of intersection (Figure 14.8). Theorem 14.1 says that $l \cap m$ exists whenever l and m are distinct (that is, when $l \neq m$). We've eliminated the exceptions caused by parallel lines in the Euclidean plane; any two lines in the extended plane intersect at a unique point, without exception.

If A and B are two points in the extended plane, we let AB be the line they determine (Figure 14.9). Theorem 14.2 says that A and B lie on a unique line AB whenever A and B are distinct (that is, when $A \neq B$).

We call points *collinear* if they all lie on one line, and *noncollinear* if they don't. We call lines *concurrent* if they all lie on one point, and *nonconcurrent* if they don't.

We can't talk about the distance between two points in the extended plane when one of the points is ideal. We also can't talk about the angles formed by two lines in the extended plane when one of the lines is ideal. *Incidence* is the relation between points and lines that specifies which points lie on each line. Statements about incidence are the only ones that we can obviously make about the extended plane without distinguishing

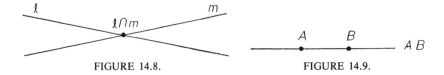

FIGURE 14.8.   FIGURE 14.9.

between ordinary and ideal points or between ordinary lines and the ideal line.

The progression from the Euclidean plane to the extended plane can be viewed as giving up something to gain something. We give up the chance to talk about most properties other than incidence (at least, if we don't want to distinguish between ordinary and ideal points). We gain the chance to study incidence effectively (since any two lines in the extended plane have a point in common, without the exceptions caused by parallel lines in the Euclidean plane).

In fact, there are interesting theorems that can be stated solely in terms of incidence, without reference to distances or angles. Pappus' Theorem, which is stated here and proved in the next section, is an important example of such a theorem.

Pappus' Theorem. In the extended plane, let l and m be two lines. Let A, B, C be three points on l other than $l \cap m$, and let A', B', C' be three points on m other than $l \cap m$. Then the points $E = AB' \cap A'B$, $F = AC' \cap A'C$, and $G = BC' \cap B'C$ are collinear.

Figure 14.10 illustrates Pappus' Theorem. We emphasize that A, B, C are any three points on l except $l \cap m$, and that they occur in any order. Likewise, A', B', and C' are any three points on m except $l \cap m$, and they occur in any order. Thus, Figure 14.11 also illustrates Pappus' Theorem. The fact that we don't consider lengths or angles in incidence theorems makes these theorems broadly applicable.

There is another sense in which a theorem about incidence in the extended plane is broadly applicable. If the theorem doesn't specify the position of

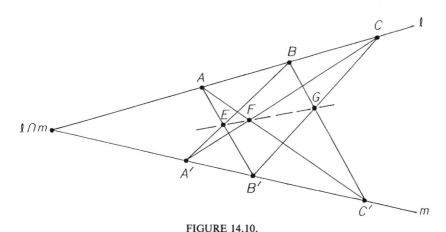

FIGURE 14.10.

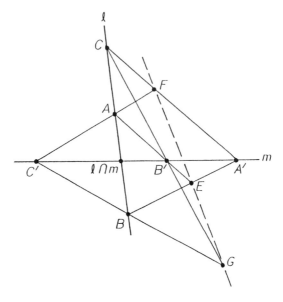

FIGURE 14.11.

the ideal line, then we can take the ideal line to lie anywhere. Thus, one theorem about the extended plane implies a number of different Euclidean results as the position of the ideal line varies. Hence, the extended plane provides a very efficient way to obtain Euclidean results.

For example, if the ideal line contains none of the points mentioned in Pappus' Theorem, we obtain the following Euclidean theorem, again illustrated by Figures 14.10 and 14.11.

Theorem. In the Euclidean plane, let l and m be two nonparallel lines. Let A, B, C be three points on l other than $l \cap m$, and let A', B', C' be three points on m other than $l \cap m$. Assume that AB' and $A'B$ aren't parallel, that AC' and $A'C$ aren't parallel, and that BC' and $B'C$ aren't parallel. Then the points $E = AB' \cap A'B$, $F = AC' \cap A'C$, and $G = BC' \cap B'C$ are collinear.

We can obtain a different Euclidean result from Pappus' Theorem by assuming that A is the one point named in Pappus' Theorem that lies on the ideal line. The corresponding Euclidean result should be stated in familiar Euclidean terms, and so the ideal point A should not appear in the Euclidean version. To accomplish this, we note that AB' is now the line p through B' parallel to l (since AB' and l intersect at the ideal point A).

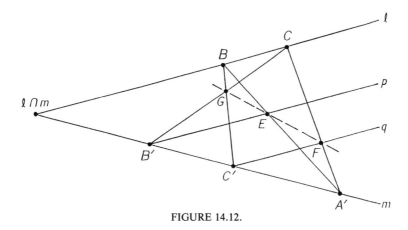

FIGURE 14.12.

(See Figure 14.12, and compare it with Figure 14.10.) Similarly, AC' is now the line q through C' parallel to l (since AC' and l intersect at the ideal point A). Thus, we obtain the following Euclidean result from Pappus' Theorem by taking A to be the one ideal point named in the theorem.

Theorem. In the Euclidean plane, let l and m be two nonparallel lines. Let B and C be two points on l other than $l \cap m$, and let A', B', C' be three points on m other than $l \cap m$. Let E be the point where $A'B$ intersects the line p through B' parallel to l, and let F be the point where $A'C$ intersects the line q through C' parallel to l. Assume that the lines BC' and $B'C$ intersect at a point G. Then the points E, F, G are collinear (Figure 14.12).

In the statement of this theorem, we must assume that BC' and $B'C$ aren't parallel, so that their point of intersection G exists in the Euclidean plane. On the other hand, E and F exist in the Euclidean plane because p and q aren't parallel to $A'B$ or $A'C$ (since p and q are parallel to l, but $A'B$ and $A'C$ aren't).

We can obtain another Euclidean result from Pappus' Theorem by assuming that BC' is the ideal line. Then the points B, C', G are all ideal (since they lie on the ideal line BC'), and so we must state the Euclidean result without mentioning these points. To do so, we note that $A'B$ is now the line p through A' parallel to l (since $A'B$ and l intersect at the ideal point B). (See Figure 14.13, and compare it with Figure 14.10.) AC' is now the line q through A parallel to m (since AC' and m intersect at the ideal point C'). The conclusion of the theorem, which states that E, F, G are collinear, becomes the statement that the lines EF and $B'C$ are parallel (since they both lie on the same ideal point G). Of course, the ideal line BC'

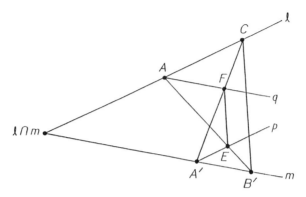

FIGURE 14.13.

doesn't appear in the Euclidean version. In short, if we take BC' to be the ideal line, Pappus' Theorem gives the following Euclidean result.

Theorem. In the Euclidean plane, let l and m be two nonparallel lines. Let A and C be two points on l other than $l \cap m$, and let A' and B' be two points on m other than $l \cap m$. Let E be the point where AB' intersects the line p through A' parallel to l, and let F be the point where $A'C$ intersects the line q through A parallel to m. Then the lines EF and $B'C$ are parallel (Figure 14.13).

The assumptions of this theorem imply that AB' and p aren't parallel (since p is parallel to l, but AB' isn't) and that $A'C$ and q aren't parallel (since q is parallel to m, but $A'C$ isn't). Thus, the points E and F exist in the Euclidean plane.

We could consider many other positions for the ideal line in order to obtain Euclidean results from Pappus' Theorem. Of course, the ideal line

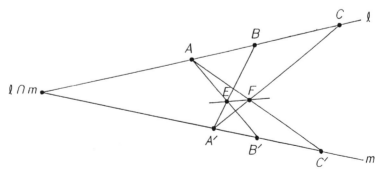

FIGURE 14.14.

can't contain noncollinear points. Thus, for example, we can't assume that the points A, B, C' in Pappus' Theorem are all ideal (since C' doesn't lie on $AB = l$ because $C' \neq l \cap m$).

As a first step toward proving Pappus' Theorem in the next section, we prove that the line EF in Pappus' Theorem exists and doesn't contain any of the points A, B, C, A', B', C' (Figure 14.14). This result is an example of a theorem that follows directly from Theorems 14.1 and 14.2. Theorem 14.1 shows that the point $l \cap m$ exists in the extended plane whenever l and m are distinct lines, and Theorem 14.2 shows that the line AB exists in the extended plane whenever A and B are distinct points.

THEOREM 14.3. In the extended plane, let l and m be two lines. Let A, B, C be three points on l other than $l \cap m$, and let A', B', C' be three points on m other than $l \cap m$.

   (i) The the line AB' exists and intersects l at the unique point A.
   (ii) The points $E = AB' \cap A'B$ and $F = AC' \cap A'C$ exist and lie on neither l nor m.
   (iii) The line EF exists and doesn't contain any of the points A, B, C, A', B', C'.

Proof: (i) B' doesn't lie on l (since $B' \neq l \cap m$, and B' lies on m). Thus, B' doesn't equal A, and so the line AB' exists (by Theorem 14.2). AB' doesn't equal l (since B' doesn't lie on l, as we've seen), and so AB' and l intersect at a unique point (by Theorem 14.1). This point is A, since A lies on both AB' and l.

(ii) The lines AB' and $A'B$ exist, and we have

$$AB' \cap l = A \neq B = A'B \cap l \tag{1}$$

(by part (i) and symmetry). Thus, AB' and $A'B$ are distinct lines, and so the point $E = AB' \cap A'B$ exists (by Theorem 14.1). AB' and $A'B$ intersect l at distinct points (by (1)), and so the point E where AB' and $A'B$ intersect doesn't lie on l. By symmetry, E also doesn't lie on m, and the point $F = AC' \cap A'C$ also exists and lies on neither l nor m.

(iii) We have

$$AB' \cap m = B' \neq C' = AC' \cap m$$

(by part (i) and symmetry). Thus, AB' and AC' are distinct lines. We have $A \neq E$ and $A \neq F$ (by part (ii)), and so the lines AE and AF exist (by Theorem 14.2). Moreover, E lies on AB', and F lies on AC', and so the two previous sentences imply that

$$AE = AB' \neq AC' = AF. \tag{2}$$

Thus, E and F are distinct points, and so line EF exists (by Theorem 14.2).

EF doesn't contain *A* (by (2)), and so it doesn't contain *A'* either (by symmetry). Neither *A* nor *B'* equals *E* (by part (ii)), and these three points are collinear, and so we have *B'E* = *AE*. Moreover, *AE* doesn't contain *F* (by (2)), and so *F* doesn't lie on *B'E*. Thus, *B'* doesn't lie on *EF*. By symmetry, none of the points *B, C, C'* lie on *EF* either. □

Theorem 14.3 shows, by symmetry, that the points *E, F, G* in Pappus' Theorem exist and are distinct. We must still prove the main assertion of Pappus' Theorem, that the points *E, F, G* lie in a line. The proof uses projection between planes, an idea we introduce in the next section.

EXERCISES

14.1. The following theorem is proved in Exercise 16.18 (Figure 14.15)).

Theorem. In the extended plane, let *l* and *m* be two lines on a point *O*. Let *A, B, C* be three points on *l* other than *O*, and let *A', B', C'* be three points on *m* other than *O*. Assume that the lines *AA', BB', CC'* are concurrent at a point *P*. Set *E* = *AB'* ∩ *A'B* and *F* = *AC'* ∩ *A'C*. Then the points *O, E, F* are collinear.

State the version of this theorem that holds in the Euclidean plane in each of the following cases. Draw a figure in the Euclidean plane to illustrate each version.

(a) *E* is the only ideal point named.
(b) *A'* is the only ideal point named.
(c) *P* is the only ideal point named.
(d) *m* is the ideal line.
(e) *AC'* is the ideal line.
(f) *BC'* is the ideal line.

14.2. The following theorem is proved in Exercise 16.19 (Figure 14.15). It is the converse of the theorem in Exercise 14.1.

Theorem. In the extended plane, let *l* and *m* be two lines on a point *O*. Let *A, B, C* be three points on *l* other than *O*, and let *A', B', C'* be three points on *m* other than *O*. Set *E* = *AB'* ∩ *A'B* and *F* = *AC'* ∩ *A'C*. Assume that the points *O, E, F* are collinear. Then the lines *AA', BB', CC'* are concurrent at a point *P*.

State the version of this theorem that holds in the Euclidean plane in each case listed in Exercise 14.1. Draw a figure in the Euclidean plane to illustrate each version.

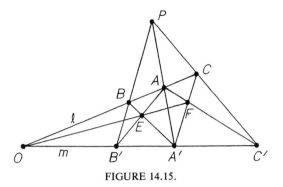

FIGURE 14.15.

14.3. The following result is proved in Exercise 16.20 (Figure 14.16).

Theorem. In the extended plane, let A, B, C be three noncollinear points. Let A' be a point on BC other than B and C, let B' be a point on CA other than C and A, and let C' be a point on AB other than A and B. Set $K = BC \cap B'C'$, $L = CA \cap C'A'$, and $M = AB \cap A'B'$. If the lines AA', BB', CC' are concurrent at a point O, then the points A', B', C' are noncollinear and the points K, L, M are collinear.

State the version of this theorem that holds in the Euclidean plane in each of the following cases. Draw a figure in the Euclidean plane to illustrate each version.

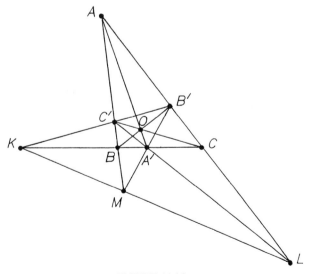

FIGURE 14.16.

(a) M is the only ideal point named.
(b) C' is the only ideal point named.
(c) C is the only ideal point named.
(d) CM is the ideal line.
(e) CC' is the ideal line.
(f) BC is the ideal line.

14.4. The following result is proved in Exercise 16.21 (Figure 14.16). It is the converse of the theorem in Exercise 14.3.

Theorem. In the extended plane, let A, B, C be three noncollinear points. Let A' be a point on BC other than B and C, let B' be a point on CA other than C and A, and let C' be a point on AB other than A and B. Set $K = BC \cap B'C'$, $L = CA \cap C'A'$, and $M = AB \cap A'B'$. If the points A', B', C' are noncollinear and the points K, L, M are collinear, then the lines AA', BB', CC' are concurrent at a point O.

State the version of this theorem that holds in the Euclidean plane in each case listed in Exercise 14.3. Draw a figure in the Euclidean plane to illustrate each version.

14.5. The following result is proved in Exercises 16.22 and 17.4 (Figure 14.17).

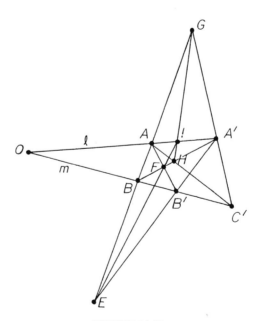

FIGURE 14.17.

Theorem. In the extended plane, let l and m be two lines on a point O. Let A and A' be two points on l other than O, and let B, B', C' be three points on m other than O. Set $E = AB \cap A'B'$, $F = AB' \cap A'B$, $G = AB \cap A'C'$, and $H = AC' \cap A'B$. Then the lines EF, GH, l are concurrent.

State the version of this theorem that holds in the Euclidean plane in each of the following cases. Draw a figure in the Euclidean plane to illustrate each version.

(a) $A'B$ is the ideal line.
(b) l is the ideal line.
(c) $C'F$ is the ideal line.
(d) B is the only ideal point named.
(e) E is the only ideal point named.

14.6. The following result is proved in Exercises 16.23 and 17.5 (Figure 14.18).

Theorem. In the extended plane, let A, B, C, D be four points, no three of which are collinear. Set $E = AB \cap CD$, $F = AC \cap BD$, $G = AD \cap BC$, $H = EF \cap AD$, $I = EF \cap BC$, and $J = BH \cap DI$. Then E, G, J are collinear.

State the version of this theorem that holds in the Euclidean plane in each of the following cases. Draw a figure in the Euclidean plane to illustrate each version.

(a) D is the only ideal point named.
(b) E is the only ideal point named.

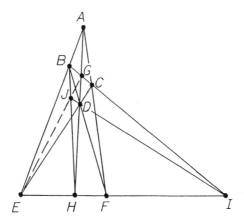

FIGURE 14.18.

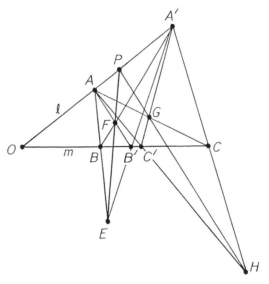

FIGURE 14.19.

(c) H is the only ideal point named.
(d) AD is the ideal time.
(e) CD is the ideal line.

14.7. The following result is proved in Exercise 17.6 (Figure 14.19).

Theorem. In the extended plane, let l and m be two lines on a point O. Let A and A' be two points on l other than O, and let B, B', C, C' be four points on m other than O. Set $E = AB \cap A'B'$, $F = AB' \cap A'B$, $G = AC \cap A'C'$, and $H = AC' \cap A'C$. Then the lines EF, GH, l are concurrent at a point P.

State the version of this theorem that holds in the Euclidean plane in each of the following cases. Draw a figure in the Euclidean plane to illustrate each version.

(a) A' is the only ideal point named.
(b) P is the only ideal point named.
(c) $A'C'$ is the ideal line.
(d) l is the ideal line.
(e) GH is the ideal line, and it doesn't contain B, B', or E.
(f) GH is the ideal line, and it contains B'.

14.8. In the extended plane, let A and B be two points (Figure 14.20). Let l and m be two lines on A other than AB, and let n and o be two lines on B other than AB.

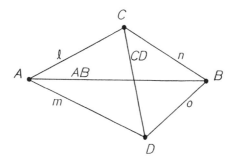

FIGURE 14.20.

(a) Prove that the points $C = l \cap n$ and $D = m \cap o$ exist and don't lie on line AB.

(b) Prove that line CD exists and that neither A nor B lies on CD.

14.9. In the extended plane, let l, m, n be three lines on a point O (Figure 14.21). Let A and B be two points on l other than O, and let C be a point on m other than O.

(a) Prove that the points $D = AC \cap n$ and $E = BD \cap m$ exist.

(b) Prove that the points O, A, B, C, D, E are distinct.

14.10. In the the extended plane, let a, b, c, d be four lines, no three of which are concurrent. Prove that the line $(a \cap b)(c \cap d)$ exists and contains none of the points $a \cap c$, $a \cap d$, $b \cap c$, $b \cap d$.

14.11. In the extended plane, let A, B, C, D, E, F be six points, no three of which are collinear. Set $G = AB \cap DE$ and $H = AC \cap DF$.

(a) Prove that the points A–H are distinct.

(b) Prove that none of the points A–F lies on line GH.

14.12. In the notation of Exercise 14.11, set $I = EB \cap FC$ and $J = ED \cap FA$.

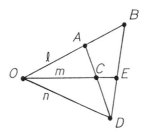

FIGURE 14.21.

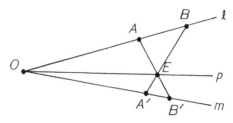

FIGURE 14.22.

(a) Prove that the points G, H, I, J are distinct.
(b) Can line IJ contain G? Can line IJ contain H? Justify your answers.

14.13. In the notation of Exercise 14.11, set $K = CF \cap BD$.

(a) Prove the points A–H, K are distinct.
(b) Which of the points A-F, H can line GK contain? Justify your answer.

14.14. In the extended plane, let l, m, p be three lines on a point O (Figure 14.22). Let A be a point on l other than O, and let A' be a point on m other than O.

(a) For any point B on l, prove that the points $E = A'B \cap p$ and $B' = AE \cap m$ exist.
(b) What points B' on m correspond to $B = A$ and $B = O$?

14.15. In the extended plane, let l and m be two lines. Let A, B, C be three points on l other than $l \cap m$, and let A' and B' be two points on m other than $l \cap m$. Set $E = AB' \cap A'B$.

(a) For any point C' on m, prove that the point $F = AC' \cap A'C$ and the line $p = EF$ exist.
(b) Prove that the map $C' \to p$ matches up the points C' on m with the lines p on E. (In other words, for every line p on E, prove that there is a unique point C' on m that maps to p.)
(c) What lines on E correspond to $C' = A'$, $C' = B'$, and $C' = l \cap m$? Justify your answers.

Section 15.

Pappus' Theorem and Projections Between Planes

We now introduce the key idea that enables us to prove Pappus' Theorem and the main results of Sections 16–22. This idea is the operation of projection between planes.

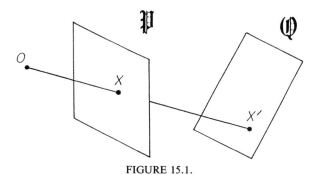

FIGURE 15.1.

We start by discussing projections informally to motivate the formal presentation later. The term *Euclidean space* means familiar three-dimensional space. In calculus, Euclidean space is called $\mathbb{R}^3$, and its points are represented as ordered triples of real numbers. We let AB denote the unique line through two points A and B in Euclidean space.

Let $\mathcal{P}$ and $\mathcal{Q}$ be two planes in Euclidean space (Figure 15.1). Let O be a point in space that doesn't lie on either plane. The projection from $\mathcal{P}$ to $\mathcal{Q}$ through O sends any point X on $\mathcal{P}$ to the point X' on $\mathcal{Q}$ where line XO intersects $\mathcal{Q}$. Conversely, any point X' on $\mathcal{Q}$ is the image of the point X on $\mathcal{P}$ where line $X'O$ intersects $\mathcal{P}$. In short, the projection maps points from $\mathcal{P}$ to $\mathcal{Q}$ along lines through O. Figure 15.2 shows several points X_i on $\mathcal{P}$ and their images X_i' on $\mathcal{Q}$. As this figure illustrates, the points O, X_i, X_i' can occur in any order. If the planes $\mathcal{P}$ and $\mathcal{Q}$ intersect in a line, each point on this line projects to itself, as shown by the fact that $X_3 = X_3'$.

We recall that two planes in Euclidean space are called *parallel* when they don't intersect. A problem arises in the previous paragraph when the planes $\mathcal{P}$ and $\mathcal{Q}$ aren't parallel, and so they intersect in a line. In this case, let $\mathcal{R}$ be the plane through O parallel to $\mathcal{Q}$ (Figure 15.3). $\mathcal{R}$ intersects $\mathcal{P}$ in a line m. If X is any point on m, the line XO lies in the plane $\mathcal{R}$, which is parallel to $\mathcal{Q}$. Thus, XO doesn't intersect $\mathcal{Q}$, and so X doesn't project to a point on $\mathcal{Q}$. We say that the points on m *vanish* under projection, and we call m the *vanishing line* on $\mathcal{P}$ of the projection. (Of course, the position of m depends on $\mathcal{Q}$ and O as well as on $\mathcal{P}$.) On the other hand, if X is any point of $\mathcal{P}$ that doesn't lie on m, then line XO intersects $\mathcal{Q}$ at a point X', and X projects to X' (Figure 15.4).

Similarly, when $\mathcal{P}$ and $\mathcal{Q}$ aren't parallel, let $\mathcal{S}$ be the plane through O parallel to $\mathcal{P}$ (Figure 15.5). $\mathcal{S}$ intersects $\mathcal{Q}$ in a line n. If X' is any point on n, the line $X'O$ lies in the plane $\mathcal{S}$, which is parallel to $\mathcal{P}$. Thus, $X'O$ doesn't intersect $\mathcal{P}$, and so no point on $\mathcal{P}$ projects to X'. We call n the *vanishing line* on $\mathcal{Q}$ for the projection. On the other hand, if X' is any point of $\mathcal{Q}$ that

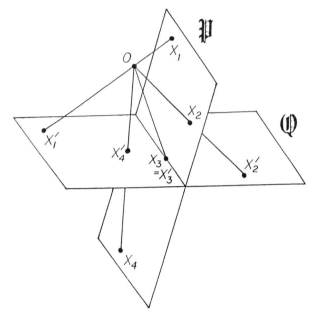

FIGURE 15.2.

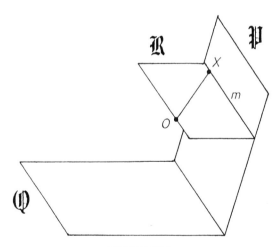

FIGURE 15.3.

doesn't lie on n, then $X'O$ intersects $\mathcal{P}$ at a point X, and X is the unique point on $\mathcal{P}$ that projects to X' (Figure 15.6).

Combining the last two paragraphs gives the following result. When the planes $\mathcal{P}$ and $\mathcal{Q}$ aren't parallel, the projection from $\mathcal{P}$ to $\mathcal{Q}$ through O

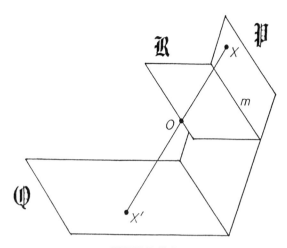

FIGURE 15.4.

matches up the points X of $\mathcal{P}$ that don't lie on the vanishing line m of $\mathcal{P}$ with the points X' of $\mathcal{Q}$ that don't lie on the vanishing line n of $\mathcal{Q}$ (Figure 15.7). The points on the vanishing line of each plane don't correspond to points on the other plane.

We can avoid treating the vanishing lines as exceptional, however. We add an ideal line to each of the planes $\mathcal{P}$ and $\mathcal{Q}$ to form extended planes $\mathcal{P}^*$ and $\mathcal{Q}^*$. We define the projection from $\mathcal{P}$ to $\mathcal{Q}$ through O so that points on the vanishing line of $\mathcal{P}^*$ project to ideal points of $\mathcal{Q}^*$, and ideal points of $\mathcal{P}^*$

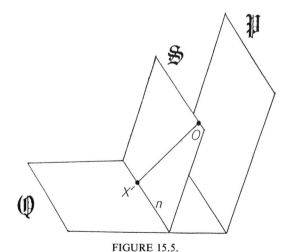

FIGURE 15.5.

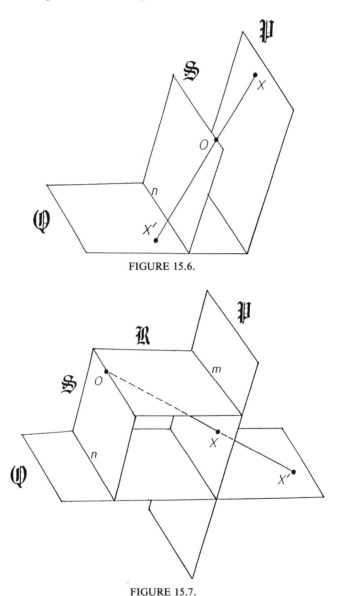

FIGURE 15.6.

FIGURE 15.7.

project to points on the vanishing line of $\mathcal{Q}^*$. In this way projection matches up the points on the extended planes $\mathcal{P}^*$ and $\mathcal{Q}^*$ without exceptions. *The points on the vanishing line of each plane correspond to the points on the ideal line of the other plane.* This shows again why it's natural to consider extended planes.

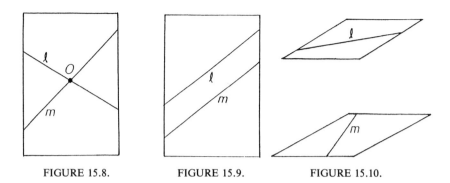

FIGURE 15.8.   FIGURE 15.9.   FIGURE 15.10.

In Figures 15.4 and 15.7, if the point X approaches the vanishing line m of $\mathcal{P}$, the image X' of X moves "off to infinity" in plane $\mathcal{Q}$. As we noted at the start of Section 14, we imagine that a point "becomes ideal" as it moves "off to infinity." This shows why it's natural to think that points on the vanishing line of $\mathcal{P}$ project to ideal points on $\mathcal{Q}$.

We call lines in Euclidean space *coplanar* if they all lie in one plane. We review the ways that two lines l and m in Euclidean space can be positioned. One possibility is that the lines intersect at a point O; in this case, the lines are coplanar (Figure 15.8). Another possibility is that the lines are coplanar but don't intersect; such lines are called *parallel* (Figure 15.9). Intuitively, two lines are parallel when they point in the same direction. The third possibility is that the lines aren't coplanar; such lines are called *skew* (Figure 15.10). Skew lines don't intersect (since intersecting lines are coplanar, as in Figure 15.8). Thus, lines in Euclidean space don't intersect when they are either parallel or skew. We also recall that a line and a plane in Euclidean space are called *parallel* when they don't intersect or when the plane contains the line.

Our goal is to formalize the definition of the projection from a plane $\mathcal{P}$ to a plane $\mathcal{Q}$ through a point O. When the planes aren't parallel, we must specify which ideal point on each plane corresponds to each point on the vanishing line of the other plane. We do so by constructing the projection in two parts. First, we map the points of the extended plane $\mathcal{P}^*$ to the lines in space on O, and then we map the lines in space on O to the points of the extended plane $\mathcal{Q}^*$. The net effect is to map the points of $\mathcal{P}^*$ to the points of $\mathcal{Q}^*$, and this map is the projection from $\mathcal{P}$ to $\mathcal{Q}$ through O. Thus, the key step in defining the projection is to match up the lines in space on a point O with the points on an extended plane that doesn't contain O.

Formally, let $\mathcal{P}$ be a plane in Euclidean space, and let O be a point in Euclidean space that doesn't lie on $\mathcal{P}$ (Figure 15.11). Map any point X of $\mathcal{P}$ to the line $l = XO$; l is the line on O that intersects $\mathcal{P}$ at X. Conversely,

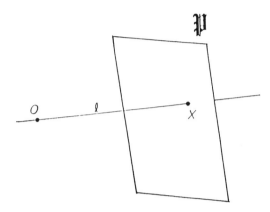

FIGURE 15.11.

if l is any line on O that intersects $\mathcal{P}$ at a point X, then X is the unique point of $\mathcal{P}$ such that $l = XO$, and so X is the unique point of $\mathcal{P}$ that maps to l. Thus, the map $X \to l = XO$ matches up the points X of $\mathcal{P}$ with the lines l on O that intersect $\mathcal{P}$. Figure 15.12 shows several points X_i of $\mathcal{P}$ and the corresponding lines l_i on O.

Form an extended plane $\mathcal{P}^*$ from $\mathcal{P}$ by adding an ideal line. Any ideal point X of $\mathcal{P}^*$ lies on a family of parallel lines in $\mathcal{P}$ (Figure 15.13). We map X to the line l on O parallel to the lines in this family. In particular, l is parallel to the plane $\mathcal{P}$. Conversely, if l is any line on O parallel to $\mathcal{P}$, there is a unique ideal point X of $\mathcal{P}^*$ that maps to l; X is the ideal point on the lines in $\mathcal{P}$ parallel to l. Thus, we've matched up the ideal points of $\mathcal{P}^*$ with the lines through O parallel to l.

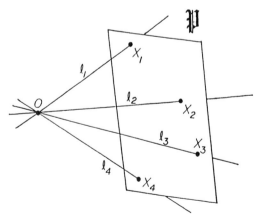

FIGURE 15.12.

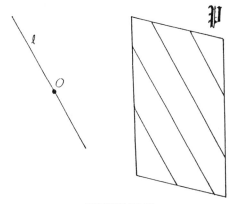

FIGURE 15.13.

The last two paragraphs match up the points of the extended plane $\mathcal{P}^*$ with the lines in Euclidean space on the point O. The ordinary points of $\mathcal{P}^*$ match up with the lines on O that intersect $\mathfrak{P}$ (Figures 15.11 and 15.12). The ideal points of $\mathcal{P}^*$ match up with the lines on O parallel to $\mathcal{P}$ (Figure 15.13). We call this matching the *elementary perspectivity* between $\mathcal{P}$ and O. We think of the elementary perspectivity as acting in either direction, either as mapping points of $\mathcal{P}^*$ to lines on O or as mapping lines on O to points of $\mathcal{P}^*$.

The key property of the elementary perspectivity between $\mathcal{P}$ and O is that it matches up collinear points of $\mathcal{P}^*$ with coplanar lines on O. For example, in Figure 15.14, the pints X_1, X_2, X_3 on the line m in $\mathcal{P}^*$ correspond to the lines l_1, l_2, l_3 on the plane $\mathcal{R}$ through O that contains m.

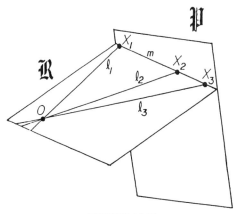

FIGURE 15.14.

THEOREM 15.1. In Euclidean space, let $\mathcal{P}$ be a plane, and let O be a point that doesn't lie on $\mathcal{P}$.

(i) Then the elementary perspectivity between $\mathcal{P}$ and O matches up the points of the extended plane $\mathcal{P}^*$ with the lines through O in Euclidean space.

(ii) The elementary perspectivity between $\mathcal{P}$ and O matches up the sets of collinear points of the extended plane $\mathcal{P}^*$ with the sets of coplanar lines through O in Euclidean space.

Proof: Part (i) holds, by the discussion accompanying Figures 15.11–15.13. We must prove part (ii).

Let m be an ordinary line of $\mathcal{P}^*$ (Figure 15.15). The ordinary points of m lie on a plane $\mathcal{R}$ through O. The elementary perspectivity maps any ordinary point X on m to the line $l = OX$, which lies in $\mathcal{R}$ and isn't parallel to $\mathcal{P}$. Conversely if l is a line on O that lies in $\mathcal{R}$ and isn't parallel to $\mathcal{P}$, then l intersects $\mathcal{P}$ in a point X; X lies on the line m where $\mathcal{R}$ intersects $\mathcal{P}$, and the elementary perspectivity maps X to l. Thus, the elementary perspectivity matches up the ordinary points of m with the lines through O that lie in $\mathcal{R}$ and aren't parallel to $\mathcal{P}$. The elementary perspectivity matches up the ideal point of m with a line l' through O parallel to $\mathcal{P}$; l' is the unique line through O that lies in $\mathcal{R}$ and is parallel to $\mathcal{P}$. The last two sentences show that the elementary perspectivity matches up the ordinary points of m and the ideal point of m with the lines through O that lie in $\mathcal{R}$.

Conversely, let $\mathcal{R}$ be any plane through O that isn't parallel to $\mathcal{P}$ (Figure 15.15). $\mathcal{R}$ intersects $\mathcal{P}$ in a line m, and the previous paragraph shows that

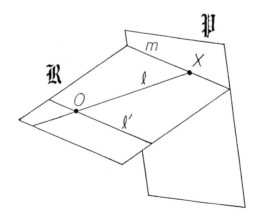

FIGURE 15.15.

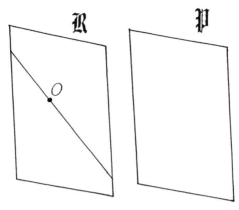

FIGURE 15.16.

the elementary perspectivity matches up the ordinary points of m and the ideal point of m with the lines through O that lie in $\mathfrak{R}$.

Next, let m be the ideal line of the extended plane $\mathcal{P}^*$, and let $\mathfrak{R}$ be the plane through O parallel to $\mathcal{P}$ (Figure 15.16). The elementary perspectivity matches up the ideal points on $\mathcal{P}^*$ with the lines through O parallel to $\mathcal{P}$ (as in the discussion accompanying Figure 15.13). In other words, the elementary perspectivity matches up the points of m with the lines through O that lie in $\mathfrak{R}$.

The three paragraphs preceding show that the elementary perspectivity matches up the sets of collinear points of $\mathcal{P}^*$ with the sets of coplanar lines through O. □

We can now formally define projections between planes. In Euclidean space, let $\mathcal{P}$ and $\mathcal{Q}$ be two planes, and let O be a point that doesn't lie on either plane. Let $\mathcal{P}^*$ and $\mathcal{Q}^*$ be the extended planes formed from $\mathcal{P}$ and $\mathcal{Q}$ by adding an ideal line to each plane. The elementary perspectivity from $\mathcal{P}$ to O maps the points of $\mathcal{P}^*$ to the lines on O, and the elementary perspectivity from O to $\mathcal{Q}$ maps the lines on O to the points of $\mathcal{Q}^*$. Combining these two maps gives a map from the points of $\mathcal{P}^*$ to the points of $\mathcal{Q}^*$. This map is called the *projection from $\mathcal{P}$ to $\mathcal{Q}$ through O*.

If X is an ordinary point on $\mathcal{P}^*$ and X' is an ordinary point on $\mathcal{Q}^*$ such that X and X' lie on a line through O in Euclidean space, then the projection from $\mathcal{P}$ to $\mathcal{Q}$ through O maps X to X' (Figure 15.1); the elementary perspectivity from $\mathcal{P}$ to O maps X to line XO, and the elementary perspectivity from O to $\mathcal{Q}$ maps line XO to X'. When the planes $\mathcal{P}$ and $\mathcal{Q}$ are parallel, a line on O intersects $\mathcal{P}$ if and only if it intersects $\mathcal{Q}$ (since $\mathcal{P}$ and $\mathcal{Q}$ are parallel to the same lines on O). Thus, the projection from $\mathcal{P}$ to $\mathcal{Q}$

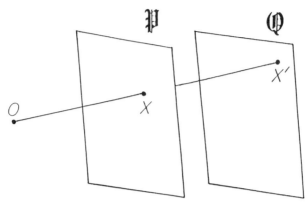

FIGURE 15.17.

through O matches up the ordinary points X on $\mathcal{P}^*$ with the ordinary points X' on $\mathcal{Q}^*$ (Figure 15.17). The projection also matches up the ideal points of $\mathcal{P}^*$ with the ideal points of $\mathcal{Q}^*$; $\mathcal{P}$ and $\mathcal{Q}$ are parallel to the same lines through O, and so these lines correspond to the ideal points of both $\mathcal{P}^*$ and $\mathcal{Q}^*$ under elementary perspectivities.

When $\mathcal{P}$ and $\mathcal{Q}$ aren't parallel, let m be the line where $\mathcal{P}$ intersects the plane $\mathcal{R}$ through O parallel to $\mathcal{Q}$, and let n be the line where $\mathcal{Q}$ intersects the plane $\mathcal{S}$ through O parallel to $\mathcal{P}$ (Figure 15.7). The projection from $\mathcal{P}$ to $\mathcal{Q}$ through O maps the points of m to the ideal points of $\mathcal{Q}^*$: The elementary perspectivity from $\mathcal{P}$ to O maps the points of m, both ordinary and ideal, to the lines through O parallel to $\mathcal{Q}$, and these lines map to the ideal points of $\mathcal{Q}^*$ under the elementary perspectivity from O to $\mathcal{Q}$ (Figure 15.3). The projection also maps the ideal points of $\mathcal{P}^*$ to the points of n: The elementary perspectivity from $\mathcal{P}$ to O maps the ideal points of $\mathcal{P}^*$ to the lines through O parallel to $\mathcal{P}$, and these lines map to the points of n, both ordinary and ideal, under the elementary perspectivity from O to $\mathcal{Q}$ (Figure 15.5). The ordinary points of $\mathcal{P}^*$ that don't lie on m project to the ordinary points of $\mathcal{Q}^*$ that don't lie on n (Figure 15.7). As before, we call m and n the *vanishing lines* of the projection. We've seen that the points on the vanishing line of each plane correspond under projection to the points on the ideal line of the other plane.

Every projection between two planes consists of two elementary perspectivities. Thus, we obtain the following result on projections by two applications of Theorem 15.1.

THEOREM 15.2. In Euclidean space, let $\mathcal{P}$ and $\mathcal{Q}$ be two planes, and let O be a point that doesn't lie on either plane. Let $\mathcal{P}^*$ and $\mathcal{Q}^*$ be the extended planes formed from $\mathcal{P}$ and $\mathcal{Q}$.

(i) The projection from $\mathcal{P}$ to $\mathcal{Q}$ through O matches up the points of $\mathcal{P}^*$ with the points of $\mathcal{Q}^*$.

(ii) The projection from $\mathcal{P}$ to $\mathcal{Q}$ through O matches up the sets of collinear points of $\mathcal{P}^*$ with the sets of collinear points of $\mathcal{Q}^*$.

Proof:

(i) The elementary perspectivity from $\mathcal{P}$ to O matches up the points of $\mathcal{P}^*$ with the lines on O (by Theorem 15.1i). The elementary perspectivity from O to $\mathcal{Q}$ matches up the lines on O with the points of $\mathcal{Q}^*$ (by Theorem 15.1i). Combining these two elementary perspectivities gives the projection from $\mathcal{P}$ to $\mathcal{Q}$ through O. Thus, the projection matches up the points of $\mathcal{P}^*$ with the points of $\mathcal{Q}^*$.

(ii) The elementary perspectivity from $\mathcal{P}$ to O matches up the sets of collinear points of $\mathcal{P}^*$ with the sets of coplanar lines through O (by Theorem 15.1ii). The elementary perspectivity from O to $\mathcal{Q}$ matches up the sets of coplanar lines through O with the sets of collinear points of $\mathcal{Q}^*$. Combining these two elementary perspectivities gives the projection from $\mathcal{P}$ to $\mathcal{Q}$ through O. Thus, the projection matches up the sets of collinear points of $\mathcal{P}^*$ with the sets of collinear points of $\mathcal{Q}^*$. $\square$

Figure 15.18 illustrates the conclusion in Theorem 15.2ii that collinear points in one plane project to collinear points in another. In this figure, n is a line on $\mathcal{P}$, $\mathcal{I}$ is the plane through n and O, and n' is the line where $\mathcal{I}$ intersects $\mathcal{Q}$. If X is any point on n, then the line XO lies in the plane $\mathcal{I}$. X projects to the point X' where XO intersects $\mathcal{Q}$, and X' lies on the line n' where $\mathcal{I}$ intersects $\mathcal{Q}$. Thus, the points of n project to the points of n', and so collinear points of $\mathcal{P}$ map to collinear points of $\mathcal{Q}$. Of course, this informal discussion neglects the possibility that one or more of the points

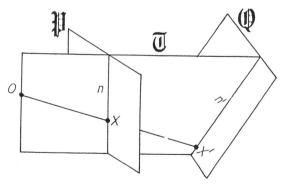

FIGURE 15.18.

X and X' and the lines n and n' may be ideal, possibilities that are included in Theorem 15.2ii.

Theorem 15.2 shows that *projections preserve points and collinearity*. In other words, *projections preserve incidence*. Thus, any statement about incidence in the extended plane is unaffected by projection between planes. Once again, we give up something to gain something; by considering only statements about incidence in the extended plane, we are free to project from one plane to another.

What can we gain by projecting between planes? Assume that the planes $\mathcal{P}$ and $\mathcal{Q}$ aren't parallel. Let m be the vanishing line of $\mathcal{P}$, and let l_1 and l_2 be two lines of $\mathcal{P}$ other than m that intersect at a point of m (Figure 15.19). l_1 and l_2 project to the lines l_1' and l_2' where $\mathcal{Q}$ intersects the planes through O and each of the lines l_1 and l_2 (as in the discussion accompanying Figure 15.18). The points on the vanishing line m of $\mathcal{P}$ project to the ideal line of the extended plane $\mathcal{Q}^*$, and projection preserves incidence. Thus, l_1' and l_2' intersect at an ideal point of $\mathcal{Q}^*$, and so l_1' and l_2' are parallel in the Euclidean plane $\mathcal{Q}$. In fact, as Figure 15.19 shows, l_1' and l_2' are both parallel to the line through O and the point of intersection of l_1 and l_2. Thus, we project lines that intersect on the vanishing line of one plane to parallel lines in the other plane. In this way we can project a figure in one plane to another figure that may be easier to analyse. Figure 15.19 shows again why it's natural to imagine that parallel lines intersect at an ideal point; the point on the

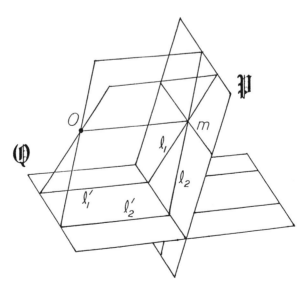

FIGURE 15.19.

vanishing line of $\mathcal{P}$ where l_1 and l_2 intersect projects to the ideal point where the parallel lines l_1' and l_2' intersect.

The next result shows that we can make any line in a Euclidean plane the vanishing line of a projection. Thus, we can project any ordinary line in one plane to the ideal line in another plane. Hence, the incidence properties of the ideal line in the extended plane are the same as those of any ordinary line (since projections preserve incidence, by Theorem 15.2). Although the ideal line differs from ordinary lines in the construction of the extended plane, there is no difference with respect to incidence properties.

THEOREM 15.3. Any line in a Euclidean plane can be projected to the ideal line in another plane.

Proof: Let m be a line in a Euclidean plane $\mathcal{P}$. We can think of $\mathcal{P}$ as lying in Euclidean space (Figure 15.3). Let $\mathcal{R}$ be a plane in Euclidean space other than $\mathcal{P}$ that contains m. Let O be a point of $\mathcal{R}$ that doesn't lie on m. Let $\mathcal{Q}$ be a plane in Euclidean space that is parallel to $\mathcal{R}$ and not equal to it. Then m is the vanishing line of the projection from $\mathcal{P}$ and $\mathcal{Q}$ through O (as in the discussion before Theorem 15.2). Thus, m projects to the ideal line on $\mathcal{Q}^*$. $\square$

We can now prove Pappus' Theorem (Figure 15.20). Since this is a theorem about incidence in the extended plane, we can project from one plane to another (by Theorem 15.2). Thus, we can assume that any line in the statement of Pappus' Theorem is ideal (by Theorem 15.3). We choose the line EF in Pappus' Theorem to be ideal. Then AB' and $A'B$ are parallel lines (since $E = AB' \cap A'B$ is ideal), and AC' and $A'C$ are parallel lines

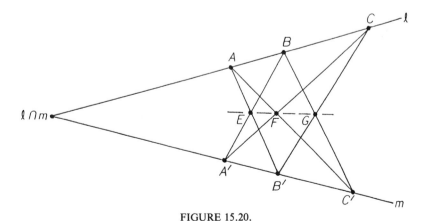

FIGURE 15.20.

(since $F = AC' \cap A'C$ is ideal). In this case, Pappus' Theorem follows directly from the results on division ratios and parallel lines in Section 1.

THEOREM 15.4 (Pappus' Theorem). In the extended plane, let l and m be two lines. Let A, B, C be three points on l other than $l \cap m$, and let A', B', C' be three points on m other than $l \cap m$. Then the points $E = AB' \cap A'B$, $F = AC' \cap A'C$, and $G = BC' \cap B'C$ are collinear (Figure 15.20).

Proof: The points E, F, G and the line EF exist, by Theorem 14.3 and symmetry. We consider two cases, depending on whether or not line EF is ideal.

Case 1: EF is the ideal line. The points A, B, C, A', B', C' are ordinary because they don't lie on the ideal line EF (by Theorem 14.3iii). l and m are ordinary lines (since they don't equal the ideal line EF, by Theorem 14.3ii). Since $E = AB' \cap A'B$ and $F = AC' \cap A'C$ are ideal points, AB' and $A'B$ are parallel lines, and AC' and $A'C$ are parallel lines. The conclusion of Pappus' Theorem is that $G = BC' \cap B'C$ lies on the ideal line EF. This is equivalent to the statement that G is ideal, which is equivalent to the statement that BC' and $B'C$ are parallel lines. The ideal line EF may or may not contain $l \cap m$, and we consider these two possibilities separately.

Subcase 1a: $l \cap m$ does not lie on the ideal line EF. Then l and m intersect at an ordinary point O, and the Euclidean version of Pappus' Theorem in this subcase is as follows (Figure 15.21):

In the Euclidean plane, let l and m be two lines that intersect at a point O. Let A, B, C be three points on l other than O, and let A', B', C' be three points on m other than O. Assume that the lines AB' and $A'B$ are parallel and that the lines AC' and $A'C$ are parallel. Then the lines BC' and $B'C$ are parallel.

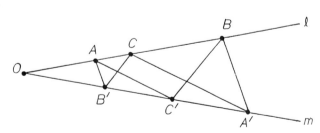

FIGURE 15.21.

To prove this version of Pappus' Theorem, note that

$$\overline{OB}/\overline{OA} = \overline{OA'}/\overline{OB'}$$

(by Theorem 1.11, since $A'B$ and AB' are parallel lines) and that

$$\overline{OA}/\overline{OC} = \overline{OC'}/\overline{OA'}$$

(by Theorem 1.11, since AC' and $A'C$ are parallel lines). It follows that

$$\frac{\overline{OB}}{\overline{OC}} = \frac{\overline{OB}}{\overline{OA}} \cdot \frac{\overline{OA}}{\overline{OC}} = \frac{\overline{OA'}}{\overline{OB'}} \cdot \frac{\overline{OC'}}{\overline{OA'}} = \frac{\overline{OC'}}{\overline{OB'}}.$$

Thus, BC' and $B'C$ are parallel lines (by Theorem 1.11), and we've proved Pappus' Theorem in this subcase.

Subcase 1b: $l \cap m$ lies on the ideal line EF. Then l and m are parallel in the Euclidean plane (since they intersect at an ideal point). The Euclidean version of Pappus' Theorem in this subcase is as follows (Figure 15.22):

In the Euclidean plane, let l and m be two parallel lines. Let A, B, C be three points on l, and let A', B', C' be three points on m. Assume that the lines AB' and $A'B$ are parallel and that the lines AC' and $A'C$ are parallel. Then the lines BC' and $B'C$ are parallel.

To prove this version of Pappus' Theorem, choose corresponding positive ends for directed distances on the parallel lines l and m (as in Definition 1.15). Then we have

$$\overline{BA} = \overline{A'B'} \quad \text{and} \quad \overline{AC} = \overline{C'A'}$$

(by Theorem 1.16, since $A'B$ and AB' are parallel lines, and since AC' and $A'C$ are parallel lines). Combining these equations with Theorem 1.10 shows that

$$\overline{BC} = \overline{BA} + \overline{AC} = \overline{A'B'} + \overline{C'A'} = \overline{C'B'}.$$

Thus, BC' and $B'C$ are parallel lines (by Theorem 1.16), and we've proved Pappus' Theorem in this subcase.

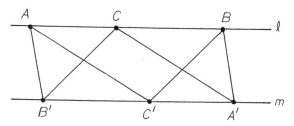

FIGURE 15.22.

Combining the two subcases completes Case 1.

Case 2: *EF* is an ordinary line. We can project to another plane so that *EF* maps to the ideal line (by Theorem 15.3). Projections preserve incidence (by Theorem 15.2), and Pappus' Theorem holds when *EF* is the ideal line (by Case 1). Combining the last two sentences shows that Pappus' Theorem holds when *EF* is an ordinary line, completing Case 2.

The two cases include all possibilities and show that Pappus' Theorem always holds. □

If A, B, C, A', B', C' are six distinct points such that the six lines

$$AB', \quad B'C, \quad CA', \quad A'B, \quad BC', \quad C'A$$

are distinct, we call the figure formed by the six points and the six lines *hexagon AB'CA'BC'* (Figure 15.23). The six points are called the *vertices* of the hexagon, and the six lines are called the *sides* of the hexagon. The six lines are formed by taking each pair of consecutive points when the vertices are listed in the order $AB'CA'BC'$, if we imagine that C' and A are also consecutive. The six solid lines in Figure 15.20 other than *l* and *m* are the six sides of hexagon $AB'CA'BC'$.

If we travel around hexagon $AB'CA'BC'$ in Figure 15.20, the vertices alternate between the points A, B, C on *l* and the points A', B', C' on *m*. Thus, the hypothesis of Pappus' Theorem is that the vertices of hexagon $AB'CA'BC'$ lie alternately on *l* and *m*, avoiding $l \cap m$. We describe this situation by saying that hexagon $AB'CA'BC'$ is *inscribed* in the lines *l* and *m*.

As Figure 15.23 suggests, the six sides of hexagon $AB'CA'BC'$ can be grouped into three pairs of *opposite sides* $\{AB', A'B\}$, $\{BC', B'C\}$, and $\{CA', C'A\}$. The three pairs of opposite sides intersect in the points

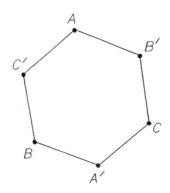

FIGURE 15.23.

$AB' \cap A'B = E$, $BC' \cap B'C = G$, and $CA' \cap C'A = F$ in Pappus' Theorem. Thus, we can state Pappus' Theorem as follows: *If a hexagon is inscribed in two lines, its three pairs of opposite sides intersect in collinear points.*

In general, whenever one proves a theorem, one should look for ways to apply it in different situations. The next example illustrates an application of Pappus' Theorem. This example also illustrates the need to check that a theorem's hypotheses are satisfied before applying the theorem.

EXAMPLE 15.5. In the extended plane, let l and m be two lines. Let A, B, C be three points on l other than $l \cap m$, and let A', B', C' be three points on m other than $l \cap m$. Set $E = AB' \cap A'B$, $G = BC' \cap B'C$, $H = C'A \cap B'C$, and $I = C'A \cap A'B$. Prove that the points $A, A'H \cap EG$, $GI \cap m$ are collinear (Figure 15.24).

Solution: The points E, G–I exist, by Theorem 14.3ii. In order to apply Pappus' Theorem, we want to find a hexagon whose three pairs of alternate sides intersect at the points $A, A'H \cap EG, GI \cap m$. The only solid lines in Figure 15.24 on $A'H \cap EG$ are $A'H$ and EG, and so we take these to be opposite sides of the hexagon. (We don't yet know that the dotted line in Figure 15.24 exists.) A' and H are the only two named points on line $A'H$, and E and G are the only two named points on line EG, and so we take these four points to be the vertices determining the opposite sides $A'H$ and EG of the hexagon. (The point $A'H \cap EG$ can't be a vertex of the hexagon, since opposite sides don't intersect at a vertex when a hexagon is inscribed in two lines, by Theorem 14.3ii.) Thus, since we can start naming the hexagon with side $A'H$, the hexagon has either the form $A'H\_\_EG\_\_$ or the form $A'H\_\_GE\_\_$. We want the hexagon to be inscribed in two lines, and so the two sets of alternate vertices are to be collinear. Thus, since A' and G are the only named points on line $A'G$, there is no natural point to fill the first blank in $A'H\_\_EG\_\_$. Hence, we consider the form $A'H\_\_GE\_\_$.

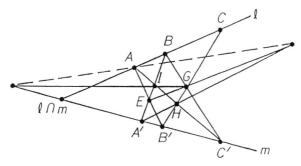

FIGURE 15.24.

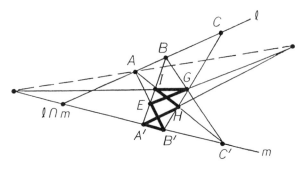

FIGURE 15.25.

The only solid lines in Figure 15.24 on $GI \cap m$ are GI and m, and so we want these to be opposite sides of the hexagon. G and I are the only points named in Figure 15.24 that lie on line GI, and so we want these to be the vertices on side GI. Thus, we replace the first blank by I and consider the form $A'HIGE\_\_$. The last vertex must be on line GH (since alternate vertices of the hexagon are to be collinear) and on line m (since GI and m are to be opposite sides of the hexagon). Thus, the last vertex is $GH \cap m = B'$.

In short, we want to consider hexagon $A'HIGEB'$. Figure 15.25 suggests that this is a hexagon that is inscribed in the lines $A'B$ and $B'C$ and whose three pairs of opposite sides intersect at the points $A'H \cap GE$, $HI \cap EB' = A$, and $IG \cap B'A' = GI \cap m$. If we verify these facts, then Pappus' Theorem shows that these three points are collinear, as desired.

We must check that the hypotheses of Pappus' Theorem 15.4 are satisfied, i.e., that hexagon $A'HIGEB'$ is inscribed in the lines $A'B$ and $B'C$. None of the points E, I, G, H lies on l or m (by Theorem 14.3ii); it follows that neither E nor I equals A', neither G nor H equals B', E doesn't equal I (since $AE = AB' \neq AC' = AI$, by Theorem 14.3i), and G doesn't equal H (since $GC' = BC' \neq AC' = HC'$, by Theorem 14.3i). Thus, A', E, I are three distinct points on line $A'B$, and B', G, H are three distinct points on line $B'C$. $A'B$ and $B'C$ are distinct lines, by Theorem 14.3i.

We must check that none of the points A', E, I, B', G, H equals $A'B \cap B'C$. In fact, $A'B \cap B'C$ doesn't lie on m (by Theorem 14.3ii), and so it doesn't equal A' or B'. Since $B'E = B'A \neq B'C$ (by Theorem 14.3i and ii), E doesn't lie on $B'C$, and so E doesn't equal $A'B \cap B'C$. Since $GB = C'B \neq A'B$ (by Theorem 14.3i and ii), G doesn't lie on $A'B$, and so G doesn't equal $A'B \cap B'C$. H and I equal $A'B \cap B'C$ when the lines $A'B, B'C, C'A$ are concurrent, however (since $H = C'A \cap B'C$ and $I = C'A \cap A'B$) (Figure 15.26).

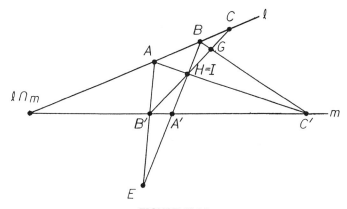

FIGURE 15.26.

First, assume that $A'B$, $B'C$, $C'A$ are not concurrent (Figure 15.25). Then $H = C'A \cap B'C$ doesn't lie on $A'B$, and $I = C'A \cap A'B$ doesn't lie on $B'C$, and so neither H nor I equals $A'B \cap B'C$. Together with the two previous paragraphs, this shows that the hypotheses of Pappus' Theorem 15.4 are satisfied, i.e., that hexagon $A'HIGEB'$ is inscribed in the lines $A'B$ and $B'C$. Thus, Pappus' Theorem shows that the opposite sides of the hexagon intersect in three collinear points. Hence, the points $A'H \cap GE$, $HI \cap EB' = A$, and $IG \cap A'B' = GI \cap m$ are collinear, as desired.

Next, assume that the lines $A'B$, $B'C$, $C'A$ are concurrent (Figure 15.26). Then we have

$$H = C'A \cap B'C = C'A \cap A'B = I,$$

and so hexagon $A'HIGEB'$ doesn't exist, and we can't apply Pappus' Theorem. In this case, however, we have

$$GI \cap m = B'C \cap m = B'$$

and

$$A'H \cap EG = A'B \cap EG = E.$$

These points are collinear with A, since E lies on AB'. Thus, the theorem holds in this case as well, even though Pappus' Theorem doesn't apply. $\square$

We end this section by emphasizing why we could use projection between planes to prove Pappus' Theorem: Pappus' Theorem is a result about incidence in the extended plane, and projections between planes preserve incidence (by Theorem 15.2). On the other hand, we cannot generally use projections between planes to prove theorems about distances and angles, since projections don't always preserve distances and angles. For example, the projection in Figure 15.2 maps points X_1 and X_2 on $\mathcal{P}$ to points X_1' and

X_2' on Q that are much farther apart, and so projections don't always preserve distances. $\angle X_1 X_4 X_2$ in this figure projects to $\angle X_1' X_4' X_2'$, which is much larger, and so projections don't always preserve angles.

The term "projective geometry" has different meanings in different contexts. For our purposes, *projective geometry* is the study of those properties of the extended plane that are preserved by projections between planes. By the previous paragraph, projective geometry includes all theorems about incidence in the extended plane, but it does not include most theorems about distances and angles. We will see in Section 22 that every projection between two planes maps the conic sections in the first plane to the conic sections in the second; thus, projective geometry includes the study of conic sections. Any ordinary line in one extended plane can be projected to the ideal line of another extended plane (by Theorem 15.3); thus, from the point of view of projective geometry, the ordinary lines and the ideal line in the extended plane are indistinguishable. The *real projective plane* is the standard name for the extended plane when it is viewed in terms of projective geometry, which means that only properties preserved by projections are considered. The word "real" in this term emphasizes the fact that the extended plane is formed from the Euclidean plane, where points can be represented as ordered pairs of real numbers.

EXERCISES

In Exercises 15.1–15.7, prove the given statements and illustrate them with figures. These exercises use the following notation (Figure 15.27). *In the extended plane, let l and m be two lines. Let A, B, C be three points on l other than $l \cap m$, and let A', B', C' be three points on m other than $l \cap m$. Set $E = AB' \cap A'B$, $F = AC' \cap A'C$, $G = BC' \cap B'C$, $H = BC' \cap A'C$, $I = AC' \cap B'C$, and $J = AC' \cap A'B$.*

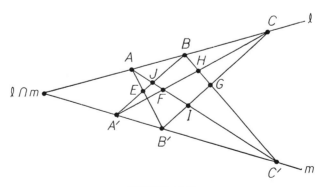

FIGURE 15.27.

15.1. The points $AA' \cap BC'$, $BB' \cap CA'$, $CC' \cap AB'$ are collinear.

15.2. The points $m \cap FG$, $A'I \cap BC'$, $AC' \cap B'H$ are collinear.

15.3. The points F, $l \cap EI$, $BC' \cap A'I$ are collinear.

15.4. The points $AC' \cap A'G$, $BA' \cap B'F$, $CB' \cap C'E$ are collinear.

15.5. The points A, $m \cap CJ$, $A'I \cap CE$ are collinear.

15.6. The lines AA', BB', HI are concurrent. (*Hint:* One possible approach is to prove that the points $AA' \cap BB'$, H, I are collinear.)

15.7. The lines $(l \cap m)F$, BI, $B'H$ are concurrent. (See the hint to Exercise 15.6.)

In Exercises 15.8–15.10, prove the given statements and illustrate them with figures. These exercises use the following notation (Figure 15.28). In the extended plane, let k, l, m be three lines on a point O. Let A and A' be two points on k other than O, let B and B' be two points on l other than O, and let C and C' be two points on m other than O. Assume that A, B, C are noncollinear and that A', B', C' are noncollinear. Set $H = AC \cap A'B'$, $L = AC \cap A'C'$, and $M = AB \cap A'B'$.

15.8. The points L, $AA' \cap B'C$, $B'C' \cap OH$ are collinear.

15.9. The points $BB' \cap CM$, $AC \cap B'C'$, $OH \cap C'M$ are collinear.

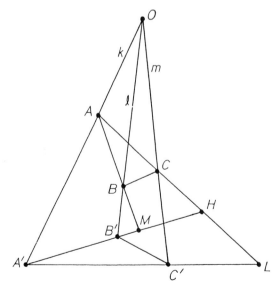

FIGURE 15.28.

15.10. The points $AA' \cap B'C$, $AB \cap B'L$, $A'C' \cap CM$ are collinear.

15.11. Prove the following theorem and illustrate it with a figure.

Theorem. In the extended plane, let D, E, F be three noncollinear points. Let A, B, C be three points on line DE other than D and E, let A', B', C' be three points on line EF other than E and F, and let A'', B'', C'' be three points on line DF other than D and F. Assume that AA', BB', CC' are concurrent at a point P, that $A'A''$, $B'B''$, $C'C''$ are concurrent at a point Q, and that D, P, Q are collinear. Then AA'', BB'', CC'' are concurrent at a point R.

(*Hint:* One possible approach is to prove that each of the lines AA'', BB'', CC'' contains the point $PF \cap QE$.)

15.12. In the Euclidean plane, let $ABCD$ be a parallelogram (Figure 15.29). Let E be a point on AB other than A and B, and let F be a point on CD other than C and D. Assume that AF intersects DE be at a point G, BF intersects CE at a point H, GH intersects AD at a point I, and GH intersects BC at a point J. Prove that the distance from A to I equals the distance from C to J.

(*Hint:* One possible approach is to use Pappus' Theorem to prove that GH contains the center of the parallelogram.)

15.13. In Euclidean space, let $\mathcal{P}$ and $\mathcal{Q}$ be two nonparallel planes, and let O be a point that doesn't lie on either plane. Let l be the line where $\mathcal{P}$ and $\mathcal{Q}$ intersect, and let X be the ideal point in the extended plane $\mathcal{P}^*$ on the lines in $\mathcal{P}$ parallel to l. What is the image of X under the projection from $\mathcal{P}$ to $\mathcal{Q}$ through O? Be specific, and explain your answer.

15.14. In Euclidean space, let $\mathcal{P}$ and $\mathcal{Q}$ be two parallel planes, and let O be a point that doesn't lie on either plane. Let X be an ideal point on the extended plane $\mathcal{P}^*$. What point on the extended plane $\mathcal{Q}^*$ is the image of X under the projection from $\mathcal{P}$ to $\mathcal{Q}$ through O? Be specific, and explain your answer.

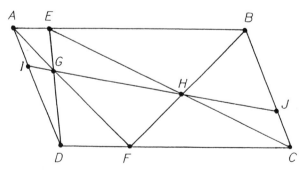

FIGURE 15.29.

Exercises 15.15–15.22 use the following terminology. In the extended plane, a *collineation* ϕ is a map that sends each point X to a point $\phi(X)$ such that the following properties hold:

(i) For every point X', there is a unique point X such that $\phi(X) = X'$; and

(ii) Points X_1, X_2, X_3 are collinear if and only if their images $\phi(X_1)$, $\phi(X_2)$, $\phi(X_3)$ are collinear.

We say that a collineation ϕ *fixes a point* X if $\phi(X) = X$. We say that ϕ *fixes a line* m if ϕ maps the points of m among themselves. The *identity map* is the collineation that fixes every point.

A *central collineation* is a collineation ϕ such that ϕ fixes all points on a line l, fixes all lines on a point G, and isn't the identity map. We call l the *axis* of ϕ, and we call G the *center* of ϕ. We note that l may or may not contain G.

15.15. In Euclidean space, let $\mathcal{P}$ and $\mathcal{Q}$ be two planes that intersect in a line l. Let N and O be two points that don't lie on $\mathcal{P}$ or $\mathcal{Q}$, and assume that line NO intersects $\mathcal{P}$ at a point G. When the projection from $\mathcal{P}$ to $\mathcal{Q}$ through N is followed by the projection from $\mathcal{Q}$ to $\mathcal{P}$ through O, prove that the net result is a central collineation of the extended plane $\mathcal{P}^*$ with axis l and center G.

15.16. In the extended plane, let ϕ and ψ be two central collineations that have the same axis l and center G. Assume further that $\phi(A) = \psi(A)$ for some point A that doesn't equal G or lie on l. Prove that $\phi = \psi$.

(*Hint:* One possible approach is as follows. For any point X that doesn't lie on GA or l, prove that $\phi(X)$ and $\psi(X)$ both lie on GX and the line through the points $AX \cap l$ and $\phi(A) = \psi(A)$.)

15.17. In the extended plane, let l be a line, and let G, A, A' be three collinear points such that neither A nor A' lies on l. Prove that there is a unique central collineation that has axis l and center G and maps A to A'.

(*Hint:* One possible approach is to combine Exercise 15.15 with projection between planes and Exercise 15.16.)

15.18. In Euclidean space, let $\mathcal{P}$ be a plane. Let $\mathcal{P}^*$ be the extended plane formed from $\mathcal{P}$, and let ϕ be a map from $\mathcal{P}^*$ to itself. Prove that the following conditions are equivalent.

(i) ϕ is a central collineation.

(ii) In Euclidean space, there is a plane $\mathcal{Q} \neq \mathcal{P}$ and there are two points N and O that don't lie on either $\mathcal{P}$ or $\mathcal{Q}$ such that ϕ consists of the projection from $\mathcal{P}$ to $\mathcal{Q}$ through N followed by the projection from $\mathcal{Q}$ to $\mathcal{P}$ through O.

(*Hint:* One possible approach is to combine Exercise 15.17 with Exercise 15.15 extended to include cases where $\mathcal{P}$ is parallel to $\mathcal{Q}$ or line *NO*.)

15.19. In the extended plane, let A, B, C be three collinear points, and let A', B', C' also be three collinear points. Prove that there is a sequence of central collineations whose combined effect is to map A to A', B to B', and C and C'. (See Exercise 15.17.)

15.20. In the extended plane, let A, B, C, D be four points, no three of which are collinear. Let A', B', C', D' also be four points, no three of which are collinear. Prove that there is a sequence of central collineations whose combined effect is to map A to A', B to B', C to C', and D to D'. (See Exercise 15.19.)

15.21. In the extended plane, let ϕ be a collineation that fixes all points on a line and is not the identity map. Prove that ϕ is a central collineation.

(*Hint:* One possible approach is to prove that ϕ fixes the lines $X\phi(X)$ for all points X that aren't fixed by ϕ. Deduce that all these lines $X\phi(X)$ lie on one point G.)

15.22. In the extended plan, let ϕ be a collineation that fixes all lines on a point G and is not the identity map. Prove that ϕ is a central collineation.

Section 16.

Desargues' Theorem and Duality

We used projections between planes to prove Pappus' Theorem in the last section. We now use the same method to prove another important result in projective geometry, Desargues' Theorem. We start by introducing the Duality Principle, which lets us take a theorem about incidence in the extended plane and obtain another theorem.

Given any statement about incidence in the extended plane, the *dual* statement is obtained by interchanging points and lines while preserving incidence. For example, the statement "*A* and *B* are points on a line *l*" (Figure 16.1a) has dual "*a* and *b* are lines on a point *L*" (Figure 16.1b). If we describe the same situations in different words, we see that the statement

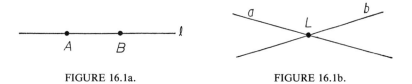

FIGURE 16.1a.   FIGURE 16.1b.

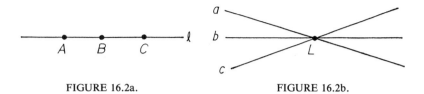

FIGURE 16.2a.   FIGURE 16.2b.

"*l* is the line *AB*" has dual "*L* is the point *a* ∩ *b*." The statement "*A*, *B*, *C* are collinear points" has dual "*a*, *b*, *c* are concurrent lines," since the first statement means that there is a line *l* on the points *A*, *B*, *C* (Figure 16.2a), and the second statement means that there is a point *L* on the lines *a*, *b*, *c* (Figure 16.2b).

We know that the statement "*A* and *B* are points on a line *l*" has dual "*a* and *b* are lines on a point *L*" (Figures 16.1a and b). If we interchange points and lines in the second statement, we return to the first. In general, if statement 2 is the dual of statement 1, then statement 1 is the dual of statement 2. Each statement comes from the other by interchanging points and lines. One interchange of points and lines takes us from statement 1 to statement 2, and a second interchange of points and lines takes us back to statement 1. Thus, statements about incidence in the extended plane occur in pairs, where each statement in a pair is the dual of the other.

Given any true statement about incidence in the extended plane, the dual statement is also true. This remarkable fact is called the *Duality Principle*, and we prove it shortly. Once we prove a theorem about incidence in the extended plane, the Duality Principle guarantees that the dual is also true. For example, if we apply the Duality Principle to Pappus' Theorem 15.4, we obtain the next result without additional work.

THEOREM 16.1. In the extended plane, let *L* and *M* be two points. Let *a*, *b*, *c* be three lines on *L* other than *LM*, and let *a'*, *b'*, *c'* be three lines on *M* other than *LM*. Then the lines $e = (a \cap b')(a' \cap b)$, $f = (a \cap c')(a' \cap c)$, and $g = (b \cap c')(b' \cap c)$ are concurrent (Figure 16.3). □

The points *E*, *F*, *G* in Pappus' Theorem 15.4 exist and are distinct, by Theorem 14.3. Thus, the lines *e*, *f*, *g* in Theorem 16.1 exist and are distinct, by the Duality Principle.

We prove the Duality Principle by proving that we can interchange the points and lines of the extended plane while preserving incidence. In Euclidean space, let 𝒫 be a plane, and let *O* be a point that doesn't lie on 𝒫 (Figure 15.14). Theorem 15.1 states that the elementary perspectivity between 𝒫 and *O* matches up the points of the extended plane 𝒫* with the lines in space through *O* in such a way that points of 𝒫* lie on a line *m* if

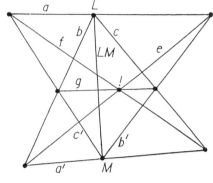

FIGURE 16.3.

and only if the corresponding lines through O lie on a plane $\mathcal{R}$. In this way the elementary perspectivity matches up the lines m of $\mathcal{P}^*$ with the planes $\mathcal{R}$ through O. Thus, we can restate Theorem 15.1 as follows.

THEOREM 16.2. In Euclidean space, let $\mathcal{P}$ be a plane, and let O be a point that doesn't lie on $\mathcal{P}$. Then the elementary perspectivity between $\mathcal{P}$ and O matches up the points of the extended plane $\mathcal{P}^*$ with the lines in space through O and it matches up the lines of the extended plane $\mathcal{P}^*$ with the planes in space through O. Moreover, if the elementary perspectivity maps a point X of $\mathcal{P}^*$ to a line l through O, and if it maps a line m of $\mathcal{P}^*$ to a plane $\mathcal{R}$ through O, then X lies on m if and only if l lies on $\mathcal{R}$ (Figure 15.15). $\square$

In the notation of Theorem 16.2, incidence is the relation that specifies which points of $\mathcal{P}^*$ lie on each line of $\mathcal{P}^*$. Similarly, we use the term *incidence* for the relation that specifies which lines in space through O lie on each plane in space through O. Thus, Theorem 16.2 shows that *elementary perspectivities preserve incidence*; that is, we can match up the points and lines of $\mathcal{P}^*$ with the lines and planes in space through O in such a way that incidence is preserved. The Duality Principle holds if we show that we can interchange the points and lines of $\mathcal{P}^*$ while preserving incidence. Thus, by the last two sentences, the Duality Principle follows if we prove that we can interchange the lines and planes in space through O while preserving incidence.

For any line l through the point O in Euclidean space, let $l^{\perp}$ be the plane through O perpendicular to l (Figure 16.4). Conversely, any plane through O has the form $l^{\perp}$ for a unique line l through O: l is the line through O perpendicular to the plane. Thus, the map $l \to l^{\perp}$ matches up the lines

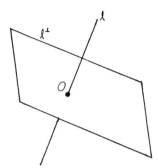

FIGURE 16.4.

through O with the planes through O. Accordingly, we can interchange the lines and planes through O as follows: We send each line l through O to the plane $l^\perp$, and we send the plane $m^\perp$ to the line m for each line m through O. The next result shows that this interchange of lines and planes through O preserves incidence.

THEOREM 16.3. In Euclidean space, let l and m be lines on a point O. Then l lies on $m^\perp$ if and only if m lies on $l^\perp$ (Figure 16.5).

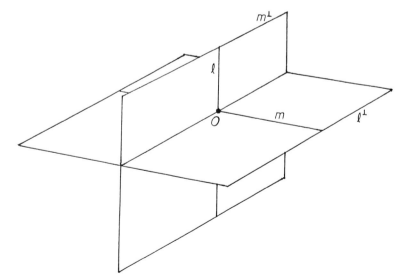

FIGURE 16.5.

Proof: *l* lies on $m^\perp$ if and only if *l* and *m* are perpendicular. Likewise, *m* lies on $l^\perp$ if and only if *l* and *m* are perpendicular. Thus, *l* lies on $m^\perp$ if and only if *m* lies on $l^\perp$. □

In Euclidean space, let $\mathcal{P}$ be a plane, and let *O* be a point that doesn't lie on $\mathcal{P}$. By Theorem 16.2, we can match up the points and lines of the extended plane $\mathcal{P}^*$ with the lines and planes through *O* in Euclidean space so that incidence is preserved. By Theorem 16.3, we can interchange the lines and planes through *O* in Euclidean space while preserving incidence. The last two sentences imply that we can interchange the points and lines of $\mathcal{P}^*$ while preserving incidence. Thus, the Duality Principle holds.

THEOREM 16.4 (Duality Principle). Given any true statement about incidence in the extended plane, the dual statement is also true. □

Theorem 16.2 provides further insight into the extended plane. In Euclidean space, let *O* be a point. Two lines through *O* lie on a unique plane through *O* (Figure 16.6), and two planes through *O* lie on a unique line through *O* (Figure 16.7). By Theorem 16.2, we can match up the lines and planes through *O* with the points and lines of an extended plane $\mathcal{P}^*$ so that incidence is preserved. The last two sentences imply that in $\mathcal{P}^*$ two points lie on a unique line, and two lines lie on a unique point. Thus, we have new proofs of Theorems 14.1 and 14.2. In fact, Theorems 14.1 and 14.2 are duals; thus, we could prove either one of them and deduce the other from the Duality Principle. We also note that any two planes in Euclidean space

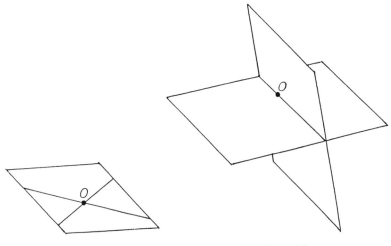

FIGURE 16.6.   FIGURE 16.7.

through O have the same properties of incidence with lines through O; thus, Theorem 16.2 provides another justification for the statement in the last paragraph of Section 15 that any two lines of the extended plane have the same properties of incidence with points.

Having proved the Duality Principle, we now proceed with Desargues' Theorem on perspective triangles. We start by introducing the necessary terminology.

In the extended plane, a *triangle ABC* consists of three noncollinear points A, B, C and the lines $a = BC$, $b = AC$, and $c = AB$ (Figure 16.8). The lines a, b, c are distinct because A, B, C are noncollinear. We have $A = b \cap c$, $B = a \cap c$, and $C = a \cap b$, and so the lines a, b, c are non-concurrent. The equations in the last sentence show that the triangle is determined by the lines a, b, c, and so we refer to Figure 16.8 as triangle abc as well as triangle ABC.

The dual of triangle ABC consists of three nonconcurrent lines a, b, c and the points $A = b \cap c$, $B = a \cap c$, and $C = a \cap b$ in the extended plane. Dualizing the results of the last paragraph shows that the points A, B, C are distinct and noncollinear. Thus, Figure 16.8 illustrates this situation as well, and the dual of a triangle is again a triangle.

We say that the triangle in Figure 16.8 has *vertices* A, B, C and *sides* a, b, c. We call a vertex and a side of a triangle *opposite* if they don't lie on each other; thus, in Figure 16.8, vertex A and side $a = BC$ are opposite, vertex B and side $b = AC$ are opposite, and vertex C and side $c = AB$ are opposite. The dual of a vertex of a triangle is a side of a triangle, by the previous paragraph. Note that, in the extended plane, a side of a triangle is a line rather than a segment.

DEFINITION 16.5. In the extended plane, triangles ABC and $A'B'C'$ are *perspective fr·m a point O* if the following conditions are satisfied (Figure 16.9):

(i) The points $A\,B, C, A', B', C', O$ are distinct.
(ii) The points A, B, C are noncollinear, and the points A', B', C' are noncollinear.

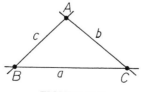

FIGURE 16.8.

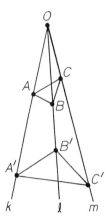

FIGURE 16.9.

(iii) The points O, A, A' lie on a line k, the points O, B, B' lie on a line l, and the points O, C, C' lie on a line m.

(iv) The lines k, l, m are distinct. □

When conditions (i)–(iv) of Definition 16.5 hold, we say that $\{A, A'\}$, $\{B, B'\}$, $\{C, C'\}$ are three pairs of *corresponding points*. Basically, two triangles are perspective from a point if their vertices can be paired so that corresponding vertices lie on three concurrent lines.

Next we state the dual of Definition 16.5.

DEFINITION 16.6. In the extended plane, triangles abc and $a'b'c'$ are *perspective from a line o* if the following conditions are satisfied (Figure 16.10):

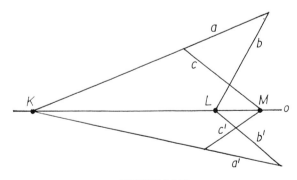

FIGURE 16.10.

(i) The lines a, b, c, a', b', c', o are distinct.

(ii) The lines a, b, c are nonconcurrent, and the lines a', b', c' are nonconcurrent.

(iii) The lines o, a, a' lie on a point K, the lines o, b, b' lie on a point L, and the lines o, c, c' lie on a point M.

(iv) The points K, L, M are distinct.   □

When conditions (i)–(iv) of Definition 16.6 hold, we say that $\{a, a'\}$, $\{b, b'\}$, $\{c, c'\}$ are three pairs of *corresponding lines*. Basically, two triangles are perspective from a line if their sides can be paired so that corresponding sides intersect in three collinear points.

Desargues' Theorem states that two triangles are perspective from a point if and only if they're perspective from a line. The next result presents technical conditions satisfied by triangles perspective from a point. This result is a first step toward proving Desargues' Theorem, just as Theorem 14.3 was a first step toward proving Pappus' Theorem.

THEOREM 16.7. In the extended plane, let triangles ABC and $A'B'C'$ be perspective from a point O. Let the lines k, l, m be as in Definition 16.5. Set $a = BC$, $b = AC$, $c = AB$, $a' = B'C'$, $b' = A'C'$, and $c' = A'B'$ (Figure 16.11).

(i) Then the points $K = a \cap a'$, $L = b \cap b'$, $M = c \cap c'$ exist and are distinct.

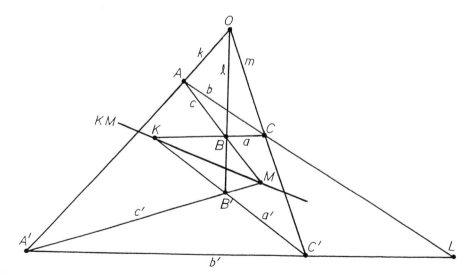

FIGURE 16.11.

(ii) Neither l nor m contains K, neither k nor m contains L, and neither k nor l contains M.

(iii) The line KM exists and doesn't contain any of the points A, B, C, A', B', C'.

(iv) The lines a, b, c, a', b', c', KM are distinct.

Proof: (i) The points K, L, M exist, by Theorem 14.3(ii) and Definition 16.5. We have $a \cap b = C \neq C' = a' \cap b'$ (by Definition 16.5i and ii), and so the lines a, b, a', b' aren't concurrent. It follows that $K = a \cap a' \neq b \cap b' = L$. Thus, the points K, L, M are distinct, by symmetry.

Part (ii) holds, by Theorem 14.3(ii) and Definition 16.5.

(iii) The line KM exists, by part (i). The lines KB, MB, MA exist, by part (ii). Thus, we have $KB = BC \neq AB = MB$ (by Definition 16.5ii), and so KM doesn't contain B. It follows that $KM \neq MB = MA$, and so KM doesn't contain A. The last two sentences show that none of the points A, B, C, A', B', C' lies on KM, by symmetry.

(iv) The lines a, b, c are distinct, and the lines a', b', c' are distinct (by Definition 16.5ii). We have $a \cap m = C \neq C' = a' \cap m = b' \cap m$ (by Theorem 14.3i and Definition 16.5), and so we have $a \neq a'$ and $a \neq b'$. Thus, none of the lines a, b, c equals any of the lines a', b', c', by symmetry. Part (iii) implies that KM doesn't equal any of the lines a, b, c, a', b', c'. Thus, we've shown that the lines a, b, c, a', b', c', KM are all distinct. $\square$

We can now prove half of Desargues' Theorem.

THEOREM 16.8. In the extended plane, if two triangles are perspective from a point, then they are perspective from a line. Corresponding sides are opposite corresponding vertices (Figure 16.12).

In the notation of Figure 16.12, Theorem 16.8 states that, if triangles ABC and $A'B'C'$ are perspective from a point O, then they are perspective from a line o. Thus, the main hypothesis of the theorem is that the lines $k = AA'$, $l = BB'$, $m = CC'$ are concurrent at a point O, and the main conclusion is that the points $K = a \cap a'$, $L = b \cap b'$, $M = c \cap c'$ lie on a line o. In other words, the main conclusion is that the points $K = BC \cap B'C'$, $L = AC \cap A'C'$, $M = AB \cap A'B'$ are collinear.

Proof: We label points and lines as in Definition 16.5 and Theorem 16.7 (Figure 16.11). The points K, L, M and the line KM exist (by Theorem 16.7i and iii). We consider two cases, depending on whether or not line KM is ideal.

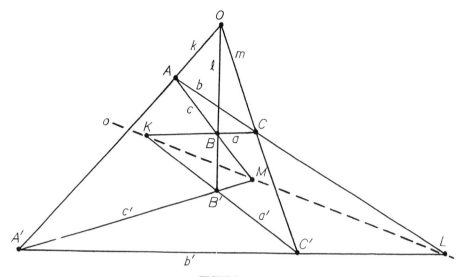

FIGURE 16.12.

Case 1: KM is the ideal line. The points A, B, C, A', B', C' are ordinary because they don't lie on the ideal line KM (by Theorem 16.7iii). Since $K = BC \cap B'C'$ and $M = AB \cap A'B'$ are ideal points, BC and $B'C'$ are parallel lines, and AB and $A'B'$ are parallel lines. The main conclusion of Theorem 16.8 is that $L = AC \cap A'C'$ lies on the ideal line KM. This is equivalent to the statement that L is ideal, which is in turn equivalent to the statement that AC and $A'C'$ are parallel lines. The ideal line KM may or may not contain O, and we consider these two possibilities separately in proving that AC and $A'C'$ are parallel.

Subcase 1a: O does not lie on the ideal line KM. Then O is an ordinary point that lies on each of the lines k, l, m. Thus, we must prove the following result in this subcase (Figure 16.13):

In the Euclidean plane, let k, l, m be three lines on a point O. Let A and A' be two points on k other than O, let B and B' be two points on l other than O, and let C and C' be two points on m other than O. Assume that the points A, B, C are noncollinear and that the points A', B', C' are noncollinear. Assume that the lines AB and $A'B'$ are parallel and that the lines BC and $B'C'$ are parallel. Then the lines AC and $A'C'$ are parallel.

To prove this result, note that

$$\overline{OA}/\overline{OA'} = \overline{OB}/\overline{OB'}$$

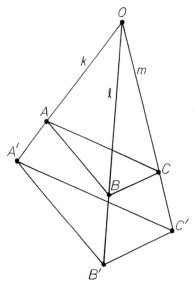

FIGURE 16.13.

(by Theorem 1.11, since AB and $A'B'$ are parallel lines) and that

$$\overline{OB}/\overline{OB'} = \overline{OC}/\overline{OC'}$$

(by Theorem 1.11, since BC and $B'C'$ are parallel lines). Thus we have

$$\overline{OA}/\overline{OA'} = \overline{OC}/\overline{OC'},$$

and so AC and $A'C'$ are parallel lines (by Theorem 1.11), as desired.

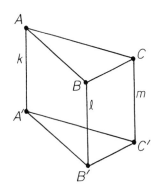

FIGURE 16.14.

Subcase 1b: O lies on the ideal line KM. Then k, l, m are parallel lines (since they lie on the same ideal point O). Thus, we must prove the following result in this subcase (Figure 16.14):

In the Euclidean plane, let k, l, m be three parallel lines. Let A and A' be two points on k, let B and B' be two points on l, and let C and C' be two points on m. Assume that the points A, B, C are noncollinear and that the points A', B', C' are noncollinear. Assume that the lines AB and $A'B'$ are parallel and that the lines BC and $B'C'$ are parallel. Then the lines AC and $A'C'$ are parallel.

To prove this result, choose corresponding positive ends for directed distances on the parallel lines k, l, m (as in Definition 1.15). Then we have

$$\overline{AA'} = \overline{BB'} \quad \text{and} \quad \overline{BB'} = \overline{CC'}$$

(by Theorem 1.16, since AB and $A'B'$ are parallel lines, and since BC and $B'C'$ are parallel lines). Thus, we have

$$\overline{AA'} = \overline{CC'},$$

and so AC and $A'C'$ are parallel lines (by Theorem 1.16), as desired.

The two subcases show that, if KM is the ideal line o, then o also contains L (Figure 16.12). Thus, the conditions of Definition 16.6 hold (by Theorem 16.7i and iv and the discussion accompanying Figure 16.8). Hence, triangles abc and $a'b'c'$ are perspective from the line o. Corresponding sides lie opposite corresponding vertices: a and a' are opposite A and A', b and b' are opposite B and B', and c and c' are opposite C and C'. Thus, we've proved that the theorem holds when KM is the ideal line.

Case 2: KM is an ordinary line. We can project to another plane so that KM maps to the ideal line (by Theorem 15.3). Projections preserve incidence (by Theorem 15.2), and the theorem holds when KM is the ideal line (by Case 1). The last two sentences show that the theorem holds when KM is an ordinary line.

Cases 1 and 2 include all possibilities and show that the theorem always holds. □

Triangles perspective from a line are dual to triangles perspective from a point. Thus, if we apply the Duality Principle (Theorem 16.4) to Theorem 16.8, we obtain the following result.

THEOREM 16.9. In the extended plane, if two triangles are perspective from a line, then they are perspective from a point. Corresponding vertices are opposite corresponding sides. □

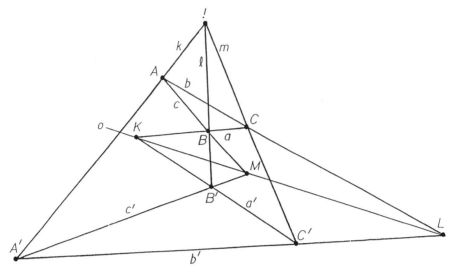

FIGURE 16.15.

This theorem is illustrated by Figure 16.15. In the notation of the figure, the main hypothesis of the theorem is that the points K, L, M are collinear, and the main conclusion of the theorem is that the lines k, l, m are concurrent. This theorem is the converse of Theorem 16.8. (Theorem 16.8 has the unusual property that its dual is its converse.) If we combine Theorem 16.8 and its converse Theorem 16.9 into a single statement, we obtain Desargues' Theorem.

THEOREM 16.10 (Desargues' Theorem). In the extended plane, two triangles are perspective from a point if and only if they are perspective from a line. Corresponding sides are opposite corresponding vertices (Figures 16.12 and 16.15). □

Just as Example 15.5 illustrates how to apply Pappus' Theorem, the next example illustrates how to apply Desargues' Theorem. The exercises contain many other applications of Desargues' Theorem. Like Example 15.5, the next example also illustrates the general principle that it's necessary to check that the hypotheses of a theorem hold before applying the theorem.

EXAMPLE 16.11. In the extended plane, let triangles ABC and $A'B'C'$ be perspective from a point O. Set $L = AC \cap A'C'$, $H = BC \cap AA'$, $I = OB \cap HC'$, and $J = AB \cap HL$. Prove that A', I, J are collinear (Figure 16.16).

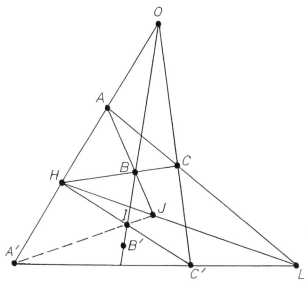

FIGURE 16.16.

Solution: If two triangles are perspective from a point, Desargues' Theorem shows that the triangles are perspective from a line, and so corresponding sides intersect in collinear points (Figure 16.12). Thus, Desargues' Theorem shows that A', I, J are collinear if there are two triangles that are perspective from a point and whose corresponding sides intersect at A', I, J. In Figure 16.16, A' lies only on the solid lines AA' and $A'C'$, I lies only on the solid lines OB and HC', and J lies only on the solid lines AB and HL. (We don't yet know that the dotted line in Figure 16.16 exists. We've also simplified Figure 16.16 by omitting the lines $A'B'$, $B'C'$, and KM and the points K and M in Figure 16.12, because these lines and points aren't mentioned in the example.) Thus, we must choose one line from each of the pairs

$$\{AA', A'C'\}, \qquad \{OB, HC'\}, \qquad \{AB, HL\} \qquad (1)$$

to form a triangle in such a way that the three remaining sides form a triangle and the two triangles are perspective from a point.

To prove that the triangles formed are perspective from a point, we would like their vertices to be points already named, and so we would like the pairs of sides of each triangle to intersect at points in Figure 16.16. Thus, we don't choose $A'C'$ to form a triangle with either OB or AB because the points $A'C' \cap OB$ and $A'C' \cap AB$ aren't named in Figure 16.16. Hence, we choose $A'C'$, HC', HL to be the sides of a triangle. Intersecting

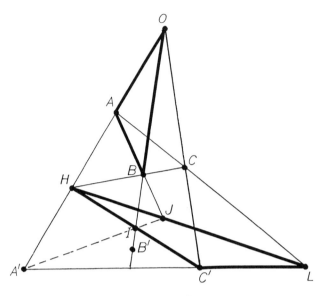

FIGURE 16.17.

these sides in pairs gives vertices $A'C' \cap HC' = C'$, $A'C' \cap HL = L$, and $HC' \cap HL = H$, and so we consider triangle $C'LH$ (Figure 16.17). The three remaining lines AA', OB, AB in (1) form the sides of another triangle. Intersecting these sides in pairs gives vertices $AA' \cap OB = O$, $AA' \cap AB = A$, and $OB \cap AB = B$, and so we consider triangle OAB. Now, (1) contains the three pairs of corresponding sides of triangles OAB and $C'LH$. These triangles have corresponding vertices $\{O, C'\}, \{A, L\}, \{B, H\}$, which are collinear with C. This suggests that triangles OAB and $C'LH$ are perspective from C.

First, assume that C', L, H are noncollinear (as shown in Figure 16.17). To prove formally that triangles OAB and $C'LH$ are perspective from C, we must check that the conditions in Definition 16.5 hold. Accordingly, we must check that the points O, A, B, C', L, H, C are distinct (so Definition 16.5i holds), that O, A, B are noncollinear and C', L, H are noncollinear (so Definition 16.5ii holds), that the three pairs of points $\{O, C'\}, \{A, L\}, \{B, H\}$ are collinear with C (so Definition 16.5iii holds), and that the lines OC, AC, BC are distinct (so Definition 16.5iv holds). The fact that all of these conditions hold follows from the given fact that triangles ABC and $A'B'C'$ are perspective from O, Definition 16.5, Theorem 16.7, and the assumption that C', L, H are noncollinear. Thus, triangles OAB and $C'LH$ are perspective from C, and so Desargues' Theorem shows that corresponding sides of these triangles intersect at collinear points. These points are

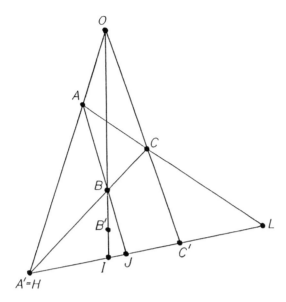

FIGURE 16.18.

$OA \cap C'L = A'$, $OB \cap C'H = I$, and $AB \cap LH = J$. Thus, we've proved that A', I, J are collinear, as desired, when C', L, H are noncollinear.

On the other hand, if C', L, H are collinear, we can't talk about "triangle $C'LH$" or apply Desargues' Theorem (Figure 16.18). In this case, we have

$$HL = HC' = C'L = A'C' \tag{2}$$

(since L lies on $A'C'$). H lies on $C'L$, I lies on HC', and J lies on HL, and so the equations in (2) show that A', I, J all lie on $A'C'$. Thus, A', I, J are collinear in this case as well, even though Desargues' Theorem doesn't apply. (As Figure 16.18 shows, H equals A' in this case, since (2) shows that H lies on $A'C'$, since H lies on AA', and since $A'C'$ and AA' intersect at A'.) $\square$

EXERCISES

16.1. State the duals of the theorems in the following exercises.

(a) Exercise 14.1
(b) Exercise 14.2
(c) Exercise 14.3
(d) Exercise 14.4

(e) Exercise 14.5
(f) Exercise 14.6
(g) Exercise 14.7
(h) Exercise 15.11

16.2. Draw figures to illustrate the duals of the theorems in exercises listed in Exercise 16.1.

16.3. The following result is proved in Exercise 16.24.

Theorem. In the extended plane, let A, B, C, A', B', C' be six points, no three of which are collinear. If the points $AB \cap A'B'$, $AC \cap A'C'$, $BC \cap B'C'$ are collinear, then the points $AB' \cap A'B$, $AC' \cap A'C$, $BC \cap B'C'$ are collinear.

(a) State the dual of this theorem.
(b) Draw a figure to illustrate the given theorem.
(c) Draw a figure to illustrate the dual of the given theorem.

16.4. The following result is proved in Exercise 16.25.

Theorem. In the extended plane, let l, m, n be three lines on a point O. Let A, B, C be three points on l other than O, let A', B', C' be three points on m other than O, and let A'', B'', C'' be three points on n other than O. If AA', BB', CC' are concurrent at a point P, and if AA'', BB'', CC'' are concurrent at a point Q, then $A'A'', B'B'', C'C''$ are concurrent at a point R. Moreover, the points P, Q, R are collinear.

(a) State the dual of this theorem.
(b) Draw a figure to illustrate the given theorem.
(c) Draw a figure to illustrate the dual of the given theorem.

16.5. The following result is proved in Exercise 16.26.

Theorem. In the extended plane, let l, m, n be three lines on a point O. Let A, B, C be three points on l other than O, let A', B', C' be three points on m other than O, and let A'', B'', C'' be three points on n other than O. If the points $AB' \cap A'B$ and $BC' \cap B'C$ are collinear with O, and if the points $AB'' \cap A''B$ and $BC'' \cap B''C$ are collinear with O, then the points $A'B'' \cap A''B'$ and $B'C'' \cap B''C'$ are collinear with O.

(a) State the dual of this theorem.
(b) Draw a figure to illustrate the given theorem.
(c) Draw a figure to illustrate the dual of the given theorem.

16.6. The following result is proved in Exercise 16.27.

Theorem. In the extended plane, let A, B, C, D be four points, no three of which are collinear. Let l be a line on A that doesn't contain any of the points B, C, D, and let m be a line on B that doesn't contain any of the points A, C, D. Then the points $l \cap BD$, $AB \cap CD$, $m \cap AC$ are collinear if and only if the points $l \cap m$, $AD \cap BC$, $AC \cap BD$ are collinear.

(a) State the dual of this theorem.
(b) Draw a figure to illustrate the given theorem.
(c) Draw a figure to illustrate the dual of the given theorem.

16.7. The following result is proved in Exercise 17.7.

Theorem. In the extended plane, let l, m, n be three lines on a point O. Let A and A' be two points on l other than O, let B and B' be two points on m other than O, and let C and C' be two points on n other than O. Set $E = AB \cap A'B'$, $F = AB' \cap A'B$, $G = AC \cap A'C'$, and $H = AC' \cap A'C$. Then the lines EF, GH, l are concurrent.

(a) State the dual of this theorem.
(b) Draw a figure to illustrate the given theorem.
(c) Draw a figure to illustrate the dual of the given theorem.

In Exercises 16.8–16.17, prove the given statements and illustrate them with figures. Exercises 16.8–16.15 are applications of Desargues' Theorem, and Exercises 16.16 and 16.17 are applications of Pappus' Theorem that use Desargues' Theorem. These exercises use the following notation (Figure 16.19): In the extended plane, let triangles ABC and $A'B'C'$ be perspective from a point O. Set $K = BC \cap B'C'$, $L = AC \cap A'C'$, $M = AB \cap A'B'$, $G = AA' \cap BC$, and $H = AC \cap A'B'$.

16.8. L, $AB' \cap A'B$, $BC' \cap B'C$ are collinear.

16.9. C, $AB \cap OH$, $C'H \cap KM$ are collinear.

16.10. L, $OK \cap B'C$, $OM \cap AB'$ are collinear.

16.11. $OA \cap KM$, $OL \cap AC'$, $OK \cap GC'$ are collinear.

16.12. $OA \cap CM$, $OM \cap AB'$, $OC \cap GB'$ are collinear.

16.13. O, $AK \cap CM$, $A'K \cap C'M$ are collinear.

16.14. C, $C'M \cap KH$, $OM \cap BH$ are collinear.

16.15. B', $AA' \cap C'H$, $AB \cap KH$ are collinear.

16.16. C, $AB \cap A'K$, $A'C' \cap GM$ are collinear.

16.17. $AB \cap OL$, $A'B' \cap OK$, $AK \cap A'C'$, $AB' \cap OC'$ are collinear.

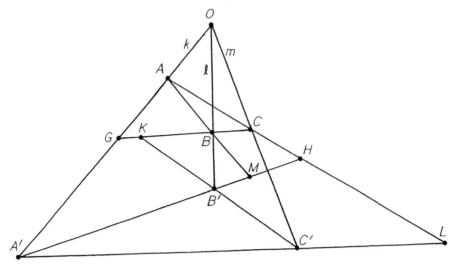

FIGURE 16.19.

In Exercises 16.18–16.27, prove the theorems in the exercises listed.

16.18. Exerise 14.1.

16.19. Exercise 14.2. (The theorems in Exercises 14.1 and 14.2 show that the line through the points E, F, G in Pappus' Theorem 15.4 contains $l \cap m$ if and only if the lines AA', BB', CC' are concurrent.)

16.20. Exercise 14.3. (*Hint:* One possible way to prove that A', B', C' are noncollinear is to use Ceva's Theorem, Menelaus' Theorem, and projection between planes.)

16.21. Exercise 14.4.

16.22. Exercise 14.5.

16.23. Exercise 14.6.

16.24. Exercise 16.3.

16.25. Exercise 16.4.

16.26. Exercise 16.5. (One possible approach is to use the theorems in Exercises 16.4, 14.1, and 14.2.)

16.27. Exercise 16.6.

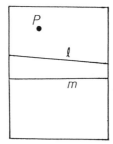

FIGURE 16.20.

16.28. Let l and m be two lines drawn on a piece of paper (Figure 16.20). Suppose that l and m aren't parallel, but that their point of intersection lies beyond the edge of the paper. Let P be a point that doesn't lie on either line. Suppose that you have an unmarked straightedge and a pencil with which to draw on the paper, but that you can't work beyond the edge of the paper. Use Desargues' Theorem to describe how to draw the line through P that would be concurrent with l and m if the lines were extended beyond the edge of the paper.

16.29. Prove the following result, and illustrate it with a figure.

Theorem. In the extended plane, let A, B, C, A', B', C' be six points. If triangles ABC and $A'B'C'$ are perspective from a point O, and if triangles ABC and $B'C'A'$ are perspective from a point P, then triangles ABC and $C'A'B'$ are perspective from a point Q.

(*Hint:* The points $A, B, C, A', B', C', O, P, Q$ can be renamed to be the nine points other than $l \cap m$ in Pappus' Theorem.)

16.30. State the dual of the theorem in Exercise 16.29. Draw a figure to illustrate the result you state.

16.31. Use Pappus' Theorem to prove its dual without using the Duality Principle. It is a special property of Pappus' Theorem that this is possible. (*Hint:* The hypotheses of the dual of Pappus' Theorem are illustrated by Figure 16.3, where we don't yet know that the lines e, f, g are concurrent. Find a way to apply Pappus' Theorem to this figure in order to prove that e, f, g are concurrent.)

16.32. Use Theorem 16.8 to prove Theorem 16.9 without using the Duality Principle.

Section 17.

Harmonic Sets

Projections between planes don't generally preserve midpoints. For example, in Figure 17.1, the projection from the plane $\mathcal{P}$ to the plane $\mathcal{Q}$ through the point O maps the points A, B, C on $\mathcal{P}$ to the points A', B', C' on $\mathcal{Q}$, where C is the midpoint of A and B, but C' is not the midpoint of A' and B'. Thus, projective geometry does not include the study of midpoints. Projective geometry does include the study of harmonic sets, however, a generalization of midpoints that involves four collinear points, rather than three.

In order to define harmonic sets in terms of incidence, we start by considering complete quadrangles.

DEFINITION 17.1. In the extended plane, a *complete quadrangle* consists of four points, no three of which are collinear, and the six lines through pairs of these points. The four points are called *vertices*, and the six lines are called *sides*. Two sides are called *opposite* if they don't lie on a common vertex. $\square$

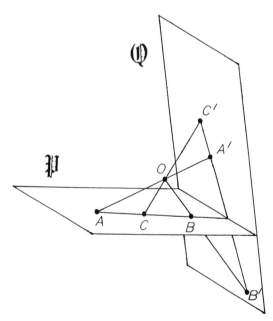

FIGURE 17.1.

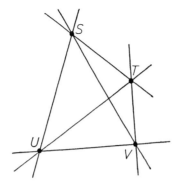

FIGURE 17.2.

Figure 17.2 illustrates Definition 17.1. The four vertices of the complete quadrangle in Figure 17.2 are labeled S, T, U, V. Thus, S, T, U, V are four points, no three of which are collinear. The six sides are the lines ST, SU, SV, TU, TV, UV through pairs of vertices. Since no three vertices are collinear, each side contains exactly two vertices, the six sides are distinct, and there are three pairs of opposite sides $\{ST, UV\}$, $\{SU, TV\}$, $\{SV, TU\}$. We name a complete quadrangle by its vertices, and so the complete quadrangle $STUV$ is shown in Figure 17.2.

We use complete quadrangles to define harmonic sets.

DEFINITION 17.2. In the extended plane, collinear points A, B, C, D form a *harmonic set $A, B; C, D$* if there is a complete quadrangle such that two opposite sides contain A, two other opposite sides contain B, a fifth side contains C, and the sixth side contains D. (See Figure 17.3, where the open dots are the vertices of the complete quadrangle that shows that $A, B; C, D$ is a harmonic set.) □

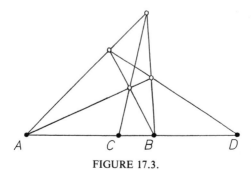

FIGURE 17.3.

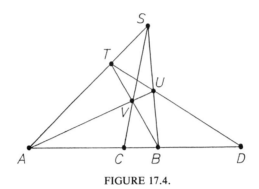

FIGURE 17.4.

Consider any complete quadrangle that gives a harmonic set $A, B; C, D$ (Figure 17.3). Let S and V be the two vertices that lie on the side containing C. Call the two remaining vertices T and U. The two pairs of opposite sides that don't include side SV are $\{ST, UV\}$ and $\{SU, TV\}$. The lines in one of these pairs contain A, and the lines in the other pair contain B. By switching the labels of T and U if necessary, we can ensure that the vertices have been labeled S, T, U, V so that the following four conditions hold (Figure 17.4):

    (i) the opposite sides ST and UV contain A;
    (ii) the opposite sides SU and TV contain B;
    (iii) the fifth side SV contains C; and
    (iv) the sixth side TU contains D.

When these four conditions hold, we say that the complete quadrangle *STUV determines* the harmonic set $A, B; C, D$. Given any complete quadrangle associated with a harmonic set $A, B; C, D$ (Figure 17.3), we've shown that the vertices can be labeled S, T, U, V so that conditions (i)–(iv) hold and the complete quadrangle *STUV* determines the harmonic set $A, B; C, D$ (Figure 17.4).

The points A and B play a symmetric role in Definition 17.2 of a harmonic set $A, B; C, D$, and so do the points C and D (Figure 17.3). Thus, if any one of the following sets is harmonic, they all are:

$$A, B; C, D \qquad B, A; C, D \qquad A, B; D, C \qquad B, A; D, C. \qquad (1)$$

Harmonic sets are defined in terms of incidence, and projections between planes preserve incidence (by Theorem 15.2). Thus, projections between planes preserve harmonic sets, a fact we record below for later reference.

THEOREM 17.3. Consider a projection from a plane $\mathcal{P}$ to a plane $\mathcal{Q}$ in Euclidean space. Let $\mathcal{P}^*$ and $\mathcal{Q}^*$ be the extended planes formed from $\mathcal{P}$ and

Q. Let A, B, C, D be points of $\mathcal{P}^*$, and let A', B', C', D' be their images on Q^* under the projection. Then $A, B; C, D$ is a harmonic set in $\mathcal{P}^*$ if and only if $A', B'; C', D'$ is a harmonic set in Q^*. $\square$

Given three collinear points A, B, C, the next result shows how to construct a point D such that $A, B; C, D$ is a harmonic set.

THEOREM 17.4. In the extended plane, let A, B, C be three collinear points. Let S and V be two points that are collinear with C and don't lie on line AB. Then the points $T = AS \cap BV$, $U = AV \cap BS$, and $D = TU \cap AB$ exist, and $A, B; C, D$ is a harmonic set determined by the complete quadrangle $STUV$ (Figure 17.4).

Proof: By assumption, the points A, B, C, S, V are distinct, and AB and SV are distinct lines that intersect at C. Thus, as in Figure 17.4, the points $T = AS \cap BV$ and $U = AV \cap BS$ exist and don't lie on either AB or SV (by Theorem 14.3ii). Then T and U don't equal S, and so we have

$$ST \cap AB = AS \cap AB = A \neq B = BS \cap AB = SU \cap AB \qquad (2)$$

(by Theorem 14.3i), which shows that $T \neq U$. The last two sentences imply that line TU exists and doesn't equal AB, and so the point $D = TU \cap AB$ exists.

We claim that S, T, U, V are four points, no three of which are collinear. We have $S \neq V$, and we've seen that T and U are distinct points that don't lie on SV, and so S, T, U, V are distinct points, S, T, V are noncollinear, and S, U, V are noncollinear. Equation 2 shows that $ST \neq SU$, and so S, T, U are noncollinear. Then T, U, V are also noncollinear, by symmetry. Thus, S, T, U, V are four points, no three of which are collinear, and so they are the vertices of a complete quadrangle. The opposite sides ST and UV contain A, two other opposite sides SU and TV contain B, the fifth side SV contains C, and the sixth side TU contains D. Thus, the complete quadrangle $STUV$ determines the harmonic set $A, B; C, D$. $\square$

In the extended plane, let A, B, C be three collinear points. By the previous theorem, there is at least one point D such that $A, B; C, D$ is a harmonic set. In fact, the point D is unique, as we see in Theorem 17.7. Thus, if a complete quadrangle $STUV$ determines a harmonic set $A, B; C, D$, and if another complete quadrangle $S'T'U'V'$ determines a harmonic set $A, B; C, D'$, then we have $D = D'$ (Figure 17.5). Hence, in the notation of Theorem 17.4, the position of the point D does not depend on the positions of the points S and V: If we replace S and V with two other points S' and V' that are collinear with C and don't lie on AB, and if we

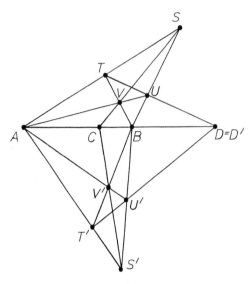

FIGURE 17.5.

set $T' = AS' \cap BV'$, $U' = AV' \cap BS'$, and $D' = T'U' \cap AB$, then D' equals the point D in Theorem 17.4 (as Figure 17.5 shows).

We start with a technical result. If A, B; C, D is a harmonic set, we do not *assume* in Definition 17.2 that the points A, B, C, D are distinct. We *prove*, however, in Theorem 17.7(ii) that these points are distinct. The next result is the first step.

THEOREM 17.5. In the extended plane, let A, B; C, D be a harmonic set determined by a complete quadrangle $STUV$ (Figure 17.6).

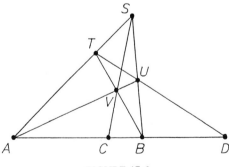

FIGURE 17.6.

(i) Then we have $A \neq B$, and line AB doesn't contain any of the points S, T, U, V.

(ii) Neither C nor D equals either A or B.

Proof:

(i) Since no three of the points S, T, U, V are collinear, A and B don't equal any of the points S, T, U, V. Then we have $AS = ST \neq SU = BS$, which shows that $A \neq B$ and that line AB doesn't contain S. By symmetry, AB also doesn't contain V. We have $AT = ST \neq AB$ (since we've seen that $A \neq T$ and that AB doesn't contain S), and so AB doesn't contain T. Then AB doesn't contain U either, by symmetry.

(ii) C doesn't lie on ST (since C lies on SV and doesn't equal $S = ST \cap SV$, by part (i)). Thus, C doesn't equal A (since ST contains A but not C). By symmetry, neither C nor D equals either A or B. $\square$

We want to prove that the fourth point D in a harmonic set $A, B; C, D$ is uniquely determined by the other three points A, B, C. By projecting between planes, we can assume that the points A–D and the vertices of the associated complete quadrangle are all ordinary. We then use Menelaus' Theorem and Ceva's Theorem to characterize D in terms of the division ratios $\overline{CA}/\overline{CB}$ and $\overline{DA}/\overline{DB}$. This characterization is surprising because projections between planes preserve harmonic sets (by Theorem 17.3) but don't generally preserve division ratios (since, for example, in Figure 17.1 we have $\overline{CA}/\overline{CB} = -1$ and $\overline{C'A'}/\overline{C'B'} > 0$, by Theorems 1.5 and 1.4).

THEOREM 17.6. In the extended plane, let $A, B; C, D$ be a harmonic set determined by a complete quadrangle $STUV$ (Figure 17.6). Assume that the points A–D and S–V are all ordinary. Then the equation $\overline{DA}/\overline{DB} = -\overline{CA}/\overline{CB}$ holds.

Proof: ABS is a triangle, and U, T, D are points that lie on the three sides of the triangle and don't equal the vertices (by Theorem 17.5 and the assumption that S, T, U are distinct). The points U, T, D are collinear, and so Menelaus' Theorem 2.4 shows that

$$\frac{\overline{UB}}{\overline{US}} \cdot \frac{\overline{TS}}{\overline{TA}} \cdot \frac{\overline{DA}}{\overline{DB}} = 1 \tag{3}$$

(since all points named are ordinary).

The points U, T, C also lie on the three sides of triangle ABS and don't equal the vertices (by Theorem 17.5 and the assumption that S, T, U are distinct). The lines AU, BT, SC are concurrent at the point V, and so

Ceva's Theorem 3.4 shows that

$$\frac{\overline{UB}}{\overline{US}} \cdot \frac{\overline{TS}}{\overline{TA}} \cdot \frac{\overline{CA}}{\overline{CB}} = -1 \tag{4}$$

(since the points A–D and S–V are all ordinary).

Combining Equations 3 and 4 shows that $\overline{DA}/\overline{DB} = -\overline{CA}/\overline{CB}$. $\square$

We can now use projections between planes to prove the main theorem on harmonic sets.

THEOREM 17.7. In the extended plane, the following results hold.

   (i) If $A, B; C, D$ and $A, B; C, D'$ are harmonic sets, then D equals D'.

  (ii) If $A, B; C, D$ is a harmonic set, then A–D are distinct collinear points.

 (iii) If A, B, C are three collinear points, then there is a unique point D such that $A, B; C, D$ is a harmonic set.

Proof:

   (i) By the discussion accompanying Figure 17.4, the harmonic set $A, B; C, D$ is determined by a complete quadrangle $STUV$, and the harmonic set $A, B; C, D'$ is determined by a complete quadrangle $S'T'U'V'$ (Figure 17.5). We can assume that the extended plane is formed from a plane in Euclidean space. Let l be a line in the extended plane that doesn't contain any of the points A-D, D', S-V, S'-V'. We can project l to the ideal line in another plane (by Theorem 15.3), and so the points in the last sentence project to ordinary points. Projection preserves incidence (by Theorem 15.2), and so we can replace the points A-D, D', S-V, S'-V' with their images under the projection. Thus, we can assume that these points are ordinary. Then we have the equations

$$\overline{DA}/\overline{DB} = -\overline{CA}/\overline{CB} \quad \text{and} \quad \overline{D'A}/\overline{D'B} = -\overline{CA}/\overline{CB}$$

(by Theorem 17.6), which show that

$$\overline{DA}/\overline{DB} = \overline{D'A}/\overline{D'B}.$$

Thus, D equals D' (by Theorem 1.9).

  (ii) For a harmonic set $A, B; C, D$, Definition 17.2 shows that A–D are collinear, and Theorem 17.5 shows that $A \neq B$ and that neither C nor D equals either A or B. Thus, all that remains to be proved is that $C \neq D$. By the proof of part (i), we can assume that A–D are ordinary points and that $\overline{DA}/\overline{DB} = -\overline{CA}/\overline{CB}$. This equation shows that $C \neq D$ (since $\overline{CA}/\overline{CB} \neq 0$, by Theorem 1.2i).

(iii) If A, B, C are three collinear points, Theorem 17.4 shows that there is at least one point D such that $A, B; C, D$ is a harmonic set. Thus, there is exactly one such point, by part (i). □

Part (ii) of the previous result shows that it's necessary to assume in part (iii) that A, B, C are distinct collinear points if there is to be a harmonic set $A, B; C, D$.

If $A, B; C, D$ is a harmonic set, we call D the *harmonic conjugate* of C with respect to A and B. This terminology reflects the fact that D is uniquely determined (by Theorem 17.7iii). We also call C and D *harmonic conjugates* with respect to A and B, since the condition that "$A, B; C, D$ is a harmonic set" is symmetric in A and B and in C and D (by (1)).

By Theorem 17.7, we can strengthen Theorem 17.6 and use division ratios to characterize harmonic conjugates with respect to two ordinary points.

THEOREM 17.8. In the extended plane, let A and B be two ordinary points.

(i) Then the midpoint of A and B and the ideal point on line AB are harmonic conjugates with respect to A and B.

(ii) Let C and D be two ordinary points on line AB other than A and B. Then $A, B; C, D$ is a harmonic set if and only if the equation $\overline{DA}/\overline{DB} = -\overline{CA}/\overline{CB}$ holds.

Proof: First, let C be any ordinary point on AB other than A and B. Let V be an ordinary point that doesn't lie on AB (Figure 17.7). Let G be

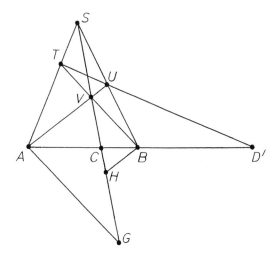

FIGURE 17.7.

the point where CV intersects the line through A parallel to BV. Let H be the point where CV intersects the line through B parallel to AV. Let S be an ordinary point on CV other than C, V, G, H. The points $T = AS \cap BV$, $U = AV \cap BS$, $D' = TU \cap AB$ exist, and $A, B; C, D'$ is a harmonic set determined by the complete quadrangle $STUV$ (by Theorem 17.4). T is ordinary (since $S \neq G$), and U is ordinary (since $S \neq H$). If D' is ordinary, then the equation $\overline{D'A}/\overline{D'B} = -\overline{CA}/\overline{CB}$ holds (by Theorem 17.6 and the fact that the points A–C, D', S–V are all ordinary).

Next, let C and D'' be two ordinary points on AB other than A and B such that the equation

$$\overline{D''A}/\overline{D''B} = -\overline{CA}/\overline{CB} \tag{5}$$

holds. Neither of these division ratios equals 1 (by Theorem 1.2i), and so Equation 5 implies that neither ratio equals -1. Thus, neither C nor D'' is the midpoint of A and B (by Theorem 1.5).

(i) Let M be the midpoint of A and B. There is a unique point N that is the harmonic conjugate of M with respect to A and B (by Theorem 17.7iii). If N were an ordinary point, then the equation $\overline{NA}/\overline{NB} = -\overline{MA}/\overline{MB}$ would hold (by the first paragraph of the proof), which would contradict the second paragraph of the proof (since M is the midpoint of A and B). Thus, N is the ideal point on line AB.

(ii) If C and D are ordinary points such that $A, B; C, D$ is a harmonic set, then the first paragraph of the proof shows that $\overline{DA}/\overline{DB} = -\overline{CA}/\overline{CB}$ (since the fourth point in a harmonic set is uniquely determined by the other three points, by Theorem 17.7i). Conversely, let C and D be two ordinary points on line AB other than A and B such that the equation

$$\overline{DA}/\overline{DB} = -\overline{CA}/\overline{CB} \tag{6}$$

holds. There is a unique point D' that is the harmonic conjugate of C with respect to A and B (by Theorem 17.7iii). Equation 6 implies that C is not the midpoint of A and B (by the second paragraph of the proof), and so D' is not ideal (by part (i)). Thus, the first paragraph of the proof shows that

$$\overline{D'A}/\overline{D'B} = -\overline{CA}/\overline{CB}.$$

Combining this equation with Equation 6 shows that

$$\overline{DA}/\overline{DB} = \overline{D'A}/\overline{D'B}.$$

Thus, D equals D' (by Theorem 1.9), and so $A, B; C, D$ is a harmonic set. $\square$

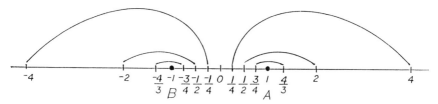

FIGURE 17.8.

Theorem 17.8(i) justifies the statement at the beginning of this section that midpoints in Euclidean geometry generalize to harmonic sets in projective geometry.

To illustrate Theorem 17.8, choose a coordinate system on an ordinary line l (as in Definition 1.6). Identify points on l by their coordinates. Let A and B be the points 1 and -1 (Figure 17.8). The pairs of ordinary points C and D that are harmonic conjugates with respect to A and B are the pairs of points such that $\overline{DA}/\overline{DB} = -\overline{CA}/\overline{CB}$ (by Theorem 17.8ii). If C and D have coordinates c and d, Theorem 1.8 shows that the equation $\overline{DA}/\overline{DB} = -\overline{CA}/\overline{CB}$ is equivalent to each of the following equations:

$$\frac{1 - d}{-1 - d} = -\frac{1 - c}{-1 - c};$$

$$(1 - d)(-1 - c) = -(1 - c)(-1 - d);$$

$$-1 - c + d + cd = 1 - c + d - cd;$$

$$2cd = 2;$$

$$d = 1/c.$$

Thus, the pairs of reciprocals c and $1/c$ are the pairs of ordinary points that are harmonic conjugates with respect to 1 and -1, as c varies over all real number other than 0, 1, and -1. The curved lines in Figure 17.8 connect several such pairs. The midpoint 0 of 1 and -1 and the ideal point on l are also harmonic conjugates with respect to 1 and -1 (by Theorem 17.8i).

In general, in the extended plane, let A and B be two ordinary points. By Theorem 17.8(ii), the pairs of ordinary points C and D that are harmonic conjugates with respect to A and B are the pairs of ordinary points on line AB such that the division ratios $\overline{DA}/\overline{DB}$ and $\overline{CA}/\overline{CB}$ have equal absolute values and opposite signs. This means that C and D lie at the same relative distances from A and B and that one of the points C and D lies between A and B and the other doesn't (by Theorem 1.4). Thus, as C approaches A from one side, D approaches A from the other side. As C approaches B from one side, D approaches B from the other side. As C approaches the

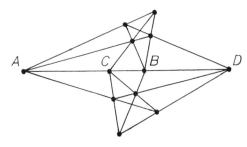

FIGURE 17.9.

midpoint of A and B, D approaches the ideal point on line AB. Figure 17.8 illustrates these properties.

Suppose that a projection from a plane $\mathcal{P}$ to a plane $\mathcal{Q}$ in Euclidean space maps four ordinary collinear points A, B, C, D on $\mathcal{P}$ to ordinary points A', B', C', D' on $\mathcal{Q}$. The equations $\overline{DA}/\overline{DB} = -\overline{CA}/\overline{CB}$ and $\overline{D'A'}/\overline{D'B'} = -\overline{C'A'}/\overline{C'B'}$ are equivalent (by Theorems 17.3 and 17.8ii), even though it's not generally true that $\overline{DA}/\overline{DB} = \overline{D'A'}/\overline{D'B'}$ and $\overline{CA}/\overline{CB} = \overline{C'A'}/\overline{C'B'}$ (as we saw before Theorem 17.6).

The next result states that $A, B; C, D$ is a harmonic set if and only if $C, D; A, B$ is a harmonic set. This is remarkable because A and B play a different role than C and D in Definition 17.2 of a harmonic set $A, B; C, D$. Collinear points A, B, C, D form a harmonic set $A, B; C, D$ when there is a complete quadrangle that has two pairs of opposite sides containing A and B and two other sides through C and D (as does the complete quadrangle above line AB in Figure 17.9). Collinear points A, B, C, D form a harmonic set $C, D; A, B$ when there is a complete quadrangle that has two pairs of opposite sides containing C and D and two other sides through A and B (as does the complete quadrangle below line AB in Figure 17.9). Thus, the next result shows that there is a complete quadrangle such as the one above line AB in Figure 17.9 if and only if there is a complete quadrangle such as the one below line AB in Figure 17.9.

THEOREM 17.9. In the extended plane, $A, B; C, D$ is a harmonic set if and only if $C, D; A, B$ is a harmonic set (Figure 17.9).

Proof: Let l be a line that doesn't contain any of the points A–D. We can assume that the extended plane is formed from a plane in Euclidean space. We can project l to the ideal line in another plane (by Theorem 15.3), and so A–D project to ordinary points. Projections preserve harmonic sets (by Theorem 17.3), and so we can replace A–D with their images. Thus, we can assume that A–D are ordinary points.

Moreover, we can assume that A–D are distinct collinear points, since this condition holds when either A, B; C, D or C, D; A, B is a harmonic set (by Theorem 17.7ii). The line m through the points A–D is ordinary (since A–D are ordinary points), and we choose a positive end for directed distances on m. Then Theorems 17.8(ii) and 1.10 show that the following conditions are equivalent:

$$A, B; C, D \text{ is a harmonic set;}$$

$$\overline{DA}/\overline{DB} = -\overline{CA}/\overline{CB};$$

$$-\overline{AD}/(-\overline{BD}) = \overline{AC}/(-\overline{BC});$$

$$\overline{AD} \cdot \overline{BC} = -\overline{AC} \cdot \overline{BD};$$

$$\overline{BC}/\overline{BD} = -\overline{AC}/\overline{AD};$$

$$C, D; A, B \text{ is a harmonic set}$$

(where the last two conditions are seen to be equivalent by interchanging A with C and B with D in Theorem 17.8ii). $\square$

Together with Equation 1, the previous theorem shows that one of the following sets is harmonic if and only if they all are:

$$A, B; C, D \qquad B, A; C, D \qquad A, B; D, C \qquad B, A; D, C$$

$$C, D; A, B \qquad D, C; A, B \qquad C, D; B, A \qquad D, C; B, A.$$

Thus, the statement that "A, B; C, D is a harmonic set" depends only on the pairs of points $\{A, B\}$ and $\{C, D\}$ and not on the order of the pairs or the order of the points in each pair. This symmetry is suggested by the semicolon in "A, B; C, D," which divides the points into two pairs.

We end this section by presenting a way to use one harmonic set to produce another.

THEOREM 17.10. In the extended plane, let l and m be two lines, and let O be a point that doesn't lie on either line. Let A, B, C, D be points on l, and let A', B', C', D' be points on m such that the points in each of the sets $\{A, A'\}$, $\{B, B'\}$, $\{C, C'\}$, $\{D, D'\}$ are collinear with O. If A, B; C, D is a harmonic set, then so is A', B', C', D' (Figure 17.10).

Proof: As in the proofs of Theorems 17.7(i) and 17.9, we can assume that the extended plane is formed from a plane $\mathcal{P}$ in Euclidean space and that the lines l and m and the points O, A–D, A'–D' are all ordinary. Let $\mathcal{Q}$ be a plane through l other than $\mathcal{P}$, and let $\mathcal{R}$ be a plane through m other than $\mathcal{P}$ (Figure 17.11). Let $\mathcal{P}^*$, $\mathcal{Q}^*$, $\mathcal{R}^*$ be the extended planes formed from

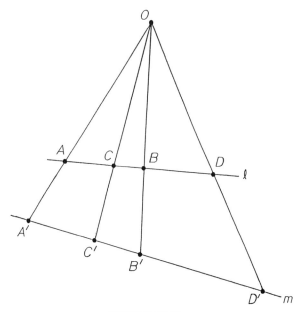

FIGURE 17.10.

$\mathcal{P}$, $\mathcal{Q}$, $\mathcal{R}$. Since $A, B; C, D$ is a harmonic set in $\mathcal{P}^*$, and since these points are all ordinary, we have $\overline{DA}/\overline{DB} = -\overline{CA}/\overline{CB}$ (by Theorem 17.8ii). This equation still holds when we consider $A–D$ as points of $\mathcal{Q}$, so $A, B; C, D$ is also a harmonic set in $\mathcal{Q}^*$ (by Theorem 17.8ii). The projection from $\mathcal{Q}$ to $\mathcal{R}$ through O maps A to A', B to B', C to C', D to D', since these pairs of points are collinear with O. The last two sentences imply that $A', B'; C', D'$ is a harmonic set in $\mathcal{R}^*$, since projections between planes preserve harmonic sets (by Theorem 17.3). Thus, we have $\overline{D'A'}/\overline{D'B'} = -\overline{C'A'}/\overline{C'B'}$ (by Theorem 17.8ii, since $A'–D'$ are ordinary points). This equation still holds when we consider $A'–D'$ as points of $\mathcal{P}^*$, and so $A', B'; C', D'$ is a harmonic set in $\mathcal{P}^*$ (by Theorem 17.8ii), as desired. $\square$

EXERCISES

17.1. In the extended plane, let S, T, U, V be four points, no three of which are collinear. Set $A = ST \cap UV$, $B = SU \cap TV$, $C = SV \cap TU$, and $D = BC \cap ST$. Prove that $S, T; A, D$ is a harmonic set. Illustrate this result with a figure.

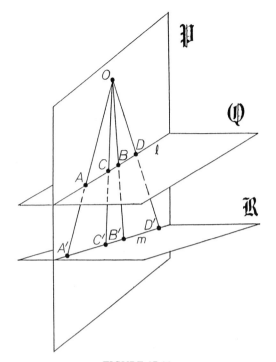

FIGURE 17.11.

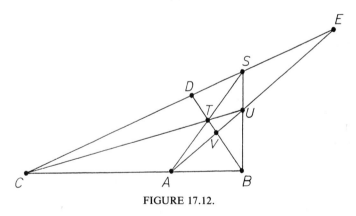

FIGURE 17.12.

17.2. In the extended plane, let S, T, U, V be four points, no three of which are collinear (Figure 17.12). Set $A = ST \cap UV$, $B = SU \cap TV$, $C = AB \cap TU$, $D = CS \cap TV$, and $E = CS \cap UV$.

(a) Prove that $S, C; D, E$ is a harmonic set.
(b) Prove that $T, B; V, D$ is a harmonic set.

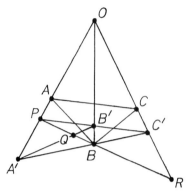

FIGURE 17.13.

17.3. In the extended plane, let triangles ABC and $A'B'C'$ be perspective from a point O. Assume that B lies on $A'C'$ (Figure 17.13). Set $P = AA' \cap B'C'$, $Q = BP \cap A'B'$, and $R = BP \cap CC'$. Prove that $B, P; Q, R$ is a harmonic set.

In Exercises 17.4–17.7, use Definition 17.2 and Theorems 17.7 and 17.9 to prove the theorems in the exercises listed.

17.4. Exercise 14.5

17.5. Exercise 14.6

17.6. Exercise 14.7

17.7. Exercise 16.7

17.8. Let A, B, C, D be four ordinary points on a line l. Let M be the midpoint of A and B. Choose a positive end on l for directed distances. Prove that $A, B; C, D$ is a harmonic set if and only if neither C nor D equals M and the relation $\overline{MC} \cdot \overline{MD} = (\overline{MA})^2$ holds. (See Theorems 17.8 and either 1.7 or 1.10.)

17.9. In the extended plane, let l be an ordinary line.

(a) Let A, B, C, D be four ordinary points on l. Prove that $A, B; C, D$ is a harmonic set if and only if

$$\frac{2}{\overline{AB}} = \frac{1}{\overline{AC}} + \frac{1}{\overline{AD}}.$$

(See Theorems 17.8ii and 1.10.)

(b) Let c and d be positive numbers. The *harmonic mean* of c and d is the reciprocal of $\frac{1}{2}(1/c + 1/d)$. Choose a coordinate system on l (in the sense of Definition 1.6), and identify points on l by their coordinates. Prove that the harmonic mean of c and d equals the harmonic conjugate of 0 with respect to c and d. (The term "harmonic mean" is the source of the term "harmonic conjugate.")

(c) Prove that part (a) remains true when one of the points B, C, D is ideal, if we set $1/\overline{AX} = 0$ and $2/\overline{AX} = 0$ when X is ideal.

17.10. Consider the following result.

Theorem. In the extended plane, let ABC be a triangle. Let A' and A'' be harmonic conjugates with respect to B and C, let B' and B'' be harmonic conjugates with respect to A and C, and let C' and C'' be harmonic conjugates with respect to A and B. Then the lines AA', BB', CC' are concurrent if and only if the points A'', B'', C'' are collinear. Moreover, when these conditions hold, we have $A'' = BC \cap B'C'$, $B'' = AC \cap A'C'$, and $C'' = AB \cap A'B'$.

(a) Illustrate this result with a figure.

(b) Prove that the theorem holds when A'' and B'' are ideal by using Theorem 17.8(i) and results from Chapter I. Deduce that the theorem always holds.

(c) Prove that the theorem holds when $AA' \cap BB'$ and all points named are ordinary by using Theorem 17.8(ii) and results from Chapter I. Deduce that the theorem always holds

(d) Without using Theorem 17.8, prove that the given theorem holds. (*Hint:* One possible approach is as follows. When AA', BB', CC' are concurrent, use Definition 17.2 to deduce that $A'' = BC \cap B'C'$, $B'' = AC \cap A'C'$, and $C'' = AB \cap A'B'$, and then use Desargues' Theorem to conclude that A'', B'', C'' are collinear. As a consequence, deduce that AA', BB', CC' are concurrent when A'', B'', C'' are collinear.)

(e) In the notation of the theorem, if AA', BB', CC' are concurrent at a point P, and if A'', B'', C'' lie on a line p, can P lie on p? Justify your answer.

17.11. In the extended plane, let a, b, c be three lines on a point O. Use Theorem 17.10 to prove that there is a unique line d through O that has the following property: For every line l that doesn't contain O, $c \cap l$ and $d \cap l$ are harmonic conjugates with respect to $a \cap l$ and $b \cap l$. (See Figure 17.14. We call $a, b; c, d$ a *harmonic set of lines* on the point O.)

17.12. In the extended plane, let harmonic sets of lines be defined as in Exercise 17.11. Let a, b, c, d be lines on a point O. Prove that $a, b; c, d$ is

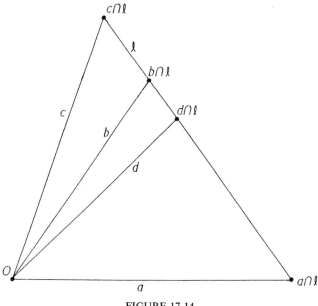

FIGURE 17.14.

a harmonic set of lines if and only if there are four lines s, t, u, v, no three of which are concurrent, such that $s \cap t$ and $u \cap v$ lie on a, $s \cap u$ and $t \cap v$ lie on b, $s \cap v$ lies on c, and $t \cap u$ lies on d.

(See Figure 17.15. By the discussion accompanying Figure 17.4, this exercise shows that a harmonic set of lines is the dual of a harmonic set of points.)

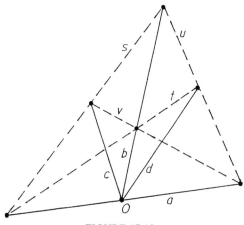

FIGURE 17.15.

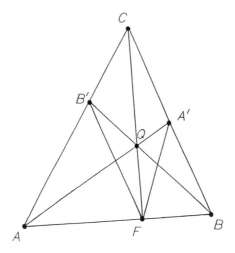

FIGURE 17.16.

17.13. In the extended plane, let harmonic sets of lines be defined as in Exercise 17.11. Let a, b; c, d be a harmonic set of lines on an ordinary point O. Prove that c and d are perpendicular if and only if c bisects a pair of angles formed by a and b.

(*Hint:* One possible approach is to choose the line l in Exercise 17.11 to be perpendicular to c.)

17.14. In the extended plane, let A, B, C, Q be four points, no three of which are collinear, such that AB and CQ are perpendicular ordinary lines (Figure 17.16). Set $A' = AQ \cap BC$, $B' = BQ \cap CA$, and $F = CQ \cap AB$. Use Exercise 17.13 to prove that AB and CQ bisect the angles formed by $A'F$ and $B'F$. (This generalizes the theorem in Exercise 5.18 and provides another proof of that result.)

Exercises 17.15–17.17 use the following notation. In the extended plane, let ABC be a triangle. Let D and E be two points, neither of which lies on any side of triangle ABC. Set $A' = AD \cap BC$, $B' = BD \cap AC$, $C' = CD \cap AB$, $A'' = AE \cap BC$, $B'' = BE \cap AC$, and $C'' = CE \cap AB$.

17.15. Prove that the following conditions are equivalent (Figure 17.17):

(i) $B, C; A', A''$ is a harmonic set.
(ii) A', B'', C'' are collinear.
(iii) A'', B', C' are collinear.

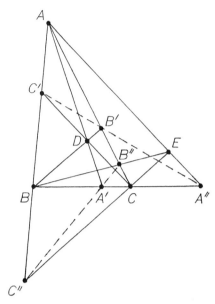

FIGURE 17.17.

17.16. Prove that any two of the following conditions imply the third.

(i) $B, C; A', A''$ is a harmonic set.
(ii) $A, C; B', B''$ is a harmonic set.
(iii) C, D, E are collinear.

Illustrate this result with a figure where conditions (i)–(iii) hold. (See Exercise 17.15.)

17.17. Prove that any two of the following conditions imply the third.

(i) A', B'', C' are collinear.
(ii) A'', B', C' are collinear.
(iii) The lines $AB, A'B', A''B''$ are concurrent.

Illustrate this result with a figure where conditions (i)–(iii) hold. (See Exercises 17.15 and 17.16.)

Section 18.

Triangles and Conic Sections

Sections 14–17 have demonstrated that projections between planes are useful in proving results about incidence of points and lines. We now use projections between planes to obtain theorems about conic sections from theorems about circles. We apply Desargues' Theorem and results on harmonic sets as well as results on circles from Section 6.

A *circle* in an extended plane is a set of ordinary points that forms a circle in the usual sense in the Euclidean plane composed of ordinary points. In particular, *all points on a circle are ordinary*.

We use projections between planes to define conic sections.

DEFINITION 18.1. A *conic section* is the figure obtained by projecting a circle from one plane in Euclidean space to another. □

Note that a projection between planes in Euclidean space maps points from one extended plane to another, and it can map ordinary points to ideal points (as in Theorem 15.3). Thus, although all points on a circle are ordinary, a conic section can have ideal points. We return to this fact shortly.

In Euclidean space, let $\mathcal{P}$ be a plane that doesn't contain a point O, and let $\mathcal{K}$ be a circle in $\mathcal{P}$ (Figure 18.1). The *cone* $\mathcal{C}$ determined by O and $\mathcal{K}$ is the set of all points that lie on lines through O and points of $\mathcal{K}$. We call O the *vertex* of $\mathcal{C}$. Note that the line through O and the center of $\mathcal{K}$ need *not* be perpendicular to $\mathcal{P}$. A *section* of the cone $\mathcal{C}$ is the intersection $\mathcal{K}'$ of $\mathcal{C}$ with a plane $\mathcal{Q}$ that doesn't contain the vertex O (Figure 18.2). The next theorem justifies the use of the term "conic section" in Definition 18.1.

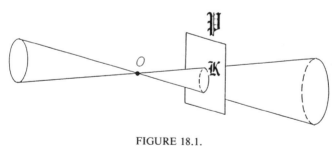

FIGURE 18.1.

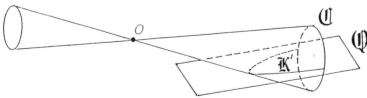

FIGURE 18.2.

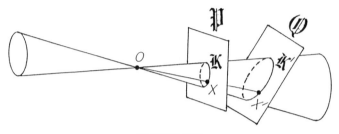

FIGURE 18.3.

THEOREM 18.2. A set of points $\mathcal{K}'$ is a section of a cone if and only if $\mathcal{K}'$ is the set of ordinary points on a conic section (Figure 18.3).

Proof: First let $\mathcal{K}'$ be the set of ordinary points on a conic section. By Definition 18.1, $\mathcal{K}'$ consists of the ordinary points in the image when a circle $\mathcal{K}$ is projected from a plane $\mathcal{P}$ to a plane $\mathcal{Q}$ through a point O in Euclidean space. Neither $\mathcal{P}$ nor $\mathcal{Q}$ contains O (by the definition of projection between planes), and so the lines through O and the points of $\mathcal{K}$ form a cone $\mathcal{C}$ whose vertex O doesn't lie on $\mathcal{Q}$. A point X' of $\mathcal{Q}$ lies on $\mathcal{K}'$ if and only if X' lies on the line through O and a point X of $\mathcal{K}$ (since points project from $\mathcal{P}$ to $\mathcal{Q}$ along lines through O, as in Figure 15.1). Thus, $\mathcal{K}'$ consists of the points where $\mathcal{Q}$ intersects the cone $\mathcal{C}$, and so $\mathcal{K}'$ is a section of a cone.

Conversely, let $\mathcal{K}'$ be a section of a cone $\mathcal{C}$. $\mathcal{C}$ is determined by a point O and a circle $\mathcal{K}$ in a plane $\mathcal{P}$ that doesn't contain O. $\mathcal{K}'$ is the intersection of $\mathcal{C}$ with a plane $\mathcal{Q}$ that doesn't contain O. We can assume that $\mathcal{P} \neq \mathcal{Q}$ (by moving $\mathcal{P}$ parallel to itself, if necessary). Then the projection from $\mathcal{P}$ to $\mathcal{Q}$ through O maps $\mathcal{K}$ to a conic section (by Definition 18.1), and $\mathcal{K}'$ consists of the ordinary points on this conic section (by the previous paragraph). □

We use the terms *Euclidean ellipse*, *Euclidean parabola*, and *Euclidean ellipse* for the curves familiar from analytic geometry (Figure 18.4). These

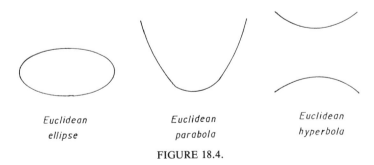

Euclidean	*Euclidean*	*Euclidean*
ellipse	*parabola*	*hyperbola*

FIGURE 18.4.

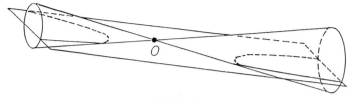

FIGURE 18.5.

curves are graphs of quadratic equations

$$Ax^2 + Bxy + Cy^2 + Dx + Ey + F = 0.$$

We review these curves and their properties in Section 21.

Euclidean ellipses, Euclidean parabolas, and Euclidean hyperbolas are sections of cones, and, conversely, every section of a cone is one of these curves. (For example, Figure 18.3 shows a Euclidean ellipse as a section of a cone, Figure 18.2 shows a Euclidean parabola, and Figure 18.5 shows a Euclidean hyperbola.) Thus, by Theorem 18.2, Euclidean ellipses, Euclidean parabolas, and Euclidean hyperbolas are exactly the sets of all ordinary points on the conic sections in Definition 18.1.

We can't use the first sentence of the previous paragraph in proofs until we justify it formally in Section 22. Even before then, however, we use Euclidean ellipses, Euclidean parabolas, and Euclidean hyperbolas in figures that illustrate results about the conic sections in Definition 18.1, although we won't formally justify this practice until Section 22.

By Definition 18.1, every conic section is the projection of a circle from one plane in Euclidean space to another. Thus, *if property of circles can be stated in terms of incidence of points and lines in the extended plane, then the property holds for all conic sections* (since projections preserve incidence, by Theorem 15.2). We consider a number of such properties in this section and in Sections 20–22. As always in projective geometry, we give up something to gain something: by limiting ourselves to incidence properties of circles, we obtain results that apply to all conic sections.

The previous paragraph shows that *all conic sections are the same as circles from the point of view of projective geometry.* It may seem surprising that Euclidean ellipses, Euclidean parabolas, and Euclidean hyperbolas all look the same when they're considered projectively. The apparent differences among these curves, however, are due to the number of ideal points we add to these curves in the extended plane. We add no ideal points to a Euclidean ellipse; its points are all visible in the Euclidean plane (Figure 18.3). We add one ideal point to a Euclidean parabola; when this point is added, the curve resembles a Euclidean ellipse (Figure 18.6). We add two ideal points to a

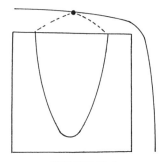

FIGURE 18.6.

Euclidean hyperbola; the two branches of the curve that approach each asymptote meet at an ideal point (Figure 18.7a). The curve resembles a Euclidean ellipse when the ideal points on both asymptotes are added (Figure 18.7b). Since each ideal point lies on a family of parallel ordinary lines, it's natural to add one ideal point to a Euclidean parabola (because

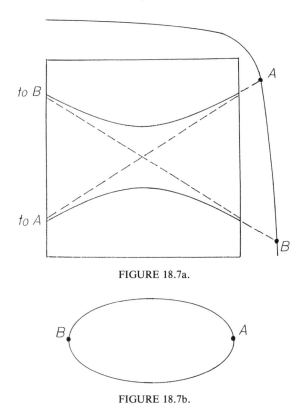

FIGURE 18.7a.

FIGURE 18.7b.

the two "ends" of the curve become parallel to the axis of symmetry) and to add two ideal points to a Euclidean hyperbola (because each of the two asymptotes is approached by two "ends" of the curve).

The informal discussion in the previous paragraph suggests the following formal definition.

DEFINITION 18.3. An *ellipse* is a conic section whose points are all ordinary. A *parabola* consists of the ordinary points on a conic section that has exactly one ideal point. A *hyperbola* consists of the ordinary points on a conic section that has exactly two ideal points.   □

The terms "ellipse," "parabola," and "hyperbola" refer to Definition 18.3, and the terms "Euclidean ellipse," "Euclidean parabola," and "Euclidean hyperbola" refer to the familiar graphs of quadratic equations. We prove in Section 22 that these two sets of terms refer to exactly the same curves.

Unless a result on conic sections refers specifically to parabolas and hyperbolas, we illustrate it with ellipses. We do so because all points on an ellipse are ordinary (by Definition 18.3), and so they appear in a figure drawn in the Euclidean plane. It should be remembered, however, that *any result we prove about all conic sections applies to all parabolas and hyperbolas as well as all ellipses.*

We already know several basic incidence properties of circles. As we've noted, these properties extend to all conic sections.

THEOREM 18.4. In the extended plane, let $\mathcal{K}$ be a conic section.

  (i) No three points of $\mathcal{K}$ are collinear, and $\mathcal{K}$ has infinitely many points.
  (ii) Let A be any point of $\mathcal{K}$. There is a unique line l through A that intersects $\mathcal{K}$ only at A. Every line through A other than l intersects $\mathcal{K}$ at exactly two points, A and one other (Figure 18.8).

Proof: These results hold in the Euclidean plane when $\mathcal{K}$ is a circle (by Theorems 4.5ii and 5.7). Thus, these properties hold in the extended plane when $\mathcal{K}$ is a circle (since all points on a circle are ordinary). Since these

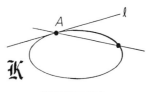

FIGURE 18.8.

properties involve only incidence, they are preserved by projections between planes (by Theorem 15.2), and so they hold in the extended plane when $\mathcal{K}$ is any conic section (by Definition 18.1). □

This theorem justifies the following definition.

DEFINITION 18.5. In the extended plane, let A be a point on a conic section $\mathcal{K}$. The *tangent* to $\mathcal{K}$ at A, written as tan A, is the unique line l that intersects $\mathcal{K}$ only at A (by Theorem 18.4ii). We also say that l is tangent to $\mathcal{K}$ at A, and we call A the *point of contact* of l (Figure 18.8). □

Suppose that a projection maps a circle $\mathcal{K}$ from one plane in Euclidean space to another. Let A be a point on $\mathcal{K}$. The projection maps $\mathcal{K}$ to a conic section $\mathcal{K}'$, and it maps A to a point A' on $\mathcal{K}'$. Because tangents are defined in terms of incidence (by Definition 18.5), the tangent to $\mathcal{K}$ at A projects to the tangent to $\mathcal{K}'$ at A' (by Theorem 15.2).

In the extended plane, let l be a line, and let $\mathcal{K}$ be a conic section. Then l is tangent to $\mathcal{K}$ if and only if l and $\mathcal{K}$ intersect at exactly one point (by Definition 18.5). If l isn't tangent to $\mathcal{K}$, then l and $\mathcal{K}$ intersect in either 0 or 2 points (by Theorem 18.4i). In particular, *the ordinary points on a conic section $\mathcal{K}$ form a parabola if and only if $\mathcal{K}$ is tangent to the ideal line*, since each of these conditions holds if an only if $\mathcal{K}$ intersects the ideal line at exactly one point (by Definitions 18.3 and 18.5) (Figure 18.6).

The ideal points on a conic section $\mathcal{K}$ are the points where $\mathcal{K}$ intersects the ideal line. There are either 0, 1, or 2 such points (by Theorem 18.4i). Thus, *the ordinary points on any conic section form either an ellipse, a parabola, or a hyperbola* (by Definition 18.3).

Let $\mathcal{K}'$ be a conic section obtained by projecting a circle $\mathcal{K}$ from a plane $\mathcal{P}$ to a plane $\mathcal{Q}$. If $\mathcal{P}$ and $\mathcal{Q}$ aren't parallel, the projection matches up the points on the vanishing line m of the extended plane $\mathcal{P}^*$ with the ideal

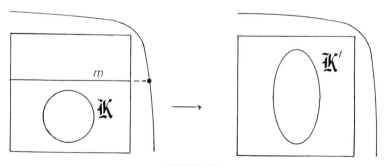

FIGURE 18.9.

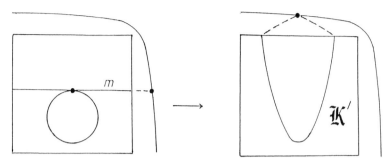

FIGURE 18.10.

points on the extended plane $\mathbb{Q}^*$ (by the second paragraph before Theorem 15.2). $\mathcal{K}$ intersects m at either 0, 1, or 2 points (by Theorem 4.5), and these points project to the points where $\mathcal{K}'$ intersects the ideal line of $\mathbb{Q}^*$. If $\mathcal{K}$ doesn't intersect m, then $\mathcal{K}'$ doesn't intersect the ideal line and is an ellipse (Figure 18.9). If $\mathcal{K}$ is tangent to m, then $\mathcal{K}'$ is tangent to the ideal line and its ordinary points form a parabola (Figure 18.10). If $\mathcal{K}$ intersects m at two points, then $\mathcal{K}'$ intersects the ideal line at two points and its ordinary points form a hyperbola (Figure 18.11). If $\mathcal{P}$ and $\mathbb{Q}$ are parallel, then the points of $\mathcal{K}$, which are all ordinary, project to ordinary points of $\mathbb{Q}^*$ (by the third paragraph before Theorem 15.2); thus, $\mathcal{K}'$ has no ideal points, and so it is an ellipse.

Let $\mathcal{K}$ be a conic section whose ordinary points form a parabola (Figure 18.12). $\mathcal{K}$ has exactly one ideal point A and is tangent to the ideal line at A. We say that a line l is *parallel to the axis of symmetry of the parabola* if l is an ordinary line that contains A. Such a line l isn't tangent to $\mathcal{K}$ (since tan A is the ideal line), and so it intersects $\mathcal{K}$ at exactly two points, the ideal point A and an ordinary point B. Of course, the ordinary lines through A form a family of parallel lines in the Euclidean plane.

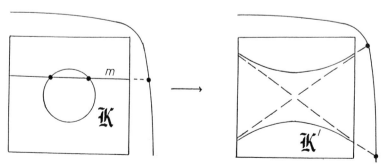

FIGURE 18.11.

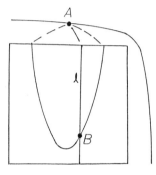
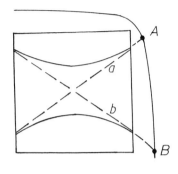

FIGURE 18.12.   FIGURE 18.13.

Next, let $\mathcal{K}$ be a conic section whose ordinary points form a hyperbola (Figure 18.13). $\mathcal{K}$ intersects the ideal line at two points A and B, and so $\mathcal{K}$ isn't tangent to the ideal line. Thus, the tangents a and b at A and B are ordinary lines, and we call these lines the *asymptotes* of the hyperbola. We prove in Section 22 that this definition of the asymptotes of a hyperbola agrees with the usual one.

We already know several basic incidence properties of tangents to circles, and these results extend to conic sections.

THEOREM 18.6. In the extended plane, let $\mathcal{K}$ be a conic section.

(i) Any point not on $\mathcal{K}$ lies on either 0 or exactly 2 tangents to $\mathcal{K}$.

(ii) A point of $\mathcal{K}$ lies on exactly one tangent to $\mathcal{K}$.

(iii) No three tangents to $\mathcal{K}$ are concurrent.

Proof: (i) First assume that $\mathcal{K}$ is a circle. Let P be a point of the extended plane that doesn't lie on $\mathcal{K}$. If P is ordinary, then P lies on either 0 or exactly 2 tangents to $\mathcal{K}$ (by Theorem 6.9i and ii). If P is ideal, let O be the center of $\mathcal{K}$, and let r be the radius (Figure 18.14). Let n be the line through

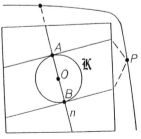

FIGURE 18.14.

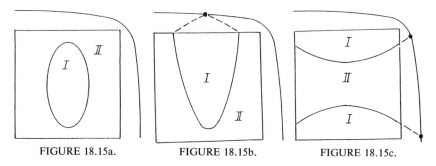

FIGURE 18.15a.   FIGURE 18.15b.   FIGURE 18.15c.

O perpendicular to the parallel ordinary lines through P. Let A and B be the two points on n that lie r units from O. Then A and B are the two points of $\mathcal{K}$ at which the tangents are perpendicular to n (by Theorem 5.7), and so A and B are the two points of $\mathcal{K}$ at which the tangents contain P.

We've proved part (i) when $\mathcal{K}$ is a circle. Then part (i) holds for every conic section $\mathcal{K}$ by projection between planes.

Part (ii) holds, by Definition 18.5. Part (iii) holds, since parts (i) and (ii) show that no point of the extended planes lies on more than two tangents to $\mathcal{K}$. $\square$

Figures 18.15a–c illustrate Theorem 18.6(i). In each of these figures, the points in region I lie on no tangents to the conic section shown, and each point in region II lies on exactly two tangents.

In the extended plane, let $\mathfrak{I}$ be a triangle inscribed in a conic section $\mathcal{K}$, and let $\mathfrak{I}'$ be the circumscribed traingle formed by the tangent lines to $\mathcal{K}$ at the vertices of $\mathfrak{I}$. Then $\mathfrak{I}$ and $\mathfrak{I}'$ are perspective from a point and a line. (See Figures 18.16 and 18.17, where $\mathfrak{I}$ is triangle ABC, and where $\mathfrak{I}'$ is triangle

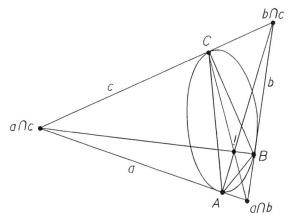

FIGURE 18.16.

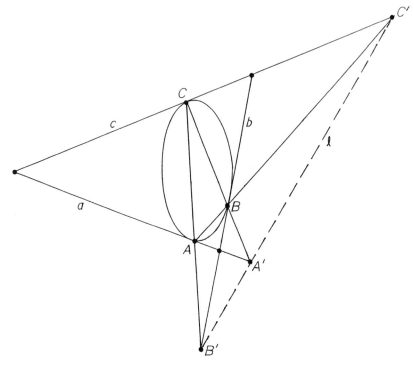

FIGURE 18.17.

abc.) This is the main result of this section, and we prove it by combining Theorem 6.8, projection between planes, and Desargues' Theorem.

THEOREM 18.7. In the extended plane, let A, B, C be three points on a conic section $\mathcal{K}$, Let a, b, c be the tangents at A, B, C, respectively. Then triangles ABC and abc are perspective from a point and a line. The corresponding vertices are A and $b \cap c$, B and $a \cap c$, C and $a \cap b$, and the corresponding sides are BC and a, AC and b, AB and c (Figures 18.16 and 18.17).

Proof: ABC is a triangle (by Theorem 18.4i), and abc is also a triangle (by Theorem 18.6iii). Corresponding sides of these triangles intersect at the points $A' = BC \cap a$, $B' = AC \cap b$, and $C' = AB \cap c$; these points exist because a, b, c intersect $\mathcal{K}$ at the unique points A, B, C, respectively (Figure 18.17).

We claim that A', B', C' are collinear when $\mathcal{K}$ is a circle. If triangle ABC is scalene, Theorem 6.8(i) shows that A', B', C' are collinear (Figures 6.26

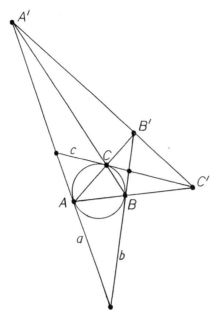

FIGURE 18.18.

and 18.18). If exactly two sides of triangle ABC have equal length, we can assume by symmetry that AC and BC are the two sides of equal length (Figure 6.27); then A' and B' are ordinary points that lie on a line parallel to both c and AB (by Theorem 6.8ii), and so $C' = c \cap AB$ is the ideal point on $A'B'$, and so A', B', C' are collinear (Figure 18.19). If triangle ABC is equilateral (Figure 6.28), each side of the triangle is parallel to the tangent through the opposite vertex (by Theorem 6.8iii); then the points A', B', C'

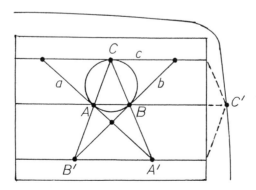

FIGURE 18.19.

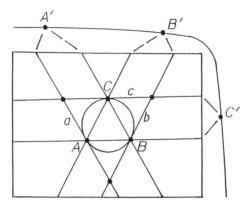

FIGURE 18.20.

are ideal, and so they all lie on the ideal line (Figure 18.20). In short, A', B', C' are collinear whenever $\mathcal{K}$ is a circle.

Now let $\mathcal{K}$ be any conic section. Since projections between planes preserve incidence (by Theorem 15.2), A', B', C' lie on a line l (by the previous paragraph and Definition 18.1) (Figure 18.17).

The lines AB, AC, BC, a, b, c are distinct (by Theorem 18.4i and Definition 18.5). $A' = BC \cap a$ doesn't equal B (since a doesn't contain B, by Definition 18.5). We have $AB \cap BC = B$ and $b \cap BC = B$ (by the first sentence of this paragraph). Since A' lies on BC, the two previous sentences imply that neither AB nor b contains A'. Then neither AB nor b equals the line l through A', B', C'. By symmetry, none of the lines AB, AC, BC, a, b, c equals l.

We can now prove that triangles ABC and abc are perspective from the line l (Figure 18.17). The previous paragraph shows that the lines AB, AC, BC, a, b, c, l are all distinct, and so condition (i) of Definition 16.6 holds. The first paragraph of the proof shows that the lines AB, AC, BC are nonconcurrent and that the lines a, b, c are nonconcurrent, and so condition (ii) of Definition 16.6 holds. The first and third paragraphs of the proof show that l, BC, a contain A', that l, AC, b contain B', and that l, AB, c contain C', and so condition (iii) of Definition 16.6 holds. A' doesn't lie on AB (by the previous paragraph), but C' does, and so we have $A' \neq C'$. Then the points A', B', C' are distinct (by symmetry), and so condition (iv) of Definition 16.6 holds. Thus, all conditions of Definition 16.6 are satisfied, and so triangles ABC and abc are perspective from the line l (Figure 18.17). Hence, these triangles are also perspective from a point, by Desargues' Theorem 16.10 (Figure 18.16). Corresponding vertices are A and $b \cap c$, B and $a \cap c$, C and $a \cap b$, and these lie opposite corresponding sides BC and a, AC and b, AB and c, respectively, as Desargues' Theorem specifies. $\square$

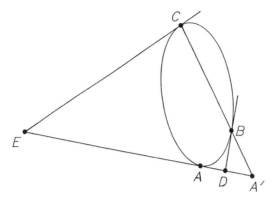

FIGURE 18.21.

The previous theorem implies that certain sets of points determined by a conic section are harmonic. The next result provides one example.

THEOREM 18.8. In the extended plane, let A, B, C be three points on a conic section. Then the point $A' = BC \cap \tan A$ is the harmonic conjugate of A with respect to $D = \tan B \cap \tan A$ and $E = \tan C \cap \tan A$ (Figure 18.21).

Proof: Set $F = \tan B \cap \tan C$ (Figure 18.22). Triangles ABC and FED are perspective from a point (by Theorem 18.7), and so the lines AF, BE, CD through the pairs of corresponding vertices are concurrent at a point G.

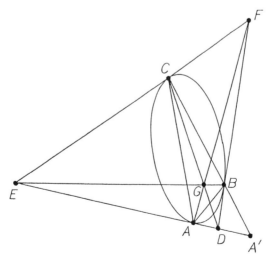

FIGURE 18.22.

Because triangles ABC and FED are perspective from G, the points B, C, F, G are distinct (by Definition 16.5i), neither B nor C lies on line FG (since the lines BG, CG, FG are distinct, by Definition 16.5iii and iv), and G doesn't lie on BC (since $BG \neq CG$, as before). Moreover, F doesn't lie on BC (since $BF = \tan B \neq BC$). Thus, B, C, F, G are four points, no three of which are collinear, and so they are the vertices of a complete quadrangle.

The complete quadrangle $BCFG$ has two opposite sides BF and CG that contain D, two other opposite sides BG and CF that contain E, a fifth side FG that contains A, and a sixth side BC that contains A'. Thus, since D, E, A, A' all lie on $\tan A$, the complete quadrangle $BCFG$ shows that $D, E; A, A'$ is a harmonic set (by Definition 17.2). $\square$

In the extended plane, let A be an ordinary point on a conic section $\mathcal{K}$. Let $\mathcal{K}'$ be the ellipse, parabola, or hyperbola formed by the ordinary points of $\mathcal{K}$. $\mathcal{K}$ determines the line $\tan A$ in the extended plane, as in Definition 18.5. The ordinary points of $\tan A$ form a line in the Euclidean plane, which we call the *tangent* to $\mathcal{K}'$ at A. We again denote this line by $\tan A$, and we call A its point of contact.

Suppose that we take line BC in the previous theorem to be the ideal line (Figure 18.21). Then the conic section has two ideal points B and C, and so its ordinary points form a hyperbola (by Definition 18.3). The tangent lines at the ideal points B and C are asymptotes of the hyperbola (by the discussion accompanying Figure 18.13). Thus, D and E are the points where $\tan A$ intersects the asymptotes. $A' = BC \cap \tan A$ is the only point of $\tan A$ that lies on the ideal line BC. Thus, since $D, E; A, A'$ is a harmonic set (by Theorem 18.8), the points D, E, A are ordinary (by Theorem 17.7ii), and A is the midpoint of D and E (by Theorems 17.7iii and 18.8i). Hence, by taking line BC in Theorem 18.8 to be the ideal line, we obtain the following notable result.

THEOREM 18.9. In the Euclidean plane, any point A on a hyperbola is the midpoint of the points D and E where the tangent at A intersects the asymptotes (Figure 18.23). $\square$

Theorem 18.7 implies another result, in addition to Theorem 18.8, about harmonic sets determined by conic sections.

THEOREM 18.10. In the extended plane, let A, B, C be three points on a conic section. Set $A' = BC \cap \tan A$, $F = \tan B \cap \tan C$, and $H = AF \cap BC$. Then A' and H are harmonic conjugates with respect to B and C (Figure 18.24).

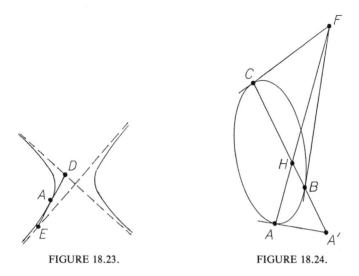

FIGURE 18.23.   FIGURE 18.24.

Proof: Set $D = \tan B \cap \tan A$ and $E = \tan C \cap \tan A$ (Figure 18.25). $F = \tan B \cap \tan C$ doesn't lie on $\tan A$ (by Theorem 18.6iii) or on BC (since $BF = \tan B \neq BC$), and so the point $H = AF \cap BC$ exists. Moreover, BC and $\tan A$ are distinct lines (by Definition 18.5), B, C, H, A' lie on BC, and F is collinear with the points in each of the pairs $\{B, D\}$, $\{C, E\}$, $\{H, A\}$, and $\{A', A'\}$. Thus, since $D, E; A, A'$ is a harmonic set (by Theorem 18.8), so is $B, C; H, A'$ (by Theorem 17.10). $\square$

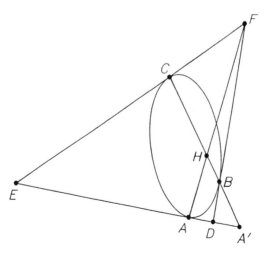

FIGURE 18.25.

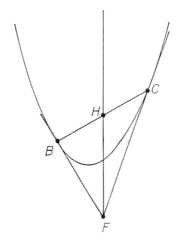

FIGURE 18.26.

In particular, suppose that we take tan A in Theorem 18.10 to be the ideal line (Figure 18.24). Then the conic section is tangent to the ideal line, and so its ordinary points form a parabola (by the discussion after Definition 18.5). Since $F = \tan B \cap \tan C$ doesn't lie on the ideal line tan A (by Theorem 18.6iii), F is an ordinary point. Then FH is an ordinary line through the only ideal point A on the conic section, and so FH is parallel to the axis of symmetry of the parabola formed by the ordinary points on the conic section (by the discussion accompanying Figure 18.12). $B, C; H, A'$ is a harmonic set (by Theorem 18.10), and $A' = BC \cap \tan A$ is the only point of BC that lies on the ideal line tan A; thus, B and C are ordinary points (by Theorem 17.7ii), and H is their midpoint (by Theorems 17.7iii and 17.8i). Thus, we obtain the following result by taking tan A in Theorem 18.10 to be the ideal line.

THEOREM 18.11. In the Euclidean plane, let B and C be two points on a parabola. Then the tangents at B and C intersect at a point F, and the midpoint H of B and C lies on the line through F parallel to the axis of symmetry of the parabola (Figure 18.26).

EXERCISES

18.1. Theorem 18.7 implies the following result (Figure 18.27):

Theorem. In the Euclidean plane, let A, B, C be three points on a conic section. Set $A_1 = \tan B \cap \tan C$, $B_1 = \tan A \cap \tan C$, and $C_1 = \tan A \cap \tan B$. Then the lines AA_1, BB_1, CC_1 are concurrent at a point O.

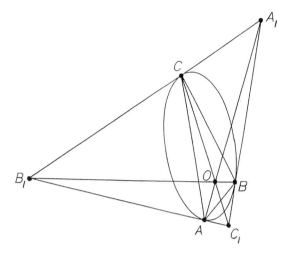

FIGURE 18.27.

State the version of this theorem that holds in the Euclidean plane in each of the following cases. Draw a figure in the Euclidean plane to illustrate each version.

(a) C_1 is the only ideal point named.
(b) C is the only ideal point named.
(c) $A_1 B_1$ is the ideal line.
(d) BC is the ideal line.
(e) CC_1 is the ideal line.
(f) All points named are ordinary.

18.2. Theorem 18.7 implies the following result (Figure 18.17):

Theorem. In the Euclidean plane, let ABC be three points on a conic section. Then the points $A' = BC \cap \tan A$, $B' = AC \cap \tan B$, and $C' = AB \cap \tan C$ are collinear.

State the version of this theorem that holds in the Euclidean plane in each of the following cases. Draw a figure in the Euclidean plane to illustrate each version.

(a) C' is the only ideal point named.
(b) C is the only ideal point named.
(c) CC' is the ideal line.
(d) BC is the ideal line.
(e) All points named are ordinary.

18.3. State the version of Theorem 18.8 that holds in the Euclidean plane in each of the following cases. Use Theorem 17.8(i) to eliminate the term "harmonic" from the statements. Draw a figure in the Euclidean plane to illustrate each version.

(a) $\tan C$ is the ideal line.
(b) CA is the ideal line.
(c) CD is the ideal line.
(d) A is the only ideal point named.
(e) E is the only ideal point named.
(f) A' is the only ideal point named.

18.4. State the version of Theorem 18.10 that holds in the Euclidean plane in each of the following cases. Use Theorem 17.8(i) to eliminate the term "harmonic" from the statements. Draw a figure in the Euclidean plane to illustrate each version.

(a) $\tan C$ is the ideal line.
(b) AC is the ideal line.
(c) AF is the ideal line.
(d) $A'F$ is the ideal line.
(e) A' is the only ideal point named.
(f) C is the only ideal point named.
(g) H is the only ideal point named.

18.5. Prove Theorem 18.10 directly from Theorem 18.7 and Definition 17.2. Do not use Theorems 17.10 and 18.8.

18.6. In the Euclidean plane, the tangents to an ellipse, parabola, or hyperbola determined by a conic section $\mathcal{K}$ in the extended plane are the lines $\tan A$ as A varies over the ordinary points of $\mathcal{K}$. Use Theorem 18.4 or Theorem 18.6 to prove that the following statements hold in the Euclidean plane.

(a) No two tangents to a parabola are parallel.
(b) If m is a line that isn't parallel to the axis of symmetry of a parabola, then m is parallel to exactly one tangent to the parabola.
(c) No tangent to a parabola is parallel to the axis of symmetry.
(d) Any tangent to an ellipse or a hyperbola is parallel to exactly one other tangent.
(e) No tangent to a hyperbola is parallel to an asymptote.
(f) There is a unique point that lies on both asymptotes of a hyperbola, and this point lies on no tangents to the hyperbola.
(g) A point lies on exactly one asymptote of a hyperbola if and only if it lies on exactly one tangent and not on the hyperbola.

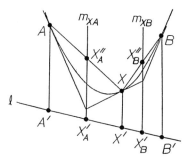

FIGURE 18.28.

(h) If a line l is parallel to and distinct from an asymptote of a hyperbola, then l intersects the hyperbola at a unique point.

18.7. In the Euclidean plane, let l be a line that isn't parallel to the axis of symmetry of a parabola $\mathcal{K}$ (Figure 18.28). Let A, B, X be three points on $\mathcal{K}$, and let A', B', X' be the points where l intersects the lines through A, B, X parallel to the axis of symmetry. Let m_{XA} and m_{XB} be the lines through $\tan X \cap \tan A$ and $\tan X \cap \tan B$ parallel to the axis of symmetry. Set $X'_A = m_{XA} \cap l$, $X'_B = m_{XB} \cap l$, $X''_A = m_{XA} \cap XA$, and $X''_B = m_{XB} \cap XB$.

(a) Prove that X'_A is the midpoint of X' and A' and that X'_B is the midpoint of X' and B'. (See Theorem 18.11.)

(b) Choose a positive end of l for directed distances. Conclude from part (a) that $\overline{X'_A X'_B} = \frac{1}{2}\overline{A'B'}$. (Thus, the directed distance from X'_A to X'_B remains the same as X varies over all points on the parabola other than A and B. In other words, the points $\tan X \cap \tan A$ and $\tan X \cap \tan B$, where a variable tangent to a parabola intersects two fixed tangents, project to a line not parallel to the axis of symmetry in segments of fixed length.)

Exercises 18.8–18.13 use the following notation. In the extended plane, a *quadrangular set* $ABC; DEF$ consists of points A, B, C, D, E, F on a line l such that there are four points S, T, U, V that satisfy the following conditions: no three of the points S, T, U, V are collinear; none of the points S, T, U, V lies on l; ST contains A; SU contains B; SV contains C; UV contains D; TV contains E; and TU contains F (Figure 18.29). Note that we do not assume that the points A–F are distinct. We say that the quadrangular set $ABC; DEF$ is *determined* by the complete quadrangle $STUV$. A, B, C lie on the three sides of the complete quadrangle that contain S, and D, E, F lie on the sides of the complete quadrangle opposite the sides containing A, B, C, respectively.

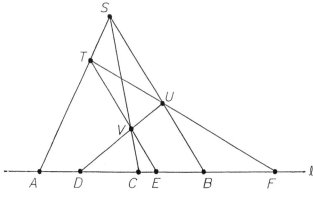

FIGURE 18.29.

18.8.

(a) In the extended plane, let $ABC; DEF$ be a quadrangular set. Prove that no equalities can hold among the points A–F except possibly $A = D$, $B = E$, $C = F$.

(b) In the extended plane, prove that $ABC; ABF$ is a quadrangular set if and only if $A, B; C, F$ is a harmonic set. (*Thus, quadrangular sets generalize harmonic sets. Harmonic sets correspond to quadrangular sets that involve only four points.*)

(c) Draw a figure that shows a quadrangular set $ABC; DEC$ determined by a complete quadrangle, where A–E are distinct points.

18.9. In the extended plane, let A, B, C, D, E be points on a line l such that no equalities holds among these points except possibly $A = D$ and $B = E$. Let S and V be two points that are collinear with C and that don't lie on l. Prove that there are points F, T, U such that $ABC; DEF$ is a quadrangular set determined by the complete quadrangle $STUV$. (See Figure 18.29. This exercise generalizes Theorem 17.4.)

18.10. In the extended plane, let $ABC; DEF$ be a quadrangular set determined by a complete quadrangle $STUV$ (Figure 18.29). Assume that the points A–F and S–V are all ordinary. Obtain three equations by applying Menelaus' Theorem 2.4 to the three sets of collinear points $\{T, U, F\}$, $\{T, V, E\}$, $\{U, V, D\}$ on the sides of triangles SAB, SAC, SBC, respectively. Combine these equations to obtain the relation

$$\frac{\overline{DB}}{\overline{DC}} \cdot \frac{\overline{EC}}{\overline{EA}} \cdot \frac{\overline{FA}}{\overline{FB}} = 1.$$

18.11. In the extended plane, let A–E be collinear points such that no equalities hold among them except possibly $A = D$ or $B = E$. Prove that there is a unique point F such that $ABC; DEF$ is a quadrangular set.

(This exercise generalizes Theorem 17.7(iii). It shows that all choices of S and V in Exercise 18.9 give the same point F. One possible approach to this exercise is to use Exercise 18.10 and Theorems 1.9, 15.2, and 15.3 to prove that there is at most one point F such that $ABC; DEF$ is a quadrangular set. Then combine this result with Exercise 18.9.)

18.12.

(a) In the extended plane, prove that $ABC; DEF$ is a quadrangular set if and only if $DEF; ABC$ is a quadrangular set. (*Hint:* One possible approach is to reduce to the case where Exercise 18.10 can be applied.)
(b) Illustrate part (a) by a figure that shows distinct points A–F, S–V, S'–V' such that $ABC; DEF$ is a quadrangular set determined by the complete quadrangle $STUV$ and such that $DEF; ABC$ is a quadrangular set determined by the complete quadrangle $S'T'U'V'$.

18.13. In the extended plane, let A–F be collinear points, and let A'–F' be collinear points such that there is a point O collinear witht he points in each of the sets $\{A, A'\}$, $\{B, B'\}$, …, $\{F, F'\}$. Prove that $ABC; DEF$ is a quadrangular set if and only if $A'B'C'; D'E'F'$ is a quadrangular set. (This exercise generalizes Theorem 17.10.)

Chapter IV
Conic Sections

INTRODUCTION AND HISTORY

In Section 18, we considered the sets of ordinary points obtained by projecting a circle to another plane. We claimed that these sets are exactly the familiar Euclidean ellipses, Euclidean parabolas, and Euclidean hyperbolas defined by quadratic equations. This chapter is devoted to proving that claim and a variety of characterizations of conic sections. In particular, we prove that there is a unique conic section through any five points, no three of which are collinear. We also prove a number of theorems about polygons inscribed in conic sections; these theorems generalize the results of Section 18 on triangles inscribed in conic sections.

The theorems in this chapter about inscribed polygons are based on cross-ratios. The cross-ratio of four collinear ordinary points is the product of two of the division ratios determined by triples of the points. Projections between nonparallel lines don't preserve distances: for example, in Figure III.3 of the introduction to Chapter III, points that lie far apart along the railroad tracks project to points much closer together on the canvas. Thus. it is remarkable that projections preserve cross-ratios, a fact we prove in Section 19. This fact, like the results on harmonic sets in Section 17, was proved by Desargues, generalizing results that Pappus had proved in the Euclidean plane without reference to projections.

The first person to study conic sections was apparently Menaechmus, a member of Plato's Academy in Athens during the fourth century B.C. Euclid wrote a four-book treatise on conic sections about 300 B.C., and Apollonius wrote the definitive classical work on conic sections about 200 B.C. Apollonius constructed any conic section $\mathcal{K}$ as the intersection of a plane and an oblique cone (as in Figures 18.2, 18.3, and 18.5). Working without analytic geometry and algebraic notation, he deduced a result equivalent to a quadratic equation of $\mathcal{K}$ by considering the distances from a variable point of $\mathcal{K}$ to two fixed lines. He proved that the midpoints of parallel chords of $\mathcal{K}$ lie on a line, called a "diameter" of $\mathcal{K}$, and he proved hundreds of theorems about diameters, tangents, and normals of conic sections. Apollonius developed the reflection properties of ellipses and hyperbolas, and he probably knew the focus–directrix properties of these conic sections, but Pappus wrote the earliest surviving treatment of the focus–directrix properties of all conic sections in the fourth century A.D.

Girard Desargues used projective geometry to study conic sections in the first half of the seventeenth century. By considering the polar of a point with respect to a conic section, he generalized and unified Apollonius' results on diameters of conic sections. (See Exercises 20.10, 20.11, and 20.16.) Desargues' work on the projective properties of conic sections was continued through the seventeenth century by Blaise Pascal and Philippe de La Hire. We use cross-ratios in Section 20 to derive Pascal's Theorem on hexagons incribed in conic sections, and we deduce corollaries on pentagons and quadrilaterals inscribed in conic sections.

In the seventeenth century also, René Descartes and Pierre de Fermat developed analytic geometry. John Wallis characterized the conic sections as the graphs of nondegenerate quadratic equations in two variables, and he deduced properties of the conic sections from these equations. In Section 21, we review the familiar classification of the graphs of these equations as Euclidean ellipses, Euclidean parabolas, and Euclidean hyperbolas. In the exercises of Section 21, we use coordinates to derive the reflection properties and other focal properties of conic sections.

At the end of the seventeenth century, Isaac Newton used the inverse-square law of gravitational attraction to deduce the laws of planetary motion that Johannes Kepler had asserted at the start of the century. In particular, Newton proved that bodies move along conic sections exactly when they are subject to an inverse-square law force. Newton also proved a number of results in which a conic section is uniquely determined by five given conditions. One result of this type is Theorem 22.8, which shows that a conic section is uniquely determinbed by five of its points. Exercises 22.3 and 22.4 contain analogous results.

Projective geometry was revived in the nineteenth century by mathe-maticians reacting against the algebraic and analytic computations that dominated geometry in the eighteenth century. They sought to base projective geometry on general principles such as duality and to develop it independently of Euclidean geometry. Jacob Steiner and Michel Chasles gave intrinsic characterizations of conic sections in the projective plane, including some of the characterizations in Section 22. Karl Georg Christian von Staudt found an intrinsic construction of coordinates on a line in the projective plane. Such work led to axiomatic treatments of projective geometry in the twentieth century, which led in turn to some of the axiomatic treatments of Euclidean geometry discussed in Chapter V.

Von Staudt's work also led in the twentieth century to generalized projective planes coordinatized by abstract algebraic systems. Geometric properties of the planes correspond to algebraic properties of the coor-dinates: for example, Desargues' Theorem corresponds roughly to the associative law of multiplication, and Pappus' Theorem corresponds roughly to the commutative law of multiplication. Planes with finitely many points link finite algebraic systems to combinatorics and coding theory, topics of considerable current interest.

Projective geometry became tied to linear algebra in the nineteenth century, when Augustus Ferdinand Möbius and Julius Plücker adapted analytic methods to projective geometry by developing homogeneous coordinates. As in Section 15, an elementary perspectivity matches up the points and lines of an extended plane with the lines and planes through a point O in Euclidean space. When O is the origin, the lines and planes through O are subspaces of Euclidean space, and so they can be studied via linear algebra. Analytic projective geometry plays a key role in computer graphics because it provides an efficient way to compute the projection of a figure onto a viewing screen. Homogeneous coordinates are also crucial in algebraic geometry because they make it possible to study the behavior of algebraic curves and surfaces at infinity in the same way as at ordinary points.

The references for Chapter III serve also for this chapter.

Section 19.

Cross-Ratios

Incidence theorems about circles generalize to all conic sections, as we've seen in Section 18. We present more of these theorems in Sections 20–22. Cross-ratios are the key to proving these theorems, and we study properties of cross-ratios in this section.

The cross-ratio of four points on an ordinary line is a number computed from the directed distances between pairs of the points.

DEFINITION 19.1. In the extended plane, let A, B, C, D be four points on an ordinary line l. If these points are ordinary, we define the cross-ratio $R(A, B; C, D)$ by

$$R(A, B; C, D) = \frac{\overline{AC} \cdot \overline{BD}}{\overline{AD} \cdot \overline{BC}}. \tag{1}$$

If one of the points A–D is the ideal point l_∞ on l, we define the cross-ratio $R(A, B; C, D)$ by omitting the directed distances that involve l_∞ from Equation 1. This gives the following equations:

$$R(l_\infty, B; C, D) = \overline{BD}/\overline{BC}; \tag{2}$$

$$R(A, l_\infty; C, D) = \overline{AC}/\overline{AD}; \tag{3}$$

$$R(A, B; l_\infty, D) = \overline{BD}/\overline{AD}; \tag{4}$$

$$R(A, B; C, l_\infty) = \overline{AC}/\overline{BC}. \quad \square \tag{5}$$

Let A, B, C be three ordinary points on a line in the extended plane. The relation

$$\frac{\overline{AC}}{\overline{BC}} = \frac{-\overline{CA}}{-\overline{CB}} = \frac{\overline{CA}}{\overline{CB}} \tag{6}$$

holds, by Theorem 1.10. Thus, the division ratio of A, B by C is $\overline{AC}/\overline{BC}$, as well as $\overline{CA}/\overline{CB}$.

Let A, B, C, D be four ordinary points on a line in the extended plane. Equation 1 shows that the cross-ratio $R(A, B; C, D)$ is the product of the ratios $\overline{AC}/\overline{AD}$ and $\overline{BD}/\overline{BC}$ in which A divides C, D and B divides D, C. The values of these division ratios don't depend on the choice of a positive end of l for directed distances (by Theorem 1.2ii). Thus, the value of $R(A, B; C, D)$ doesn't depend on the choice of a positive end of l when A–D are ordinary.

Equation 6 shows that the right sides of Equations 4 and 5 are division ratios, as are the right sides of Equations 2 and 3. Thus, the value of $R(A, B; C, D)$ doesn't depend on the choice of a positive end of l when one of the points A–D is ideal (by Theorem 1.2ii).

Let A, B, C, D be four points on an ordinary line l in the extended plane. Since A–D are distinct, at most one of them is the ideal point on l, and so $R(A, B; C, D)$ is given by one of the Equations 1–5. Thus, the two previous paragraphs show that the value of $R(A, B; C, D)$ doesn't depend on the choice of a positive end of l for directed distances. Hence, given any coordinate system on l (in the sense of Definition 1.6), we can use coordinates to compute the directed distances in Equations 1–5 (by Theorem 1.7). Accordingly, we can restate Definition 19.1 as follows.

THEOREM 19.2. In the extended plane, let A, B, C, D be four points on an ordinary line l. Let $X \rightarrow x$ be a coordinate system on l (in the sense of Definition 1.6). If A–D are all ordinary points, then we have

$$R(A, B; C, D) = \frac{(c - a)(d - b)}{(d - a)(c - b)}. \tag{7}$$

If one of the points A–D is the ideal point l_∞, $R(A, B; C, D)$ is obtained by omitting the factors on the right-hand side of Equation 7 that involve the coordinate of the ideal point. This gives the following equations:

$$R(l_\infty, B; C, D) = \frac{d - b}{c - b}; \tag{8}$$

$$R(A, l_\infty; C, D) = \frac{c - a}{d - a}; \tag{9}$$

$$R(A, B; l_\infty, D) = \frac{d - b}{d - a}; \tag{10}$$

$$R(A, B; C, l_\infty) = \frac{c - a}{c - b}. \quad \square \tag{11}$$

This theorem makes it easy to evaluate cross-ratios. For example, we can use Theorem 19.2 to compute the following cross-ratios of points on the x axis, where we refer to ordinary points on the x axis by their x coordinates, and where we refer to the ideal point on the x axis as ∞:

$$R(3, 1; -2, 7) = \frac{(-2 - 3)(7 - 1)}{(7 - 3)(-2 - 1)} = \frac{-5 \cdot 6}{4(-3)} = \frac{5}{2}; \tag{12}$$

$$R(-2, 3; 1, 7) = \frac{(1 - (-2))(7 - 3)}{(7 - (-2))(1 - 3)} = \frac{3 \cdot 4}{9(-2)} = -\frac{2}{3}; \tag{13}$$

$$R(1, \infty; -2, 3) = \frac{(-2 - 1)(3 - \infty)}{(3 - 1)(-2 - \infty)} = \frac{-2 - 1}{3 - 1} = -\frac{3}{2}. \tag{14}$$

We obtained Equation 14 from Equation 7 by substituting 1, ∞, -2, 3 for a, b, c, d, respectively, and by omitting the factors that involve ∞. This procedure makes it unnecessary to use Equations 8–11, since they also arise from Equation 7 by omitting factors that involve ∞.

Equations 12 and 13 demonstrate that the value of the cross-ratio of four points depends on the order in which the points are listed. The discussion after Equation 2 of Section 2 shows why it's natural to omit factors involving ideal points in Definition 19.1 and Theorem 19.2.

We reviewed the definition of $\sin(\theta)$ for $0° < \theta < 180°$ in the discussion accompanying Figures 0.21a–c. In the extended plane, let a and b be two lines that intersect at an ordinary point (Figure 19.1). The lines a and b form supplementary angles θ and ψ (by Theorem 0.6), and so we have $\sin(\theta) = \sin(\psi)$ (by Equation 4 of Section 0). Thus, we can define the sine of a and b to be the sine of either of the angles θ and ψ formed by the lines.

DEFINITION 19.3. In the extended plane, let a and b be two lines that intersect at an ordinary point. We define the *sine of a and b*, written $\sin(ab)$, to be the sine of either angle between $0°$ and $180°$ formed by the lines (Figure 19.1). $\square$

Since $\sin(\theta)$ is positive for $0° < \theta < 180°$, the sine of two lines a and b is positive. The relation

$$\sin(ba) = \sin(ab) \tag{15}$$

holds because Definition 19.3 is symmetric in a and b.

We can now define the cross-ratio of four lines on an ordinary point. Theorem 19.6 shows that cross-ratios of points and lines are closely related to each other. This relationship makes cross-ratios significant.

DEFINITION 19.4. In the extended plane, let a, b, c, d be four lines on an ordinary point. We define their *cross-ratio* $R(a, b; c, d)$ by

$$R(a, b; c, d) = \pm \frac{\sin(ac)\sin(bd)}{\sin(ad)\sin(bc)}, \tag{16}$$

where the minus sign is used when a and b lie in different pairs of vertical angles formed by c and d. $\square$

Figure 19.2 shows an example where a and b lie in different pairs of vertical angles formed by c and d, so that $R(a, b; c, d)$ is negative (by Definition 19.4). Figure 19.3 shows an example where a and b lie in the same pair of vertical angles formed by c and d, so that $R(a, b; c, d)$ is positive.

The value of $R(a, b; c, d)$ is generally affected when two of the lines a, b, c, d are interchanged. For example, in Figure 19.2, $R(a, b; c, d)$ is negative (since a and b lie in different angles determined by c and d), and $R(a, c; b, d)$ is positive (since a and c lie in the same angles determined by

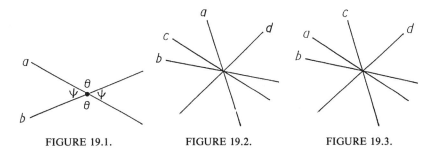

FIGURE 19.1.   FIGURE 19.2.   FIGURE 19.3.

b and *d*). The next theorem, however, shows that the cross-ratio of four lines is unchanged when we switch the lines in two pairs. This result is useful in proving Theorem 19.6.

THEOREM 19.5. In the extended plane, let *a*, *b*, *c*, *d* be four lines on an ordinary point. Then the value of the cross-ratio $R(a, b; c, d)$ is unaffected when the lines are interchanged in two pairs. Specifically, we have the following results:

    (i) interchanging *a* with *b* and *c* with *d* gives

$$Rb, a; d, c) = R(a, b; c, d); \qquad (17)$$

    (ii) interchanging *a* with *c* and *b* with *d* gives

$$R(c, d; a, b) = R(a, b; c, d); \qquad (18)$$

    (iii) interchanging *a* with *d* and *b* with *c* gives

$$R(d, c; b, a) = R(a, b; c, d).$$

Proof:

    (i) Interchanging *a* with *b* and *c* with *d* in Equation 16 shows that

$$|R(b, a; d, c)| = \frac{\sin(bd)\sin(ac)}{\sin(bc)\sin(ad)}$$

$$= \frac{\sin(ac)\sin(bd)}{\sin(ad)\sin(bc)}$$

$$= |R(a, b; c, d)| \qquad \text{(by Equation 16)}.$$

Both sides of Equation 16 are negative if and only if *a* and *b* lie in different pairs of vertical angles formed by *c* and *d* (by Definition 19.4). Thus, both sides of Equation 17 have the same sign as well as the same absolute value, and so they are equal.

(ii) Interchanging a with c and b with d in Equation 16 shows that

$$|R(c, d; a, b)| = \frac{\sin(ca) \sin(db)}{\sin(cb) \sin(da)}$$

$$= \frac{\sin(ac) \sin(bd)}{\sin(ad) \sin(bc)} \qquad \text{(by Equation 15)}$$

$$= |R(a, b; c, d)| \qquad \text{(by Equation 16).} \qquad (19)$$

The lines c and d lie in different pairs of vertical angles formed by a and b if and only if a and b lie in different pairs of vertical angles formed by c and d (Figures 19.2 and 19.3). Thus, $R(c, d; a, b)$ is negative if and only if $R(a, b; c, d)$ is negative (by Definition 19.4). Together with Equation 19, this shows that both sides of Equation 18 have the same sign as well as the same absolute value, and so they are equal.

(iii) Interchanging a with b and c with d in Equation 18 shows that $R(d, c; b, a) = R(b, a; d, c)$. We also have $R(b, a; d, c) = R(a, b; c, d)$ (by part (i)). The last two sentences show that $R(d, c; b, a) = R(a, b; c, d)$. $\square$

We can now prove the main theorem of this section, which relates cross-ratios of points and lines. This result shows that Definitions 19.1 and 19.4 are not arbitrary. The key to the proof of this result is Theorem 0.15, the Law of Sines. Because the Law of Sines relates ratios of distances with ratios of sines of angles, it provides a natural way to relate Definitions 19.1 and 19.4.

THEOREM 19.6. In the extended plane, let O be an ordinary point, and let l be an ordinary line that doesn't contain O. Let A, B, C, D be four points on l, and let a, b, c, d be four lines on O that contain A, B, C, D, respectively. Then the equation

$$R(A, B; C, D) = R(a, b; c, d)$$

holds (Figure 19.4).

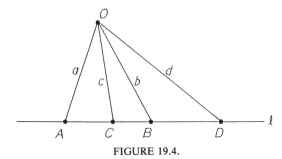

FIGURE 19.4.

Proof: There are five cases, depending on which of the points A–D, if any, is the ideal point l_∞ of l. Let XY be the distance between points X and Y.

Case 1: The points A–D are all ordinary (Figure 19.4). We have

$$|R(A, B; C, D)| = \frac{AC \cdot BD}{AD \cdot BC} \qquad \text{(by Equation 1)}$$

$$= \frac{(AC/OA)(BD/OB)}{(AD/OA)(BC/OB)}$$

$$= \frac{[\sin(ac)/\sin(lc)] \cdot [\sin(bd)/\sin(ld)]}{[\sin(ad)/\sin(ld)] \cdot [\sin(bc)/\sin(lc)]}$$

(by Theorem 0.15, the Law of Sines)

$$= \frac{\sin(ac)\sin(bd)}{\sin(ad)\sin(bc)}$$

$$= |R(a, b; c, d)| \qquad \text{(by Equation 16).} \qquad (20)$$

$R(A, B; C, D)$ is the product of the division ratios $\overline{AC}/\overline{AD}$ and $\overline{BD}/\overline{BC}$ (by Equation 1), and $R(A, B; C, D)$ is negative if and only if these division ratios have opposite signs. This occurs if and only if exactly one of the points A and B lies between C and D (by Theorem 1.4). This happens exactly when a and b lie in different pairs of vertical angles formed by c and d (since a, b, c, d contain A, B, C, D, respectively), which occurs exactly when $R(a, b; c, d)$ is negative (by Definition 19.4). In short, $R(A, B; C, D)$ is negative if and only if $R(a, b; c, d)$ is negative.

The previous paragraph shows that $R(A, B; C, D)$ and $R(a, b; c, d)$ have the same signs. These cross-ratios also have equal absolute values (by Equation 20), and so they are equal.

Case 2: A is the ideal point l_∞ (Figure 19.5). We have

$$|R(l_\infty, B; C, D)| = \frac{BD}{BC} \qquad \text{(by Equation 2)}$$

$$= \frac{BD/OB}{BC/OB}$$

$$= \frac{\sin(bd)/\sin(ld)}{\sin(bc)/\sin(lc)}$$

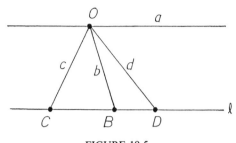

FIGURE 19.5.

(by Theorem 0.15, the Law of Sines)

$$= \frac{\sin(lc) \sin(bd)}{\sin(ld) \sin(bc)}$$

$$= \frac{\sin(ac) \sin(bd)}{\sin(ad) \sin(bc)}$$

(by Theorem 0.7, since a and l are parallel)

$$= |R(a, b; c, d)| \qquad \text{(by Equation 16).} \qquad (21)$$

Since $R(l_\infty, B; C, D) = \overline{BD}/\overline{BC}$ (by Equation 2), $R(l_\infty, B; C, D)$ is negative if and only if B lies between C and D (by Theorem 1.4). This occurs if and only if a and b lie in different pairs of vertical angles formed by c and d (since a and l are parallel, and since b, c, d contain B, C, D, respectively). This happens if and only if $R(a, b; c, d)$ is negative (by Definition 19.4). In short, $R(l_\infty, B; C, D)$ is negative if and only if $R(a, b; c, d)$ is negative.

The previous paragraph shows that $R(l_\infty, B; C, D)$ and $R(a, b; c, d)$ have the same signs. These cross-ratios also have the same absolute values (by Equation 21), and so they are equal.

Case 3: B is the ideal point l_∞. We have

$$R(A, l_\infty; C, D) = \overline{AC}/\overline{AD} \qquad \text{(by Equation 3)}$$

$$= R(l_\infty, A; D, C)$$

(by replacing B with A and interchanging C with D in Equation 2)

$$= R(b, a; d, c) \qquad \text{(by Case 2)}$$

$$= R(a, b; c, d) \qquad \text{(by Theorem 19.5i).}$$

Case 4: C is the ideal point l_∞. We have

$$R(A, B; l_\infty, D) = \overline{BD}/\overline{AD} \qquad \text{(by Equation 4)}$$

$$= \overline{DB}/\overline{DA} \qquad \text{(by Equation 6)}$$

$$= R(l_\infty, D; A, B)$$

(by replacing C with A and interchanging B with D in Equation 2)

$$= R(c, d; a, b) \qquad \text{(by Case 2)}$$

$$= R(a, b; c, d) \qquad \text{(by Theorem 19.5ii)}.$$

Case 5: D is the ideal point l_∞. We have

$$R(A, B; C, l_\infty) = \overline{AC}/\overline{BC} \qquad \text{(by Equation 5)}$$

$$= \overline{CA}/\overline{CB} \qquad \text{(by Equation 6)}$$

$$= R(l_\infty, C; B, A)$$

(by replacing D with A and interchanging B with C in Equation 2)

$$= R(d, c; b, a) \qquad \text{(by Case 2)}$$

$$= R(a, b; c, d) \qquad \text{(by Theorem 19.5iii)}. \qquad \square$$

It follows from the previous theorem that we can use slopes to compute the cross-ratio of four lines on an ordinary point. We use this fact in Section 22 when we prove that the ordinary points on the conic sections in Definition 18.1 are the familiar Euclidean conic sections.

THEOREM 19.7. In the extended plane, let a, b, c, d be four lines on an ordinary point O. Let these lines have slopes α, β, γ, δ, respectively, where a vertical line has slope ∞. Then we have

$$R(a, b; c, d) = \frac{(\gamma - \alpha)(\delta - \beta)}{(\delta - \alpha)(\gamma - \beta)},$$

where factors involving ∞ are omitted.

Proof: Let l be the vertical line one unit to the right of O (Figure 19.6). Let Q be the point of l that lies on the horizontal line through O. Consider the coordinate system on l (as in Definition 1.6) such that Q has coordinate 0 and such that the points with positive coordinates lie above Q. We follow the convention that the ideal point of l has coordinate ∞.

Let X be a point of l, and let x be its coordinate. Let m be the slope of line OX.

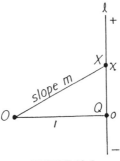

FIGURE 19.6.

If X is ordinary, we can evaluate the slope of OX as the rise over the run and obtain

$$|m| = \frac{|x - 0|}{1} = |x| \tag{22}$$

(by Definition 1.6). Moreover, m is positive if and only if X lies above Q, and this occurs if and only if x is positive. Thus, m and x have the same sign. Together with Equation 22, this shows that $m = x$ when X is ordinary.

When X is the ideal point of l, OX is the line through O parallel to l. Then OX is vertical, and so its slope m is ∞. Since X is ideal, its coordinate x is ∞, and so we also have $m = x$ when X is ideal.

The three previous paragraphs show that $m = x$ for every point X on l. In other words, the coordinate of any point X on l is the slope of the line through X and O.

The points $A = a \cap l$, $B = b \cap l$, $C = c \cap l$, $D = d \cap l$ exist (since O lies on a, b, c, d but not on l) (Figure 19.7). We have $a = OA$, $b = OB$, $c = OC$, and $d = OD$, so that the coordinates of A–D are the slopes α, β, γ, δ of the lines a–d (by the previous paragraph). Thus, we have

$$R(A, B; C, D) = \frac{(\gamma - \alpha)(\delta - \beta)}{(\delta - \alpha)(\gamma - \beta)} \tag{23}$$

(by Theorem 19.2), where factors involving ∞ are omitted. Since the lines a–d contain the points A–D, we also have

$$R(A, B; C, D) = R(a, b; c, d) \tag{24}$$

(by Theorem 19.6). Combining Equations 23 and 24 shows that

$$R(a, b; c, d) = \frac{(\gamma - \alpha)(\delta - \beta)}{(\delta - \alpha)(\gamma - \beta)},$$

where factors involving ∞ are omitted. $\square$

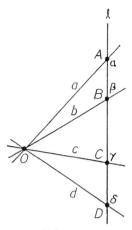

FIGURE 19.7.

It's now easy to compute the cross-ratio of four lines through an ordinary point by using their slopes. For example, let a be the line $y = 3x + 1$, let b be $x = 0$, let c be $y = -x + 1$, and let d be $y = 1$ (Figure 19.8). These lines all contain the point $(0, 1)$, and so their cross-ratio is defined. The lines have respective slopes 3, ∞, -1, 0, and so the previous theorem shows that

$$R(a, b; c, d) = \frac{(-1 - 3)(0 - \infty)}{(0 - 3)(-1 - \infty)} = \frac{-1 - 3}{0 - 3} = \frac{4}{3}.$$

Theorem 19.6 also implies that cross-ratios are preserved when a projection between two planes in Euclidean space maps points from one ordinary line to another.

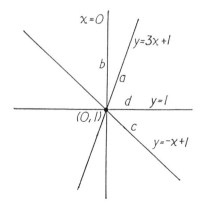

FIGURE 19.8.

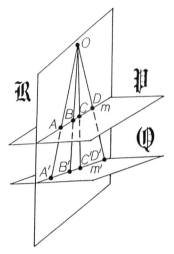

FIGURE 19.9.

THEOREM 19.8. In Euclidean space, let $\mathcal{P}$ and $\mathcal{Q}$ be two planes, and let O be a point that doesn't lie on either plane. Let $\mathcal{P}^*$ and $\mathcal{Q}^*$ be the extended planes formed from $\mathcal{P}$ and $\mathcal{Q}$, respectively. Let m be an ordinary line of $\mathcal{P}^*$ such that the projection from $\mathcal{P}$ to $\mathcal{Q}$ through O maps m to a ordinary line m' of $\mathcal{Q}^*$. If A, B, C, D are any four points of m, then we have

$$R(A, B; C, D) = R(A', B'; C', D'),$$

where A', B', C', D' are the images of A, B, C, D under the projection (Figure 19.9).

Proof: The elementary perspectivity between $\mathcal{P}$ and O matches up the points on m in $\mathcal{P}^*$ with the lines through O that lie in the plane through O and m in Euclidean space (by the second paragraph of the proof of Theorem 15.1). Likewise, the elementary perspectivity between $\mathcal{Q}$ and O matches up the points of m' with the lines through O that lie in the plane through O and m' in Euclidean space. Combining these two elementary perspectivities gives the projection from $\mathcal{P}$ to $\mathcal{Q}$ through O (by the first paragraph after the proof of Theorem 15.1), which maps m to m'. Thus, m and m' lie on the same plane $\mathcal{R}$ through O in Euclidean space (as in Figure 19.9).

Let $\mathcal{R}^*$ be the extended plane formed from $\mathcal{R}$. We consider m to have the same ideal point m_∞ in $\mathcal{R}^*$ and $\mathcal{P}^*$, and we consider m' to have the same ideal point m'_∞ in $\mathcal{R}^*$ and $\mathcal{Q}^*$.

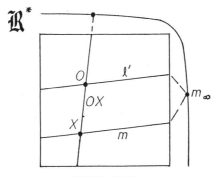

FIGURE 19.10.

If X is any ordinary point of m, the elementary perspectivity between $\mathcal{P}$ and O maps X to the line OX in $\mathcal{R}^*$ (by the discussion accompanying Figure 15.11) (Figure 19.10). The elementary perspectivity between $\mathcal{P}$ and O maps the ideal point m_∞ on m to the line l' through O parallel to m (by the second paragraph of the proof of Theorem 15.1); since l' and m are parallel lines in $\mathcal{R}$, l' contains the ideal point m_∞ on m, and we have $l' = Om_\infty$ in $\mathcal{R}^*$ (Figure 19.10). The last two sentences show that the elementary perspectivity between $\mathcal{P}$ and O maps each point X on m to the line OX in $\mathcal{R}^*$. By symmetry, the elementary perspectivity between $\mathcal{Q}$ and O maps each point Y on m' to the line OY in $\mathcal{R}^*$.

If we take the elementary perspectivity from $\mathcal{P}$ to O, and if we follow it by the elementary perspectivity from O to $\mathcal{Q}$, we obtain the projection from $\mathcal{P}$ to $\mathcal{Q}$ through O. The elementary perspectivity from $\mathcal{P}$ to O maps A, B, C, D to OA, OB, OC, OD in $\mathcal{R}^*$ (by the previous paragraph) (Figure 19.9). Similarly, the lines OA', OB', OC', OD' in $\mathcal{R}^*$ are the unique lines through O in Euclidean space that map to A', B', C', D' under the elementary perspectivity from O to $\mathcal{Q}$ (by the previous paragraph and Theorem 15.1). The projection from $\mathcal{P}$ to $\mathcal{Q}$ through O maps A, B, C, D to A', B', C', D'. Combining the last four sentences shows that $OA = OA'$, $OB = OB'$, $OC = OC'$, and $OD = OD'$ in $\mathcal{R}^*$ (as Figure 19.9 shows).

Theorem 19.6 shows that

$$R(A, B; C, D) = R(OA, OB; OC, OD) \qquad (25)$$

and

$$R(A', B'; C', D') = R(OA', OB'; OC', OD') \qquad (26)$$

in $\mathcal{R}^*$ (since OA–OD and OA'–OD' contain A–D and A'–D'). We have $OA = OA'$, $OB = OB'$, $OC = OC'$, and $OD = OD'$ (by the previous paragraph), and so Equations 25 and 26 show that

$$R(A, B; C, D) = R(A', B'; C', D') \tag{27}$$

in $\mathfrak{R}^*$. The values of the cross-ratios in Equation 27 don't depend on whether we consider A–D as points of $\mathfrak{R}^*$ or $\mathcal{P}^*$ or whether we consider A'–D' as points of $\mathfrak{R}^*$ or $\mathfrak{Q}^*$ (by Definition 19.1). Thus, Equation 27 still holds when we consider A–D as points of $\mathcal{P}^*$ and consider A'–D' as points of $\mathfrak{Q}^*$. □

We now know that a projection between planes preserves cross-ratios when it maps points from one ordinary line to another. This fact makes it easy to prove that we can use cross-ratios to determine the location of a point on an ordinary line.

THEOREM 19.9. In the extended plane, let A, B, C be three points on an ordinary line l. Then the map $X \to R(A, B; C, X)$ matches up the points on l other than A, B, C with the real numbers other than 0 and 1.

Proof: We can assume that the extended plane in question is the extended plane $\mathcal{P}^*$ formed from a plane $\mathcal{P}$ in Euclidean space. Let m be a line in $\mathcal{P}$ that contains C and doesn't equal l. There is a projection from $\mathcal{P}$ to another plane in Euclidean space so that m projects to the ideal line (by Theorem 15.3). Then l projects to an ordinary line, and C projects to an ideal point. We can replace A, B, C, l with their images under the projection, since a projection between planes preserves cross-ratios when it maps points from one ordinary line to another (by Theorem 19.8).

Thus, in addition to the theorem's hypotheses, we can assume that C is ideal. Then, for any point X on l other than A, B, C, we have

$$R(A,B; C, X) = \overline{BX}/\overline{AX} = \overline{XB}/\overline{XA}$$

(by Equations 4 and 6). The map $X \to \overline{XB}/\overline{XA}$ matches up the ordinary points X on l other than A and B with the real numbers other than 0 and 1 (by Theorem 1.9). The ordinary points on l other than A and B are the points on l other than A, B, C (since C is the ideal point on l). Combining the last three sentences shows that the map $X \to R(A, B; C, X)$ matches up the points on l other than A, B, C with the real numbers other than 0 and 1. □

We use the next theorem in Section 20 to derive results on incidence from results on cross-ratios. Theorems 19.6, 19.8, and 19.9 are all used to prove the next theorem.

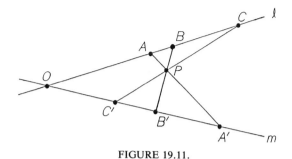

FIGURE 19.11.

THEOREM 19.10. In the extended plane, let l and m be two ordinary lines on a point O. Let A, B, C be three points on l other than O, and let A', B', C' be three points on m other than O. Then the lines AA', BB', CC' are concurrent if and only if the equation

$$R(O, A; B, C) = R(O, A'; B', C')$$

holds (Figure 19.11).

Proof: The point $P = AA' \cap BB'$ exists and lies on neither l nor m (by Theorem 14.3ii).

First, assume that the lines AA', BB', CC' are concurrent and that P is ordinary. Since P is the unique point where AA' and BB' intersect, the fact that AA', BB', CC' are concurrent implies that P also lies on CC'. P doesn't lie on l or m (by the previous paragraph), and P, l, m are all ordinary, and so Theorem 19.6 shows that

$$R(O, A; B, C) = R(PO, PA; PB, PC) \tag{28}$$

and

$$R(O, A'; B', C') = R(PO, PA'; PB', PC'). \tag{29}$$

The relations $PA = PA'$, $PB = PB'$, $PC = PC'$ hold (since AA', BB', CC' contain P), and so Equations 28 and 29 show that

$$R(O, A: B, C) = R(O, A'; B', C').$$

Conversely, assume that

$$R(O, A; B, C) = R(O, A'; B', C') \tag{30}$$

and that P is ordinary. Since P doesn't lie on l or m (by the first paragraph of the proof), the line PC exists and intersects m at a unique point C'' (Figure 19.12). The lines AA', BB', CC'' are concurrent at the ordinary point P, and so the previous paragraph shows that

$$R(O, A; B, C) = R(O, A'; B', C''). \tag{31}$$

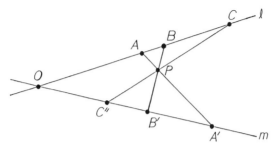

FIGURE 19.12.

Equation 30 also holds, by assumption. Combining Equations 30 and 31 shows that

$$R(O, A'; B', C') = R(O, A'; B', C'').$$

It follows that $C' = C''$ (by Theorem 19.9). Thus, since AA', BB', CC'' are concurrent at P (Figure 19.12), so are AA', BB', CC' (Figure 19.11).

The two previous paragraphs show that the theorem holds when P is ordinary. If P is ideal, we can project to another plane so that P projects to an ordinary point and so that l and m project to ordinary lines (by Theorem 15.3). We can replace P, A–D, A'–D' by their images under the projection (since Theorems 15.2 and 19.8 show that a projection preserves incidence and that it preserves cross-ratios when it maps points from one ordinary line to another). Thus, we can assume that P is ordinary, and we've already proved that the theorem holds in this case. ☐

EXERCISES

19.1. Identify ordinary points on the x axis by their x coordinates, and identify the ideal point on the x axis as ∞. Evaluate the following cross-ratios of points on the x axis.

(a) $R(3, -1; 5, 2)$

(b) $R(3, 1; -2, 6)$

(c) $R(0, 5; 2, 1)$

(d) $R(\infty, 3; 1, 4)$

(e) $R(0, 3; \infty, -2)$

(f) $R(-4, 2; 1, \infty)$

19.2. Four lines a, b, c, d on an ordinary point O are given in each part of this exercise. Evaluate $R(a, b; c, d)$.

(a) a is $y = x$, b is $y = 2x$, c is $y = 0$, d is $y = 4x$, O is $(0, 0)$.

(b) a is $y = -x$, b is $x = 0$, c is $y = 3x$, d is $y = 2x$, O is $(0, 0)$.

(c) a is $y = 2$, b is $y = 3x + 2$, c is $x = 0$, d is $y = -x + 2$, O is $(0, 2)$.

(d) a is $y = -2x + 2$, b is $y = 0$, c is $y = x - 1$, d is $x = 1$, O is $(1, 0)$.

(e) a is $y = x - 3$, b is $y = \frac{1}{2}x - 2$, c is $y = -3x + 5$, d is $y = -x + 1$, O is $(2, -1)$.

19.3. In the extended plane, let A, B, C, D be four points on an ordinary line. Prove that A, B; C, D is a harmonic set if and only if $R(A, B; C, D) = -1$.

In Exercises 19.4–19.9, let A, B, C, D be four points on an ordinary line, and set $r = R(A, B; C, D)$. These exercises show that the cross-ratio generally takes 6 different values when A, B, C, D are substituted in $R(\_, \_; \_, \_)$ in the 24 possible ways.

19.4. Prove that $R(B, A; D, C)$, $R(C, D; A, B)$, and $R(D, C; B, A)$ all equal r. (Thus, we can interchange the points in $R(A, B; C, D)$ in two pairs without affecting the value of the cross-ratio.)

19.5.
(a) Use Definition 19.1 to prove that $R(B, A; C, D) = 1/r$.
(b) Use part (a) and Exercise 19.4 to prove that $R(A, B; D, C)$, $R(C, D; B, A)$, and $R(D, C; A, B)$ all equal $1/r$.

19.6.
(a) Prove that $R(A, C; B, D) = 1 - r$. (See Theorem 19.2 and possibly Theorem 19.8.)
(b) Use Exercise 19.4 to find three other ways to substitute A, B, C, D into $R(\_, \_; \_, \_)$ to obtain the value $1 - r$.

19.7.
(a) Use Exercises 19.5(a) and 19.6(a) to prove that
$$R(C, A; B, D) = 1/(1 - r).$$
(b) Use Exercise 19.4 to find three other ways to substitute A, B, C, D into $R(\_, \_; \_, \_)$ to obtain the value $1/(1 - r)$.

19.8.
(a) Use Exercises 19.5(a) and 19.6(a) to prove that
$$R(B, C; A, D) = (r - 1)/r.$$
(b) Use Exercise 19.4 to find three other ways to substitute A, B, C, D into $R(\_, \_; \_, \_)$ to obtain the value $(r - 1)/r$.

19.9.
(a) Use Exercises 19.5(a) and 19.8(a) to prove that
$$R(C, B; A, D) = r/(r - 1).$$
(b) Use Exercise 19.4 to find three other ways to substitute A, B, C, D into $R(\_, \_; \_, \_)$ to obtain the value $r/(r - 1)$.

19.10. In Euclidean space, let $\mathcal{P}$ and $\mathcal{Q}$ be two planes, and let O be a point that doesn't lie on either plane. Let $\mathcal{P}^*$ and $\mathcal{Q}^*$ be the extended planes formed from $\mathcal{P}$ and $\mathcal{Q}$. Let S be an ordinary point of $\mathcal{P}^*$ such that the

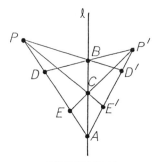

FIGURE 19.13.

projection from $\mathcal{P}$ to $\mathcal{Q}$ through O maps S to an ordinary point S' of $\mathcal{Q}^*$. If a, b, c, d are four lines in $\mathcal{P}$ through S, prove that

$$R(a, b; c, d) = R(a', b'; c', d'),$$

where a', b', c', d' are the images of a, b, c, d under the projection. (See Theorems 19.6 and 19.8.)

19.11. In the extended plane, let A, B, C, D, E be five points on an ordinary line. Prove that

$$R(A, B; C, E) = R(A, B; C, D)R(A, B; D, E).$$

19.12. In the extended plane, let A, B, C be three points on a line l, and let P and P' be two points that don't lie on l (Figure 19.13). Assume that none of the points A, B, C lies on line PP'. Set $D = AP \cap BP'$, $D' = AP' \cap BP$, $E = AP \cap CP'$, and $E' = AP' \cap CP$. Assume that AP and AP' are ordinary lines. Prove that $R(A, P; D, E) = R(A, P'; D', E')$.

(*Hint:* Possible approaches use Theorem 19.6 or 19.10. These theorems require certain points and lines to be ordinary, and this condition can be met by using Theorems 15.2, 15.3, and 19.8.)

19.13. In the extended plane, let A, B, C, D be four points, no three of which are collinear (Figure 19.14). Set $E = AC \cap BD$ and $F = AB \cap CD$. Let l be a line on F that doesn't contain any of the points A, B, C, D, E. Set $G = AC \cap l$, $H = BD \cap l$, and $I = BC \cap l$. Assume that BD and l are ordinary lines. Prove that $R(B, D; E, H) = R(F, I; H, G)$. (See the hint to Exercise 19.12.)

19.14. In the extended plane, let l and m be two ordinary lines. Let A, B, C, D be four points on l other than $l \cap m$, and let A', B', C', D' be four points on m other than $l \cap m$. Prove that $R(A, B; C, D) = R(A', B'; C', D')$ if and only if $AB' \cap A'B$, $AC' \cap A'C$, $AD' \cap A'D$ are collinear. (See the hint to Exercise 19.12.)

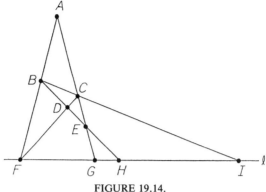

FIGURE 19.14.

19.15. In the extended plane, let A–F be six points on a circle (Figure 19.15). The points $L = EF \cap AB$, $M = AB \cap CD$, $N = CD \cap EF$ exist (by Theorem 4.5ii). Prove that

$$\frac{\overline{AL} \cdot \overline{BL} \cdot \overline{CM} \cdot \overline{DM} \cdot \overline{EN} \cdot \overline{FN}}{\overline{AM} \cdot \overline{BM} \cdot \overline{CN} \cdot \overline{DN} \cdot \overline{EL} \cdot \overline{FL}} = 1,$$

when directed distances involving ideal points are omitted. (See Theorem 6.5.)

19.16. In the extended plane, let LMN be a triangle whose sides are ordinary lines (Figure 19.16). Let A and B be points on line LM, let C and D points on line MN, and let E and F be points on line LN, such that the only possible equality among the points A–D, L–N is $A = B$.

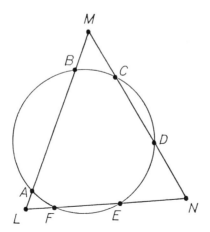

FIGURE 19.15.

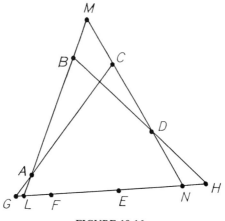

FIGURE 19.16.

Set $G = AC \cap LN$ and $H = BD \cap LN$. Assume also that the only possible equality among the points E-H, L-N is $G = H$. Prove that

$$R(E, G; L, N) = R(F, H; N, L) \tag{32}$$

if and only if the equation

$$\frac{\overline{AL} \cdot \overline{BL} \cdot \overline{CM} \cdot \overline{DM} \cdot \overline{EN} \cdot \overline{FN}}{\overline{AM} \cdot \overline{BM} \cdot \overline{CN} \cdot \overline{DN} \cdot \overline{EL} \cdot \overline{FL}} = 1 \tag{33}$$

holds when directed distances involving ideal points are omitted.

(*Hint:* When A-H and L-N are ordinary, one possible way to relate Equations 32 and 33 is to apply Menelaus' Theorem to the two sets of collinear points $\{A, C, G\}$ and $\{B, D, H\}$ on the sides of triangle LMN. Cases where A-H and L-N are not all ordinary need separate consideration, and Theorem 2.2 and Exercise 2.11 may be helpful.)

19.17. Prove the following result, known as *Carnot's Theorem* (Figure 19.17).

Theorem. In the extended plane, let A-F be six points on a conic section $\mathcal{K}$ such that the lines AB, CD, EF are ordinary. Set $L = EF \cap AB$, $M = AB \cap CD$, and $N = CD \cap EF$. Then the equation

$$\frac{\overline{AL} \cdot \overline{BL} \cdot \overline{CM} \cdot \overline{DM} \cdot \overline{EN} \cdot \overline{FN}}{\overline{AM} \cdot \overline{BM} \cdot \overline{CN} \cdot \overline{DN} \cdot \overline{EL} \cdot \overline{FL}} = 1 \tag{34}$$

holds when directed distances involving ideal points are omitted.

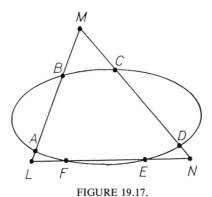

FIGURE 19.17.

(*Hint:* One possible approach is as follows. The conic section $\mathcal{K}$ is the image of a circle $\mathcal{K}'$ under a projection in Euclidean space from a plane $\mathcal{P}$ to a plane $\mathcal{Q}$ through a point O. Let A'-F' and L'-N' be the points of $\mathcal{P}$ that project to A-F and L-N in $\mathcal{Q}$. Set $G' = A'C' \cap L'N'$ and $H' = B'D' \cap L'N'$. Use Exercises 19.15 and 19.16 to deduce that

$$R(E', G'; L', N') = R(F', H'; N', L'),$$

and then use Theorem 19.8 and Exercise 19.16 to conclude that Equation 34 holds.)

19.18. In the extended plane, let A, C, D, E, F be five points on a conic section $\mathcal{K}$ such that the lines $\tan A$, CD, EF are ordinary (Figure 19.18). Set $L = EF \cap \tan A$, $M = \tan A \cap CD$, and $N = CD \cap EF$. Prove that

$$\frac{(\overline{AL})^2 \cdot \overline{CM} \cdot \overline{DM} \cdot \overline{EN} \cdot \overline{FN}}{(\overline{AM})^2 \cdot \overline{CN} \cdot \overline{DN} \cdot \overline{EL} \cdot \overline{FL}} = 1, \tag{35}$$

where directed distances involving ideal points are omitted.

(*Hint:* One possible approach is as follows. First use Theorems 6.5 and 6.6 to derive Equation 35 when $\mathcal{K}$ is a circle. Then prove that Equation 35 holds for every conic section $\mathcal{K}$ by adapting the hint to Exercise 19.17.)

19.19. In the Euclidean plane, let n be an asymptote of a hyperbola $\mathcal{K}$ (Figure 19.19). Let C-F be four points on $\mathcal{K}$ such that CD and EF are parallel lines. Prove that CD intersects n at a point M and that EF intersects n at a point L. Use Exercise 19.18 with an appropriate choice of the ideal line to prove that $\overline{CM} \cdot \overline{DM} = \overline{EL} \cdot \overline{FL}$. (Thus, the value of the expression $\overline{CM} \cdot \overline{DM}$ remains constant as C and D vary over all pairs of points on a hyperbola such that CD is parallel to a fixed line. For each choice of C and D, M is the point where CD intersects a specified asymptote.)

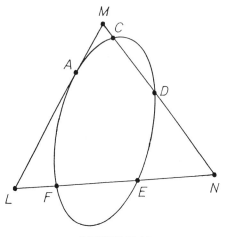

FIGURE 19.18.

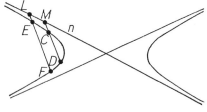

FIGURE 19.19.

Section 20.

Polygons Inscribed in Conic Sections

A triangle $\mathfrak{I}$ inscribed in a conic section is perspective from a point and a line with the triangle of tangents at the vertices of $\mathfrak{I}$ (by Theorem 18.7) (Figures 18.16 and 18.17). Hexagons, pentagons, and quadrilaterals inscribed in conic sections also have striking properties. We develop some of these properties in this section by using cross-ratios.

We start by relating cross-ratios and circles. If A–D are four points on a circle $\mathfrak{K}$, we consider the cross-ratio of the four lines a–d that join A–D to another point E on $\mathfrak{K}$ (Figure 20.1). We prove that the value of this cross-ratio doesn't depend on the choice of E; we obtain the same value for the cross-ratio of the lines a'–d' that join A–D to F for any point F on $\mathfrak{K}$ other than A–E.

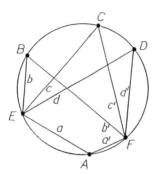

FIGURE 20.1.

THEOREM 20.1. In the extended plane, let A–F be six points on a circle $\mathcal{K}$. Then we have

$$R(EA, EB; EC, ED) = R(FA, FB; FC, FD)$$

(Figure 20.1).

Proof: As in Figure 20.1, we set $a = EA$, $b = EB$, $c = EC$, $d = ED$, $a' = FA$, $b' = FB$, $c' = FC$, and $d' = FD$. We must prove that

$$R(a, b; c, d) = R(a', b'; c', d'). \tag{1}$$

No three points on a circle are collinear (by Theorem 4.5ii), and all points on a circle are ordinary. Thus, a–d are distinct lines on the ordinary point E, and a–d' are distinct lines on the ordinary point F, and so the cross-ratios in Equation 1 are determined by Definition 19.4.

$\angle AEC$ and $\angle AFC$ are either equal or supplementary (by Theorem 6.2), and so we have

$$\sin(ac) = \sin(a'c') \tag{2}$$

(by Definition 19.3 and Equation 4 of Section 0). By symmetry, we also have

$$\sin(bd) = \sin(b'd'), \tag{3}$$

$$\sin(ad) = \sin(a'd'), \tag{4}$$

$$\sin(bc) = \sin(b'c'). \tag{5}$$

It follows that

$$|R(a, b; c, d)| = \frac{\sin(ac)\sin(bd)}{\sin(ad)\sin(bc)} \quad \text{(by Definition 19.4)}$$

$$= \frac{\sin(a'c')\sin(b'd')}{\sin(a'd')\sin(b'c')} \quad \text{(by Equations 2–5)}$$

$$= |R(a', b'; c', d')| \tag{6}$$

(by Definition 19.4).

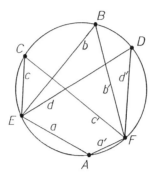

FIGURE 20.2.

$R(a, b; c, d)$ is negative if and only if a and b lie in different pairs of vertical angles formed by c and d, by Definition 19.4. (Figure 20.1 illustrates the case where $R(a, b; c, d)$ is positive, and Figure 20.2 illustrates the case where $R(a, b; c, d)$ is negative.) Thus, $R(a, b; c, d)$ is negative if and only if A and B lie on different arcs of $\mathcal{K}$ having C and D as endpoints (since a–d contain A–D). By symmetry, $R(a', b'; c', d')$ is also negative if and only if A and B lie on different arcs of $\mathcal{K}$ having C and D as endpoints. Thus, $R(a, b; c, d)$ and $R(a', b'; c', d')$ have the same sign. Since these cross-ratios also have the same absolute value (by Equation 6), they are equal. $\square$

The next result provides the link between the previous theorem and hexagons. We state this result in general form for use in Section 22.

THEOREM 20.2. In the extended plane, let A–F be six points such that no three are collinear and A–E are ordinary. Set $L = AB \cap DE$, $M = BC \cap EF$, and $N = CD \cap FA$. Then L, M, N are collinear if and only if the equation

$$R(CA, CB; CD, CF) = R(EA, EB; ED, EF) \qquad (7)$$

holds (Figure 20.3).

Proof: Since no three of the points A–F are collinear, the points L, M, N exist, and so do the points $O = AF \cap BC$ and $P = AB \cap EF$ (Figure 20.4).

Since no three of the points A–F are collinear, C doesn't lie on AF, and the lines CA, $CO = CB$, $CN = CD$, CF exist and are distinct. Thus, the points A, O, N, F are distinct, and we have

$$R(A, O; N, F) = R(CA, CB; CD, CF), \qquad (8)$$

by Theorem 19.6 and the facts that AF and C are ordinary (since A–E are).

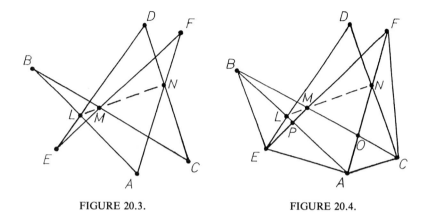

FIGURE 20.3.   FIGURE 20.4.

Similarly, since no three of the points A–F are collinear, E doesn't lie on AB, and the lines EA, EB, $EL = ED$, $EP = EF$ exist and are distinct. Thus, the points A, B, L, P are distinct, and we have

$$R(A, B; L, P) = R(EA, EB; ED, EF), \tag{9}$$

by Theorem 19.6 and the facts that AB and E are ordinary (since A–E are). Equations 8 and 9 show that Equation 7 holds if and only if

$$R(A, O; N, F) = R(A, B; L, P). \tag{10}$$

AF and AB are distinct ordinary lines (since A is ordinary and A, B, F are noncollinear), and so Equation 10 holds if and only if the lines OB, NL, FP are concurrent (by Theorem 19.10). We have $M = BC \cap EF = OB \cap FP$; thus, OB, NL, FP are concurrent if and only if M lies on NL. The last three sentences show that Equation 7 holds if and only if L, M, N are collinear.   □

In the proof of Theorem 20.2, we used Theorem 19.6 to obtain Equations 8 and 9. This use of Theorem 19.6 depends on the fact that the points C and E and the lines AF and AB are ordinary, which follows from the assumption that the points A–E are all ordinary. On the other hand, we note that each of the points F, L, M, N, O, P can be either ordinary or ideal.

The points L, M, N in Theorem 20.2 are the intersections of the three pairs of opposite sides of hexagon $ABCDEF$—$\{AB, DE\}$, $\{BC, EF\}$, $\{CD, FA\}$. (See Figure 20.5, which shows the same solid lines as Figure 20.3.) Thus, we can restate Theorem 20.2 as follows: In the extended plane, if A–F are six points such that no three are collinear and A–E are ordinary, then the three pairs of opposite sides of hexagon $ABCDEF$ intersect in three collinear points if and only if Equation 7 holds.

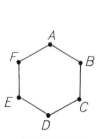

FIGURE 20.5.   FIGURE 20.6.

We say that a hexagon is *inscribed* in a conic section if the vertices of the hexagon lie on the conic section. By combining Theorems 20.1 and 20.2, we can prove that, *if a hexagon is inscribed in a conic section, the three pairs of opposite sides intersect in collinear points* (Figure 20.6). The result is called *Pascal's Theorem*. It is analogous to Pappus' Theorem, which states that, if a hexagon is inscribed in two lines, the three pairs of opposite sides intersect in collinear points (Figure 15.20).

THEOREM 20.3 (Pascal's Theorem).
In the extended plane, let A–F be six points on a conic section $\mathcal{K}$. Then the points $L = AB \cap DE$, $M = BC \cap EF$, $N = CD \cap FA$ are collinear (Figure 20.6).

Proof: First assume that $\mathcal{K}$ is a circle. We have

$$R(CA, CB; CD, CF) = R(EA, EB; ED, EF), \qquad (11)$$

by replacing F with C, C with D, and D with F in Theorem 20.1 (Figure 20.7). Since A–F lie on the circle $\mathcal{K}$, they are ordinary points, and no three of them are collinear (by Theorem 4.5ii). Thus, Equation 11 implies that L, M, N are collinear (by Theorem 20.2).

We've proved that the theorem holds when $\mathcal{K}$ is a circle. Since the theorem is stated in terms of incidence, projection between planes shows that the theorem holds for any conic section $\mathcal{K}$ (by Definition 18.1 and Theorem 15.2) (Figure 20.6). □

Let A–F be six points on a conic section. If we arrange the points in any order to form a hexagon, Pascal's Theorem shows that there is a line through the points where the three pairs of opposite sides of the hexagon intersect. Of course, this line depends on the order in which the points are arranged into a hexagon. For example, in Figure 20.8, one dotted line contains the three points $L = AB \cap DE$, $M = BC \cap EF$, $N = CD \cap FA$

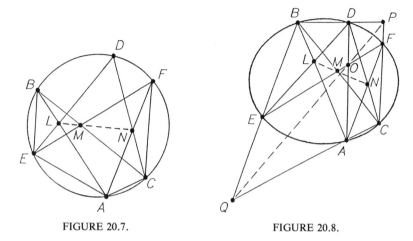

FIGURE 20.7.　　　　　　　　　FIGURE 20.8.

where the opposite sides of hexagon $ABCDEF$ intersect, and another dotted line contains the three points $O = AD \cap EF$, $P = DB \cap FC$, $Q = BE \cap CA$ where the opposite sides of hexagon $ADBEFC$ intersect.

If A–F are six points on a conic section $\mathcal{K}$, the three pairs of opposite sides of hexagon $ABCDEF$ intersect in the points

$$L = AB \cap DE, \qquad M = BC \cap EF, \qquad N = CD \cap FA, \qquad (12)$$

and Pascal's Theorem 20.3 states that these points are collinear (Figure 20.9a). Imagine that F moves around $\mathcal{K}$ until it approaches A. Then the line EF in (12) approaches EA, and the line FA in (12) approaches $\tan A$. This suggests that the points

$$L = AB \cap DE, \qquad M = BC \cap EA, \qquad N = CD \cap \tan A \qquad (13)$$

are collinear (Figure 20.9b). Instead of the inscribed hexagon $ABCDEF$, we now consider the inscribed pentagon $ABCDE$ and the tangent at the vertex A of the pentagon.

We follow the convention that AA represents $\tan A$. Then we obtain (13) be replacing F with A in Equation 12. Thus, we can remember the points in (13) as the intersections of the three pairs of opposite sides of "hexagon" $ABCDEA$—$\{AB, DE\}$, $\{BC, EA\}$, $\{CD, AA\}$ (Figure 20.10).

We prove that the points in (13) are collinear by matching up the lines on A with the lines on E so that AF matches up with EF for every point F on $\mathcal{K}$ other than A and E. We use the fact that the points in (12) are collinear to define this matching in terms of the points A–E. The fact that EA corresponds to $\tan A$ under this matching implies that the points in (13) are collinear.

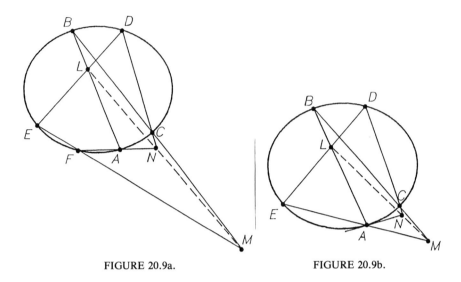

FIGURE 20.9a.   FIGURE 20.9b.

THEOREM 20.4. In the extended plane, let A–E be five points on a conic section $\mathcal{K}$ (Figure 20.11). Set $L = AB \cap DE$. For any line m on E, the points $M = m \cap BC$ and $N = LM \cap CD$ and the line $n = AN$ exist. The correspondence $m \to n$ matches up the lines on E with the lines on A. The lines $m = EX$ and $n = AX$ correspond for each point X on $\mathcal{K}$ other than A and E. The lines $m = EA$ and $n = \tan A$ also correspond, as do $m = \tan E$ and $n = AE$.

Proof: No three points on $\mathcal{K}$ are collinear (by Theorem 18.4i). Thus, $L = AB \cap DE$ exists and doesn't equal $B = AB \cap BC$ or $D = DE \cap CD$. Hence, L doesn't lie on BC or CD.

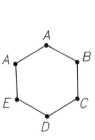

FIGURE 20.10.

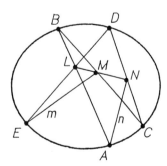

FIGURE 20.11.

If m is any line on E, we have $m \neq BC$ (since E lies on m but not BC, by Theorem 18.4i), and so $M = m \cap BC$ exists. Since L doesn't lie on BC or CD (by the previous paragraph), line LM exists and intersects CD at a unique point N. Since A doesn't lie on CD (by Theorem 18.4i), the line $n = AN$ exists.

Conversely, let n be any line on A. We have $n \neq CD$ (since A lies on n but not CD), and so n and CD intersect at a unique point N. L doesn't lie on BC or CD (by the first paragraph of the proof), and so LN exists and intersects BC at a unique point M. E doesn't lie on BC, and so line EM exists and is the unique line m that maps to n under the given correspondence.

The two previous paragraphs show that the map $m \to n$ matches up the lines m on E with the lines n on A.

Let F be a point on $\mathcal{K}$ other than A–E, and set $m = EF$ (Figure 20.12). Then we have $M = m \cap BC = BC \cap EF$. It follows that $N = LM \cap CD = CD \cap FA$ (since $L = AB \cap DE$, $M = BC \cap EF$, and $CD \cap FA$ are collinear, by Theorem 20.3). Accordingly, we have $n = AN = AF$. Thus, the correspondence $m \to n$ maps EF to AF for any point F on $\mathcal{K}$ other than A–E.

If we set $m = EB$ (Figure 20.13), then we have $M = m \cap BC = B$, $N = LM \cap CD = AB \cap CD$ (since both L and M lie on AB), and $n = AN = AB$. Thus, the correspondence $m \to n$ maps EB to AB.

If we set $m = EC$ (Figure 20.14), then we have $M = m \cap BC = C$, $N = LM \cap CD = C$, and $n = AN = AC$. Thus, the correspondence $m \to n$ maps EC to AC.

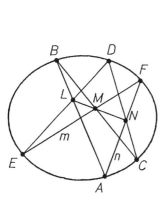

FIGURE 20.12.

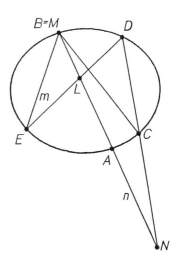

FIGURE 20.13.

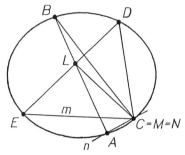

FIGURE 20.14.

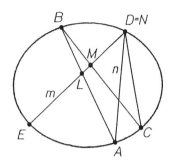

FIGURE 20.15.

If we set $m = ED$ (Figure 20.15), then we have $M = m \cap BC = ED \cap BC$, $N = LM \cap CD = ED \cap CD = D$ (since both L and M lie on ED), and $n = AN = AD$. Thus, the correspondence $m \to n$ maps ED to AD.

The correspondence $m \to n$ matches up the lines on E with the lines on A (by the fourth paragraph of the proof). This correspondence matches up the lines EX and AX for every point X on $\mathcal{K}$ other than E and A (by the last four paragraphs). Every line through E other than $\tan E$ has the form EX for a unique point X on $\mathcal{K}$ other than E, and every line through A other than $\tan A$ has the form AX for a unique point X on $\mathcal{K}$ other than A (by Theorem 18.4ii and Definition 18.5).

The previous paragraph implies that the correspondence $m \to n$ matches up the lines $\tan E$ and EA on E with the lines $\tan A$ and AE on A in some order.

If we set $m = EA$, then we have $M = m \cap BC = EA \cap BC$ (Figure 20.16). We have $E = DE \cap EA$ (by Theorem 18.4i), and E doesn't equal $L = DE \cap AB$ (since A, B, E are noncollinear, by Theorem 18.4i), and so L doesn't lie on EA. The last two sentences imply that

$$M = LM \cap AE. \tag{14}$$

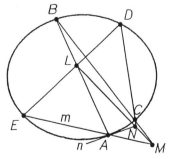

FIGURE 20.16.

$M = EA \cap BC$ doesn't equal $C = CD \cap BC$ (by Theorem 18.4i), and so M doesn't lie on CD. Thus, $N = LM \cap CD$ doesn't equal M, and so N doesn't lie on AE (by Equation 14). Hence, we have $EA \neq AN = n$, and so the previous paragraph implies that $m = EA$ maps to $n = \tan A$ (Figure 20.16) and that $m = \tan E$ maps to $n = AE$. □

As in the second paragraph before Theorem 20.4, we let AA denote $\tan A$, and let EE denote $\tan E$, for points A and E on a conic section. Thus, we can combine the last two sentences in the statement of Theorem 20.4 by saying that the correspondence $m \to n$ matches up $m = EX$ with $n = AX$ for every point X on $\mathcal{K}$; the second-to-last sentence of Theorem 20.4 describes the case where X doesn't equal A or E, and the last sentence describes the cases $X = A$ (where $m = EA$ maps to $n = AA = \tan A$), and $X = E$ (where $m = EE = \tan E$ maps to $n = AE$). Theorem 18.4(ii) implies that every line on E has the form EX for a unique point X on $\mathcal{K}$ (where $X = E$ gives $EE = \tan E$) and that every line on A has the form AX for a unique point X on $\mathcal{K}$ (where $X = A$ gives $AA = \tan A$).

In particular, Theorem 20.4 states that $m = EA$ maps to $n = \tan A$ (Figure 20.16). In this case, we have $M = m \cap BC = BC \cap EA$. The fact that $N = LM \cap CD$ lies on $AN = n = \tan A$ implies that $N = CD \cap \tan A$ lies on LM (where $CD \neq \tan A$, by Theorem 18.4i). Thus, we've proved the following theorem, which states the points in (13) are collinear.

THEOREM 20.5. In the extended plane, let A–E be five points on a conic section. Then the points $L = AB \cap DE$, $M = BC \cap EA$, $N = CD \cap \tan A$ are collinear (Figure 20.17). □

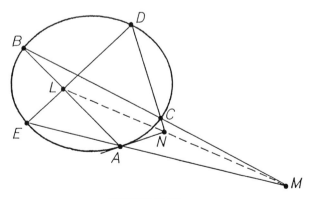

FIGURE 20.17.

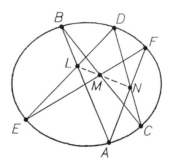

FIGURE 20.18.

As we noted after (13), the previous result is the analogue of Pascal's Theorem 20.3 where, instead of a hexagon inscribed in a conic section, we consider an inscribed pentagon and the tangent at one of the vertices.

Theorem 20.4 also implies the next result, which uses Pascal's Theorem 20.3 to determine exactly which points lie on a conic section.

THEOREM 20.6. In the extended plane, let A–E be five points on a conic section $\mathcal{K}$. Then $\mathcal{K}$ consists of A–E and all points F such that F is not collinear with any two of the points A–E and such that the points $L = AB \cap DE$, $M = BC \cap EF$, $N = CD \cap FA$ are collinear (Figure 20.18).

Proof: If F is a point on $\mathcal{K}$ other than A–E, then F is not collinear with any two of the points A–E (by Theorem 18.4i), and the points $L = AB \cap DE$, $M = BC \cap EF$, $N = CD \cap FA$ are collinear (by Theorem 20.3).

Conversely, let F be a point that is not collinear with any two of the points A–E and is such that the points $L = AB \cap DE$, $M = BC \cap EF$, $N = CD \cap FA$ are collinear. If we set $m = EF$, we have $m \cap BC = BC \cap EF = M$, $LM \cap CD = N$ (since L, M, N are collinear), and $AN = AF$ (since $N = CD \cap FA$). Thus, the correspondence in Theorem 20.4 maps $m = EF$ to $n = AF$.

Neither EF nor AF equals EA (since F is not collinear with E and A), and so Theorem 20.4 shows that there is a point G on $\mathcal{K}$ other than E and A such that $EF = m = EG$ and $AF = n = AG$. It follows that

$$G = EG \cap AG = EF \cap AF = F$$

(by Theorem 18.4i). Thus, F lines on $\mathcal{K}$, since G does. $\square$

If A–E are five points on a conic section $\mathcal{K}$, the three pairs of opposite sides of "hexagon" $ABCDEA$ intersect at the points

$$L = AB \cap DE, \quad M = BC \cap EA, \quad N = CD \cap AA = CD \cap \tan A \quad (15)$$

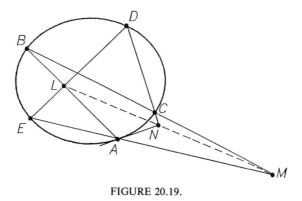

FIGURE 20.19.

(Figure 20.19). These points are collinear, by Theorem 20.5. Imagine that C moves around $\mathcal{K}$ until it approaches D. Then the line BC in (15) approaches BD, and CD approaches $DD = \tan D$. This suggests that the points

$$L = AB \cap DE, \qquad M = BD \cap EA, \qquad N = \tan D \cap \tan A \qquad (16)$$

are collinear (Figure 20.20). Instead of the inscribed hexagon in Pascal's Theorem, we now consider a quadrilateral inscribed in a conic section and the tangents at two of the vertices. The points in (16) can be remembered as the intersections of the three pairs of opposite sides of "hexagon" $ABDDEA$—$\{AB, DE\}$, $\{BD, EA\}$, $\{DD, AA\}$ (Figure 20.21).

We prove that the points in (16) are collinear by matching up the lines on B with the lines on D so that BC matches up with DC for every point C on $\mathcal{K}$ other than A and D. We use the fact that the points in (15) are collinear in order to define this matching in terms of A, B, D, E, and $\tan A$. The fact that BD corresponds to $\tan D$ under this matching implies that the points in (16) are collinear.

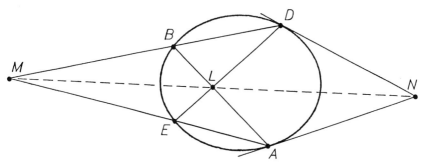

FIGURE 20.20.

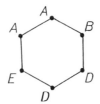

FIGURE 20.21.

THEOREM 20.7. In the extended plane, let A, B, D, E be four points on a conic section $\mathcal{K}$ (Figure 20.22). Set $L = AB \cap DE$. For any line m on B, the points $M = m \cap EA$ and $N = LM \cap \tan A$ and the line $n = DN$ exist. The correspondence $m \to n$ matches up the lines on B with the lines on D. This correspondence matches up $m = BX$ with $n = DX$ for every point X on $\mathcal{K}$ other than B and D. The correspondence also matches up $m = BD$ with $n = \tan D$, and it matches up $m = \tan B$ with $n = DB$.

Proof: No three points of $\mathcal{K}$ are collinear (by Theorem 18.4i). Thus, $L = AB \cap DE$ exists and doesn't equal A or B. It follows that L doesn't lie on EA, $\tan A$, or BD, since we have $A = AB \cap EA$, $A = AB \cap \tan A$, and $B = AB \cap BD$ (by Theorem 18.4i and Definition 18.5).

If m is any line on B, we have $m \neq EA$ (since B lies on m but not EA, by Theorem 18.4i), and so $M = m \cap EA$ exists. Since L doesn't lie on EA or $\tan A$ (by the previous paragraph), the line LM and the point $N = LM \cap \tan A$ exist. We have $D \neq N$ (since $\tan A$ contains N but not D, by Definition 18.5), and so the line $n = DN$ exists.

Conversely, let n be any line on D. We have $n \neq \tan A$ (since D lies on n but not $\tan A$), and so n and $\tan A$ intersect at a unique point N. L doesn't lie on $\tan A$ or EA (by the first paragraph of the proof), and so the line LN

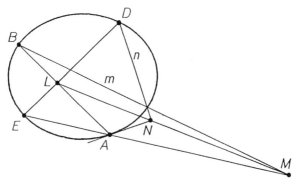

FIGURE 20.22.

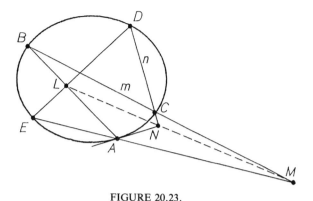

FIGURE 20.23.

exists and intersects EA at a unique point M. We have $B \neq M$ (since EA contains M but not B), and so BM is the unique line m that maps to n under the given correspondence.

The last two paragraphs show that the map $m \rightarrow n$ matches up the lines m on B with the lines n on D.

Let C be a point on $\mathcal{K}$ other than A, B, D, E, and set $m = BC$ (Figure 20.23). Then we have $M = m \cap EA = BC \cap EA$. It follows that $N = LM \cap \tan A = CD \cap \tan A$ (since $L = AB \cap DE$, $M = BC \cap EA$, and $CD \cap \tan A$ are collinear, by Theorem 20.5). Accordingly, we have $n = DN = DC$. Thus, the correspondence $m \rightarrow n$ maps BC to DC for any point C on $\mathcal{K}$ other than A, B, D, E.

If we set $m = BA$ (Figure 20.24), we have $M = m \cap EA = A$, $N = LM \cap \tan A = A$, and $n = DN = DA$. Thus, the correspondence $m \rightarrow n$ maps BA to DA.

If we set $m = BE$ (Figure 20.25), we have $M = m \cap EA = E$, $N = LM \cap \tan A = DE \cap \tan A$ (since both L and M lie on DE), and $n = DN = DE$. Thus, the correspondence $m \rightarrow n$ maps BE to DE.

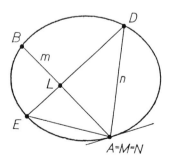

FIGURE 20.24.

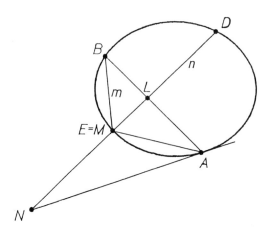

FIGURE 20.25.

The correspondence $m \to n$ matches up the lines on B with the lines on D (by the fourth paragraph of the proof). This correspondence matches up the lines BX and DX for every point X on $\mathcal{K}$ other than B and D (by the last three paragraphs). Every line through B other than tan B has the form BX for a unique point X on $\mathcal{K}$ other than B, and every line through D other than tan D has the form DX for a unique point X on $\mathcal{K}$ other than D (by Theorem 18.4ii and Definition 18.5).

The previous paragraph implies that the correspondence $m \to n$ matches up tan B and BD with tan D and DB in some order.

If we set $m = BD$, then we have $M = m \cap EA = BD \cap EA$ (Figure 20.26). M doesn't equal A (since A doesn't lie on BD, by Theorem 18.4i), and we have $A = \tan A \cap EA$ (by Definition 18.5). The two previous sentences imply that M doesn't lie on tan A, and so $M \neq N = LM \cap \tan A$. L doesn't lie on BD (by the first paragraph of the proof), and so we have

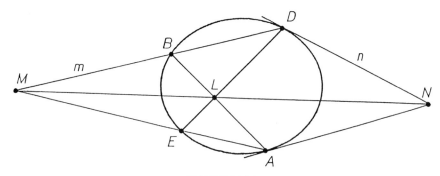

FIGURE 20.26.

$M = LM \cap BD$. The two previous sentences imply that N doesn't lie on BD, and so we have $n = DN \neq DB$. Thus, the previous paragraph implies that the correspondence $m \to n$ matches up $m = BD$ with $n = \tan D$ (Figure 20.26) and matches up $m = \tan B$ with $n = DB$. $\Box$

As usual, we set $XX = \tan X$ when X is a point on a conic section. We can combine the last two sentences in the statement of Theorem 20.7 by saying that the correspondence $m \to n$ matches up $m = BX$ with $n = DX$ for every point X on $\mathcal{K}$; the second-to-last sentence of the theorem describes the case where X doesn't equal B or D, and the last sentence describes the cases $X = D$ (where $m = BD$ maps to $n = DD = \tan D$) and $X = B$ (where $m = BB = \tan B$ maps to $n = DB$). Every line on B equals BX for a unique point X on $\mathcal{K}$ (where $BB = \tan B$), and every line on D equals DX for a unique point X on $\mathcal{K}$ (where $DD = \tan D$), by Theorem 18.4(ii) and Definition 18.5.

In particular, Theorem 20.7 states that $m = BD$ maps to $n = \tan D$ (Figure 20.26). In this case, we have $M = m \cap EA = BD \cap EA$. The fact that $N = LM \cap \tan A$ lies on $DN = n = \tan D$ implies that $N = \tan D \cap \tan A$ lies on LM (since $\tan D \neq \tan A$, by Definition 18.5). Thus, we've proved the following theorem, which states that the points in (16) are collinear.

THEOREM 20.8. In the extended plane, let A, B, D, E be four points on a conic section. Then the points $L = AB \cap DE$, $M = BD \cap EA$, $N = \tan D \cap \tan A$ are collinear (Figure 20.27). $\Box$

As we noted after (16), this result is analogous to Pascal's Theorem 20.3 where, instead of a hexagon inscribed in a conic section, we consider an inscribed quadrilateral and the tangents at two of the vertices.

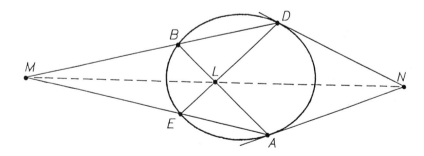

FIGURE 20.27.

EXERCISES

20.1. State the version of Pascal's Theorem 20.3 that holds in the Euclidean plane in each of the following cases. Illustrate each version with a figure in the Euclidean plane.

(a) F is the only ideal point named and is the only ideal point on $\mathcal{K}$.
(b) F is the only ideal point named, and $\mathcal{K}$ has a second ideal point.
(c) EF is the ideal line.
(d) DF is the ideal line.
(e) LM is the ideal line.
(f) L is the only ideal point named.

20.2. State the version of Theorem 20.5 that holds in the Euclidean plane in each of the following cases. Illustrate each version with a figure in the Euclidean plane.

(a) C is the only ideal point named and is the only ideal point on the conic section.
(b) M is the only ideal point named.
(c) N is the only ideal point named.
(d) A is the only ideal point named.
(e) AN is the ideal line.
(f) BC is the ideal line.

20.3. State the version of Theorem 20.8 that holds in the Euclidean plane in each of the following cases. Illustrate each version with a figure in the Euclidean plane.

(a) B is the only ideal point named and the only ideal point on the conic section.
(b) B is the only ideal point named, and the conic section has a second ideal point.
(c) M is the only ideal point named.
(d) A is the only ideal point named.
(e) AN is the ideal line.
(f) AD is the ideal line.
(g) AB is the ideal line.

20.4. Use Theorem 20.7 to prove the following result. Illustrate it with a figure.

Theorem. In the extended plane, let A, B, D, E be four points on a conic section. Then the points $L = AB \cap DE$, $M = \tan B \cap EA$, $N = BD \cap \tan A$ are collinear.

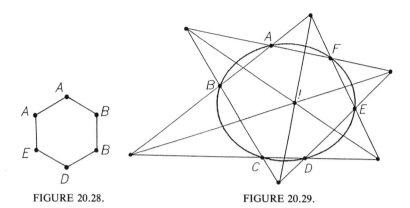

FIGURE 20.28.   FIGURE 20.29.

(This result corresponds to Pascal's Theorem when the hexagon *ABCDEF* is replaced by "hexagon" *ABBDEA*, as in Figure 20.28. Like Theorem 20.8, this result concerns a quadrilateral inscribed in a conic section and the tangents at two of the vertices. Comparing Figures 20.21 and 20.28 shows the difference between these theorems: the two tangents are opposite sides of the "hexagon" in Figure 20.21 but not in Figure 20.28. Accordingly, we consider the intersection of two tangents in Theorem 20.8 but not in the preceding result.)

20.5. Prove the following result (Figure 20.29):

Theorem. In the extended plane, let *A–F* be six points on a conic section. Assume that *AB*, *CD*, *EF* are the sides of a triangle and that *DE*, *FA*, *BC* are the sides of a triangle. Then these triangles are perspective from a point.

20.6. In the extended plane, let *A–E* be five points on a conic section. Assume that tan *A*, *BC*, *DE* are the sides of a triangle. Prove that this triangle is perspective from a point with the triangle that has *CD*, *EA*, *AB* as sides. Illustrate this result with a figure.

20.7. In the Euclidean plane, let *A*, *B*, *C* be three points on a hyperbola $\mathcal{K}$, and assume that *ABC* is not a right triangle. Let *H* be the orthocenter of triangle *ABC*. Prove that *H* lies on $\mathcal{K}$ if and only if the asymptotes of $\mathcal{K}$ are perpendicular.

(See Figure 20.30. A *rectangular hyperbola* is a hyperbola whose asymptotes are perpendicular. This exercise implies that a rectangular hyperbola contains the orthocenter of every inscribed triangle. One possible approach to this exercise is as follows. Let *D* and *E* be the ideal points on the asymptotes. Set $M = BC \cap EH$ and $N = CD \cap HA$. By replacing *F* with *H* in Theorem 20.6, deduce that *H* lies on $\mathcal{K}$ if and only if *MN* is parallel to *AB*. Then apply Exercise 4.17.)

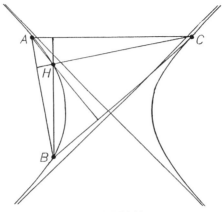

FIGURE 20.30.

20.8. In the extended plane, let A–E be five points on a circle $\mathcal{K}$. Set $X' = \tan X \cap \tan E$ for each point X of $\mathcal{K}$ other than E. Prove that

$$R(A', B'; C', D') = R(EA, EB; EC, ED). \qquad (17)$$

(*Hint:* One possible way to establish Equation 17 is to prove that each side of this equation equals

$$R(TA', TB'; TC', TD'),$$

where T is the center of $\mathcal{K}$.)

20.9. In the extended plane, let A–F be six points on a conic section $\mathcal{K}$. Set $X' = \tan X \cap \tan E$ and $X'' = \tan X \cap \tan F$ for each point X of $\mathcal{K}$ other than E and F. Assume that $\tan E$ and $\tan F$ are ordinary lines. Prove that

$$R(A', B'; C', D') = R(A'', B''; C'', D''). \qquad (18)$$

(*Hint:* One possible approach is to combine two applications of Exercise 20.8 with Theorem 20.1 and projection between planes.)

20.10. Prove the following result.

Theorem. In the extended plane, let P be a point that doesn't lie on a conic section $\mathcal{K}$. Then there is a unique line l that has the following three properties.

  (i) l contains the intersection of the tangent lines at any two points on $\mathcal{K}$ collinear with P.
  (ii) l contains the harmonic conjugate of P with respect to any two points on $\mathcal{K}$ collinear with P.
  (iii) l contains $AC \cap BD$ and $AD \cap BC$ for any two pairs $\{A, B\}$ and $\{C, D\}$ of points on $\mathcal{K}$ collinear with P.

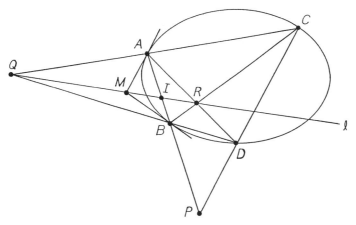

FIGURE 20.31.

(See Figure 20.31. By the theorem, as $\{A, B\}$ and $\{C, D\}$ vary over all pairs of points on $\mathcal{K}$ collinear with P, the same line l contains the points $M = \tan A \cap \tan B$, the harmonic conjugate I of P with respect to A and B, $Q = AC \cap BD$, and $R = AD \cap BC$. One possible approach to this exercise is as follows. Choose two points S and T on $\mathcal{K}$ collinear with P. Prove that there is a unique line l that contains $\tan S \cap \tan T$ and the harmonic conjugate of P with respect to S and T. If U and V are any other two points of $\mathcal{K}$ collinear with P, use Theorem 20.8 to prove that l contains $SU \cap TV$ and $SV \cap TU$. Conclude that l is the unique line satisfying conditions (i)–(iii).)

20.11. State the version of the theorem in Exercise 20.10 that holds in the Euclidean plane when P is an ideal point. Use Theorem 17.8(i) to eliminate the term "harmonic" from the statement. Draw a figure in the Euclidean plane to illustrate the result you state. (The line l is called a *diameter* of $\mathcal{K}$ when P is ideal.)

20.12.

(a) In the extended plane, let A–D be four points on a conic section $\mathcal{K}$ (Figure 20.32). Set $P = AB \cap CD$, and let m be a line on P that doesn't contain any of the points A–D. Assume that m intersects $\mathcal{K}$ at two points E and F, and that m doesn't contain the point $AC \cap BD$. Set $G = AC \cap m$ and $H = BD \cap m$. Prove that P has the same harmonic conjugate with respect to E and F as with respect to G and H. (This result is called the *Butterfly Theorem*, as Figure 20.32 suggests. One possible approach is to combine Exercise 20.10 with Theorem 17.10.)

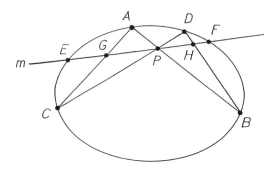

FIGURE 20.32.

(b) In the Euclidean plane, let $\mathcal{K}$ be an ellipse, a parabola, or a hyperbola. Let E and F be two points on $\mathcal{K}$, and let P be their midpoint. Let $\{A, B\}$ and $\{C, D\}$ be two pairs of points on $\mathcal{K}$ other than $\{E, F\}$ that lie on lines through P. Assume that EF is not parallel to both AC and BD. Prove that EF intersects AC and BD at distinct points G and H that have P as their midpoint. Illustrate this result with a figure.

20.13.

(a) In the extended plane, let A, B, C, D be four points on a conic section. Let E, F, G be the points where CD intersects $\tan A$, $\tan B$, AB, respectively. If $E \neq F$, prove that G has the same harmonic conjugate with respect to C and D as with respect to E and F. (See Exercise 20.10.)

(b) In the Euclidean plane, let m be a line that intersects a hyperbola at two points C and D and that doesn't contain the point of intersection of the asymptotes (Figure 20.33). Prove that m intersects the asymptotes at two points E and F that have the same midpoint M as do C and D.

20.14.

(a) Let the notation be as in Exercise 20.10. For any point G of $\mathcal{K}$, prove that G lies on l if and only if $\tan G$ contains P. (See part (ii) of the theorem in Exercise 20.10 and also Theorems 18.8 and 17.10.)

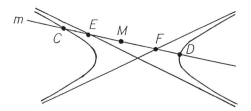

FIGURE 20.33.

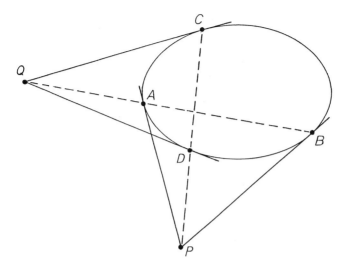

FIGURE 20.34.

(b) In the notation of Exercise 20.11, prove that a diameter of a parabola is parallel to the axis of symmetry.

20.15. In the extended plane, let A, B, C, D be four points on a conic section (Figure 20.34). Set $P = \tan A \cap \tan B$ and $Q = \tan C \cap D$. Prove that P lies on CD if and only if Q lies on AB. (See Exercises 20.10 and 20.14.)

20.16. In the extended plane, let $\mathcal{K}$ be a conic section. For any point P, we define the *polar* of P with respect to $\mathcal{K}$ as follows: it is the line l in Exercise 20.10 if P doesn't lie on $\mathcal{K}$, and it is $\tan P$ if P lies on $\mathcal{K}$.

(a) Let S and T be any two points. Prove that S lies on the polar of T if and only if T lies on the polar of S. (*Hint:* One possible approach is as follows. If neither S or T lies on $\mathcal{K}$, prove that one of the points lies on the polar of the other if and only if there are four points A–D on $\mathcal{K}$ such that $AB \cap CD = S$ and $AC \cap BD = T$. Use Exercise 20.14 if S or T lies on $\mathcal{K}$.)

(b) Prove that every line in the plane is the polar of exactly one point.

20.17. Use Pascal's Theorem and Exercise 20.16 to prove the following result (Figure 20.35):

Brianchon's Theorem. In the extended plane, let a–f be six tangents of a conic section. Then the lines $l = (a \cap b)(d \cap e)$, $m = (b \cap c)(e \cap f)$, and $n = (c \cap d)(f \cap a)$ are concurrent.

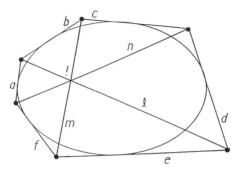

FIGURE 20.35.

(*Hint:* Let A–F be the respective points of contact of a–f. The points $AB \cap DE$, $BC \cap EF$, and $CD \cap FA$ lie on a line t, by Pascal's theorem. There is a point P whose polar is t, by Exercise 20.16(b). Use Exercise 20.16(a) to prove that P lies on each of the lines l, m, and n.)

20.18. In the extended plane, let a–e be five tangents of a conic section, and let A be the point of contact of a. Use Theorem 20.5 and Exercise 20.16 to prove that the lines $l = (a \cap b)(d \cap e)$, $m = (b \cap c)(e \cap a)$, and $n = (c \cap d)A$ are concurrent. Illustrate this result with a figure.

20.19. In the extended plane, let A–D be four points, no three of which are collinear. Let l be a line through two of the points $AB \cap CD$, $AC \cap BD$, $AD \cap BC$. Use Exercises 20.10, 20.14, and possibly 20.16 to prove that no conic section contains A–D and is tangent to l.

Section 21.

Graphs of Quadratic Equations

In this section, we review the characterizations of Euclidean ellipses, Euclidean parabolas, and Euclidean hyperbolas by quadratic equations. We prove that these curves are all sections of right circular cones. It follows in the next section that these curves are the sets of ordinary points on conic sections.

We start by using quadratic equations to define Euclidean ellipses, Euclidean parabolas, and Euclidean hyperbolas. We prove in the next section that these are the same curves as in Definition 18.3.

DEFINITION 21.1. In the Euclidean plane, a *Euclidean ellipse* is the graph of an equation of the form

$$x^2/a^2 + y^2/b^2 = 1 \tag{1}$$

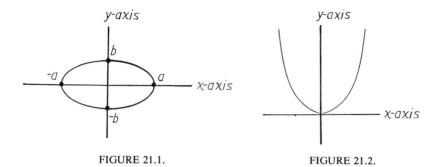

FIGURE 21.1.   FIGURE 21.2.

for $a \geq b > 0$ and any choice of coordinate axes (Figure 21.1). A *Euclidean parabola* is the graph of an equation of the form

$$ay = x^2 \tag{2}$$

for $a > 0$ and any choice of coordinate axes; the *axis of symmetry* in this coordinate system is the y axis (Figure 21.2). A *Euclidean hyperbola* is the graph of an equation of the form

$$x^2/a^2 - y^2/b^2 = 1 \tag{3}$$

for $a > 0$, $b > 0$, and any choice of coordinate axes; the *asymptotes* in this coordinate system are the lines $y = (b/a)x$ and $y = -(b/a)x$. (See Figure 21.3, where the dotted lines are the asymptoytes.) □

   In the previous definition, the coordinate axes are chosen so that the equations of the curves are as simple as possible. In order to determine the equations of these curves in any coordinate system, we consider how equations change when we move the coordinate axes.
   Translating the coordinate axes affects equations as follows. Suppose that the x and y axes are translated to x' and y' axes that intersect at a

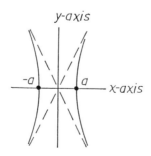

FIGURE 21.3.

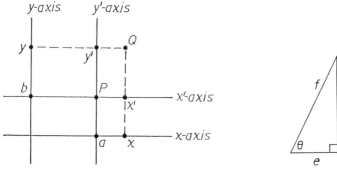

FIGURE 21.4.   FIGURE 21.5.

point P having coordinates (a, b) in the xy system (Figure 21.4). The xy coordinates of a point Q are obtained by adding a and b to the $x'y'$ coordinates. Thus, the coordinates (x, y) and (x', y') of Q in the two systems are related by the equations

$$x = x' + a \quad \text{and} \quad y = y' + b. \tag{4}$$

We can rewrite these equations as

$$x' = x - a \quad \text{and} \quad y' = y - b. \tag{5}$$

Consider a right triangle that has a hypotenuse of length f (Figure 21.5). Let d be the length of a leg opposite an angle θ, and let e be the length of the leg adjacent to θ. Equations 5 and 6 of Section 0 state that

$$\sin \theta = d/f \quad \text{and} \quad \cos \theta = e/f.$$

We rewrite these equations as

$$d = f \cdot \sin \theta \quad \text{and} \quad e = f \cdot \cos \theta. \tag{6}$$

We use these equations to determine the effects of rotating the coordinate axes about the origin.

THEOREM 21.2. In the Euclidean plane, let the x' and y' axes be obtained by rotating the x and y axes counterclockwise through an angle θ about the origin, where $0° < \theta < 90°$ (Figure 21.6). If a point has coordinates (x, y) and (x', y') in the two coordinate systems, then these coordinates are related by the equations

$$x = x' \cos \theta - y' \sin \theta, \tag{7}$$

$$y = x' \sin \theta + y' \cos \theta. \tag{8}$$

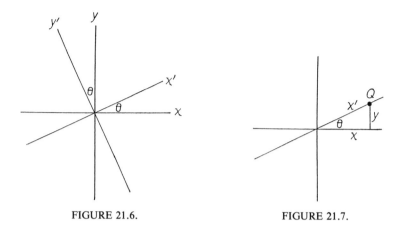

FIGURE 21.6.   FIGURE 21.7.

Proof: We claim that the point $Q = (x', 0)$ in the $x'y'$ system has xy coordinates given by

$$x = x' \cos \theta \quad \text{and} \quad y = x' \sin \theta. \tag{9}$$

When $x' > 0$, these equations follow from those in (6) (Figure 21.7). Then the equations in (9) also hold when $x' < 0$, since reflecting Q across the origin multiplies x', x, and y by -1. Finally, the equations in (9) hold when $x' = 0$, since x and y are also zero in this case.

Next we claim that the point $R = (0, y')$ in the $x'y'$ system has xy coordinates given by

$$x = -y' \sin \theta \quad \text{and} \quad y = y' \cos \theta. \tag{10}$$

When $y' > 0$, the equations in (6) show that both sides of each equation in (10) have the same absolute value (Figure 21.8). Thus, the equations in (10) hold when $y' > 0$, because $x < 0$ and $y > 0$ in this case. It follows that the

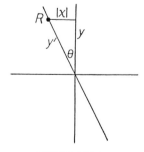

FIGURE 21.8.

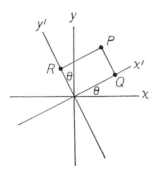

FIGURE 21.9.

equations in (10) also hold when $y' < 0$, since reflecting R across the origin multiplies y', x, and y by -1. Finally, the equations in (10) hold when $y' = 0$, since x and y are also zero in this case.

Now let P be any point in the plane, and let Q and R be the feet of the perpendiculars drawn from P to the x' and y' axes (Figure 21.9). If P has coordinates (x', y') in the $x'y'$ system, then Q has coordinates $(x', 0)$ and R has coordinates $(0, y')$ in this system. Then the equations in (9) give the xy coordinates of Q, and the equations in (10) give the xy coordinates of R. We can obtain the xy coordinates of P by taking the xy coordinates of Q and adding the changes of the xy coordinates in moving from Q to P. These changes are the same as those in moving from the origin to R, and so the xy coordinates of P are the sums of the xy coordinates of Q and R. Thus, we obtain Equations 7 and 8 by adding the right-hand sides of corresponding equations in (9) and (10). □

We could use trigonometric identities to extend the previous theorem to all angles θ. The only additional case we need, however, is $\theta = 90°$, which is easy to handle directly. If we rotate the coordinate axes $90°$ counterclockwise about the orgin, we obtain x' and y' axes, where the positive x' axis coincides with the positive y axis, and where the positive y' axis coincides with the negative x axis (Figure 21.10). Thus, the xy and $x'y'$ coordinates of a point are related by the equations

$$x = -y' \quad \text{and} \quad y = x'. \tag{11}$$

A *quadratic equation* in two variables x and y is an equation of the form

$$Ax^2 + Bxy + Cy^2 + Dx + Ey + F = 0, \tag{12}$$

where A–F are real numbers. A quadratic equation is called *nondegenerate* if it doesn't consist of the entire plane, one or two lines, one point, or the

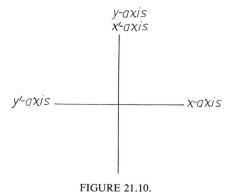

FIGURE 21.10.

empty set. We now prove that we can simplify any nondegenerate quadratic equation into one of the equations in Definition 21.1 by moving the coordinate axes.

THEOREM 21.3. In the Euclidean plane, the graph $\mathcal{K}$ of any nondegenerate quadratic equation is a Euclidean ellipse, a Euclidean parabola, or a Euclidean hyperbola.

Proof: Suppose that we rotate the coordinate axes counterclockwise through an angle θ about the origin, where $0° < \theta < 90°$. Then Equation 12 is transformed by substituting the expressions for x and y in Equations 7 and 8. If we perform this substitution and collect terms with like powers of x' and y', Equation 12 becomes

$$A'x'^2 + B'x'y' + C'y'^2 + D'x' + E'y' + F' = 0,$$

where A'–F' are real numbers and

$$B' = B(\cos^2 \theta - \sin^2 \theta) + 2(C - A) \cos \theta \sin \theta. \tag{13}$$

If $B \neq 0$, we claim that θ can be chosen so that $B' = 0$. By Equation 13, it's enough to show that there is an angle θ between $0°$ and $90°$ such that

$$\frac{\cos^2 \theta - \sin^2 \theta}{\cos \theta \sin \theta} = \frac{2(A - C)}{B}. \tag{14}$$

For $0° < \theta < 90°$, $\cos \theta$ and $\sin \theta$ are positive, and so the left side of Equation 14 is a continuous function of θ that takes arbitrarily large positive values (since $\cos \theta$ nears 1 and $\sin \theta$ nears 0 as θ approaches 0) and that takes arbitrarily negative values (since $\cos \theta$ nears 0 and $\sin \theta$ nears 1 as θ approaches 90°). Thus, the left side of Equation 14 assumes every real value as θ varies between $0°$ and $90°$, and so Equation 14 holds for one of these angles θ.

In short, by the previous paragraph, we can assume that the equation of $\mathcal{K}$ has the form

$$Ax^2 + Cy^2 + Dx + Ey + F = 0. \tag{15}$$

If A and C were both zero, then $\mathcal{K}$ would be either a line (if $D \neq 0$ or $E \neq 0$), the empty set (if $D = E = 0 \neq F$), or the entire plane (if $D = E = F = 0$). Thus, since $\mathcal{K}$ is the graph of a nondegenate quadratic equation, either A or C is nonzero.

First assume that $C = 0$. Then $\mathcal{K}$ has the equation

$$Ax^2 + Dx + Ey + F = 0,$$

where $A \neq 0$ (by the previous paragraph). Dividing through by A, we can assume that $A = 1$ and that $\mathcal{K}$ has the equation

$$x^2 + Dx + Ey + F = 0.$$

Completing the square, we can rewrite this equation as

$$(x + D/2)^2 = -Ey + G \tag{16}$$

for $G = -F + D^2/4$. If E were zero, Equation 16 would become

$$(x + D/2)^2 = G,$$

and its graph would be either the two lines $x = -D/2 \pm \sqrt{G}$ (if $G > 0$), the one line $x = -D/2$ (if $G = 0$), or the empty set (if $G < 0$), contradicting the assumption that $\mathcal{K}$ is the graph of a nondegenerate quadratic equation. Thus, E is nonzero, and we can rewrite Equation 16 as

$$(x + D/2)^2 = -E(y - G/E).$$

Accordingly, if we translate the coordinate axes $-D/2$ units horizontally and G/E units vertically and use the equations in (5), we can assume that $\mathcal{K}$ has the equation

$$x^2 = -Ey, \tag{17}$$

where $E \neq 0$. If E is negative, Equation 17 shows that $\mathcal{K}$ is a Euclidean parabola (by Definition 21.1). If E is positive, rotating the axes 180° about the origin switches the ends of each coordinate axis; this replaces x and y with $-x$ and $-y$, and so Equation 17 becomes $x^2 = Ey$ in the new coordinate system, and $\mathcal{K}$ is once again a Euclidean parabola.

Next assume that $A = 0$ in Equation 15. Let x' and y' axes be obtained by rotating the x and y axes 90° counterclockwise about the origin. Substituting the equations in (11) into Equation 15 with $A = 0$ shows that $\mathcal{K}$ has the equation

$$Cx'^2 - Dy' + Ex' + F = 0$$

in the $x'y'$ system. Thus the previous paragraph shows that $\mathcal{K}$ is a Euclidean parabola.

It only remains to consider Equation 15 when A and C are both nonzero. In this case, we can complete the squares in both x and y and rewrite Equation 15 as

$$A(x + D/2A)^2 + C(y + E/2C)^2 = G$$

for $G = -F + D^2/4A + E^2/4C$. By translating the coordinate axes $-D/2A$ horizontally and $-E/2C$ vertically and using the equations in (5), we can assume that $\mathcal{K}$ has the equation

$$Ax^2 + Cy^2 = G, \tag{18}$$

where A and C are both nonzero.

We claim that G is also nonzero. If G were zero, and if A and C were both positive or both negative, the graph of Equation 18 would be the one point $(0, 0)$. If G were zero, and if A were positive and C negative, we could rewrite Equation 18 as

$$u^2x^2 - v^2y^2 = 0, \tag{19}$$

where $u = \sqrt{A}$ and $v = \sqrt{|C|}$ are positive numbers; we can rewrite Equation 19 as

$$(ux + vy)(ux - vy) = 0,$$

and so the graph of Equation 18 would be the two lines $ux + vy = 0$ and $ux - vy = 0$. Similarly, if G were zero, and if A were negative and C positive, the graph of Equation 18 would again consist of two lines. Since $\mathcal{K}$ is the graph of a nontrivial quadratric equation, the last three sentences show that G is nonzero, as claimed.

Since A, C, G are all nonzero, we can rewrite Equation 18 as

$$\frac{x^2}{s} + \frac{y^2}{t} = 1, \tag{20}$$

where $s = G/A$ and $t = G/C$ are nonzero real numbers. We can assume that $s \geq t$; if $s < t$, the equations in (11) show that a 90° rotation of the coordinate axes transforms Equation 20 into

$$\frac{y'^2}{s} + \frac{x'^2}{t} = 1,$$

and so the roles of s and t are reversed. The assumption that $s \geq t$ implies that s is positive; if s were negative, then t would also be negative, and the graph of Equation 20 would be the empty set, contradicting the assumption that $\mathcal{K}$ is the graph of a nontrivial quadratic equation.

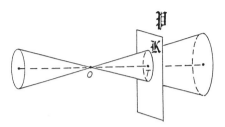

FIGURE 21.11.

In short, we can assume that $\mathcal{K}$ is the graph of Equation 20, where $s > 0$, $t \neq 0$, and $s \geq t$. Since $s > 0$, we have $s = a^2$ for $a = \sqrt{s} > 0$. If $t > 0$, we have $t = b^2$ for $b = \sqrt{t} > 0$; then we can rewrite Equation 20 as $x^2/a^2 + y^2/b^2 = 1$, where $a \geq b$ (since $s \geq t$), and so $\mathcal{K}$ is a Euclidean ellipse (by Definition 21.1). If $t < 0$, we have $t = -b^2$ for $b = \sqrt{|t|} > 0$; then we can rewrite Equation 20 as $x^2/a^2 - y^2/b^2 = 1$, and so $\mathcal{K}$ is a Euclidean hyperbola (by Definition 21.1). □

In the discussion after Definition 18.1, we defined the cone $\mathcal{C}$ determined by a point O and a circle $\mathcal{K}$ in Euclidean space, and we defined sections of $\mathcal{C}$. We call $\mathcal{C}$ *right circular* if the line through O and the center T of $\mathcal{K}$ is perpendicular to the plane $\mathcal{P}$ containing $\mathcal{K}$. For example, the cone in Figure 21.11 is right circular, but the cone in Figure 18.1 is not.

The last theorem in this section states that Euclidean ellipses, Euclidean parabolas, and Euclidean hyperbolas are all sections of right circular cones. We prove in the next section that, conversely, every section of a cone is a Euclidean ellipse, a Euclidean parabola, or a Euclidean hyperbola, whether or not the cone is right circular (Figures 18.2, 18.3, and 18.5).

In Euclidean space, we set up x, y, and z axes along three mutually perpendicular lines through a point O (Figure 21.12). We set up coordinate systems on the three axes (in the sense of Definition 1.6) so that O has coordinate 0 on each axis. A point in Euclidean space is assigned coordinates (a, b, c) if it lies on the plane perpendicular to the x axis through the point on the axis with coordinate a, on the plane perpendicular to the y axis through the point on the axis with coordinate b, and on the plane perpendicular to the z axis through the point on the axis with coordinate c. In this way we match up the points of Euclidean space with the ordered triples (a, b, c) of real numbers.

When a point $(0, b, c)$ in the yz plane is rotated about the z axis, the z coordinate is unchanged, and we obtain a point (a', b', c) (Figure 21.13). Both points lie at the same distance from the z axis, and so we have

$$\sqrt{a'^2 + b'^2} = |b|. \tag{21}$$

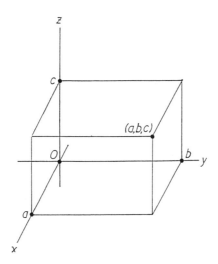

FIGURE 21.12.

Theorem 21.2 describes the effect of rotating the Euclidean plane about the origin. We use this result in the proof of the next theorem to compute the effect of rotating Euclidean space about a coordinate axis.

THEOREM 21.4. Every Euclidean ellipse, Euclidean parabola, and Euclidean hyperbola is a section of a right circular cone.

Proof: Let d and e be positive numbers. In the yz plane, the graph of the equation $y = dz + e$ is a line (Figure 21.14). Setting $z = 0$ in this equation gives $y = e$, and setting $y = 0$ gives $z = -e/d$. Thus, the y intercept is e

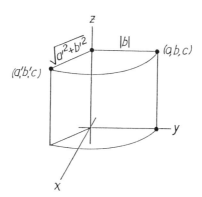

FIGURE 21.13.

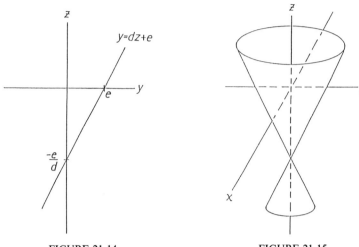

FIGURE 21.14.   FIGURE 21.15.

and the z intercept is $-e/d$, and so the assumption that d and e are positive ensures that both intercepts are nonzero. Thus, rotating the line about the z axis gives a right circular cone $\mathcal{C}$ whose vertex lies on the z axis and has nonzero z coordinate (Figure 21.15). By Equation 21, the equation of $\mathcal{C}$ is

$$\sqrt{x^2 + y^2} = |dz + e|.$$

This equation is equivalent to

$$x^2 + y^2 = (dz + e)^2. \tag{22}$$

Let x', y', and z' axes be obtained by rotating the x, y, and z axes through an angle θ about the x axis, where $0° < \theta < 90°$; specifically, the direction of rotation is chosen so that a $90°$ rotation would take the positive y axis to the positive z axis. (See Figure 21.16, where the square marks the plane of the y, z, y', and z' axes.) A point's coordinates (x, y, z) and (x', y', z') in the two coordinate systems are related by the equations

$$x = x',$$
$$y = y' \cos \theta - z' \sin \theta,$$
$$z = y' \sin \theta + z' \cos \theta,$$

by Theorem 21.2 and the fact that the rotation fixes the x axis. Substituting these equations in Equation 22 shows that the cone $\mathcal{C}$ has the equation

$$x'^2 + (y' \cos \theta - z' \sin \theta)^2 = [d(y' \sin \theta + z' \cos \theta) + e]^2 \tag{23}$$

in the $x'y'z'$ system.

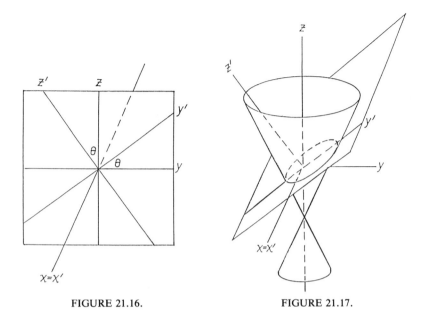

FIGURE 21.16.   FIGURE 21.17.

The vertex of $\mathcal{C}$ does not lie in the $x'y'$ plane (since it lies on the z axis and has nonzero z coordinate, and since $0° < \theta < 90°$) (Figure 21.17). Thus, the intersection of $\mathcal{C}$ and the $x'y'$ plane is a section of the right circular cone $\mathcal{C}$. We find the equation of this intersection by setting $z' = 0$ in Equation 23 (since $z' = 0$ for points on the $x'y'$ plane). Hence, the intersection has the equation

$$x'^2 + (y' \cos \theta)^2 = [dy' \sin \theta + e]^2$$

in the $x'y'$ plane. Multiplying this equation out and collecting terms gives

$$x'^2 + (\cos^2 \theta - d^2 \sin^2 \theta)y'^2 - 2de(\sin \theta)y' = e^2.$$

In short, the graph of

$$x^2 + (\cos^2 \theta - d^2 \sin^2 \theta)y^2 - 2de(\sin \theta)y = e^2 \qquad (24)$$

in the xy plane is a section of a right circular cone for any positive numbers d and e and any angle θ such that $0° < \theta < 90°$. Thus, to prove that every Euclidean ellipse, Euclidean parabola, and Euclidean hyperbola is a section of a right circular cone, it's enough to prove that these curves are all given by Equation 24 for some choices of d, e, and θ.

If we set $\theta = 45°$ and $d = 1$ in Equation 24, we obtain

$$x^2 - 2e\left(\frac{\sqrt{2}}{2}\right)y = e^2.$$

We rewrite this equation as

$$x^2 = \sqrt{2}\, e(y + e/\sqrt{2}).$$

Translating the coordinate axes $-e/\sqrt{2}$ units vertically transforms this equation into

$$x^2 = \sqrt{2}\, ey \tag{25}$$

(by the equations in (5)). Any Euclidean parabola $\mathcal{K}$ has the equation $ay = x^2$ for some positive number a and some coordinate system (by Definition 21.1). Hence, setting $e = a/\sqrt{2}$ in Equation 25 shows that $\mathcal{K}$ is a section of a right circular cone.

If we choose d and θ so that $\cos^2 \theta - d^2 \sin^2 \theta$ is nonzero, we can rewrite Equation 24 as

$$x^2 + (\cos^2 \theta - d^2 \sin^2 \theta)\left(y^2 - \frac{2de \sin \theta}{\cos^2 \theta - d^2 \sin^2 \theta}\, y\right) = e^2.$$

Completing the square gives

$$x^2 + (\cos^2 \theta - d^2 \sin^2 \theta)\left(y - \frac{de \sin \theta}{\cos^2 \theta - d^2 \sin^2 \theta}\right)^2$$

$$= e^2 + \frac{d^2 e^2 \sin^2 \theta}{\cos^2 \theta - d^2 \sin^2 \theta}$$

$$= \frac{e^2 \cos^2 \theta}{\cos^2 \theta - d^2 \sin^2 \theta}.$$

We can use a vertical translation to transform this equation into

$$x^2 + (\cos^2 \theta - d^2 \sin^2 \theta)y^2 = \frac{e^2 \cos^2 \theta}{\cos^2 \theta - d^2 \sin^2 \theta}$$

(by the equations in (5)). If we rotate the axes 90° counterclockwise about the origin, we transform the previous equation by replacing x with $-y$ and y with x (by the equations in (11)), and so we obtain

$$(\cos^2 \theta - d^2 \sin^2 \theta)x^2 + y^2 = \frac{e^2 \cos^2 \theta}{\cos^2 \theta - d^2 \sin^2 \theta}. \tag{26}$$

A Euclidean hyperbola $\mathcal{K}$ has the equation $x^2/a^2 - y^2/b^2 = 1$ for positive numbers a and b (by Definition 21.1). Multiplying by $-b^2$, we can rewrite this equation as

$$-b^2 x^2/a^2 + y^2 = -b^2. \tag{27}$$

On the other hand, if we set $\theta = 45°$ in Equation 26, we obtain

$$\frac{(1 - d^2)}{2} x^2 + y^2 = \frac{e^2}{1 - d^2}. \tag{28}$$

As d varies over all numbers greater than 1, $(1 - d^2)/2$ varies over all negative numbers, and so we can choose $d > 1$ such that

$$(1 - d^2)/2 = -b^2/a^2. \tag{29}$$

Then we can choose a positive number e such that

$$\frac{e^2}{1 - d^2} = -b^2 \tag{30}$$

(since the fact that $1 - d^2 < 0$ implies that the left side of Equation 30 varies over all negative numbers as e varies over all positive numbers). Substituting Equations 29 and 30 in Equation 28 gives Equation 27, so $\mathcal{K}$ is a section of a right circular cone.

A Euclidean ellipse $\mathcal{K}$ has the equation $x^2/a^2 + y^2/b^2 = 1$ for $a \geq b > 0$ (by Definition 21.1). Multiplying by b^2, we can rewrite this equation as

$$\frac{b^2}{a^2} x^2 + y^2 = b^2. \tag{31}$$

If $a = b$, Equation 31 becomes $x^2 + y^2 = b^2$, whose graph is a circle (since it consists of all points (x, y) whose distance $\sqrt{x^2 + y^2}$ from the origin is b). A circle is obviously a section of a right circular cone (Figure 21.11), and so we can assume that $a > b$. Then $0 < b/a < 1$, and so there is an angle θ such that $0° < \theta < 90°$ and $b/a < \cos\theta < 1$. Thus, there is a positive number d such that

$$\cos^2\theta - d^2 \sin^2\theta = b^2/a^2 \tag{32}$$

(since the fact that $\sin\theta > 0$ implies that the left side of Equation 32 varies over all numbers less than $\cos^2\theta$ as d varies over all positive numbers). We can then choose a positive number e such that

$$\frac{e^2 \cos^2\theta}{\cos^2\theta - d^2 \sin^2\theta} = b^2 \tag{33}$$

(since Equation 32 shows that $\cos^2\theta - d^2 \sin^2\theta > 0$, and so the left side of Equation 33 varies over all positive numbers as e varies over all positive numbers). Substituting Equations 32 and 33 in Equation 26 gives Equation 31, and so $\mathcal{K}$ is a section of a right circular cone. $\square$

EXERCISES

21.1. Graph each of the following nondegenerate quadratic equations by writing it as one of the Equations 1–3 in an appropriate coordinate system. Show the axes of all coordinate systems considered. (See the proof of Theorem 21.3.)

(a) $2x^2 - 12x + y^2 + 4y = 3$
(b) $4x^2 + 8x - y^2 + 6y = 9$
(c) $x^2 - 6x - y^2 + 8y = 3$
(d) $y^2 - 3y + 2x = 4$
(e) $3x^2 + 10xy + 3y^2 = 8$
(f) $13x^2 - 6\sqrt{3}\,xy + 7y^2 = 16$
(g) $x^2 + \sqrt{3}\,xy = 6$
(h) $x^2 + 2xy + y^2 - 2\sqrt{2}\,x + 6\sqrt{2}\,y + 18 = 0$
(i) $5x^2 - 6xy + 5y^2 + 2\sqrt{2}\,x - 14\sqrt{2}\,y + 18 = 0$

Exercises 21.2–21.13 use the following terminology. In the Euclidean plane, a set of points $\mathcal{K}$ is determined by the *focus–directrix property* with *focus F, directrix d,* and *eccentricity e* if F is a point, d is a line that doesn't contain F, e is a positive number, and $\mathcal{K}$ consists of all points A such that the distance from A to F is e times the distance from A to d. Thus, if B is the foot of the perpendicular drawn from any point A in the plane to d, A lies on $\mathcal{K}$ if and only if $AF = e(AB)$, where AF and AB are the distances from A to F and B. (See Figure 21.18, where the case $e = 0.88$ is shown.)

21.2. Let the focus–directrix property be defined as in the preceding. In the Euclidean plane, let $\mathcal{K}$ be the Euclidean parabola with equation $ay = x^2$ for $a > 0$. Set $p = a/4$, and so we can rewrite the equation of $\mathcal{K}$ as $4py = x^2$. Prove that $\mathcal{K}$ is determined by the focus–directrix property with focus $(0, p)$, directrix $y = -p$, and eccentricity $e = 1$ (Figure 21.19).

21.3. In the notation of Exercise 21.2, let (x, y) be a point on $\mathcal{K}$ other than $(0, 0)$ (Figure 21.19).

(a) Defining tangent lines as in calculus, use differentiation to prove that the tangent line to $\mathcal{K}$ at (x, y) has slope $x/2p$. Conclude that the tangent line has y intercept $(0, -y)$ and that the normal at (x, y) has slope $-2p/x$ and y intercept $(0, y + 2p)$, where the normal is the line through (x, y) perpendicular to the tangent line at (x, y).

(b) Conclude from part (a) that the focus $(0, p)$ of $\mathcal{K}$ is the midpoint of the y intercepts of the tangent line and the normal at (x, y).

(c) Let f be the line through (x, y) and the focus $(0, p)$ of $\mathcal{K}$. Let g be the line through (x, y) parallel to the y axis. Use part (b) and Exercises 17.11 and 17.13 to conclude that the tangent line and the normal at (x, y) are the angle bisectors determined by f and g.

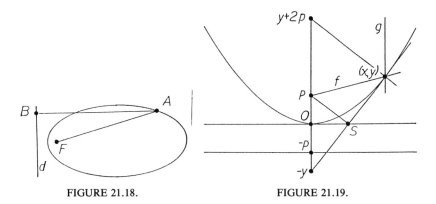

FIGURE 21.18.   FIGURE 21.19.

(Light rays reflect off a curve so that the incoming and outgoing rays form equal angles with the normal. Thus, part (c) establishes the *reflection property of parabolas*: Light rays emitted in any direction from the focus reflect off a parabola parallel to the axis of symmetry, and light rays entering a parabola parallel to the axis of symmetry reflect to pass through the focus. See Figure 21.20.)

21.4. In the notation of Exercises 21.2 and 21.3, let S be the foot of the perpendicular drawn from the focus $(0, p)$ to the tangent line at (x, y) (Figure 21.19). Deduce from Theorem 1.13 and parts (a) and (b) of Exercise 21.3 that S lies on the x axis. (Thus, if we call $(0, 0)$ the *vertex* of $\mathcal{K}$, *the tangent line at the vertex contains the feet of the perpendiculars from the focus to all tangent lines.*)

21.5. Let the focus–directrix property be defined as before Exercise 21.2. In the Euclidean plane, let $\mathcal{K}$ be the Euclidean ellipse with equation $x^2/a^2 + y^2/b^2 = 1$ for $a > b > 0$. Set $c = \sqrt{a^2 - b^2}$. Prove that $\mathcal{K}$ is determined by the focus–directrix property with focus $(c, 0)$, directrix

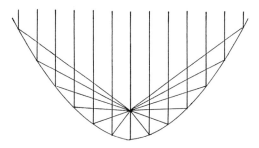

FIGURE 21.20.

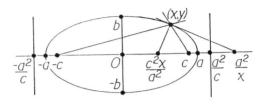

FIGURE 21.21.

$x = a^2/c$, and eccentricity $e = c/a$. Prove that $\mathcal{K}$ is also determined by the focus–directrix property with focus $(-c, 0)$, directrix $x = -a^2/c$, and eccentricity $e = c/a$ (Figure 21.21).

21.6. In the notation of Exercise 21.5, let (x, y) be a point on $\mathcal{K}$ other than $(\pm a, 0)$ and $(0, \pm b)$ (Figure 21.21).

(a) Defining tangent lines as in calculus, use implicit differentiation to prove that the tangent line to $\mathcal{K}$ at (x, y) has slope $-b^2x/a^2y$. Conclude that the tangent line has x intercept $(a^2/x, 0)$ and that the normal at (x, y) has slope a^2y/b^2x and x intercept $(c^2x/a^2, 0)$, where the normal is the line through (x, y) perpendicular to the tangent line at (x, y).
(b) Use part (a) and Theorem 17.8(ii) to prove that the x intercepts of the tangent line and the normal at (x, y) are harmonic conjugates with respect to the foci $(\pm c, 0)$.
(c) Use part (b) and Exercises 17.11 and 17.13 to prove that the tangent line and the normal at (x, y) are the angle bisectors determined by the two lines that each contain (x, y) and one of the foci $(\pm c, 0)$.

(Light rays reflect off a curve so that incoming and outgoing rays form equal angles with the normal. Thus, part (c) establishes the *reflection property of ellipses*: Light rays emitted in any direction from one focus reflect off an ellipse to pass through the other focus. See Figure 21.22.)

21.7. Let the focus–directrix property be defined as before Exercise 21.2. In the Euclidean plane, let $\mathcal{K}$ be the Euclidean hyperbola with equation $x^2/a^2 - y^2/b^2 = 1$ for positive numbers a and b. Set $c = \sqrt{a^2 + b^2}$.

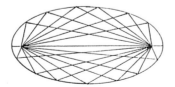

FIGURE 21.22.

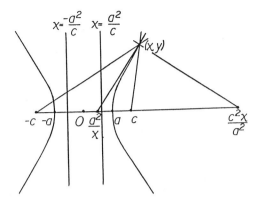

FIGURE 21.23.

Prove that $\mathcal{K}$ is determined by the focus–directrix property with focus $(c, 0)$, directrix $x = a^2/c$, and eccentricity $e = c/a$. Prove that $\mathcal{K}$ is also determined by the focus–directrix property with focus $(-c, 0)$, directrix $x = -a^2/c$, and eccentricity $e = c/a$ (Figure 21.23).

21.8. In the notation of Exercise 21.7, let (x, y) be a point on $\mathcal{K}$ other than $(\pm a, 0)$ (Figure 21.23).

(a) Defining tangent lines as in calculus, use implicit differentiation to prove that the tangent line to $\mathcal{K}$ at (x, y) has slope b^2x/a^2y. Conclude that the tangent line has x intercept $(a^2/x, 0)$ and that the normal at (x, y) has slope $-a^2y/b^2x$ and x intercept $(c^2x/a^2, 0)$, where the normal is the line through (x, y) perpendicular to the tangent line at (x, y).

(b) Use part (a) and Theorem 17.8(ii) to prove that the x intercepts of the tangent line and the normal at (x, y) are harmonic conjugates with respect to the foci $(\pm c, 0)$.

(c) Use part (b) and Exercises 17.11 and 17.13 to prove that the tangent line and the normal at (x, y) are the angle bisectors determined by the two lines that each contain (x, y) and one of the foci $(\pm c, 0)$. (This property is called the *reflection property of hyperbolas*: Light rays emitted from a focus reflect off either branch of a hyperbola on paths pointed directly away from the other focus.)

21.9. Let the notation be as in Exercise 21.5 (Figure 21.21). Use the focus–directrix property to deduce that the sum of the distances from the two foci $(c, 0)$ and $(-c, 0)$ to any point P of $\mathcal{K}$ remains constant as P varies over all points of $\mathcal{K}$. Deduce that this constant is $2a$, by setting $P = (a, 0)$.

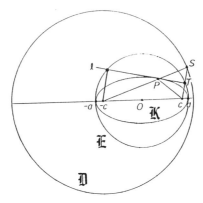

FIGURE 21.24.

21.10. Let the notation be as in Exercise 21.5. Let P be a point on $\mathcal{K}$ other than $(\pm a, 0)$ (Figure 21.24).

(a) Consider the line through the focus $(c, 0)$ perpendicular to the tangent line l at P, and consider the line through the other focus $(-c, 0)$ and P. Conclude from Exercise 21.6(c) that these two lines intersect at a point S such that S and $(c, 0)$ are equidistant from P.

(b) Use part (a) and Exercise 21.9 to conclude that S lies on the circle $\mathfrak{D}$ of radius $2a$ about the focus $(-c, 0)$.

(c) Let T be the foot of the perpendicular from $(c, 0)$ to l. Prove that T is the midpoint of $(c, 0)$ and S.

(d) Prove that T lies on the circle $\mathcal{E}$ having $(a, 0)$ and $(-a, 0)$ as the endpoints of a diameter.

(e) Prove that $\mathcal{E}$ contains the feet of the perpendiculars from both foci $(\pm c, 0)$ to all tangent lines of $\mathcal{K}$.

21.11. Let the notation be as in Exercise 21.5. As l varies over all tangent lines of $\mathcal{K}$, prove that the product of the distances from l to the foci $(\pm c, 0)$ takes the constant value b^2.

(*Hint:* One possible approach is as follows. As in Figure 21.25, let T and U be the feet of the perpendiculars from $(c, 0)$ and $(-c, 0)$ to a tangent line l of $\mathcal{K}$. Let V be the point such that $(0, 0)$ is the midpoint of V and U. Prove that V is the foot of the perpendicular from $(c, 0)$ to a tangent m of $\mathcal{K}$ parallel to l and that the distance from $(c, 0)$ to V equals the distance from $(-c, 0)$ to U. Use Exercise 21.10(e) and Theorem 6.5 to conclude that the product of the distances from the foci to l remains constant as l varies over all tangents of $\mathcal{K}$. Prove that this constant is b^2 by considering the case where l is perpendicular to the x axis.)

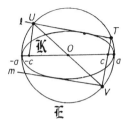

FIGURE 21.25.

21.12. Let the notation be as in Exercise 21.7 (Figure 21.23).

(a) Prove that the absolute value of the difference between the distances from the foci ($\pm c$, 0) to any point P of $\mathcal{K}$ takes the constant value $2a$ as P varies over all points of $\mathcal{K}$.

(b) Let P be any point of $\mathcal{K}$ other than ($\pm a$, 0). Consider the line through the focus (c, 0) perpendicular to the tangent at P, and consider the line through the other focus ($-c$, 0) and P. Prove that these two lines intersect at a point S on the circle of radius $2a$ about ($-c$, 0). Illustrate this result with two figures, one where P lines on each of the two branches of $\mathcal{K}$.

(c) Let $\mathcal{E}$ be the circle having ($\pm a$, 0) as the endpoints of a diameter. Prove that $\mathcal{E}$ contains the feet of the perpendiculars from both foci ($\pm c$, 0) to all tangent lines of $\mathcal{K}$. Illustrate this result with a figure.

(d) As l varies over all tangent lines of $\mathcal{K}$, prove that the product of the distances from l to the foci ($\pm c$, 0) takes the constant value b^2.

21.13. Conclude from Exercises 21.2, 21.5, and 21.7 that a set of points in the Euclidean plane is determined by the focus–directrix property if and only if it is either a Euclidean parabola, a Euclidean ellipse other than a circle, or a Euclidean hyperbola.

Section 22.

Characterizations of Conic Sections

We now complete the proof that Euclidean ellipses, Euclidean parabolas, and Euclidean hyperbolas are exactly the sets of all ordinary points on conic sections. This shows that Definitions 18.3 and 21.1 give the same curves. It justifies taking a theorem about conic sections, choosing a position for the ideal line, and interpreting the special case that results as a theorem in the Euclidean plane about Euclidean ellipses, Euclidean parabolas, and Euclidean hyperbolas. We also prove that there is a unique conic section through any five points, no three of which are collinear.

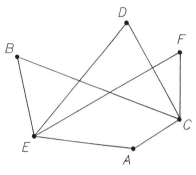

FIGURE 22.1.

We start by defining a set of points. We will prove at the end of the section that this set of points is a conic section.

DEFINITION 22.1. In the extended plane, let A–E be five ordinary points, no three of which are collinear. Define $\mathcal{K}_1(A, B, C, D, E)$ to be the set of points that consists of A–E and all points F that are not collinear with any two of the points A–E and are such that the equation

$$R(CA, CB; CD, CF) = R(EA, EB; ED, EF) \tag{1}$$

holds (Figure 22.1). $\square$

The next result relates the previous definition to conic sections. This result follows directly from Theorems 20.2 and 20.6, which connect cross-ratios and conic sections.

THEOREM 22.2. Every conic section $\mathcal{K}$ equals $\mathcal{K}_1(A, B, C, D, E)$, where A–E are any five ordinary points of $\mathcal{K}$ (Figure 22.2).

Proof: At most two points of $\mathcal{K}$ lie on the ideal line, and so $\mathcal{K}$ has infintely many ordinary points (by Theorem 18.4i). If A–E are five ordinary points of $\mathcal{K}$, no three of these points are collinear (by Theorem 18.4i). By Theorem 20.6, $\mathcal{K}$ consists of A–E and all points F such that F is not collinear with any two of the points A–E and such that the points $L = AB \cap DE$, $M = BC \cap EF$, $N = CD \cap FA$ are collinear (Figure 20.18). Thus, $\mathcal{K}$ equals $\mathcal{K}_1(A, B, C, D, E)$ (by Theorem 20.2). $\square$

Now that we've related conic sections to cross-ratios, we can relate them to nondegenerate quadratic equations. By the previous result, it's enough to relate sets of the form $\mathcal{K}_1(A, B, C, D, E)$ to nondegenerate quadratic equations. We do so in Theorem 22.4, but we require the following preliminary result.

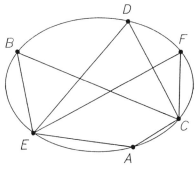

FIGURE 22.2.

THEOREM 22.3. In the extended plane, let A–F be six ordinary points such that the lines CA, CB, CD, CF are distinct, the lines EA, EB, ED, EF are distinct, and the equation

$$R(CA, CB; CD, CF) = R(EA, EB; ED, EF) \qquad (2)$$

holds. If three of the points A, B, D, F lie on a line l, and if the fourth point doesn't lie on line CE, then the fourth point also lies on l.

Proof: We can interchange F with any one of the points A, B, D if we also interchange the remaining two points (since Equation 2 still holds after such interchanges, by Theorem 19.5). Thus, we can assume that A, B, D lie on l and that F doesn't lie on CE (Figure 22.3).

C doesn't lie on l (since $CA \neq CB$), and the lines CA, CB, CD, CF are distinct, and so these lines intersect l at distinct points A, B, D, $CF \cap l$.

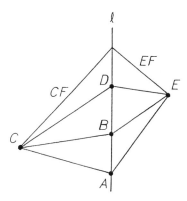

FIGURE 22.3.

Thus, we have

$$R(CA, CB; CD, CF) = R(A, B; D, CF \cap l) \qquad (3)$$

(by Theorem 19.6). By symmetry, we also have

$$R(EA, EB; ED, EF) = R(A, B; D, EF \cap l). \qquad (4)$$

Combining Equations 2–4 shows that

$$R(A, B; D, CF \cap l) = R(A, B; D, EF \cap l).$$

It follows that $CF \cap l = EF \cap l$ (by Theorem 19.9), and so we have $CF \cap l = CF \cap EF = F$ (since F doesn't lie on CE). Thus, F lies on l.   □

We can now relate the sets $\mathcal{K}_1(A, B, C, D, E)$ to nondegenerate quadratic equations.

THEOREM 22.4. In the extended plane, let A–E be five ordinary points, no three of which are collinear. Then the ordinary points of $\mathcal{K}_1(A, B, C, D, E)$ are the graph of a nondegenerate quadratic equation.

Proof: We can obtain any rotation about the origin by rotating one or more times through an angle between $0°$ and $90°$. Thus, the first paragraph of the proof of Theorem 21.3 shows that any rotation of the coordinate axes about the origin transforms one quadratic equation into another. If we substitute the expressions $x' + a$ and $y' + b$ for x and y, we transform a quadratic equation in x and y into a quadratic equation in x' and y'. Thus, a translation of the coordinate axes also transforms one quadratic equation into another (by the equations in (4) of Section 21). Moreover, any pair of coordinate axes can be transformed into any other pair by a rotation about the origin and a translation. Hence, if a set of points is the graph of a quadratic equation in one coordinate system, the same holds in every coordinate system.

We write $\mathcal{K}_1(A, B, C, D, E)$ as $\mathcal{K}_1$. Definition 22.1 of $\mathcal{K}_1$ doesn't depend on the choice of coordinate axes, and neither does the property of being the graph of a nondegenerate quadratic equation (by the previous paragraph). Thus, we can choose the coordinate axes so that C is the origin $(0, 0)$ and E is a point $(0, e)$ on the y axis, where $e \neq 0$. Since no three of the points A–E are collinear, the points A, B, D don't lie on the y axis CE, and the lines CA, CB, CD, EA, EB, ED have respective slopes a, b, d, a', b', d', where a, b, d are distinct real numbers, and so are a', b', d'.

Let $F = (x, y)$ be an ordinary point on $\mathcal{K}_1$ other than A–E. F doesn't lie on the y axis CE (by Definition 22.1), and so CF has slope y/x, and EF has slope $(y - e)/x$. We can use slopes to compute cross-ratios (by Theorem 19.7), and so Equation 1 becomes

$$\frac{(d - a)(y/x - b)}{(y/x - a)(d - b)} = \frac{(d' - a')((y - e)/x - b')}{((y - e)/x - a')(d' - b')}. \tag{5}$$

Cross-multiplying gives

$$(d - a)(d' - b')\left(\frac{y}{x} - b\right)\left(\frac{y - e}{x} - a'\right)$$

$$= (d - b)(d' - a')\left(\frac{y}{x} - a\right)\left(\frac{y - e}{x} - b'\right), \tag{6}$$

and multiplying both sides of this equation by x^2 gives

$$(d - a)(d' - b')(y - bx)(y - e - a'x)$$

$$= (d - b)(d' - a')(y - ax)(y - e - b'x). \tag{7}$$

The graph of Equation 7 contains all ordinary points on $\mathcal{K}_1$ other than A–E (by the previous paragraph). We claim that the graph also contains A–E. In fact, the graph contains C and E, since substituting $(0, 0)$ or $(0, e)$ for (x, y) makes a factor on each side of Equation 7 equal 0. The coordinates (x, y) of A are such that $y/x = a$ and $(y - e)/x = a'$ (since a and a' are the slopes of the lines through A and either $C = (0, 0)$ or $E = (0, e)$); these coordinates make one factor on each side of Equation 6 equal 0, and so they satisfy Equation 7. Similarly, the coordinates (x, y) of B are such that $y/x = b$ and $(y - e)/x = b'$, and so they make one factor on each side of Equation 6 equal 0, and thus they satisfy Equation 7. The coordinates (x, y) of D are such that $y/x = d$ and $(y - e)/x = d'$, and so they make each side of Equation 6 equal

$$(d - a)(d' - b')(d - b)(d' - a'),$$

and thus they satisfy Equation 7.

In short, the graph of Equation 7 contains all ordinary points of $\mathcal{K}_1$. Equation 7 is a quadratic equation, as we see by multiplying it out and collecting terms in like powers of x and y. We claim that this quadratic equation is nondegenerate, as defined before Theorem 21.3. In fact, we've seen that the graph of Equation 7 contains the five points A–E, no three of which are collinear, and so the graph doesn't consist of one or two lines, one point, or the empty set. If we substitute ax for y in Equation 7, we obtain

$$(d - a)(d' - b')((a - b)x)(ax - e - a'x) = 0;$$

this equation doesn't hold for all real numbers x (since $d \neq a$, $d' \neq b'$, $a \neq b$, and $e \neq 0$), and so the graph of Equation 7 is not the whole plane. In short, Equation 7 is a nondegenerate quadratic equation whose graph contains all ordinary points of $\mathcal{K}_1$.

We must prove that, conversely, $\mathcal{K}_1$ contains all points on the graph of Equation 7. First assume that $F = (x, y)$ is a point on the graph of Equation 7 such that $x \neq 0$, y/x doesn't equal any of the numbers a, b, d, and $(y - e)/x$ doesn't equal any of the numbers a', b', d'. These assumptions ensure that F doesn't lie on any of the lines CA, CB, CD, CE, EA, EB, ED and that we can reverse the steps in the second paragraph of the proof. Thus, we can obtain Equation 5 from Equation 7 and deduce that Equation 1 holds. Then Theorem 22.3 shows that no three of the points A, B, D, F are collinear (since A, B, D are noncollinear, by assumption). The last three sentences show that F is not collinear with any two of the points A–E and that Equation 1 holds. Hence, F lies on $\mathcal{K}_1$, as desired.

Next assume that (x, y) lies on the graph of Equation 7 and that $x \neq 0$ and $y/x = a$. Dividing both sides of Equation 7 by x^2 gives Equation 6, and it follows that

$$(d - a)(d' - b')(a - b)\left(\frac{y - e}{x} - a'\right) = 0,$$

and so we have $(y - e)/x = a'$ (since $d \neq a$, $d' \neq b'$, and $a \neq b$). Thus, (x, y) lies on the line of slope a' through $(0, e)$, and it also lies on the line of slope a through $(0, 0)$ (since $y/x = a$). Hence, (x, y) equals A, so it lies on $\mathcal{K}_1$. Similarly, if (x, y) satisfies Equation 7 and if $x \neq 0$, then the relation $(y - e)/x = a'$ implies that $y/x = a$ and $(x, y) = A$, the relation $y/x = b$ implies that $(y - e)/x = b'$ and $(x, y) = B$, and the relation $(y - e)/x = b'$ implies that $y/x = b$ and $(x, y) = B$; thus, (x, y) lies on $\mathcal{K}_1$ in these cases also.

Next assume that (x, y) satisfies Equation 7 and that $x \neq 0$ and $y/x = d$. Dividing both sides of Equation 7 by x^2 gives Equation 6, which shows that

$$(d - a)(d' - b')(d - b)\left(\frac{y - e}{x} - a'\right)$$
$$= (d - b)(d' - a')(d - a)\left(\frac{y - e}{x} - b'\right).$$

Since a, b, d are distinct, and since a', b', d' are distinct, it follows that

$$(d' - b')\left(\frac{y - e}{x} - a'\right) = (d' - a')\left(\frac{y - e}{x} - b'\right);$$

$$[(d' - b') - (d' - a')]\frac{y - e}{x} = (d' - b')a' - (d' - a')b';$$

$$(a' - b')\frac{y - e}{x} = d'a' - d'b';$$

$$\frac{y - e}{x} = d'.$$

Thus, (x, y) lies on the line of slope d' through $(0, e)$, and it also lies on the line of slope d through $(0, 0)$ (since $y/x = d$). Hence, (x, y) equals D, and so it lies on $\mathcal{K}_1$. Similarly, if (x, y) satisfies Equation 7 and if $x \neq 0$ and $(y - e)/x = d'$, it follows that $y/x = d$, and so (x, y) equals D and lies on $\mathcal{K}_1$.

If we multiply out Equation 7, all terms that don't contain a factor of x arise from multiplying out expressions of the form $ry(y - e)$ for real numbers r. Thus, we can write Equation 7 in the form

$$Ax^2 + Bxy + Dx + Gy(y - e) = 0 \tag{8}$$

for numbers A, B, D, G. If G were zero, we could rewrite Equation 8 as

$$x(Ax + By + D) = 0,$$

and its graph would be either the entire plane (if A, B, D are all zero), the one line $x = 0$ (if B is zero and if exactly one of the numbers A and D is zero), or two lines (if B is nonzero or if both A and D are nonzero). Thus, G is nonzero, since we've seen that Equation 7 is a nondegenerate quadratic equation. If we set $x = 0$ in Equation 8, we obtain

$$Gy(y - e) = 0.$$

Hence, since $G \neq 0$, $C = (0, 0)$ and $E = (0, e)$ are the only points on the graph of Equation 7 that have x coordinate zero, and these points lie on $\mathcal{K}_1$.

The last four paragraphs show that $\mathcal{K}_1$ contains the graph of Equation 7. We've already seen that, conversely, the graph of Equation 7 contains the ordinary points of $\mathcal{K}_1$. Thus, the ordinary points of $\mathcal{K}_1$ are the graph of Equation 7, a nondegenerate quadratic equation. $\square$

Theorem 21.4 states that every Euclidean ellipse, Euclidean parabola, and Euclidean hyperbola is a section of a right circular cone. We have now proved enough to deduce that, conversely, every section of a cone is a Euclidean ellipse, Euclidean parabola, or Euclidean hyperbola, whether or not the cone is right circular. Accordingly, we can characterize the ordinary points on a conic section as follows.

THEOREM 22.5. In the Euclidean plane, let $\mathcal{K}$ be a set of points. Then the following conditions are equivalent.

   (i) $\mathcal{K}$ is the set of ordinary points on a conic section, and so $\mathcal{K}$ consists of the ordinary points obtained by projecting a circle from one extended plane to another.

  (ii) $\mathcal{K}$ is the graph of a nondegenerate quadratic equation.

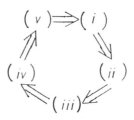

FIGURE 22.4.

(iii) $\mathcal{K}$ is a Euclidean ellipse, Euclidean parabola, or a Euclidean hyperbola.

(iv) $\mathcal{K}$ is a section of a right circular cone.

(v) $\mathcal{K}$ is a section of a cone.

Proof: We establish the implications in Figure 22.4. Thus, moving clockwise around the circle shows that any one of the statements (i)–(v) implies all of the others.

Condition (i) implies (ii), since any conic section equals $\mathcal{K}_1(A, B, C, D, E)$ for five points A–E (by Theorem 22.2), and so its ordinary points are the graph of a nondegenerate quadratic equation (by Theorem 22.4). (ii) implies (iii), by Theorem 21.3. (iii) implies (iv), by Theorem 21.4. (iv) implies (v) because a right circular cone is a special type of cone. Finally, (v) implies (i), by Theorem 18.2. □

If $\mathcal{K}$ is a set of points in the Euclidean plane, the previous theorem shows that $\mathcal{K}$ is a Euclidean ellipse, a Euclidean parabola, or a Euclidean hyperbola if and only if $\mathcal{K}$ is the set of ordinary points on a conic section $\mathfrak{M}$ in the extended plane. We can now reconcile Definitions 21.1 and 18.3. If $\mathcal{K}$ is a Euclidean ellipse (as in Definition 21.1 and Figure 21.1), every family of parallel ordinary lines contains a line that intersects $\mathcal{K}$ in two points. Then $\mathfrak{M}$ has no ideal points (by Theorem 18.4i), and so $\mathcal{K} = \mathfrak{M}$ is an ellipse (by Definition 18.3). If $\mathcal{K}$ is a Euclidean parabola (as in Definition 21.1 and Figure 21.2), every ordinary line parallel to the axis of symmetry intersects $\mathcal{K}$ in exactly one point. These lines can't all be tangent to $\mathfrak{M}$ (by Theorem 18.6iii), and so $\mathfrak{M}$ contains the ideal point A on all these lines. Each ordinary line on A intersects $\mathcal{K}$ at an ordinary point, and so the tangent at A is the ideal line (Figure 18.12), and $\mathcal{K}$ is a parabola (by Definition 18.3). If $\mathcal{K}$ is a Euclidean hyperbola (as in Definition 21.1 and Figure 21.3), the asymptotes don't intersect $\mathcal{K}$, and each ordinary line parallel to and distinct from an asymptote intersects $\mathcal{K}$ in exactly one point. These lines can't all be tangent to $\mathfrak{M}$ (by Theorem 18.6iii), and so $\mathfrak{M}$ contains the ideal points A and B on the two asymptotes, and $\mathcal{K}$ is a hyperbola (by Definition 18.3). The asymptotes intersect $\mathfrak{M}$ only at A and B, and so the asymptotes are the tangents at A and B (Figure 18.13).

In short, the previous paragraph shows that Euclidean ellipses, Euclidean parabolas, and Euclidean hyperbolas are exactly the same as ellipses, parabolas, and hyperbolas, respectively, and so Definitions 21.1 and 18.3 describe the same curves. The preceding paragraph also justifies the discussion accompanying Figures 18.12 and 18.13. Thus, we can obtain theorems about Euclidean ellipses, Euclidean parabolas, and Euclidean hyperbolas by taking a theorem about conic sections and choosing a position for the ideal line.

In the rest of this section, we use Theorem 22.5 to deduce that any five points, no three of which are collinear, lie on a conic section. The next definition is helpful.

DEFINITION 22.6. In the extended plane, let A–E be five points, no three of which are collinear. Define $\mathcal{K}_2(A, B, C, D, E)$ to be the set of points that consists of A–E and all points F that are not collinear with any two of the points A–E and are such that the points $L = AB \cap DE$, $M = BC \cap EF$, $N = CD \cap FA$ are collinear (Figure 22.5). $\square$

This definition is analogous to Definition 22.1 of $\mathcal{K}_1(A, B, C, D, E)$. We consider $\mathcal{K}_2(A, B, C, D, E)$ as well as $\mathcal{K}_1(A, B, C, D, E)$, because the former is defined solely in terms of incidence and doesn't require A–E to be ordinary points.

If A–E are five points on a conic section $\mathcal{K}$, then $\mathcal{K}$ equals $\mathcal{K}_2(A, B, C, D, E)$ (by Theorems 20.6 and 18.4i). Conversely, if we're given five points A–E, no three of which are collinear, we prove that $\mathcal{K}_2(A, B, C, D, E)$ is a conic section. We start with the next result.

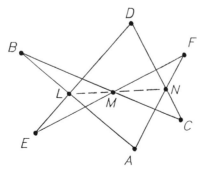

FIGURE 22.5.

THEOREM 22.7. In the extended plane, let A–E be five points, no three of which are collinear.

(i) Then $\mathcal{K}_2(A, B, C, D, E)$ contains infinitely many points, no three of which are collinear.

(ii) If A'–E' are any five points on $\mathcal{K}_2(A, B, C, D, E)$, we have

$$\mathcal{K}_2(A, B, C, D, E) = \mathcal{K}_2(A', B', C', D', E').$$

Proof: Because Definition 22.6 and the statements to be proved involve only incidence, we can assume that A–E are ordinary (by projecting between planes and using Theorems 15.2 and 15.3). The ordinary points of $\mathcal{K}_1(A, B, C, D, E)$ are the graph of a nondegenerate quadratic equation (by Theorem 22.4), and so they lie on a conic section $\mathcal{K}$ (by Theorem 22.5). Thus, A–E, which lie on $\mathcal{K}_1(A, B, C, D, E)$, lie on the conic section $\mathcal{K}$. Then $\mathcal{K}_2(A, B, C, D, E)$ equals $\mathcal{K}$ (by Theorem 20.6), and so part (i) holds (by Theorem 18.4i). If A'–E' are five points on $\mathcal{K}_2(A, B, C, D, E) = \mathcal{K}$, then we also have $\mathcal{K}_2(A', B', C', D', E') = \mathcal{K}$ (by Theorem 20.6), and so part (ii) holds.   □

We can now prove the result toward which we've been heading.

THEOREM 22.8. In the extended plane, there is a unique conic section through any five points, no three of which are collinear.

Proof: Let A–E be five points, no three of which are collinear. Theorem 22.7(i) shows that $\mathcal{K}_2(A, B, C, D, E)$ contains at most two points on the ideal line and infinitely many other points. Thus, $\mathcal{K}_2(A, B, C, D, E)$ contains five ordinary points A'–E', and we have

$$\mathcal{K}_2(A, B, C, D, E) = \mathcal{K}_2(A', B', C', D', E') \tag{9}$$

(by Theorem 22.7ii). The ordinary points on $\mathcal{K}_1(A', B', C', D', E')$ are the graph of a nondegenerate quadratic equation (by Theorem 22.4), and so they are the ordinary points on a conic section $\mathcal{K}$ (by Theorem 22.5). Thus, A'–E' lie on $\mathcal{K}$, and so we have

$$\mathcal{K} = \mathcal{K}_2(A', B', C', D', E') = \mathcal{K}_2(A, B, C, D, E)$$

(by Theorem 20.6 and Equation 9). Hence, $\mathcal{K}$ contains A–E.

Thus, A–E lie on at least one conic section. On the other hand, no more than one conic section can contain A–E, since any such conic section must equal $\mathcal{K}_2(A, B, C, D, E)$ (by Theorem 20.6).   □

In the Euclidean plane, let A–E be five points, no three of which are collinear. Theorems 22.5 and 22.8 show that A–E lie on a unique curve $\mathcal{K}$ that is either a Euclidean ellipse, a Euclidean parabola, or a Euclidean

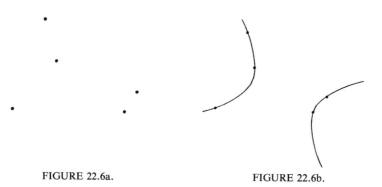

FIGURE 22.6a.   FIGURE 22.6b.

hyperbola. As an illustration, we can choose five points at random, no three of which are collinear, and draw the curve $\mathcal{K}$ they determine. For example, the five points in Figure 22.6a determine the Euclidean hyperbola in Figure 22.6b. It's interesting to vary the positions of the five points and observe the corresponding changes in $\mathcal{K}$.

In the extended plane, consider five points A–E, no three of which are collinear. These points lie on a unique conic section (by the previous theorem), which equals $\mathcal{K}_2(A, B, C, D, E)$ (by Theorem 20.6). Conversely, any conic section contains infinitely many points and equals $\mathcal{K}_2(A, B, C, D, E)$ for any five of these points (by Theorems 18.4i and 20.6). Thus, we have the following result.

THEOREM 22.9. In the extended plane, the conic sections are the sets $\mathcal{K}_2(A, B, C, D, E)$ as A–E vary over all choices of five points, no three of which are collinear. □

Of course, there are infintely many choices of points A–E that give the same set $\mathcal{K}_2(A, B, C, D, E)$, as Theorem 22.7 shows.

The sets $\mathcal{K}_2(A, B, C, D, E)$ are defined solely in terms of incidence, and so projection between planes maps such sets to other such sets (by Theorem 15.2). Thus, Theorem 22.9 has the following consequence.

THEOREM 22.10. Projections between planes map conic sections to other conic sections. □

By Definition 18.1, conic sections are the sets obtained by projecting a circle to another plane. Thus, Theorem 22.10 shows that we don't obtain anything new if we apply a sequence of projections to a circle instead of just one projection.

In the extended plane, five ordinary points A–E, no three of which are collinear, lie on a unique conic section (by Theorem 22.8), which equals $\mathcal{K}_1(A, B, C, D, E)$ (by Theorem 22.2). Conversely, any conic section has the form $\mathcal{K}_1(A, B, C, D, E)$ (by Theorem 22.2). Thus, we have the following companion result to Theorem 22.9.

THEOREM 22.11. In the extended plane, the conic sections are the sets $\mathcal{K}_1(A, B, C, D, E)$ as A–E vary over all choices of five ordinary points, no three of which are collinear. $\square$

EXERCISES

22.1. In each part of this exercise, we give five ordinary points A–E, no three of which are collinear. These points lie on a unique curve $\mathcal{K}$ that is either a Euclidean ellipse, a Euclidean parabola, or a Euclidean hyperbola (by Theorems 22.5 and 22.8). Draw a figure that shows A–E and $\mathcal{K}$.

(a) $A = (1, 1)$, $B = (-1, 0)$, $C = (0, 0)$, $D = (1, -1)$, $E = (0, 2)$.
(b) $A = (0, 0)$, $B = (1, -1)$, $C = (-2, -1)$, $D = (-1, 0)$, $E = (0, 2)$.
(c) $A = (-1, 0)$, $B = (0, -1)$, $C = (0, 1)$, $D = (1, 1)$, $E = (1, 0)$.
(d) $A = (0, 1)$, $B = (0, 3)$, $C = (2, 0)$, $D = (2, -1)$, $E = (-2, -1)$.
(e) $A = (2, 0)$, $B = (0, 4)$, $C = (0, 1)$, $D = (1, 3)$, $E = (1, 0)$.
(f) $A = (2, -2)$, $B = (1, 0)$, $C = (1, 1)$, $D = (0, 1)$, $E = (0, 0)$.
(g) $A = (1, 0)$, $B = (2, 1)$, $C = (0, 3)$, $D = (3, 3)$, $E = (-1, 1)$.

22.2. In each part of Exercise 22.1, $\mathcal{K}$ is the graph of a nondegenerate quadratic equation (by Theorem 22.5). Find the equation of $\mathcal{K}$.

(*Hint:* One possible approach is as follows: Set $F = (x, y)$ in Equation 1, use Theorem 19.7 to evaluate each side of Equation 1, and then rewrite the resulting equation as a quadratic equation in x and y. Theorem 22.2 justifies this procedure.)

22.3. In the extended plane, let A–D be four points, no three of which are collinear, and let l be a line on A that doesn't contain any of the points B, C, D. Prove that there is a unique conic section that contains A–D and is tangent to l.

(*Hint:* One possible approach is as follows. Choose a line m on A other than l that doesn't contain any of the points B, C, D. As Figure 22.7 suggests, one can deduce from Theorem 20.5 that there is a unique point E on m such that a conic section containing A–E is tangent to l. Then apply Theorem 22.8.)

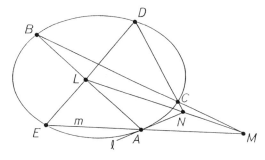

FIGURE 22.7.

22.4. In the extended plane, let A, B, C be three noncollinear points. Let l be a line on A that doesn't contain B or C, and let m be a line on B that doesn't contain A or C. Prove that there is a unique conic section that contains A, B, C and is tangent to l and m.

(*Hint:* One possible approach is as follows. Choose a line n on A other than l, AB, AC. As Figure 22.8 suggests, one can deduce from Theorem 20.8 that there is a unique point D on n such that a conic section containing A–D and tangent to l is also tangent to m. Then apply Exercise 22.3.)

22.5. In the extended plane, let A–F be six points, no three of which are collinear. Assume that AB, CD, EF are the sides of a triangle and that DE, FA, BC are the sides of a triangle. If these triangles are perspective from a point, prove that A–F lie on a conic section (Figure 20.29).

22.6. In the extended plane, let A–E be five points, no three of which are collinear. Let l be a line through A that doesn't contain any of the points B–E. Assume that l, BC, DE are the sides of a triangle and that this triangle is perspective from a point with the triangle that has CD, EA, AB as sides. Prove that A–E lie on a conic section tangent to l. Illustrate this result with a figure.

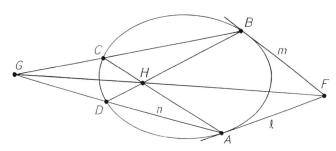

FIGURE 22.8.

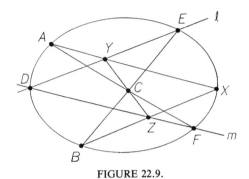

FIGURE 22.9.

22.7. Prove the following result (Figure 22.9):

Theorem. In the extended plane, let A, B, C be three noncollinear points. Let l and m be two lines that don't contain any of the points A, B, C and such that $l \cap m$ doesn't lie on any of lines AB, BC, AC.

(i) Set $D = l \cap m$, $E = BC \cap l$, and $F = AC \cap m$. Then the five points A, B, D, E, F lie on a unique conic section $\mathcal{K}$.

(ii) The conic section $\mathcal{K}$ in part (i) contains all points X such that there is a triangle XYZ that has vertex Y on l, vertex Z on m, side XY containing A, side XZ containing B, and side YZ containing C.

(Roughly, part (ii) states that, if the sides of a triangle revolve around three fixed points while two of the vertices move along fixed lines, then the third vertex traces out a conic section. Theorem 20.6 may be of use in proving part (ii).)

22.8. Prove the following result, and illustrate it with a figure analogous to Figure 22.9.

Theorem. In the extended plane, let A, B, C be three noncollinear points. Let l and m be two lines that don't contain any of the points A, B, C and such that $l \cap m$ lies on line AC.

(i) Set $D = l \cap m$ and $E = BC \cap l$. Then there is a unique conic section $\mathcal{K}$ that contains the four points A, B, D, E and is tangent to m.

(ii) The conic section $\mathcal{K}$ in part (i) contains all points X such that there is a triangle XYZ that has vertex Y on l, vertex Z on m, side XY containing A, side XZ containing B, and side YZ containing C.

(See Exercise 22.3 and Theorems 20.5 and 22.8.)

22.9. In the extended plane, let *C–F* be four points, no three of which are collinear. Let *l* be a line that contains $CD \cap EF$ and doesn't contain $CE \cap DF$, $CF \cap DE$, *C*, *D*, *E*, or *F*. Prove that there is a unique conic section $\mathcal{K}$ that contains *C–F* and is tangent to *l*. Prove that $\mathcal{K}$ is tangent to *l* at the harmonic conjugate of $CD \cap EF$ with respect to $CE \cap l$ and $DF \cap l$. (See Exercises 20.10 and 20.14 and Theorem 22.8.)

22.10. In the extended plane, let *L*, *M*, *N* be three noncollinear ordinary points (Figure 19.17). Let *A* and *B* be two points on line *LM* other than *L* and *M*, let *C* and *D* be two points on line *MN* other than *M* and *N*, and let *E* and *F* be two points on line *LN* other than *L* and *N*. Prove that the six points *A–F* lie either on a conic section or on two lines if and only if the equation

$$\frac{\overline{AL} \cdot \overline{BL} \cdot \overline{CM} \cdot \overline{DM} \cdot \overline{EN} \cdot \overline{FN}}{\overline{AM} \cdot \overline{BM} \cdot \overline{CN} \cdot \overline{DN} \cdot \overline{EL} \cdot \overline{FL}} = 1,$$

holds when directed distances involving ideal points are omitted.
   (See Exercises 19.17 and 19.18, Theorem 22.8, Menelaus' Theorem, and Theorem 2.2.)

22.11. In the extended plane, let *LMN* be a triangle. Let *A* and *B* be two points on line *LM* other than *L* and *M*, let *C* and *D* be two points on line *MN* other than *M* and *N*, and let *E* and *F* be two points on line *LN* other than *L* and *N*. Assume that the lines *AN*, *CL*, *EM* lie on a common point *O* and that the lines *BN*, *DL*, *FM* lie on a common point *P*. Then the six points *A–F* lie either on a conic section or on two lines. Illustrate this result with two figures, one showing each of the possibilities in the last sentence.
   (*Hint:* One possible approach is to reduce to the case where all points named are ordinary and then apply Exercise 22.10 and Ceva's Theorem.)

22.12. In the notation of Exercise 22.10, prove that Equation 35 of Section 19 holds if and only if *C–F* lie either on a conic section tangent to *LM* at *A* or on two lines through *A*.

22.13. In the extended plane, let *C–F* be four points, no three of which are collinear. Let *l* be a line that doesn't contain any of the points *C–F*, $CD \cap EF$, $CE \cap DF$, $CF \cap DE$. Prove that there are either 0 or 2 conic sections that contain *C–F* and are tangent to *l*. (See Exercise 22.12.)

22.14. In the extended plane, let a_1–a_5 be five lines, no three of which are concurrent. Let $\mathcal{K}'$ be a conic section.

   (a) Prove that there is a conic section $\mathcal{L}$ that contains points B_1–B_5 such that a_i is the polar of B_i with respect to $\mathcal{K}'$ for each *i*. (See Exercise 20.16 and Theorem 22.8.)

(b) Let b_1–b_5 be the tangents to $\mathcal{L}$ at B_1–B_5. Prove that there is a conic section $\mathcal{K}$ that contains points A_1–A_5 such that b_i is the polar of A_i with respect to $\mathcal{K}'$ for each i.
(c) Prove that a_1–a_5 are tangent to $\mathcal{K}$. (See Exercises 20.16 and 20.18 and Theorem 20.5.)

22.15. In the extended plane, prove that any five lines, no three of which are concurrent, are tangent to a unique conic section. (See Exercises 20.18 and 22.14 and Theorem 18.6.)

22.16. In the Euclidean plane, let e and f be two lines on a point T. Choose coordinate systems on e and f (as in Definition 1.6) so that T has coordinate 0 on each line. Identify points on e and f by their coordinates. Let p and q be nonzero real numbers.

(a) Let a be the ideal line, let b be the line through the point q/p on e and the point $2q$ on f, and let c be the line through the point $2q/p$ on e and the point $3q$ on f. Use Exercise 22.14 to prove that there is a conic section $\mathcal{K}$ tangent to a, b, c, e, f.
(b) Let d be a tangent of $\mathcal{K}$ other than a, b, c, e, f. Prove that d intersects e and f at points x and y such that $y = px + q$. (*Hint:* One possible approach is to apply Exercise 20.9, where A–F are the respective points of contact of a–f. Use Theorem 19.2 to evaluate Equation 18 of Section 20.)
(c) Prove that the ordinary points of $\mathcal{K}$ form a parabola $\mathcal{K}'$ in the Euclidean plane that is tangent to e at $-q/p$ and tangent to f at q. Prove that the tangents of $\mathcal{K}'$ are exactly the lines that join the point x on e with the point $px + q$ on f as x varies over all real numbers. (See Theorem 18.6.)

22.17. Figure 22.10 illustrates Exercise 22.16 for $p = -1$ and $q = 4$, where e and f are the x and y axes. The figure shows the lines through the points $(x, 0)$ and $(0, -x + 4)$ as x varies over all integers from -2 through 6. Draw analogous figures to illustrate Exercise 22.16 in the following cases, where c and e are the x and y axes. In each figure, sketch the parabola tangent to the lines drawn.

(a) $p = 1$ and $q = 5$
(b) $p = -2$ and $q = 7$

22.18. Prove that we can obtain any parabola in the Euclidean plane by using Exercise 22.16, taking e and f to be any two tangents, and choosing p and q appropriately.

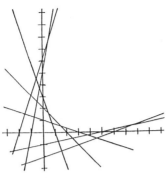

FIGURE 22.10.

22.19. In the Euclidean plane, let e and f be two lines on a point T. Choose coordinate systems on e and f so that T has coordinate 0 on each line. Let p, q, s be nonzero real numbers such that $q \neq ps$. Identify points on e and f by their coordinates.

(a) Let a be the line parallel to f through the point $-s$ on e, and let b be the line parallel to e through the point p on f. Let u be a nonzero real number other than $-s$ and $-q/p$, and let c be the line through the point u on e and the point $(pu + q)/(u + s)$ on f. Use Theorem 19.10 to prove that a, b, c are not concurrent. Then use Exercise 22.14 to prove that there is a conic section $\mathcal{K}$ tangent to a, b, c, e, f.

(b) Let d be a tangent of $\mathcal{K}$ other than a, b, c, e, f. Prove that d intersects e and f at points x and y such that $y = (px + q)/(x + s)$. (*Hint:* One possible approach is to apply Exercise 20.9, where A–F are the respective points of contact of a–f. Use Theorem 19.2 to evaluate Equation 18 of Section 20.)

(c) Prove that $\mathcal{K}$ is tangent to e at the point $-q/p$ and tangent to f at the point q/s. Prove that the ordinary points of $\mathcal{K}$ form an ellipse or a hyperbola $\mathcal{K}'$ in the Euclidean plane whose tangents (and asymptotes, if $\mathcal{K}'$ is a hyperbola) are a, b, and the lines through the point x on e and the point $(px + q)/(x + s)$ on f as x varies over all real numbers other than $-s$. (See Theorem 18.6.)

22.20. In each part of this exercise, draw a figure analogous to Figure 22.10 to illustrate Exercise 22.19 for the given values of p, q, and s, where e and f are the x and y axes. In each figure, sketch the ellipse of hyperbola tangent to the lines drawn.

(a) $p = -2$, $q = 4$, $s = -3$
(b) $p = -2$, $q = 3$, $s = 1$
(c) $p = 2$, $q = 4$, $s = 4$

22.21. Prove that we can obtain any ellipse or hyperbola in the Euclidean plane by using Exercise 22.19, taking e and f to be any two tangents that aren't parallel, and choosing p, q, and s appropriately.

22.22.

(a) In the Euclidean plane, let e and f be two lines on a point T. Choose coordinate systems on e and f so that T has coordinate 0 on each line. Let q be a nonzero real number. Identify points on e and f by their coordinates. Prove that there is a hyperbola whose asymptotes are e and f and whose tangents are exactly the lines through the point x on e and the point q/x on f as x varies over all nonzero real numbers.

(b) Illustrate part (a) with a figure analogous to Figure 22.10, where $q = 12$ and where e and f are the x and y axes.

Chapter V

Axiomatic Geometry

INTRODUCTION AND HISTORY

Geometers from Africa, Asia, and Europe worked for over two thousand years to correct an apparent flaw in Euclid's *Elements*. Their work ultimately led to a revolution in mathematics that proved Euclid had been right all along.

The supposed flaw in the *Elements* was the Fifth Postulate. A postulate is a fundamental assumption used to prove other results. The Fifth Postulate states that two lines forming acute interior angles on one side of a transversal intersect on that side. In other words, if A, B, C, D are four points such that A and D lie on the same side of line BC, and if $\angle ABC$ and $\angle BCD$ are less than $90°$, then the rays $\overrightarrow{BA}$ and $\overrightarrow{CD}$ intersect (Figure V.1). It can be shown that this postulate is equivalent to the assumption that every point lies on a unique line parallel to a given line.

Euclid's critics thought that it was unnecessary to assume the Fifth Postulate. They tried and failed for two millennia to deduce the Fifth Postulate from Euclid's other axioms. They often proved the Fifth Postulate, but only after adding an extra assumption to Euclid's other axioms. Knowingly or unknowingly, they simply replaced the Fifth Postulate with an equivalent assumption.

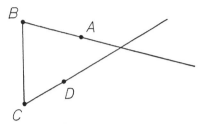

FIGURE V.1.

Nevertheless, much of this work was ultimately valuable. Geometers tried to show that Euclid's other axioms would be contradicted if the Fifth Postulate were false. They never found a contradiction, but they discovered many properties that would have to hold in a plane that satisfied all of Euclid's axioms except the Fifth Postulate. Among those who obtained notable results were Omar Khayyam (the well-known poet who wrote the *Rubaiyat*) in the eleventh century, and Girolamo Saccheri and Johann Heinrich Lambert in the eighteenth century.

The hyperbolic plane is a plane where the Fifth Postulate doesn't hold but Euclid's other axioms do. In the first half of the nineteenth century, a few mathematicians began to believe in the logical possibility of the hyperbolic plane. This idea was developed in largely unpublished work by Carl Friedrich Gauss—one of the greatest mathematicians of all time—and his correspondents Friedrich Ludwig Wachter, Ferdinand Carl Schweikart, and Franz Adolf Taurinus. Nicolai Ivanovitch Lobachevsky and Janos Bolyai asserted publicly that the hyperbolic plane was logically possible, and they published analyses of its properties.

The assertions of Lobachevsky and Bolyai were justified in the second half of the nineteenth century. Eugenio Beltrami, Felix Klein, and Henri Poincaré used Euclidean geometry to construct models of the hyperbolic plane. These models proved that the hyperbolic plane was as logically consistent as the Euclidean plane. Because the hyperbolic plane satisfied all of Euclid's axioms except the Fifth Postulate, the logical consistency of the hyperbolic plane proved that the Fifth Postulate did not follow from the other axioms of Euclidean geometry. Thus, Euclid had been right to include the Fifth Postulate among his axioms.

The discovery of the hypercolic plane freed mathematicians from the assumption that geometry was limited to Euclidean space. They began to use remarkable new geometries throughout mathematics. Differential geometers used analysis to study curvature in multidimensional spaces, and their ideas ultimately led to Albert Einstein's General Theory of Relativity. Complex analysts used Riemann surfaces to study multivalued functions geometrically. Functional analysts studied spaces whose points are functions. Algebraic geometers radically extended their study of algebraic curves to multidimensional spaces derived from abstract algebra. Topology, the study of abstract spaces, became a major branch of mathematics.

Geometers used axiom systems to illuminate the basic properties of known geometries and to point the way to new geometries. At the end of the nineteenth century, Moritz Pasch, Giuseppe Peano, and Mario Pieri revised and supplemented Euclid's axioms to make explicit his unstated assumptions. For example, Pasch added the following axiom, which Euclid had taken for granted: If a line passes through the interior of a triangle and contains a vertex, then the line intersects the side of the triangle opposite the vertex. David Hilbert—the greatest mathematician of the twentieth century—produced a definitive set of axioms for Euclidean geometry. Alternatives to Hilbert's axioms were developed by such noted mathematicians as George David Birkhoff and Oswald Veblen.

In the first section of this chapter, we study inversions in circles. An inversion in a circle is a map of the plane that interchanges points inside and outside of the circle, excluding the center. It is the circular analogue of a reflection in a line. Circular inversions were apparently first studied in the third century B.C. by Apollonius, the brilliant successor to Euclid and Archimedes. We use inversions in circles in Section 24 to develop one of Poincaré's constructions of the hyperbolic plane.

We present axioms for absolute geometry in Section 25. Absolute geometry is the study of the properties common to both Euclidean and hyperbolic geometry. The axioms of absolute geometry do not include the Fifth Postulate or its equivalents, and so the theorems of absolute geometry apply to both the Euclidean and the hyperbolic planes. The axioms we present are essentially due to George David Birkhoff. They make it possible to derive basic properties quickly and easily by using coordinate systems on lines and degree measures of angles.

In Section 26, we use the axioms of absolute geometry to prove the equivalence of a number of alternative forms of the Fifth Postulate. Each of these equivalent properties characterizes the difference between Euclidean and hyperbolic geometry.

Much of this chapter is based on the following books, which provide further reading about absolute and hyperbolic geometries.

Greenberg, Marvin Jay, *Euclidean and Non-Euclidean Geometries*, Freeman, San Francisco, 1980.

Kelly, Paul, and Matthews, Gordon, *The Non-Euclidean, Hyperbolic Plane*, Springer-Verlag, New York, 1981.

Martin, George E., *The Foundations of Geometry and the Non-Euclidean Plane*, Springer-Verlag, New York, 1975.
Wolfe, Harold E., *Introduction to Non-Euclidean Geometry*, Holt, Rinehart, and Winston, New York, 1945.

Section 23.

Inversion in Circles

Our goal in this section and the next is to construct the hyperbolic plane. Although the hyperbolic and the Euclidean planes have many similarities, they also have dramatic differences. The construction of the hyperbolic plane depends on certain properties of circles in the Euclidean plane, and we develop these properties in this section. In particular, we study inversion in a circle, a map of the Euclidean plane that interchanges the points outside the circle with those inside except the center. We prove that inversion preserves a surprising number of properties of the plane.

This section depends on two small, self-contained portions of Chapter II. Readers who skipped Chapter II should read Section 7 through the discussion accompanying Figure 7.1 and Section 13 through the proof of Theorem 13.3.

DEFINITION 23.1. In the Euclidean plane, let $\mathcal{K}$ be a circle with center P and radius r. The *inversion* $\mathcal{I}$ in $\mathcal{K}$ maps each point X in the plane other than P to the point X' on line PX such that

$$\overline{PX} \cdot \overline{PX'} = r^2. \tag{1}$$

$\mathcal{I}(X)$ denotes the image X' of X under $\mathcal{I}$. □

In this definition, the inversion $\mathcal{I}$ maps the points in the plane other than P among themselves. The image of P under $\mathcal{I}$ is undefined. Switching the positive end of line PX replaces each of the quantities $\overline{PX}$ and $\overline{PX'}$ with its negative, and so Equation 1 remains unchanged. Thus, Definition 23.1 doesn't depend on the choice of a positive end on line PX.

Let t be the set of points except P on a ray originating at P (Figure 23.1). Let $X \to x$ be a coordinate system on the line l containing t such that P has coordinate 0 and t consists of the points with positive coordinates. For every point X on l other than P, if x is the coordinate of X, and if $X' = \mathcal{I}(X)$ has coordinate x', Equation 1 becomes

$$xx' = r^2 \tag{2}$$

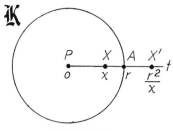

FIGURE 23.1.

(by Theorem 1.7, since we've seen that Definition 23.1 doesn't depend on the choice of a positive end of line PX), or, equivalently,

$$x' = r^2/x. \tag{3}$$

Since $x > 0$ implies that $x' > 0$ (by Equation 3), $\mathcal{I}$ maps the points of t among themselves. If $\mathcal{I}$ maps X to X', it also maps X' to X (since Equation 2 is symmetric in x and x'), and so $\mathcal{I}$ interchanges the points of t among themselves in pairs.

As X moves along t away from P, $X' = \mathcal{I}(X)$ moves along t toward P (since, as x increases, $x' = r^2/x$ decreases). One of the points X and X' lies inside $\mathcal{K}$ exactly when the other lies outside (since Equation 2 pairs positive numbers less than r with positive numbers greater than r). Let A be the point on t with coordinate r where t intersects $\mathcal{K}$. As X moves from the inside of $\mathcal{K}$ to the outside, X' moves from the outside of $\mathcal{K}$ to the inside, and both X and X' pass through the point A simultaneously (since Equation 2 holds with $x = r = x'$). Accordingly, when we say that $\mathcal{I}$ interchanges the points of t in pairs, the point A is paired with itself.

Any point in the plane except P lies on a unique ray originating at P. Thus, we've proved the following result.

THEOREM 23.2. In the Euclidean plane, let $\mathcal{I}$ be the inversion in a circle $\mathcal{K}$ having center P. Then $\mathcal{I}$ fixes the points of $\mathcal{K}$. It interchanges in pairs the points outside $\mathcal{K}$ with those inside $\mathcal{K}$ except P. The points except P on each ray originating at P are interchanged among themselves. □

We use the following convention in this section. Let $\mathcal{I}$ be the inversion in a circle $\mathcal{K}$ with center P, and set $X' = \mathcal{I}(X)$ for any point X other than P. If X doesn't lie on $\mathcal{K}$, then X' doesn't equal X, and the perpendicular bisector m of X and X' is determined by Definition 4.1, as usual (Figure 23.2). Since P, X, X' lie on a line (by Definition 23.1), m is perpendicular to line PX. If X is a point of $\mathcal{K}$, however, then X' equals X, and Definition 4.1 doesn't apply; in this case, we adopt the convention

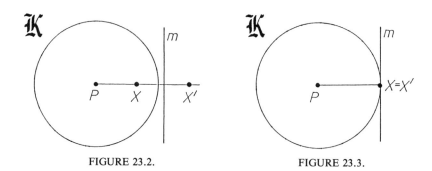

FIGURE 23.2.   FIGURE 23.3.

that the perpendicular bisector of X and X' is the line m through X perpendicular to line PX (Figure 23.3). Accordingly, for any point X except P, the perpendicular bisector m of X and X' is a line perpendicular to line PX, and the reflection σ_m in m interchanges X and X' (by Definition 7.2) (Figures 23.2 and 23.3).

If X and Y are two points in the Euclidean plane, we let $[X, Y]$ denote the segment with endpoints X and Y. We use this notation to summarize the discussion before Theorem 23.2 in the following form.

THEOREM 23.3. In the Euclidean plane, let $\mathcal{I}$ be the inversion in a circle $\mathcal{K}$ with center P. Let l be a line through P.

(i) Then $\mathcal{I}$ interchanges the points of l other than P among themselves.

(ii) Let A and B be two points on l such that $[A, B]$ doesn't contain P. If $\mathcal{I}$ maps A to A' and B to B', then $\mathcal{I}$ maps $[A, B]$ onto $[A', B']$. If m is the perpendicular bisector of A and A', the reflection σ_m interchanges the rays $\overrightarrow{AB}$ and $\overrightarrow{A'B'}$ (Figure 23.4).

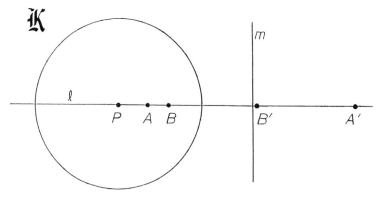

FIGURE 23.4.

Proof: Definition 23.1 gives part (i).

(ii) Let t be the set of points other than P on the ray $\overrightarrow{PA}$. $[A, B]$ lies on t (since it lies on l and doesn't contain P). As X varies over the points of t in one direction, $X' = \mathcal{I}(X)$ varies over the points of t in the opposite direction (by the discussion accompanying Figure 23.1). It follows that $\mathcal{I}$ maps $[A, B]$ onto $[A', B']$ and that the rays $\overrightarrow{AB}$ and $\overrightarrow{A'B'}$ point in opposite directions. The latter fact implies that the reflection σ_m interchanges $\overrightarrow{AB}$ and $\overrightarrow{A'B'}$ (since it interchanges A and A' and reverses the direction of rays on l). $\square$

Our next goal is to find the analogue of the previous theorem that holds when l doesn't contain P. We use the following basic property of inversions.

THEOREM 23.4. In the Euclidean plane, let PAB be a triangle. Let A' and B' be the respective images of A and B under inversion in a circle $\mathcal{K}$ with center P. Then the triangles PAB and $PB'A'$ are similar (Figure 23.5).

Proof: Let XY denote the distance between points X and Y. If $\mathcal{K}$ has radius r, we have

$$PA \cdot PA' = r^2 = PB \cdot PB'$$

(by Definition 23.1), which shows that

$$PB'/PA = PA'/PB. \tag{4}$$

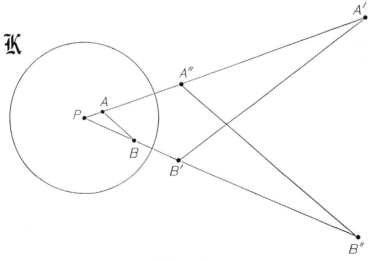

FIGURE 23.5.

Let A'' be the point on ray $\overrightarrow{PA}$ such that $PA'' = PB'$, and let B'' be the point on ray $\overrightarrow{PB}$ such that $PB'' = PA'$. Triangles $PA''B''$ and $PB'A'$ are congruent (by the SAS Property 0.2), and so Equation 4 implies that

$$\overline{PA''}/\overline{PA} = \overline{PB''}/\overline{PB} \tag{5}$$

(since both sides of this equation are positive). Then AB and $A''B''$ are parallel lines (by Theorem 1.11), and so they form equal corresponding angles with PA and PB (by Property 0.5). Thus, we have

$$\angle PAB = \angle PA''B'' \quad \text{and} \quad \angle PBA = \angle PB''A''$$

(since $\overrightarrow{PA}$ contains A'' and $\overrightarrow{PB}$ contains B''). Thus, triangles PAB and $PA''B''$ are similar. Hence, PAB and $PB'A'$ are similar (since $PA''B''$ and $PB'A'$ are congruent triangles, as we've seen). □

In order to study the angles formed by intersecting circles, we consider tangent rays to arcs.

DEFINITION 23.5. In the Euclidean plane, let $\mathcal{K}$ be a circle. An *arc s of* $\mathcal{K}$ consists of two points A and B on $\mathcal{K}$ and all points on $\mathcal{K}$ that lie on one side of line AB. A and B are the *endpoints* of s. The *tangent ray* to s at A is the ray t that originates at A, lies on the tangent line to $\mathcal{K}$ at A, and contains points that lie on the same side of line AB as points of s (Figure 23.6). □

By Theorem 23.3, the inversion $\mathcal{I}$ in a circle $\mathcal{K}$ with center P maps each line through P onto itself, excluding P. The next result shows that $\mathcal{I}$ interchanges lines that don't contain P with circles through P (excluding P itself).

THEOREM 23.6. In the Euclidean plane, let $\mathcal{I}$ be the inversion in a circle $\mathcal{K}$ with center P.

FIGURE 23.6.

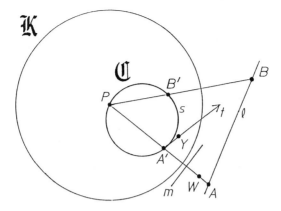

FIGURE 23.7.

(i) If *l* is a line that doesn't contain *P*, then 𝔍 interchanges the points of *l* with the points other than *P* of a circle 𝒞 through *P*. The correspondence *l* ↔ 𝒞 in part (i) matches up the lines *l* that don't contain *P* with the circles 𝒞 through *P*.

(ii) Let *l* and 𝒞 correspond as in part (i). Let *A* and *B* be two points of *l*, and let *A'* and *B'* be their respective images under 𝔍. Then 𝔍 matches up the points of the segment [*A*, *B*] of *l* with the points of the arc *s* of 𝒞 that has *A'* and *B'* as endpoints and doesn't contain *P*. Moreover, if *m* is the perpendicular bisector of *A* and *A'*, the reflection σ_m interchanges the ray $\overrightarrow{AB}$ and the tangent ray *t* to *s* at *A'* (Figure 23.7).

Proof: (i) Let *F* be the foot of the perpendicular from *P* to *l* (Figure 23.8). *F* doesn't equal *P* (since *l* doesn't contain *P*), and so we can set *F'* = 𝔍(*F*). Since *F'* ≠ *P*, there is a circle 𝒞 that has *F'* and *P* as the endpoints of a diameter.

Let *X* and *X'* be points that are interchanged by 𝔍 and that don't lie on line *PF*. *X* lies on *l* if and only if ∠*PFX* = 90°. This happens if and only if ∠*PX'F'* = 90° (by Theorem 23.4), which occurs if and only if *X'* lies on 𝒞 (by Theorem 4.7ii). Thus, 𝔍 interchanges the points *X* of *l* other than *F* with the points *X'* of 𝒞 other than *F'* and *P*. Since 𝔍 also interchanges *F* and *F'*, it interchanges the points of *l* with the points of 𝒞 other than *P*.

Conversely, let 𝒞 be any circle through *P* (Figure 23.8). Let *F'* be the other endpoint of the diameter of 𝒞 through *P*. Let *F* be the point that 𝔍 interchanges with *F'*. Let *l* be the line through *F* perpendicular to line *PF'*. Since *P* ≠ *F*, the last two paragraphs show that *l* is the unique line that 𝔍 maps to 𝒞. Thus the correspondence *l* ↔ 𝒞 matches up the lines *l* that don't contain *P* with the circles 𝒞 through *P*.

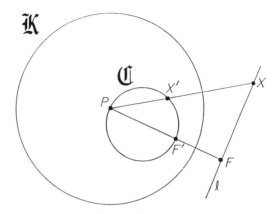

FIGURE 23.8.

(ii) For any point X on l, its image X' under $\mathfrak{J}$ is the point other than P where line PX intersects $\mathcal{C}$ (by part (i) and Definition 23.1) (Figure 23.8). As X moves along l in one direction, X' moves around $\mathcal{C}$ in one direction. Thus, the segment $[A, B]$ on l maps to the arc s on $\mathcal{C}$ that has $A' = \mathfrak{J}(A)$ and $B' = \mathfrak{J}(B)$ as endpoints and that doesn't contain P (Figure 23.7).

Let W be a point on line PA' that lies on the side of A' opposite P (Figure 23.7). Let Y be a point other than A' on the tangent ray t to s at A'. We have

$$\angle PAB = \angle PB'A' \qquad \text{(by Theorem 23.4)}$$

$$= 180° - \angle PA'Y \qquad \text{(by Theorem 6.4ii)}$$

$$= \angle WA'Y. \tag{6}$$

The reflection σ_m in the perpendicular bisector m of A and A' interchanges A and A' and switches the ends of line PA. Thus, σ_m interchanges the rays $\overrightarrow{AP}$ and $\overrightarrow{A'W}$. B and Y lie on the same side of line AA', and points remain on the same side of line AA' after reflection in m. Since reflections preserve angles, Equation 6 and the last two sentences imply that σ_m interchanges the rays $\overrightarrow{AB}$ and $\overrightarrow{A'Y} = t$. $\square$

Theorems 23.3 and 23.6 show how the inversion $\mathfrak{J}$ in a circle with center P acts on all lines and on the circles through P. The next theorem discusses the effect of $\mathfrak{J}$ on circles that don't contain P.

THEOREM 23.7. In the Euclidean plane, let $\mathfrak{J}$ be the inversion in a circle $\mathcal{K}$ having center P and radius r.

    (i) Then $\mathfrak{J}$ interchanges circles that don't pass through P among themselves.

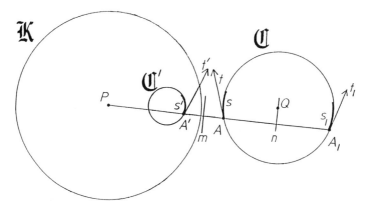

FIGURE 23.9.

(ii) Let $\mathcal{C}$ be a circle that doesn't pass through P, and set $\mathcal{C}' = \mathcal{I}(\mathcal{C})$. Let A be an endpoint of an arc s of $\mathcal{C}$. Then $\mathcal{I}$ maps s to an arc s' of $\mathcal{C}'$ that has $A' = \mathcal{I}(A)$ as an endpoint. If t and t' are the tangent rays to s and s' at A and A', respectively, then t' is the reflection of t in the perpendicular bisector m of A and A' (Figure 23.9).

Proof: Let $\mathcal{C}$ be a circle that doesn't pass through P. For any point X of $\mathcal{C}$, if the line PX intersects $\mathcal{C}$ at a second point, call this point X_1 (Figure 23.10). On the other hand, if PX is tangent to $\mathcal{C}$ at X, set $X_1 = X$. Then the quantity $\overline{PX} \cdot \overline{PX_1}$ takes a constant value k as X varies over all points of $\mathcal{C}$ (by Theorems 6.5 and 6.6). If X' is the image of X under $\mathcal{I}$, the equation $\overline{PX} \cdot \overline{PX'} = r^2$ holds, by Definition 23.1. Combining this equation with the relation $\overline{PX} \cdot \overline{PX_1} = k$ shows that $\overline{PX'}/\overline{PX_1} = r^2/k$.

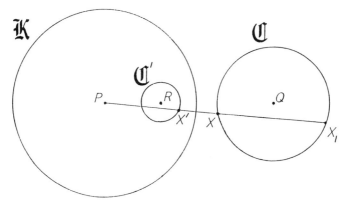

FIGURE 23.10.

Thus, the dilation δ with center P and ratio r^2/k maps X_1 to X' for every point X on $\mathcal{C}$ (by Definition 13.1). Let Q be the center of $\mathcal{C}$, and set $R = \delta(Q)$. Since δ fixes P and multiplies all distances by $|r^2/k|$ (by Theorem 13.2i), it maps $\mathcal{C}$ onto a circle $\mathcal{C}'$ that has center R and doesn't pass through P. As X varies over all points of $\mathcal{C}$, so does X_1, and so $\delta(X_1) = X' = \mathcal{I}(X)$ varies over all points of $\delta(\mathcal{C}) = \mathcal{C}'$. In short, $\mathcal{I}$ maps $\mathcal{C}$ onto $\mathcal{C}'$. Part (i) follows because $\mathcal{I}$ interchanges points in pairs (by Theorem 23.2).

As X moves around $\mathcal{C}$ in one direction, X_1 moves around $\mathcal{C}$ in one direction, and $\delta(X_1) = X'$ moves around $\mathcal{C}'$ in one direction, regardless of whether P lies on the inside or outside of $\mathcal{C}$. Thus, as X moves over an arc s of $\mathcal{C}$ with endpoint A, X_1 moves over an arc s_1 of $\mathcal{C}$ with endpoint A_1, and X' moves over an arc s' of $\mathcal{C}'$ with endpoint A' (Figure 23.9). Let t, t_1, t' be the tangent rays to s, s_1, s' at A, A_1, A', respectively. We must prove that t' is the reflection of t in the perpendicular bisector m of A and A'.

First assume that $A \neq A_1$ (as in Figure 23.9). Then line PA isn't tangent to $\mathcal{C}$, and so it doesn't contain t. Let l, l_1, l' be the lines through the rays t, t_1, t', respectively. The perpendicular bisector n of A and A_1 contains the center Q of $\mathcal{C}$ (by Theorem 4.2), and so σ_n maps $\mathcal{C}$ onto itself. Thus, since σ_n maps A to A_1, it maps l to l_1. Then $l_1 = \sigma_n(l)$ is parallel to $\sigma_m(l)$ (since m and n are both perpendicular to PA). It follows that $\delta(l_1) = \sigma_m(l)$, since δ maps each line to a parallel line (by Theorem 13.3) and since $\delta(A_1) = A'$ lies on $\sigma_m(l)$. On the other hand, we have $\delta(l_1) = l'$ (since δ maps $\mathcal{C}$ to $\mathcal{C}'$ and A_1 to A'), and so the previous sentence gives $l' = \sigma_m(l)$. Thus, we have $t' = \sigma_m(t)$, as desired, since t' and t are rays on l' and l, respectively, that each contain points on the same side of line PA.

Finally, assume that $A = A_1$ (Figure 23.11). Then the line PA is tangent to $\mathcal{C}$, and so it is also tangent to $\mathcal{C}'$ (since $\mathcal{I}$ matches up the points of $\mathcal{C}$ and

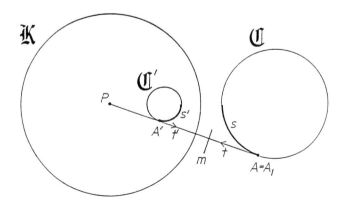

FIGURE 23.11.

$\mathcal{C}'$ along lines through P, by part (i) and Definition 23.1). Thus, t and t' both lie on line PA. Moreover, t and t' point in opposite directions on PA (since X and $X' = \mathcal{I}(X)$ moves around $\mathcal{C}$ and $\mathcal{C}'$ in opposite directions, because P lies on the outside of $\mathcal{C}$). Hence, σ_m maps t to t' (since it maps A to A' and switches the ends of line PA). $\square$

In analogy with Definition 23.5, we say that the *tangent ray* at A of a segment $[A, B]$ is the ray $\overrightarrow{AB}$. This convention lets us combine results about arcs and segments in a single statement.

Let $\mathcal{I}$ be the inversion in a circle with center P. Theorems 23.3, 23.6, and 23.7 show that $\mathcal{I}$ interchanges all lines and circles among themselves, excluding the point P. Specifically, $\mathcal{I}$ maps each line through P to itself, interchanges lines that don't contain P with circles containing P, and interchanges circles that don't contain P among themselves. Moreover, as the next theorem shows, our results on tangent rays imply that $\mathcal{I}$ maps arcs and segments that don't contain P among themselves and preserves the angles at which these curves intersect.

THEOREM 23.8. In the Euclidean plane, let $\mathcal{I}$ be the inversion in a circle $\mathcal{K}$ with center P. Let $\mathcal{S}$ be the set of all arcs that don't contain P and all segments that don't contain P. Let A be any point other than P, and set $A' = \mathcal{I}(A)$.

(i) Then $\mathcal{I}$ interchanges the elements of $\mathcal{S}$ among themselves. It interchanges the elements of $\mathcal{S}$ having endpoint A with the elements of $\mathcal{S}$ having endpoint A'.

(ii) Let s_1 and s_2 be two elements of $\mathcal{S}$ that have a common endpoint A, and let s_1' and s_2' be their respective images under $\mathcal{I}$. Let t_1 and t_2 be the respective tangent rays to s_1 and s_2 at A, and let t_1' and t_2' be the respective tangent rays to s_1' and s_2' at A'. Then the reflection in the perpendicular bisector m of A and A' interchanges t_1 and t_2 with t_1' and t_2', respectively. The angle formed by t_1 and t_2 at A equals the angle formed by t_1' and t_2' at A' (Figure 23.12).

Proof: Since $\mathcal{I}$ interchanges points in pairs (by Theorem 23.2), Theorems 23.3, 23.6, and 23.7 show that part (i) holds. Accordingly, in part (ii), s_1' and s_2' are elements of $\mathcal{S}$ that have endpoint A', and so their tangent rays t_1' and t_2' at A' are defined. If m is the perpendicular bisector of A and A' Theorems 23.3, 23.6, and 23.7 show that the reflection σ_m interchanges t_1 and t_2 with t_1' and t_2', respectively. Thus, t_1 and t_2 form the same angle as t_1' and t_2', since reflections preserve angles. $\square$

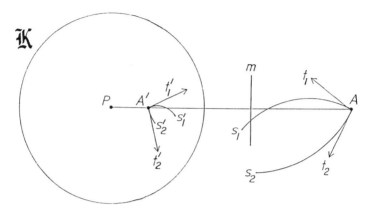

FIGURE 23.12.

We've seen that the inversion $\mathcal{I}$ in a circle $\mathcal{C}$ with center P interchanges lines and circles among themselves, excluding the point P. Accordingly, it's natural to ask which of these curves is mapped to itself by $\mathcal{I}$. In fact, $\mathcal{I}$ maps each line through P to itself (excluding P), by Theorem 23.3. By Theorem 23.6, lines that don't contain P are interchanged with circles through P (excluding P), and so none of these curves maps to itself. By Theorem 23.7, $\mathcal{I}$ interchanges circles that don't contain P among themselves, and so the remaining question is to determine which of these circles are mapped to themselves by $\mathcal{I}$. To answer this question, we consider orthogonal circles, circles that intersect at right angles.

DEFINITION 23.9. In the Euclidean plane, two circles are *orthogonal at a point A* if both circles contain A and their tangents at A are perpendicular. Two circles are *orthogonal* if they are orthogonal at a point (Figure 23.13). □

The next result presents basic properties of orthogonal circles.

THEOREM 23.10. In the Euclidean plane, let $\mathcal{C}$ and $\mathcal{D}$ be two circles. (Figure 23.14)

(i) $\mathcal{C}$ and $\mathcal{D}$ are orthogonal at a point A if and only if both circles contain A and the tangent to either circle at A contains the center of the other circle.

(ii) If $\mathcal{C}$ and $\mathcal{D}$ are orthogonal, neither circle contains the center of the other.

(iii) If $\mathcal{C}$ and $\mathcal{D}$ are orthogonal, they are orthogonal at exactly two points.

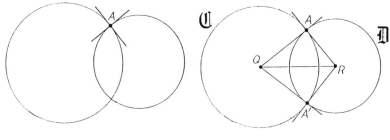

FIGURE 23.13.   FIGURE 23.14.

Proof: Let Q and R be the centers of $\mathcal{C}$ and $\mathcal{D}$, respectively.

(i) Assume that $\mathcal{C}$ and $\mathcal{D}$ both contain a point A. Line QA is perpendicular to the tangent to $\mathcal{C}$ at A (by Theorem 5.7). Thus, the tangent to $\mathcal{D}$ at A is perpendicular to the tangent to $\mathcal{C}$ at A if and only if the tangent to $\mathcal{D}$ at A contains Q. By symmetry, the tangents to $\mathcal{C}$ and $\mathcal{D}$ are also perpendicular if and only if the tangent to $\mathcal{C}$ at A contains R.

(ii) Let $\mathcal{C}$ and $\mathcal{D}$ be orthogonal at A. Q lies on the tangent line to $\mathcal{D}$ at A (by part (i)) and doesn't equal A (since $\mathcal{C}$ contains A but not Q), and so Q doesn't lie on $\mathcal{D}$. By symmetry, R doesn't lie on $\mathcal{C}$.

(iii) Let $\mathcal{C}$ and $\mathcal{D}$ be orthogonal at A. RA and QA are the tangents to $\mathcal{C}$ and $\mathcal{D}$ at A (by parts (i) and (ii)), and so they are perpendicular (by Definition 23.9). Thus, line QR exists and doesn't contain A, and so A doesn't equal its reflection A' across QR. Since $\mathcal{C}$ and $\mathcal{D}$ are orthogonal at A, they are also orthogonal at A' (because σ_{QR} fixes Q and R and preserves distances and angles). $\mathcal{C}$ and $\mathcal{D}$ aren't orthogonal at a third point because three points lie on at most one circle (by Theorems 4.3ii and 4.5).   □

The next result links orthogonal circles and inversion.

THEOREM 23.11. In the Euclidean plane, let $\mathcal{I}$ be the inversion in a circle $\mathcal{K}$ with center P. Let B be a point that doesn't lie on $\mathcal{K}$ or equal P, and set $B' = \mathcal{I}(B)$. Let $\mathcal{C}$ be a circle that contains B. Then $\mathcal{C}$ is orthogonal to $\mathcal{K}$ if and only if it contains B' (Figure 23.15).

Proof: Let r be the radius of $\mathcal{K}$. We have

$$\overline{PB} \cdot \overline{PB'} = r^2 \qquad (7)$$

(by Definition 23.1).

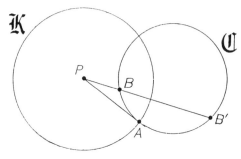

FIGURE 23.15.

First assume that $\mathcal{C}$ and $\mathcal{K}$ are orthogonal at a point A. Equation 7 becomes

$$\overline{PB} \cdot \overline{PB'} = (\overline{PA})^2. \tag{8}$$

Line PA is tangent to $\mathcal{C}$ at A (by Theorem 23.10i and ii), and so B' lies on $\mathcal{C}$ (by Theorems 6.6 and 6.9ii and the fact that Equation 8 determines B' uniquely).

Conversely, assume that B' lies on $\mathcal{C}$. Then P lies outside of $\mathcal{C}$ (since B and B' lie on a ray originating at P, by Theorem 23.2). Thus, P lies on the tangent to $\mathcal{C}$ at a point A of $\mathcal{C}$ (by Theorem 6.9ii). Equation 8 holds (by Theorem 6.6), and so we have $(\overline{PA})^2 = r^2$ (by Equations 7 and 8), and A lies on $\mathcal{K}$. The last two sentences imply that $\mathcal{C}$ and $\mathcal{K}$ are orthogonal at A (by Theorem 23.10i).   □

We can now determine which circles are mapped to themselves by an inversion.

THEOREM 23.12. In the Euclidean plane, let $\mathcal{I}$ be the inversion in a circle $\mathcal{K}$. Then $\mathcal{I}$ maps a circle $\mathcal{C}$ onto itself if and only if $\mathcal{C}$ and $\mathcal{K}$ are orthogonal or equal.

Proof: $\mathcal{I}$ maps $\mathcal{K}$ onto itself because it fixes every point of $\mathcal{K}$. Let $\mathcal{C}$ be a circle that is orthogonal to $\mathcal{K}$. $\mathcal{C}$ doesn't contain the center P of $\mathcal{K}$ (by Theorem 23.10ii). If B is a point of $\mathcal{C}$ that doesn't lie on $\mathcal{K}$, then $B \neq P$ (since $\mathcal{C}$ doesn't contain P), and $\mathcal{I}(B)$ lies on $\mathcal{C}$ (by Theorem 23.11). Moreover, $\mathcal{I}$ fixes the points where $\mathcal{C}$ equals $\mathcal{K}$, and so it maps every point of $\mathcal{C}$ back to $\mathcal{C}$. Accordingly, since $\mathcal{I}$ interchanges points in pairs (by Theorem 23.2), it maps $\mathcal{C}$ onto itself.

Conversely, let $\mathcal{C}$ be any circle that is mapped to itself by $\mathcal{I}$. If $\mathcal{C} \neq \mathcal{K}$, $\mathcal{C}$ contains a point B that doesn't lie on $\mathcal{K}$. Since $\mathcal{I}$ maps $\mathcal{C}$ to itself, $\mathcal{C}$

doesn't contain P (by Theorem 23.6). Thus, B doesn't equal P, and so $B' = \mathcal{I}(B)$ is defined. B' lies on $\mathcal{C}$ (since $\mathcal{I}$ maps $\mathcal{C}$ to itself), and so $\mathcal{C}$ and $\mathcal{K}$ are orthogonal (by Theorem 23.11). In short, $\mathcal{C}$ and $\mathcal{K}$ are either equal or orthogonal. $\square$

Finally, we consider the effect of inversion on distances. By Theorem 23.2 and the preceding discussion, the inversion $\mathcal{I}$ in a circle $\mathcal{K}$ with center P interchanges points near P with points far away from P, and it fixes points of $\mathcal{K}$. Thus, $\mathcal{I}$ multiplies distances both by arbitrarily large and by arbitrarily small positive amounts. Accordingly, it's remarkable that the distances among any four points other than P determine a ratio that is preserved by $\mathcal{I}$. We let XY denote the distance between points X and Y in the next result.

THEOREM 23.13. In the Euclidean plane, let $\mathcal{I}$ be the inversion in a circle with center P. Let A, B, C, D be four points other than P, and let A', B', C', D' be their respective images under $\mathcal{I}$. Then we have

$$\frac{AC \cdot BD}{BC \cdot AD} = \frac{A'C' \cdot B'D'}{B'C' \cdot A'D'}. \tag{9}$$

Proof: We claim that

$$A'C' = \frac{AC \cdot PA'}{PC}. \tag{10}$$

If P, A, C don't lie on a line, triangles PAC and $PC'A'$ are similar (by Theorem 23.4), and so we have

$$A'C'/AC = PA'/PC$$

(by Property 0.4), and Equation 10 follows. If P, A, C lie on a line l, choose a coordinate system on l so that P has coordinate 0. If A and C have respective coordinates a and c, then A' and C' are points of l with respective coordinates r^2/a and r^2/c (by Equation 3). It follows from Definition 1.6 that

$$A'C' = \left| \frac{r^2}{a} - \frac{r^2}{c} \right| = \left| \frac{r^2(c-a)}{ac} \right|$$

$$= \left| \frac{(c-a)(r^2/a)}{c} \right| = \frac{AC \cdot PA'}{PC}.$$

Thus, Equation 10 holds in every case.

If we combine Equation 10 with corresponding results for $B'D'$, $B'C'$, and $A'D'$, we see that

$$\frac{A'C' \cdot B'D'}{B'C' \cdot A'D'} = \frac{\left(\dfrac{AC \cdot PA'}{PC}\right)\left(\dfrac{BD \cdot PB'}{PD}\right)}{\left(\dfrac{BC \cdot PB'}{PC}\right)\left(\dfrac{AD \cdot PA'}{PD}\right)} = \frac{AC \cdot BD}{BC \cdot AD}. \quad \square$$

Readers of Section 19 will note that, if the points A, B, C, D in the previous theorem lie on a line, the left side of Equation 9 is the absolute value of the cross-ratio $R(A, B; C, D)$. Accordingly, we may think of Theorem 23.13 as roughly analogous to Theorem 19.8, which shows that cross-ratios are preserved by projections between planes.

If we compare isometries and dilations with inversions in circles, we find both similarities and differences. Isometries and dilations are maps of the plane defined for every point, and an inversion is a map of the plane defined on all points but one. Isometries and dilations map lines among themselves (by Theorems 7.7iii and 13.3), and inversions map lines and circles among themselves (by Theorem 23.8i). Isometries preserve distances, dilations multiply distances by a common factor, and inversions multiply distances by varying factors (by the discussion before Theorem 23.13) but preserve the ratios in Equation 9. Isometries preserve angles (by the SSS Property 0.1), and so do dilations (by Theorem 13.3), and inversions preserve the angles of intersection of arcs and segments (by Theorem 23.8ii).

EXERCISES

23.1. In the Euclidean plane, let A and B be two points of a circle $\mathcal{K}$. If A and B are not the endpoints of a diameter of $\mathcal{K}$, prove that A and B lie on a unique circle orthogonal to $\mathcal{K}$. If A and B are the endpoints of a diameter of $\mathcal{K}$, prove that A and B lie on no circles orthogonal to $\mathcal{K}$.

23.2. In the Euclidean plane, let $\mathcal{K}$ be a circle with center P. Let A be a point on $\mathcal{K}$, and let B be a point that doesn't lie on $\mathcal{K}$. If line AB doesn't contain P, prove that A and B lie on a unique circle orthogonal to $\mathcal{K}$. If line AB contains P, prove that A and B lie on no circle orthogonal to $\mathcal{K}$.

23.3. In the Euclidean plane, let $\mathcal{K}$ be a circle with center P. Let A and B be two points. Prove that A and B lie on a unique circle orthogonal to $\mathcal{K}$ if and only if line AB doesn't contain P. (*Hint:* See Exercises 23.1 and 23.2 and Theorem 23.11.)

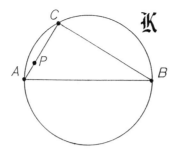

FIGURE 23.16.

23.4. In the Euclidean plane, let $\mathfrak{g}$ be the inversion in a circle $\mathcal{K}$ with center P. As in Exercise 9.9, we say that two circles are tangent at P if they both contain P and are tangent to the same line there. Prove that $\mathfrak{g}$ interchanges circles tangent at P with parallel lines that don't contain P.

Exercises 23.5–23.7 use the following consequence of Theorem 4.7(ii) (Figure 23.16):

Theorem A. In the Euclidean plane, let A, B, C be three points that don't lie on a line. Then C lies on the circle $\mathcal{K}$ with diameter AB if and only if AC and BC are perpendicular lines.

Let $\mathfrak{g}$ be the inversion in any circle whose center P is a point on line AC other than A and C (Figure 23.16). Let A', B', C' be the respective images of A, B, C under $\mathfrak{g}$ (Figure 23.17). Since AC contains P but not B, $\mathfrak{g}$ interchanges the points of AC other than P among themselves (by Theorem 23.3i), and so $A'C'$ contains P but not B'. Since the lines AB and BC don't contain P, $\mathfrak{g}$ maps AB to a circle $\mathcal{K}_1$ through A', B', and P, and it maps BC to a circle $\mathcal{K}_2$ through B', C', and P (by Theorem 23.6i). Since $\mathcal{K}$ contains A and B but not P, $\mathfrak{g}$ maps $\mathcal{K}$ to a circle $\mathcal{K}'$ through A' and B' (by Theorem 23.7i). Since AB is perpendicular to the tangents to $\mathcal{K}$ at A and B, applying $\mathfrak{g}$ shows tht $\mathcal{K}_1$ and $\mathcal{K}'$ have perpendicular tangents at A' and B' (by Theorem 23.8ii), and so $\mathcal{K}_1$ and $\mathcal{K}'$ are orthogonal. If we apply $\mathfrak{g}$ to the conclusion in Theorem A that C lies on $\mathcal{K}$ if and only if AC and BC are perpendicular, we obtain the statement that C' lies on $\mathcal{K}'$ if and only if $A'C'$ is perpendicular to the tangent to $\mathcal{K}_2$ at C' (by Theorem 23.8ii). This happens if and only if $A'C'$ contains the center of $\mathcal{K}_2$ (by Theorem 5.7), which occurs if and only if P and C' are the endpoints of a diameter of $\mathcal{K}_2$ (since P and C' are points of $\mathcal{K}_2$ on $A'C'$). As ABC varies over all triangles and $\mathfrak{g}$ varies over the inversions in all circles whose centers are points on AC other than A and C, $A'B'C'$ varies over all triangles and P varies over all points on $A'C'$ other than A' and C'. Thus, we've obtained the following result from Theorem A.

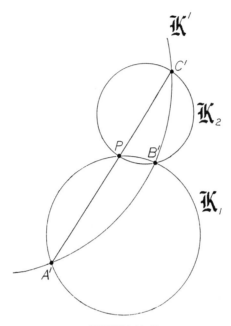

FIGURE 23.17.

Theorem. In the Euclidean plane, let $A'B'C'$ be a triangle, and let P be a point on line $A'C'$ other than A' and C'. Let $\mathcal{K}_1$ be the circle through A', B', and P, and let $\mathcal{K}_2$ be the circle through B', C', and P. Let $\mathcal{K}'$ be a circle through A' and B' orthogonal to $\mathcal{K}_1$. Then $\mathcal{K}'$ contains C' if and only if P and C' are the endpoints of a diameter of $\mathcal{K}_2$ (Figure 23.17).

Similarly, in each of the following cases, state the result that follows from Theorem A by inverting in every circle whose center P satisfies the given condition. Illustrate the result you state with a figure.

23.5. P lies on line AB and doesn't equal A or B.

23.6. P lies on $\mathcal{K}$ and doesn't equal A, B, or C.

23.7. P doesn't lie on $\mathcal{K}$, AB, BC, or AC.

Exercises 23.8–23.12 are based on the following consequence of Theorem 6.2 (Figures 6.7 and 6.8):

Theorem. In the Euclidean plane, let A, B, C, D be four points on a circle $\mathcal{K}$. Then the angles formed by the lines AC and BC equal the angles formed by the lines AD and BD.

In each of the following cases, state the result that follows from this theorem by inverting in every circle whose center P satisfies the given condition. Illustrate the result you state with a figure. (See the discussion before Exercise 23.5.)

23.8. P equals A.

23.9. P equals C.

23.10. P lies on $\mathcal{K}$ and doesn't equal A, B, C, or D.

23.11. P doesn't lie on $\mathcal{K}$ or any of the lines AC, BC, AD, BD.

23.12. P lies on AC and doesn't equal A or C.

Exercises 23.13–23.15 are based on the following consequences of Theorem 6.4 (Figures 6.15 and 6.16):

Theorem. In the Euclidean plane, let A, B, C be three points on a circle $\mathcal{K}$. Then the angles formed by the tangent at A and line AB equal the angles formed by the lines AC and BC.

In each of the following cases, state the result that follows from this theorem by inverting in every circle whose center Q satisfies the given condition. Illustrate the result you state with a figure. (See the discussion before Exercise 23.5.)

23.13. Q lies on AC and doesn't equal A or C.

23.14. Q lies on the tangent at A and doesn't equal A.

23.15. Q doesn't lie on $\mathcal{K}$, AB, BC, or AC.

Exercises 23.16–23.18 are based on the following consequence of Theorem 4.11(i):

Theorem. In the Euclidean plane, let ABC be a triangle without a right angle. Then there is a point H such that AH is perpendicular to BC, BH is perpendicular to AC, and CH is perpendicular to AB.

In each of the following cases, state the result that follows from this theorem by inverting in every circle whose center P satisfies the given condition. Illustrate the result you state with a figure. (See the discussion before Exercise 23.5.)

23.16. P equals A.

23.17. P lies on AB but not CH and doesn't equal A or B.

23.18. P lies on AH but not BC and doesn't equal A or H.

Exercises 23.19–23.22 are based on the theorem in Exercise 6.1 (Figure 6.32). In each of the following cases, state the result that follows from this theorem by inverting in every circle whose center P satisfies the given condition. Illustrate the result you state with a figure. (See Exercise 23.4 and the discussion before Exercise 23.5.)

23.19. P equals W.

23.20. P equals U.

23.21. P doesn't lie on l, m, RS, VW, $\mathcal{K}$, or $\mathcal{K}'$.

23.22. P lies on VW and doesn't equal V or W.

Exercises 23.23–23.28 are based on the theorem in Exercise 6.2 (Figure 6.33). In each of the following cases, state the result that follows from this theorem by inverting in every circle whose center P satisfies the given condition. Illustrate the result you state with a figure. (See Exercise 23.4 and the discussion before Exercise 23.5.)

23.23. P equals R.

23.24. P equals S.

23.25. P equals T.

23.26. P lies on $\mathcal{K}$ and doesn't equal R, S, or T.

23.27. P lies on $\mathcal{K}'$ and doesn't equal S, T, V, or W.

23.28. P lies on m and doesn't equal S, W, or Z.

23.29. In the Euclidean plane, let ABC be a triangle without a right angle at C. Let H be the orthocenter of the triangle. Prove that the circle with diameter CH is orthogonal to the circle with diameter AB.

23.30. Let $\mathcal{K}$ and $\mathcal{K}'$ be two circles that intersect at two points A and B. Let $\mathcal{K}$ and $\mathcal{K}'$ have respective centers P and P' and respective radii r and r'. Let s be the distance from A to B, and let t be the distance from P to P'. Let M be the midpoint of A and B. Prove that the following conditions are equivalent.

   (i) $\mathcal{K}$ and $\mathcal{K}'$ are orthogonal.
   (ii) $2rr' = st$.
   (iii) Inversion in $\mathcal{K}$ interchanges M and P'.
   (iv) Inversion in $\mathcal{K}'$ interchanges M and P.

(*Hint:* One possible approach is to prove that each condition is equivalent to the condition that triangles AMP and $P'AP$ are similar.)

23.31. In the Euclidean plane, let $\mathcal{K}$ be a circle with center P. Let X be a point other than P that doesn't lie on $\mathcal{K}$. Prove tht inversion in $\mathcal{K}$ maps X to its harmonic conjugate with respect to the two points where line PX intersects $\mathcal{K}$. (See Exercise 17.8.)

Exercises 23.32–23.37 provide a proof of the following result. As in Exercise 9.9, we say that two circles are tangent if they are tangent to the same line at the same point.

Feuerbach's Theorem. For any triangle in the Euclidean plane, the nine-point circle is tangent to or equal to the incircle and is tangent to the three excircles. The nine-point circle equals the incircle only for an equilateral triangle.

23.32. In the Euclidean plane, let ABC be a triangle (Figure 23.18). Assume that sides AC and BC are of different lengths. Let $\mathcal{K}'$ be the incircle, and let I be the incenter. Let $\mathcal{K}''$ be the excircle whose center L lies on the internal angle bisector at C. Let Z be the point where the lines CI and AB intersect. Let F, C', C'' be the feet of the perpendiculars to AB through C, I, L, respectively.

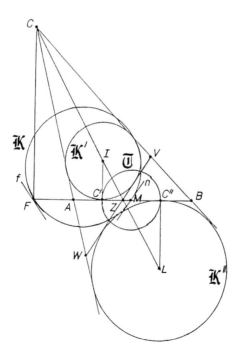

FIGURE 23.18.

(a) Let r' and r'' be the radii of $\mathcal{K}'$ and $\mathcal{K}''$, respectively. Use Property 0.4 to prove that

$$\overline{CI}/\overline{CL} = r'/r'' = -\overline{ZI}/\overline{ZL}.$$

(b) Prove that $I, L; C, Z$ is a harmonic set.
(c) Prove that $C', C''; F, Z$ is a harmonic set. (See Theorem 17.10.)

23.33. Let the notation be as in Exercise 23.32. In addition, let $\mathcal{K}$ be the nine-point circle of triangle ABC. Let M be the midpoint of A and B.

(a) Prove that M is the center of the circle $\mathcal{J}$ that has C' and C'' as the endpoints of a diameter. (See Exercise 6.10b.)
(b) Let $\mathcal{I}$ be the inversion in $\mathcal{J}$. Prove that $\mathcal{I}$ interchanges F and Z. (See Exercises 23.31 and 23.32.)
(c) Prove that F and M are distinct points of $\mathcal{K}$. If f and n are the tangents to $\mathcal{K}$ at F and M, prove that the reflection in the perpendicular bisector of F and M maps f to n.
(d) Prove that $\mathcal{I}$ maps the points of $\mathcal{K}$ other than M to the points on the line through Z parallel to n. (See parts (b) and (c) and Theorem 23.6ii.)
(e) Prove that $\mathcal{I}$ maps the points on each of the circles $\mathcal{K}'$ and $\mathcal{K}''$ among themselves.

23.34. Let the notation be as in Exercises 23.32 and 23.33. Let V and W be the respective images of A and B under the reflection in the internal angle bisector at C.

(a) Prove that V is a point on line BC other than B and C and that W is a point on line AC other than A and C.
(b) Prove that VW is a line through Z other than AB that is tangent to $\mathcal{K}'$ and $\mathcal{K}''$.
(c) Prove that the perpendicular bisector of A and B intersects line CI. Conclude that the four points A, B, V, W lie on a circle.

23.35. Let the notation be as in Exercises 23.32–23.34. In addition, let D and E be the feet of the altitudes on A and B, respectively.

(a) If triangle ABC doesn't have a right angle, use Theorem 4.7(ii) and Exercises 6.1 and 23.34(c) to prove that VW and DE are parallel lines. Then conclude from Exercise 5.3(b) that VW and n are parallel lines.
(b) If triangle ABC has a right angle at A, prove that A, B, and D lie on a circle tangent to AC. Then use Exercises 6.2 and 23.34(c) to deduce that VW and AD are parallel lines. Conclude from Exercise 5.3(b) that VW and n are parallel, and illustrate this result with a figure.

(c) If triangle ABC has a right angle at C, use Theorems 4.6 and 5.2(ii) to prove that the circumcircle of triangle ABC has a tangent at C parallel to n. Then use Exercises 6.3 and 23.34(c) to conclude that VW and n are parallel lines. Illustrate this result with a figure.

23.36. Let the notation be as in Exercises 23.32–23.34.

(a) Prove that $\mathcal{I}$ interchanges the points on $\mathcal{K}$ except M with the points on line VW. (See Exercises 23.33d, 23.34b, and 23.35.)

(b) Prove that $\mathcal{K}$ is tangent to $\mathcal{K}'$ and $\mathcal{K}''$. (See part (a) and Exercises 23.33e and 23.34b.)

23.37. Prove Feuerbach's Theorem. (*Hint:* In order to extend Exercise 23.36b to the case where AC and BC are of equal length, see Exercise 5.6.)

23.38. Let the notation be as in Exercises 23.32–23.34. In addition, let line VW be tangent to $\mathcal{K}'$ at a point P and tangent to $\mathcal{K}''$ at a point Q (by Exercise 23.34b). Prove that the line MP intersects $\mathcal{K}'$ at a second point R and that $\mathcal{K}$ is tangent to $\mathcal{K}'$ at R. Prove that the line MQ intersects $\mathcal{K}''$ at a second point S and that $\mathcal{K}$ is tangent to $\mathcal{K}''$ at S. Illustrate these results with a figure.

23.39. Let the notation be as in Exercises 23.32–23.34. Prove that the circles $\mathcal{K}$ and $\mathcal{I}$ and the line VW have two points in common.

Section 24.

The Hyperbolic Plane

We can now construct the hyperbolic plane by using results from the previous section. There are both strong similarities and striking differences between the hyperbolic and the Euclidean planes.

Throughout this section, we let $\mathcal{K}$ be a fixed circle in the Euclidean plane with center P and radius r. We use the prefix ''h-'' to denote objects in the hyperbolic plane. Geometric terms without this prefix refer to objects in the Euclidean plane.

We define an *h-point* to be a point of the Euclidean plane that lies inside the circle $\mathcal{K}$. Accordingly, the hyperbolic plane is the interior of the circle $\mathcal{K}$ in the Euclidean plane.

DEFINITION 24.1. An *h-line* is either the set of h-points on a circle orthogonal to $\mathcal{K}$ or the set of h-points on a line through P. □

Figure 24.1 shows two h-lines l and m. It must be remembered that points on or outside of $\mathcal{K}$ are not part of the hyperbolic plane.

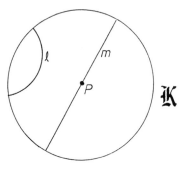

FIGURE 24.1.

One of the most basic properties of the Euclidean plane is that any two points lie on a unique line. We prove that the same property holds in the hyperbolic plane. This result begins to justify the term h-line and provides the first similarity between the hyperbolic and the Euclidean planes.

THEOREM 24.2. Any two h-points lie on a unique h-line.

Proof: Let A and B be two h-points. At most one of them equals P, and so we can assume that $B \neq P$. Then the inversion in $\mathfrak{K}$ maps B to a point B', and the points A, B, B' are distinct (since B' lies outside $\mathfrak{K}$, by Theorem 23.2). To prove that A and B lie on a unique h-line, we consider two cases, depending on whether or not line AB contains P.

First, assume that line AB doesn't contain P (Figure 24.2). A doesn't lie on line $PB = BB'$, and so A, B, and B' lie on a unique circle $\mathcal{C}$ (by Theorem 4.3ii). $\mathcal{C}$ is the unique circle through A and B orthogonal to $\mathfrak{K}$ (by Theorem 23.11). Thus, the points of $\mathcal{C}$ inside $\mathfrak{K}$ form the unique h-line through A and B (by Definition 24.1).

Next assume that line AB contains P (Figure 24.3). B' lies on line $PB = AB$, and so no circle contains A, B, and B' (by Theorem 4.5ii). Thus, there is no circle through A and B orthogonal to $\mathfrak{K}$ (by Theorem 23.11). Therefore, the points inside $\mathfrak{K}$ on line AB form the unique h-line through A and B (by Definition 24.1). $\square$

In the Euclidean plane, a coordinate system on a line matches up the points on the line with the real numbers. We now extend this to the hyperbolic plane by defining coordinate systems on h-lines.

Let l be an h-line. Every circle orthogonal to $\mathfrak{K}$ intersects $\mathfrak{K}$ at two points (by Theorem 23.10iii), and every line through P intersects $\mathfrak{K}$ at two points. Thus, there are two points A and B on $\mathfrak{K}$ that form an arc or a segment in the Euclidean plane when they are added to l (Figures 24.4 and 24.5).

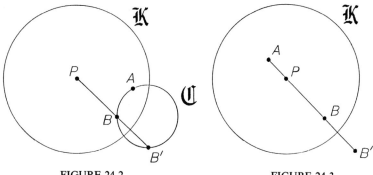

FIGURE 24.2. FIGURE 24.3.

We call A and B the *limit points* of l. We emphasize that the limit points are not h-points.

For any h-point X on l, let AX and BX be the distances in the Euclidean plane from X to the limit points A and B of l (Figures 24.4 and 24.5). These distances are positive because A and B are not h-points. As X moves along l from A to B, AX increases continuously from arbitrarily small positive values, and BX decreases continuously to arbitrarily small positive values, and so the ratio AX/BX increases and takes each positive value exactly once. As the graph of the natural logarithm function $y = \ln(x)$ in Figure 24.6 illustrates, when x increases through all positive real numbers, $\ln(x)$ increases and takes each real value exactly once. The last two sentences imply that, as X moves along l from A to B, the quantity $y = \ln(AX/BX)$ increases and takes each real value exactly once. We call the map $X \rightarrow \ln(AX/BX)$ an *h-coordinate system* on l, and we call $\ln(AX/BX)$ the *coordinate* of the h-point X of l. We've proved the following result.

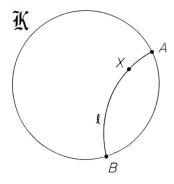

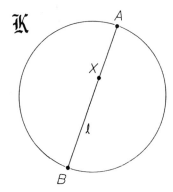

FIGURE 24.4. FIGURE 24.5.

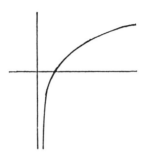

FIGURE 24.6.

THEOREM 24.3. Let l be an h-line, and let A and B be the limit points of l. Then the h-coordinate system $X \to \ln(AX/BX)$ on l matches up the h-points X of l with the real numbers. □

If we interchange the limit points A and B of an h-line l, we replace $\ln(AX/BX)$ with $\ln(BX/AX) = -\ln(AX/BX)$, and the coordinate of each point of l is multiplied by -1. In effect, switching the limit points corresponds to switching the positive end of l.

In analogy with Definition 1.6, we use h-coordinate systems to define h-distances. Two h-points C and D lie on a unique h-line l (by Theorem 24.2). We define the *h-distance* from C to D to be $|c - d|$, where c and d are the coordinates of C and D in an h-coordinate system on l. This definition is symmetric in C and D, since $|c - d| = |d - c|$. Switching the limit points of l replaces coordinates with their negatives, and so $|c - d|$ becomes

$$|-c - (-d)| = |-(c - d)| = |c - d|$$

and remains unchanged. Thus, the h-distance between C and D depends only on the h-points C and D themselves and not on the order in which the limit points of l are taken.

The next result gives a convenient formula for computing h-distances. As Theorem 23.13 suggests, this formula is an important link between inversion in circles and the hyperbolic plane.

THEOREM 24.4. Let C and D be two h-points (Figure 24.7). Let l be the unique h-line through C and D, and let A and B be the limit points of l. Then the h-distance from C to D is

$$\left| \ln\left(\frac{AC \cdot BD}{BC \cdot AD} \right) \right|.$$

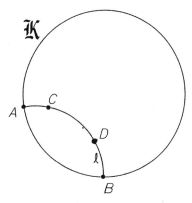

FIGURE 24.7.

Proof: C and D have respective coordinates $c = \ln(AC/BC)$ and $d = \ln(AD/BD)$ in an h-coordinate system on l. Then the h-distance from C to D is

$$|c - d| = |\ln(AC/BC) - \ln(AD/BD)|$$

$$= \left| \ln\left(\frac{AC \cdot BD}{BC \cdot AD} \right) \right|. \quad \square$$

Next we use h-coordinate systems to transfer segments and rays from the real number line to h-lines. Let A and B be two h-points. They lie on a unique h-line l (by Theorem 24.2). For any h-point X of l, let a, b, x be the respective coordinates of A, B, X in an h-coordinate system on l. We define the *h-segment* $[A, B]$ to consist of all h-points X on l such that x lies between a and b, inclusive (Figures 24.8a and b). We define the *h-ray* $\overrightarrow{AB}$ to consist

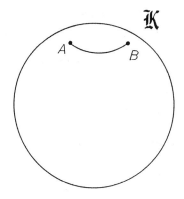

FIGURE 24.8a.

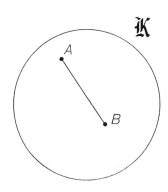

FIGURE 24.8b.

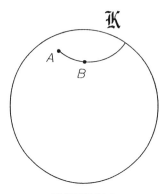

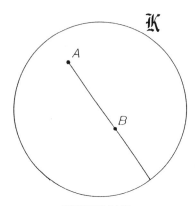

FIGURE 24.9a.   FIGURE 24.9b.

of A and all h-points X on l such that $x - a$ has the same sign as $b - a$ (Figures 24.9a and b). These definitions don't depend on the choice of an h-coordinate system on l, since the two choices differ by a factor of -1. We say that the h-segment $[A, B]$ has *endpoints* A and B and that the h-ray $\overrightarrow{AB}$ *originates* at A.

As X moves along l in one direction, its coordinate x increases (by the discussion before Theorem 24.3). It follows that the h-segment $[A, B]$ is traced out by an h-point X as it moves along l from A to B, and so an h-segment is an arc or a segment in the Euclidean plane (Figures 24.8a and b). It also follows that h-ray $\overrightarrow{AB}$ is traced out by an h-point X as it moves along l from A in the direction that makes it pass through B, and so an h-ray becomes an arc or a segment in the Euclidean plane when a point of $\mathcal{K}$ is added (Figures 24.9a and b).

We defined a tangent ray to an arc or a segment in the Euclidean plane in Definition 23.5 and the paragraph after the proof of Theorem 23.7. We define the *tangent ray* to an h-ray $\overrightarrow{AB}$ to be the tangent ray at A to the arc or segment in the Euclidean plane formed by adding a point of $\mathcal{K}$ to $\overrightarrow{AB}$ (Figures 24.10a and b).

If A is an h-point on an h-line l, we define the *tangent line* m to l at A as follows. If l lies on a circle $\mathcal{C}$ in the Euclidean plane, we take m to be the tangent to $\mathcal{C}$ at A in the Euclidean plane (Figure 24.11a). If l lies on a line through P in a Euclidean plane, we take m to be this line (Figure 24.11b).

Let A be an h-point on an h-line l. There are two h-rays s_1 and s_2 that lie on l and originate at A (Figures 24.12a and b). The union of these h-rays is l, and their intersection is the one h-point A. If t_1 and t_2 are the tangent rays to s_1 and s_2, t_1 and t_2 are the two rays in the Euclidean plane that originate at A and lie on the tangent line to l.

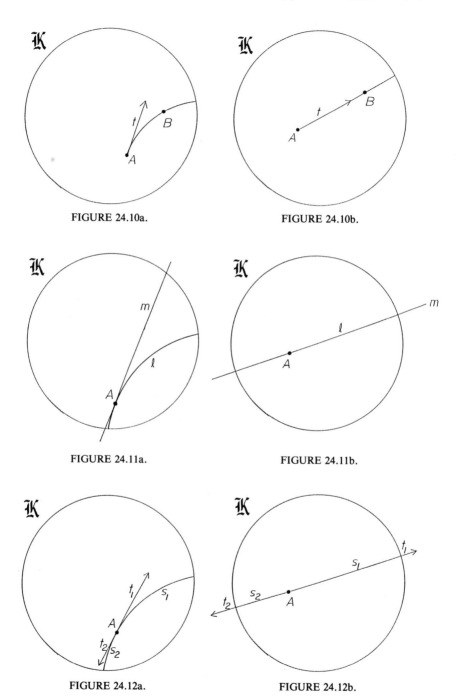

FIGURE 24.10a.

FIGURE 24.10b.

FIGURE 24.11a.

FIGURE 24.11b.

FIGURE 24.12a.

FIGURE 24.12b.

We can now set up a correspondence between the h-lines and tangent lines through an h-point and a correspondence between the h-rays and tangent rays originating at an h-point. These correspondences are important in considering h-angles and in determining how h-points are separated by an h-line.

THEOREM 24.5. Let A be a point inside the circle $\mathcal{K}$ in the Euclidean plane.

 (i) For any line m through A in the Euclidean plane, there is a unique h-line l that has m as its tangent line at A (Figures 24.11a and b).
 (ii) For any ray t in the Euclidean plane originating at A, there is a unique h-ray $\overrightarrow{AB}$ that has t as its tangent ray at A (Figures 24.10a and b).

Proof: (i) First assume that m doesn't contain the center P of $\mathcal{K}$ (Figure 24.13). A doesn't equal P (since m contains A but not P), and so inversion in $\mathcal{K}$ maps A to a point A'. A' doesn't equal A (by Theorem 23.2, since A lies inside $\mathcal{K}$), and the lines AA' and m are distinct (since P lies on AA' but not m). Let d be the line through A perpendicular to m, and let e be the perpendicular bisector of A and A'. The lines d and e aren't parallel (since they are perpendicular to distinct lines m and AA' through A), and so they intersect at a point Q. Q is the unique point that is equidistant from A and A' and is such that line AQ is perpendicular to m (by Theorem 4.2). Thus, Q is the center of the unique circle $\mathcal{C}$ through A and A' that is tangent to m (by Theorem 5.7), and $\mathcal{C}$ is the unique circle through A that is orthogonal to $\mathcal{K}$ and tangent to m (by Theorem 23.11). The points of $\mathcal{C}$ inside $\mathcal{K}$ form the unique h-line through A tangent to m (since any h-line

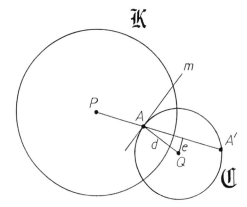

FIGURE 24.13.

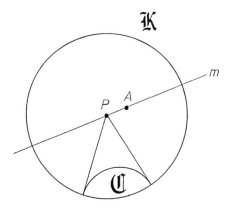

FIGURE 24.14.

tangent to m must lie on a circle orthogonal to $\mathcal{K}$, by Definition 24.1, the discussion accompanying Figure 24.11b, and the assumption that m doesn't contain P).

Next assume that m contains P (Figure 24.14). If a circle $\mathcal{C}$ is orthogonal to $\mathcal{K}$, P lies on the tangents to $\mathcal{C}$ at the two points where $\mathcal{C}$ intersects $\mathcal{K}$ (by Theorem 23.10i and iii), and so P doesn't lie on the tangent to $\mathcal{C}$ at any point of $\mathcal{C}$ inside $\mathcal{K}$ (by Theorem 6.9ii). Accordingly, since m contains P, no circle orthogonal to $\mathcal{K}$ is tangent to m at a point inside $\mathcal{K}$. Thus, the points of m inside $\mathcal{K}$ form the unique h-line through A tangent to m at A (by Definition 24.1 and the discussion accompanying Figures 24.11a and b).

Part (ii) follows from part (i) and the discussion accompanying Figures 24.12a and b. $\square$

We now consider the separation of points by a line. In the Euclidean plane, let l be a line (Figure 24.15). Let $\mathcal{S}_1$ and $\mathcal{S}_2$ be the two sets that each consist of all points in the plane lying on the same side of l. Every point in the plane belongs to exactly one of the sets l, $\mathcal{S}_1$, $\mathcal{S}_2$. If A and B are

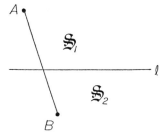

FIGURE 24.15.

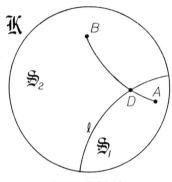

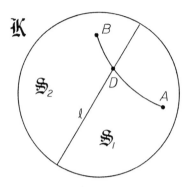

FIGURE 24.16.   FIGURE 24.17.

two points that don't lie on l, the segment $[A, B]$ intersects l if and only if A and B lie on different sides of l.

To extend the results in the last paragraph to the hyperbolic plane, let l be an h-line. If l lies on a circle $\mathcal{C}$ in the Euclidean plane, let $\mathcal{S}_1$ be the set of all h-points lying inside $\mathcal{C}$, and let $\mathcal{S}_2$ be the set of all h-points lying outside $\mathcal{C}$ (Figure 24.16). If l lies on a line m in the Euclidean plane, let $\mathcal{S}_1$ and $\mathcal{S}_2$ be the two sets of h-points lying on the two sides of m in the Euclidean plane (Figure 24.17). In both cases, we call $\mathcal{S}_1$ and $\mathcal{S}_2$ the *h-sides* of l. Every h-point belongs to exactly one of the sets l, $\mathcal{S}_1$, $\mathcal{S}_2$.

In the Euclidean plane, let s be an arc or a segment. If s contains points inside and outside of a circle $\mathcal{C}$, then s must intersect $\mathcal{C}$. If s contains points on both sides of a line m, then s must intersect m. Therefore, if A and B are h-points that lie on different sides of an h-line l, the h-segment $[A, B]$ intersects l at an h-point D (since the h-segment $[A, B]$ is an arc or a segment inside $\mathcal{K}$ in the Euclidean plane) (Figures 24.16 and 24.17).

Conversely, let A and B be two h-points that don't lie on an h-line l and are such that the h-segment $[A, B]$ intersects l at an h-point D (Figures 24.16 and 24.17). The h-line containing $[A, B]$ and the h-line l have different tangent lines at D (by Theorem 24.5i). It follows that $[A, B]$ contains points on both h-sides of l. Thus, since $[A, B]$ and l don't intersect at any points other than D (by Theorem 24.2), the previous paragraph implies that A and B lie on different h-sides of l.

In short, we've established the following result.

THEOREM 24.6. Let $\mathcal{S}_1$ and $\mathcal{S}_2$ be the h-sides of an h-line l. Then every h-point belongs to exactly one of the sets l, $\mathcal{S}_1$, $\mathcal{S}_2$. For any two h-points A and B that don't lie on l, the h-segment $[A, B]$ intersects l if and only if A and B lie on different h-sides of l (Figures 24.16 and 24.17). □

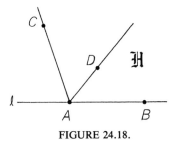

FIGURE 24.18.

Our next goal is to extend results on angles from the Euclidean plane to the hyperbolic plane. In the Euclidean plane, let $\overrightarrow{AB}$ be a ray on a line l (Figure 24.18). A *halfplane* $\mathcal{H}$ of l is the union of l and the set of all points on one side of l. For any ray $\overrightarrow{AC}$ in $\mathcal{H}$, we let $\angle BAC$ be the degree measure of the angle formed by the rays $\overrightarrow{AB}$ and $\overrightarrow{AC}$. The map taking $\overrightarrow{AC}$ to $\angle BAC$ matches up the rays AC in $\mathcal{H}$ originating at A with the real numbers $\angle BAC$ from 0 through 180, inclusive. We emphasize that, for us, $\angle BAC$ is a number rather than the union of the rays $\overrightarrow{AB}$ and $\overrightarrow{AC}$. If D is a point in the halfplane $\mathcal{H}$ such that B and C lie on different sides of line AD, we have

$$\angle BAC = \angle BAD + \angle DAC. \tag{1}$$

We extend these results to the hyperbolic plane. We define the *h-angle* $\mathrm{h}\angle BAC$ formed by the h-rays $\overrightarrow{AB}$ and $\overrightarrow{AC}$ to be the angle formed in the Euclidean plane by their tangent rays. That is, if $\overrightarrow{AB'}$ and $\overrightarrow{AC'}$ are the tangent rays to the h-rays $\overrightarrow{AB}$ and $\overrightarrow{AC}$, we set

$$\mathrm{h}\angle BAC = \angle B'AC' \tag{2}$$

(Figure 24.19). We emphasize that $\mathrm{h}\angle BAC$ is a number.

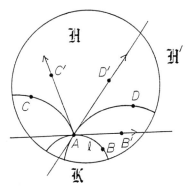

FIGURE 24.19.

Let l be the h-line that contains the h-ray $\overrightarrow{AB}$. An *h-halfplane* $\mathcal{H}$ of l is the union of l and one of its h-sides. By Theorem 24.5(ii), the map that takes each h-ray $\overrightarrow{AC}$ to its tangent ray $\overrightarrow{AC'}$ matches up the h-rays $\overrightarrow{AC}$ that originate at A and lie in the h-halfplane $\mathcal{H}$ with the rays $\overrightarrow{AC'}$ that originate at A and lie in a halfplane $\mathcal{H}'$ of line AB' in the Euclidean plane. By the previous paragraph, the map taking $\overrightarrow{AC'}$ to $\angle B'AC'$ matches up the rays $\overrightarrow{AC'}$ in the halfplane $\mathcal{H}'$ of the Euclidean plane with real numbers $\angle B'AC'$ from 0 through 180, inclusive. Combining the last two sentences with Equation 2 shows that the map taking the h-ray $\overrightarrow{AC}$ to $h\angle BAC$ matches up the h-rays $\overrightarrow{AC}$ in the h-halfplane $\mathcal{H}$ with the real numbers $h\angle BAC$ from 0 through 180, inclusive.

Let C and D be h-points in $\mathcal{H}$ such that B and C lie on different h-sides of the h-line through the h-ray $\overrightarrow{AD}$ (Figure 24.19). If $\overrightarrow{AB'}$, $\overrightarrow{AC'}$, $\overrightarrow{AD'}$ are the respective tangent rays to the h-rays $\overrightarrow{AB}$, $\overrightarrow{AC}$, $\overrightarrow{AD}$, then B' and C' lie on different sides of the line AD' in the Euclidean plane. Thus, we have

$$\angle B'AC' = \angle B'AD' + \angle D'AC' \tag{3}$$

(by Equation 1), and combining Equations 2 and 3 gives

$$h\angle BAC = h\angle BAD + h\angle DAC.$$

The last two paragraphs establish the following result.

THEOREM 24.7. Let $\overrightarrow{AB}$ be an h-ray, and let $\mathcal{H}$ be an h-halfplane of the h-line containing the h-ray $\overrightarrow{AB}$. Then the map taking the h-ray $\overrightarrow{AC}$ to $h\angle BAC$ matches up the h-rays $\overrightarrow{AC}$ in $\mathcal{H}$ originating at A with the real numbers $h\angle BAC$ from 0 through 180, inclusive. If C and D are h-points in $\mathcal{H}$ such that B and C lie on different h-sides of the h-line containing A and D, we have

$$h\angle BAC = h\angle BAD + h\angle DAC. \quad \square$$

Inversion in circles provides an analogue in the hyperbolic plane to reflections in lines in the Euclidean plane. As we shall see in the next section, this analogy lets us extend triangle congruence theorems from the Euclidean plane to the hyperbolic plane.

Specifically, let l be a line in the Euclidean plane. The reflection σ_l in l fixes each point of l and interchanges each point that doesn't lie on l with a point on the opposite side of l. Reflection preserves distances and angles: If A, B, C are three points in the plane, and if A', B', C' are their images under σ_l, then the distance from A to B equals the distance from A' to B', and the angles $\angle BAC$ and $\angle B'A'C'$ are equal. Analogously, for any h-line l, we show that there is a map of the hyperbolic plane that fixes the h-points

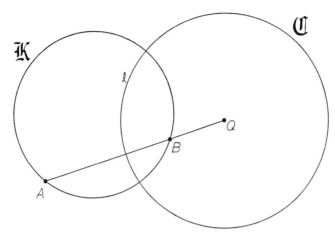

FIGURE 24.20.

of l, interchanges each h-point that doesn't lie on l with an h-point on the opposite h-side of l, and preserves h-distances and h-angles.

First, let l be an h-line that lies on a circle $\mathfrak{C}$ in the Euclidean plane (Figure 24.20). Suppose that a line through the center Q of $\mathfrak{C}$ intersects $\mathfrak{K}$ in two points A and B. Q lies outside $\mathfrak{K}$ (since it lies on two tangents to $\mathfrak{K}$, by Theorem 23.10i and iii), and so it doesn't lie on the segment $[A, B]$. $\mathfrak{C}$ and $\mathfrak{K}$ are orthogonal (by Definition 24.1), and so the inversion $\mathfrak{D}$ in $\mathfrak{C}$ maps $\mathfrak{K}$ onto itself (by Theorem 23.12, with the roles of $\mathfrak{C}$ and $\mathfrak{K}$ interchanged). It follows that $\mathfrak{I}$ interchanges A and B and that, as a point X varies over the segment $[A, B]$ from A to B, $\mathfrak{I}(X)$ varies over the same segment from B to A (by Theorem 23.3). Hence, $\mathfrak{I}$ maps the points inside $\mathfrak{K}$ among themselves, and so restricting $\mathfrak{I}$ to these points gives a map σ_l of h-points. σ_l fixes the h-points on l because $\mathfrak{I}$ fixes the points on $\mathfrak{C}$ (by Theorem 23.2). σ_l interchanges each h-point that doesn't lie on l with an h-point on the other h-side of l because $\mathfrak{I}$ interchanges points outside and inside $\mathfrak{C}$, except Q, in pairs (by Theorem 23.2).

Let C and D be two h-points on an h-line m, and let A and B be the limit points of m (Figure 24.21). Let A', B', C', D', m' be the respective images of A, B, C, D, m under $\mathfrak{I}$. Adding A and B to m gives an arc or a segment in the Euclidean plane whose tangent rays at A and B are perpendicular to the tangent lines to $\mathfrak{K}$ at these points (by Definitions 24.1 and 23.9 and Theorem 5.7). Thus, adding A' and B' to m' gives an arc or a segment in the Euclidean plane whose tangent rays at A' and B' are perpendicular to the tangent lines to $\mathfrak{K}$ at these points (by Theorem 23.8 and the fact that $\mathfrak{I}$ maps $\mathfrak{K}$ onto itself). It follows that m' is an h-line having A' and B' as its limit

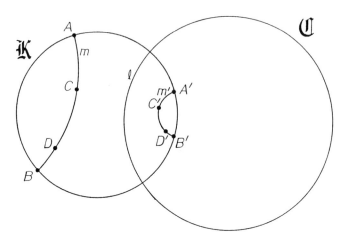

FIGURE 24.21.

points (by Definitions 24.1 and 23.9 and Theorem 5.7). We have

$$\left|\ln\left(\frac{AC \cdot BD}{BC \cdot AD}\right)\right| = \left|\ln\left(\frac{A'C' \cdot B'D'}{B'C' \cdot A'D'}\right)\right| \tag{4}$$

(by Theorem 23.13), and so the h-distance from C to D equals the h-distance from $C' = \sigma_l(C)$ to $D' = \sigma_l(D)$ (by Theorem 24.4). Thus, σ_l preserves h-distances.

In the Euclidean plane, $\mathcal{I}$ interchanges arcs and segments that don't contain Q among themselves and preserves the angles formed by their tangent rays (by Theorem 23.8). Since $\mathcal{I}$ also maps $\mathcal{K}$ to itself and preserves h-lines (by the last two paragraphs), it follows that σ_l interchanges h-rays among themselves and preserves the angles formed by their tangent rays. Thus, σ_l preserves h-angles (since, by definition, the h-angle formed by two h-rays is the angle formed by their tangent rays). That is, if σ_l maps three h-points A, B, C to A', B', C', the h-angles h$\angle BAC$ and h$\angle B'A'C'$ are equal.

Next, assume that l is an h-line that lies on a line n in the Euclidean plane through the center P of $\mathcal{K}$ (Figure 24.22). The reflection σ_n of the Euclidean plane fixes P and preserves distances. Thus, σ_n maps points inside $\mathcal{K}$ among themselves, and so its restriction to these points gives a map σ_l of h-points. σ_l fixes the h-points on l because σ_n fixes the points on n. σ_l interchanges each h-point that doesn't lie on l with an h-point on the other h-side of l because σ_n interchanges points on opposite sides of n. Let C and D be two h-points on an h-line m, and let A and B be the limit points of m. Let A',

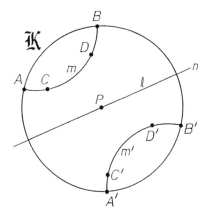

FIGURE 24.22.

B', C', D', m' be the respective images of A, B, C, D, m under σ_n. σ_n preserves distances and angles and fixes P, and so it interchanges points of $\mathcal{K}$ among themselves, circles orthogonal to $\mathcal{K}$ among themselves, and lines through P among themselves. Thus, σ_l interchanges h-lines among themselves, and m' is an h-line with limit points A' and B'. Equation 4 holds (since σ_n preserves distances), and so the h-distance from C to D equals the h-distance from $C' = \sigma_l(C)$ to $D' = \sigma_l(D)$, and σ_l preserves h-distances. σ_n interchanges arcs and segments among themselves, preserves the angles formed by their tangent rays, and maps $\mathcal{K}$ to itself. It follows that σ_l interchanges h-rays among themselves and preserves the angles formed by their tangent rays. Hence, σ_l preserves h-angles: If σ_l maps three h-points A, B, C to A', B', C', the h-angles $h\angle BAC$ and $h\angle B'A'C'$ are equal.

Every h-line l lies either on a circle orthogonal to $\mathcal{K}$ or on a line through P (by Definition 24.1). Thus, we've defined a map σ_l for every h-line l. We call σ_l the *h-reflection* in l. We've proved the following result.

THEOREM 24.8. For every h-line l, the h-reflection σ_l fixes the h-points on l and interchanges each h-point that doesn't lie on l with an h-point on the other h-side of l. σ_l preserves h-distances and h-angles: If A', B', C' are the images of three h-points A, B, C under σ_l, the h-distance from A to B equals the h-distance from A' to B', and the h-angles $h\angle BAC$ and $h\angle B'A'C'$ are equal. □

The *hyperbolic plane* is the set of h-points with the structures we've defined: h-lines, h-coordinate systems, h-distances, h-segments, h-rays, h-sides, h-angles, and h-reflections. Theorems 24.2, 24.3, and 24.6–24.8

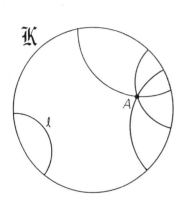

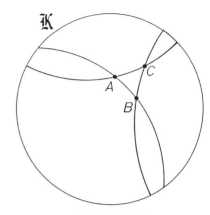

FIGURE 24.23.   FIGURE 24.24.

show that the hyperbolic and Euclidean planes share many basic properties. In the next section, we obtain more properties that the planes have in common.

Hyperbolic planes also have properties spectacularly different from those of the Euclidean plane, however. The hyperbolic plane is important because it shows that such unfamiliar properties can coexist with the familiar properties shared by the hyperbolic and Euclidean planes.

For example, suppose we call two h-lines *parallel* if they don't intersect. As Figure 24.23 illustrates, there is an h-point A and an h-line l such that A lies on infinitely many h-lines parallel to l. This is very different from the Euclidean plane, where every point lies on a unique line parallel to a given line.

Figure 24.24 illustrates another surprising property of the hyperbolic plane. The h-angle h$\angle BAC$ is less than the angle $\angle BAC$ in the Euclidean plane (since the tangent rays to the h-rays $\overrightarrow{AB}$ and $\overrightarrow{AC}$ lie between the rays $\overrightarrow{AB}$ and $\overrightarrow{AC}$ in the Euclidean plane). Similarly, the h-angle h$\angle ABC$ is less than the Euclidean angle $\angle ABC$, and the h-angle h$\angle ACB$ is less than the Euclidean angle $\angle ACB$. Thus, the fact that

$$\angle BAC + \angle ABC + \angle ACB = 180$$

in the Euclidean plane (by Theorem 0.8) implies that

$$h\angle BAC + h\angle ABC + h\angle ACB < 180$$

in the hyperbolic plane. In other words, although the angles of every triangle in the Euclidean plane sum to 180 degrees, the angles of a triangle in the hyperbolic plane can sum to less than 180 degrees.

The last two paragraphs show that it is impossible to use the properties shared by the Euclidean and hyperbolic planes—such as those in Theorems 24.2, 24.3, and 24.6–24.8—to prove that each point lies on a unique parallel to a given line or that the angles in a triangle sum to 180 degrees: Such a proof would imply that the latter properties hold in the hyperbolic plane, but they don't. In general, it is impossible to use the properties shared by the Euclidean and hyperbolic planes to prove any property that holds in only one of the planes.

In Section 25, we obtain more properties shared by the hyperbolic and Euclidean planes. In Section 26, we consider more properties that differentiate these planes.

EXERCISES

The exercises in this section are based on the following definition. In the Euclidean plane, let $\mathcal{K}$ be a circle with center P and radius r. If X is any point in the plane, the *power* of X with respect to $\mathcal{K}$ is $d^2 - r^2$, where d is the distance from X to P. (Of course, we take $d = 0$ when $X = P$.)

24.1. In the Euclidean plane, let $\mathcal{K}$ be a circle, and let X be a point. Prove that the power of X with respect to $\mathcal{K}$ is $\overline{XA} \cdot \overline{XB}$, where A and B are any two points of $\mathcal{K}$ that lie on a line through X. (We follow the convention that $\overline{XY} = 0$ if $X = Y$. One possible approach is to use Theorem 1.7 when line AB contains the center of $\mathcal{K}$ and X doesn't lie on $\mathcal{K}$. See Theorem 6.5.)

24.2. In the Euclidean plane, let I be the incenter of triangle ABC (Figure 24.25). The internal angle bisector CI at C intersects the circumcircle of triangle ABC at a point P on the arc intercepted by $\angle ACB$.

(a) Prove that $\angle ACP = \angle BAP$. (*Hint:* One possible approach is to prove that each side of this equation equals $\angle BCP$.)
(b) Prove that $\angle AIP = \angle ACP + \angle CAI$ and $\angle IAP = \angle BAP + \angle CAI$.
(c) Use parts (a) and (b) to prove that P is equidistant from A and I.
(d) Prove that P is the center of the circumcircle of triangle ABI.
(e) Let L be the excenter on the internal angle bisector at C. Prove that the circle with diameter IL contains A and B and has center P. Illustrate this result with a figure. (*Thus, an internal angle bisector l of a triangle intersects the circumcircle at the center of the circle that has the equicenters on l as the endpoints of a diameter. The circle contains the two vertices of the triangle that don't lie on l.*)

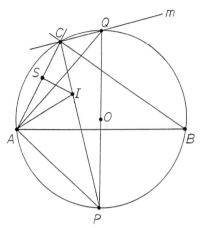

FIGURE 24.25.

24.3. Let the notation be as in Exercise 24.2 (Figure 24.25). In addition, let the circumcircle of triangle ABC have radius R and center O. Let Q be the other endpoint of the diameter OP of the circumcircle. Let r be the radius of the incircle, and let S be the foot of the perpendicular from I to line AC. Let d be the distance from O to I. Let XY denote the distance between any two points X and Y.

(a) Prove that $R^2 - d^2 = IP \cdot IC$. (See Exercise 24.1.)

(b) Prove that triangles CSI and QAP are similar. Conclude that $AP \cdot IC = QP \cdot SI$.

(c) Use parts (a) and (b) and Exercise 24.2(c) to conclude that $R^2 - d^2 = 2Rr$. (We can rewrite this equation as

$$d^2 = R^2 - 2Rr,$$

which is known as *Euler's theorem*. It expresses the distance between the centers of the circumcircle and the incircle of a triangle in terms of the radii of the circles. In Figure 24.25, d is the distance from O to I, R is the distance from O to P, and r is the distance from I to S.)

(d) Prove that $R = 2r$ if triangle ABC is equilateral and that $R > 2r$ otherwise.

24.4. Let the notation be as in Exercises 24.2 and 24.3. In addition, let m be the external angle bisector at C (Figure 24.25)

(a) Prove that Q lies on m. (See Theorem 4.7ii.)

(b) Prove that PQ is the perpendicular bisector of A and B, and conclude that Q is the unique point on m equidistant from A and B. (See Exercise 24.2(d).)

(c) Let J and K be the two excenters on the external angle bisector at C. Prove that the circle with diameter JK contains A and B and has center Q. Illustrate this result with a figure. (*Thus, an external angle bisector m of a triangle intersects the circumcircle at the center of the circle that has the excenters on m as the endpoints of a diameter. The circle contains the two vertices of the triangle that don't lie on m.*)

24.5. Let the notation be as in Exercises 24.2 and 24.3. In addition, let r' be the radius of the excircle with center L, and let S' be the foot of the perpendicular from L to line AC. Let d' be the distance from O to L. Let XY denote the distance between any two points X and Y.

(a) Prove that $d'^2 - R^2 = LP \cdot LC$.
(b) Prove that triangles $CS'L$ and QAP are similar. Conclude that $AP \cdot LC = QP \cdot S'L$.
(c) Use parts (a) and (b) and Exercise 24.2(e) to conclude that $d'^2 = R^2 + 2Rr'$.

24.6. Let the notation be as in Exercises 24.2 and 24.3. In addition, assume that sides AC and BC of triangle ABC are of different lengths. Let F be the foot of the altitude on C (Figure 24.26).

(a) Prove that $\angle FCP = \angle CPQ$. (See Exercise 24.4b.)
(b) Prove that the internal angle bisector of triangle ABC at C bisects the angle at C between the altitude CF and the radius CO of the circumcircle.
(c) Prove that the external angle bisector m at C of triangle ABC bisects two of the angles formed by the lines CF and CO.

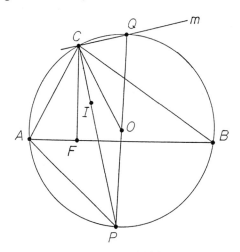

FIGURE 24.26.

24.7. In the Euclidean plane, let ABC be a triangle, and let H be its orthocenter. Let $\mathcal{K}$ be a circle that has a diameter whose endpoints are either A and a point on line BC, B and a point on line AC, or C and a point on line AB. Prove that the power of H with respect to $\mathcal{K}$ equals the quantities in (3) of Section 6. (See Exercise 24.1.)

24.8. In the Euclidean plane, let $\mathcal{K}_1$ and $\mathcal{K}_2$ be circles whose centers P_1 and P_2 are distinct. Let r_1 and r_2 be the respective radii of $\mathcal{K}_1$ and $\mathcal{K}_2$.

(a) Let X be any point in the plane, and let F be the foot of the perpendicular from X to line P_1P_2. Prove that X has the same power with respect to $\mathcal{K}_1$ and $\mathcal{K}_2$ if and only if

$$(\overline{P_1F})^2 - (\overline{P_2F})^2 = r_1^2 - r_2^2.$$

(b) Choose a coordinate system on line P_1P_2 such that P_1 has coordinate 0. If P_2 has coordinate c, let l be the line perpendicular to P_1P_2 through the point on P_1P_2 with coordinate

$$\frac{1}{2c}(c^2 + r_1^2 - r_2^2).$$

Prove that the points of l are exactly the points in the plane that have the same power with respect to $\mathcal{K}_1$ and $\mathcal{K}_2$.

24.9. In the notation of Exercise 24.8, prove that the points of intersection of $\mathcal{K}_1$ and l are exactly the points of intersection of $\mathcal{K}_1$ and $\mathcal{K}_2$.

Exercises 24.10–24.16 use the following notation. If $\mathcal{K}_1$ and $\mathcal{K}_2$ are circles with distinct centers, the line l in Exercise 24.8(b) is called the *radical axis* of $\mathcal{K}_1$ and $\mathcal{K}_2$.

In the Euclidean plane, let a, b, c, d be four lines such that no two are parallel and no three lie on the same point. The *associated triangles* are the four triangles whose sides lie on three of the lines a, b, c, d; that is, the sides of each associated triangle lie on the lines in one of the sets

$$\{a, b, c\}, \quad \{a, b, d\}, \quad \{a, c, d\}, \quad \{b, c, d\}.$$

The *diagonal circles* determined by a, b, c, d are the three circles that have the points in one of the sets

$$\{a \cap b, c \cap d\}, \quad \{a \cap c, b \cap d\}, \quad \{a \cap d, b \cap c\}$$

as the endpoints of a diameter.

24.10. Let H be the orthocenter of one of the associated triangles. Use Exercise 24.7 to prove that H has the same power with respect to all three diagonal circles.

24.11. Prove that the three diagonal circles have distinct centers. (*Hint:* One possible approach is to assume that two of the diagonal circles have the same center and obtain a contradiction to the assumption that no two of the lines a, b, c, d are parallel.)

24.12. Let $\mathfrak{I}_1$ be the associated triangle whose sides lie on the lines a, b, c, and let $\mathfrak{I}_2$ be the asssociated triangle whose sides lie on the lines a, b, d. Prove that $\mathfrak{I}_1$ and $\mathfrak{I}_2$ have the same orthocenter if and only if a and b are perpendicular. Conclude that the orthocenters of the four associated triangles are not all equal.

24.13. Use Exercises 24.10–24.12 to prove that any two diagonal circles have the same radical axis and that this line contains the orthocenters of the four associated triangles.

24.14. If two of the diagonal circles intersect at two points A and B, prove that all three diagonal circles contain A and B and that line AB contains the orthocenters of the four associated triangles. Illustrate this result with a figure. (See Exercises 24.9 and 24.13.)

24.15. If a line l is tangent to two of the diagonal circles at a point A, prove that l is tangent to all three diagonal circles at A and that l contains the orthocenters of the four associated triangles. Illustrate this result with a figure. (See Exercises 24.9, 24.13, and 9.9.)

24.16. If two of the diagonal circles don't intersect, prove that no two of the diagonal circles intersect and that the orthocenters of the four associated triangles lie on a line that doesn't intersect any of the diagonal circles. Illustrate this result with a figure. (See Exercises 24.9 and 24.13.)

Section 25.
Absolute Geometry

We have found five basic properties that hold in both the hyperbolic and the Euclidean planes, Theorems 24.2, 24.3, 24.6, 24.7, and 24.8. We study the logical consequences of these propeties in this section. In this way we obtain more results that hold in both the hyperbolic and Euclidean planes: since the five given properties hold in both planes, so do their consequences. We also highlight the elegance of elementary geometry by showing that many results follow from just a few assumptions.

In general, the axiomatic method is the study of the logical consequences of given assumptions called axioms. The previous paragraph illustrates two of the advantages of the axiomatic method. First, conclusions drawn from the axioms hold in every system that satisfies the axioms; by reasoning from the axioms, we simultaneously obtain results in all systems where the axioms hold. Second, the axiomatic method emphasizes the structure of a subject by showing how results follow from basic properties.

We start by listing the axioms, the five properties we've found that hold in both the hyperbolic and Euclidean planes. We start with a set of objects called *points*. Certain sets of points are called *lines*.

AXIOM A1.
   (i) Any two points lie on a unique line.
   (ii) There are at least two points.   □

No two points lie on more than one line, by Axiom A1(i), and so *two lines intersect in at most one point*.

AXIOM A2. The points on each line *l* can be matched up with the real numbers.   □

The matching in Axiom A2 is called the *coordinate system* of *l*. The number associated with each point X of *l* is called the *coordinate* of X as a point of *l*. We emphasize that one coordinate system is designated for each line.

Let A and B be two points. We define AB to be the unique line through A and B (by Axiom A1(i)). Let A and B have respective coordinates a and b in the coordinate system on AB. We define $d(A, B)$, the *distance* from A to B, by setting

$$d(A, B) = |a - b|. \tag{1}$$

We also set $d(A, A) = 0$. We define the *segment* $[A, B]$ to consist of all points X on AB such that the coordinate x of X lies between a and b, inclusive. We define the *ray* $\overrightarrow{AB}$ to consist of A and all points X on line AB such that $x - a$ has the same sign as $b - a$. We say that the ray $\overrightarrow{AB}$ *originates* at A.

AXIOM A3. For each line *l*, there are two sets S_1 and S_2 of points such that the following two properties hold (Figure 24.15):

   (i) Every point belongs to exactly one of the sets *l*, S_1, S_2.
   (ii) If A and B are two points that don't lie on *l*, then $[A, B]$ intersects *l* if and only if S_1 and S_2 each contain one of the points A and B.   □

The sets S_1 and S_2 in Axiom A3 are called the *sides* of *l*. A *halfplane* of a line *l* is the union of *l* and one of its sides.

AXIOM A4. A number $\angle BAC$ is assigned to each pair of rays $\overrightarrow{AB}$ and $\overrightarrow{AC}$ (not necessarily distinct) orginating at the same point A. If $\mathcal{K}$ is a halfplane of line AB, the following two properties hold (Figure 24.18):

(i) The map taking $\overrightarrow{AC}$ to $\angle BAC$ matches up the rays $\overrightarrow{AC}$ in $\mathcal{K}$ originating at A with the real numbers $\angle BAC$ from 0 through 180, inclusive.

(ii) If C and D are points of $\mathcal{K}$ such that B and C lie on different sides of line AD, then we have

$$\angle BAC = \angle BAD + \angle DAC. \quad \square$$

We refer to the quantity $\angle BAC$ in Axiom A4 as an *angle*. We emphasize that, for us, an angle is a number. The first sentence of Axiom A4 implies that the value of $\angle BAC$ depends only on the rays $\overrightarrow{AB}$ and $\overrightarrow{AC}$ and not on the order they are listed. Thus, we have $\angle BAC = \angle CAB$ and, if $\overrightarrow{AC} = \overrightarrow{AD}$, we have $\angle BAC = \angle BAD$.

AXIOM A5. For any line l, there is a map σ_l from the plane to itself that has the following two properties.

(i) σ_l maps each point of l to itself, and it interchanges each point that doesn't lie on l with a point on the other side of l.

(ii) If A', B', C' are the images under σ_l of three points A, B, C, then the equations $d(A, B) = d(A', B')$ and $\angle BAC = \angle B'A'C'$ hold. $\quad \square$

We call the map σ_l in Axiom A5 the *reflection* in l. We say that σ_l *fixes* the points of l, meaning that σ_l maps each point of l to itself. Part (ii) of Axiom A5 says that σ_l preserves distances and angles.

An *absolute plane* is a plane that satisfies Axioms A1–A5 and the accompanying definitions. The Euclidean and hyperbolic planes are examples of absolute planes, as the last section shows. Specifically, Theorems 24.2, 24.3, 24.6, 24.7, and 24.8 show that a hyperbolic plane satisfies Axiom A1–A5.

Absolute geometry is the study of the logical consequences of Axioms A1–A5 and the accompanying definitions. These consequences hold in all absolute planes, and so they hold in both the Euclidean and the hyperbolic planes. We emphasize that *all results in absolute geometry are to follow directly from Axioms A1–A5 and the acccompanying definitions*. We cannot use any additional information about the Euclidean and the hyperbolic planes in studying absolute geometry.

The results of absolute geometry that we obtain are obviously true in the Euclidean plane. Because we derive these results from the axioms, however, they also hold in the hyperbolic plane, where many of them are quite surprising. It is also notable that all of these results follow from just five axioms.

Throughout this section, we illustrate results in absolute geometry with figures in the Euclidean plane. This approach makes the results seem familiar and intuitively appealing. It's worthwhile, however, to illustrate the results with figures in the hyperbolic plane as well.

Segments, rays, and distances have been defined via coordinates on each line in the absolute plane in the same way that they are defined on the real number line. Accordingly, all facts about segments, rays, and distances on the real number line also hold on every line in the absolute plane. This observation gives the next two theorems. We say that points A and B are *equidistant* from a point C if $d(A, C)$ equals $d(B, C)$.

THEOREM 25.1. In an absolute plane, let A be a point on a line l. Then there are exactly two rays originating at A that lie on l. These are the rays $\overrightarrow{AB}$ and $\overrightarrow{AC}$ if and only if B and C are two points of l other than A such that $[B, C]$ contains A (Figure 25.1). $\square$

THEOREM 25.2. In an absolute plane, let B and C be two points (Figure 25.1).

(i) If A is a point on $[B, C]$ other than B and C, then $[A, B]$ doesn't contain C.

(ii) The correspondence $A \leftrightarrow d(B, A)$ matches up the points A of $\overrightarrow{BC}$ other than B with the positive real numbers.

(iii) For any point A on l other than B, A lies on $\overrightarrow{BC}$ if and only if $\overrightarrow{BA}$ equals $\overrightarrow{BC}$.

(iv) The segment $[B, C]$ consists of all points A on $\overrightarrow{BC}$ such that $d(B, A) \leq d(B, C)$.

(v) $[B, C]$ is the intersection of $\overrightarrow{BC}$ and $\overrightarrow{CB}$.

(vi) There is a unique point M on line BC that is equidistant from B and C. M is a point of $[B, C]$ other than B and C, and the distance from B to C is twice the distance from B to M.

(vii) There is a unique point N on line BC other than B such that C is equidistant from N and B. $\square$

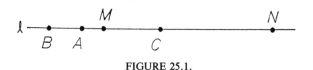

FIGURE 25.1.

We call the point M in Theorem 25.2(vi) the *midpoint* of B and C. Theorems 25.2(vi) and (vii) show that B and C have a unique midpoint and that there is a unique point N such that C is the midpoint of N and B.

We call two rays *opposite* if they originate at the same point and lie on one line. Theorem 25.1 shows that every ray is opposite exactly one other.

Axiom A4 leads to other familiar results about angles that generalize from the Euclidean plane to all absolute planes. The next theorem contains several such results.

THEOREM 25.3. In an absolute plane, let A, B, C be three points such that $[B, C]$ contains A. Let D be any point that doesn't lie on line BC, and let $\overrightarrow{AE}$ be the ray opposite $\overrightarrow{AD}$ (Figure 25.2). Then the following relations hold:

$$\angle BAC = 180: \tag{2}$$

$$\angle BAD + \angle DAC = 180: \tag{3}$$

$$\angle BAD = \angle CAE. \tag{4}$$

Moreover, if F is a point other than A and B, we have

$$0 < \angle BAF < 180 \tag{5}$$

if and only if F doesn't lie on line AB.

Proof: B and C lie on different sides of line AD (by Axiom A3), and so we have

$$\angle BAD + \angle DAC = \angle BAC \tag{6}$$

(by Axiom A4(ii)). It follows that

$$\angle BAD \le \angle BAC \le 180 \tag{7}$$

(by Axiom A4(i)). As $\overrightarrow{AD}$ varies over all rays that lie in a halfplane of line AB but not on AB itself, $\angle BAD$ takes all values from 0 through 180 except two (by Axiom A4(i) and Theorem 25.1). Thus, the inequalities in (7) imply that Equation 2 holds. Equation 3 follows from Equations 2 and 6.

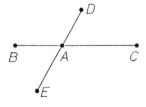

FIGURE 25.2.

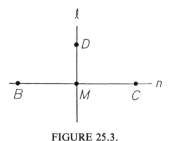

FIGURE 25.3.

A is a point of $[D, E]$ other than D and E (by Theorem 25.1), and so Equation 3 shows that

$$\angle DAC + \angle CAE = 180.$$

Thus, Equation 4 follows from Equation 3. Equation 2 and Axiom A4(i) show that $\angle BAD$ and $\angle DAC$ are less than 180, and so Equation 3 implies that $0 < \angle BAD < 180$.

There is a ray $\overrightarrow{AG}$ in a halfplane of line AB such that $\angle BAG = 0$ (by Axiom A4(i)). G lies on AB (by the last sentence of the previous paragraph), and $\overrightarrow{AG}$ is not opposite to $\overrightarrow{AB}$ (by Theorem 25.1 and Equation 2). Thus, $\overrightarrow{AG}$ equals $\overrightarrow{AB}$. Together with Equation 2, Theorem 25.1, and the last sentence of the previous paragraph, this shows that the inequalities in (5) hold if and only if F doesn't lie on line AB.   □

In the notation of Theorem 25.3, if one of the angles $\angle BAD$, $\angle DAC$, $\angle CAE$, $\angle EAB$ is 90, they all are (by Equations 3 and 4). In this case, we say that BC and DE are *perpendicular* lines.

The *perpendicular bisector* l of two points B and C is the line l that contains the midpoint M of B and C and is perpendicular to line BC (Figure 25.3). Any two points B and C have a unique perpendicular bisector (by Theorem 25.2vi, Axiom A4(i), and the last sentence of Theorem 25.3).

The next result links perpendicular bisectors and reflections.

THEOREM 25.4. In an absolute plane, let B and C be two points, and let l be a line. Then σ_l maps B to C if and only if l is the perpendicular bisector of B and C (Figure 25.3).

Proof: First assume that σ_l maps B to C. Since $B \neq C$, B and C lie on different sides of l (by Axiom A5), and so the segment $[B, C]$ intersects l at a point M (by Axiom A3). Let D be a point of l other than M. The reflection σ_l maps B to C and fixes M and D, and so we have

$$d(B, M) = d(C, M) \quad \text{and} \quad \angle BMD = \angle CMD$$

(by Axiom A5). The first of these equations shows that M is the midpoint of B and C, and the second shows that $\angle BMD = 90$ (since $\angle BMD$ and $\angle CMD$ sum to 180, by Equation 3). Thus, l is the perpendicular bisector of B and C.

Conversely, assume that l is the perpendicular bisector of B and C. Let M be the midpoint of B and C, and let D be a point on l other than M. Set $C' = \sigma_l(B)$. Since σ_l fixes points of l and preserves angles and distances (by Axiom A5), we have

$$\angle C'MD = \angle BMD = 90 = \angle CMD \tag{8}$$

and

$$d(C', M) = d(B, M) = d(C, M). \tag{9}$$

C' lies on a different side of l than B (by Axiom A5(i)), and so we have $\overrightarrow{MC'} = \overrightarrow{MC}$ (by (8) and Axiom A4(i)). It follows that $C' = C$ (by (9) and Theorem 25.2ii and iii), and so σ_l maps B to C. $\square$

The previous result implies that we can draw a unique perpendicular to any line through any point in an absolute plane.

THEOREM 25.5. In an absolute plane, let B be any point, and let l be any line. Then there is a unique line through B perpendicular to l (Figure 25.3).

Proof: The theorem holds when B lies on l, by Axiom A4(i) and the last sentence of Theorem 25.3. Thus, we can assume that B doesn't lie on l (as in Figure 25.3).

If we set $C = \sigma_l(B)$, we have $B \neq C$ (by Axiom A5(i)), and so BC is a line through B perpendicular to l (by Theorem 25.4). Conversely, if n is any line through B perpendicular to l, let n and l intersect at a point M. Let C be the point on n such that M is the midpoint of B and C (by Theorem 25.2vii). We have $n = BC$ (by Axiom A1(i)). Since l is the perpendicular bisector of B and C, we have $C = \sigma_l(B)$ (by Theorem 25.4). Thus, by the first sentence of the paragraph, the unique line through B perpendicular to l is $n = BC$ for $C = \sigma_l(B)$. $\square$

In an absolute plane, we call two lines *parallel* if they don't intersect. Note that we no longer consider a line to be parallel to itself. Let B be a point that doesn't lie on a line l. By the previous theorem, B lies on a unique line perpendicular to l. On the other hand, B can lie on infinitely many lines parallel to l (by the discussion accompanying Figure 24.23).

The next theorem lets us deduce results about parallel lines from results about perpendicular lines.

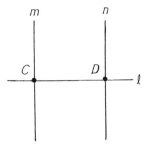

FIGURE 25.4.

THEOREM 25.6. In an absolute plane, if m and n are lines perpendicular to a line l at distinct points C and D, then m and n are parallel (Figure 25.4).

Proof: Any two lines intersect in at most one point (by Axiom A1(i)), and so C and D are the unique points where m and n intersect l. Thus, the assumption that $C \neq D$ implies that $m \neq n$. Any point B in the plane lies on a unique line perpendicular to l (by Theorem 25.5), and so it doesn't lie on both m and n. Hence, m and n don't intersect at any point. □

In an absolute plane, let C be a point that doesn't lie on a line n. C might lie on a unique line parallel to n, or it might lie on infinitely many lines parallel to n, since the Euclidean and the hyperbolic plane are both examples of absolute planes. The last two theorems, however, imply that C always lies on at least one line parallel to n.

THEOREM 25.7. In an absolute plane, let C be a point that doesn't lie on a line n. Then C lies on at least one line parallel to n.

Proof: Let l be the line through C perpendicular to n, and let m be the line through C perpendicular to l, by Theorem 25.5 (Figure 25.4). The perpendicular lines l and n intersect at a point D, and D doesn't equal C (since C doesn't lie on n). Thus, m is a line through C parallel to n, by Theorem 25.6. □

The construction of the hyperbolic plane in the last section makes it seem unlikely that the SAS Congruence Property 0.2 extends to the hyperbolic plane. Our next goal, however, is to extend the SAS property to all absolute planes. The following theorem is useful in doing so. Part (i) is a basic property of the separation of points by lines. Part (ii) is roughly the converse of Axiom A4(ii).

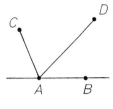

FIGURE 25.5.

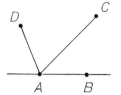

FIGURE 25.6.

THEOREM 25.8. In an absolute plane, let A and B be two points, and let C and D be two points on one side of line AB such that $\overrightarrow{AC} \neq \overrightarrow{AD}$.

(i) Then either B and C lie on different sides of AD, or B and D lie on different sides of AC (Figures 25.5 and 25.6).

(ii) If $\angle BAD < \angle BAC$, then B and C lie on different sides of line AD, and we have

$$\angle BAC = \angle BAD + \angle DAC \tag{10}$$

(Figure 25.5).

Proof: (i) $[C, D]$ doesn't contain A (by Axiom A3(ii)), and it's assumed that $\overrightarrow{AC} \neq \overrightarrow{AD}$, and so line AC doesn't contain D (by Theorem 25.1). AC doesn't contain B either (since C isn't on AB), and so B and D lie on sides of AC (by Axiom A3(i)). If they lie on different sides of AC, we're done (Figure 25.6). Thus, we can assume that B and D lie on the same side of AC (Figure 25.5).

Let $\overrightarrow{AE}$ be the ray opposite $\overrightarrow{AB}$ (Figure 25.7). $[B, E]$ contains A (by Theorem 25.1), and so B and E lie on different sides of AC (by Axiom A3). Thus, since B and D lie on the same side of AC (by assumption), D and E lie on different sides of AC. Hence, $[D, E]$ intersects AC at a point F other than D and E (by Axiom A3).

$[E, F]$ doesn't contain D (by Theorem 25.2i). DE doesn't equal AD (since D doesn't lie on $AB = AE$), and so D is the only point of DE that lies on AD (by Axiom A1(i)). The last two sentences show that E and F lie on the same side of AD (by Axiom A3). Thus, since B and E lie on different sides of AD (by Axiom A3), B and F lie on different sides of AD.

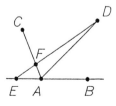

FIGURE 25.7.

$[D, F]$ doesn't contain E (by Theorem 25.2i). E is the only point of DE that lies on AB (by Axiom A1(i), since the fact that D doesn't lie on AB implies that $DE \neq AB$). The last two sentences show that D and F lie on the same side of AB (by Axiom A3). Thus, since C and D lie on the same side of AB (by assumption), so do C and F. Hence, either C and F are equal or $[C, F]$ doesn't contain A (by Axiom A3). AC doesn't equal AD (by the first sentence of the proof), and so A is the only point of AC that lies on AD (by Axiom A1(i)). The last two sentences imply that F and C lie on the same side of AD (by Axiom A3).

The last two paragraphs show that B and F lie on different sides of AD and that C and F lie on the same side. Thus, B and C lie on different sides of AD, as desired.

(ii) If B and D lay on different sides of AC (Figure 25.6), we would have

$$\angle BAD = \angle BAC + \angle CAD$$

(by Axiom A4(ii)), which would contradict the assumption that $\angle BAD < \angle BAC$. Thus, B and C lie on opposite sides of AD (by part (i)), and Equation 10 holds (by Axiom A4(ii)) (Figure 25.5). $\square$

Reflections play a key role in extending the SAS Congruence Property to absolute planes. Theorem 25.4 relates reflections and perpendicular bisectors. The next theorem, which extends Theorem 4.2 to absolute planes, relates perpendicular bisectors and distances between points.

THEOREM 25.9. In an absolute plane, let A, B, C be three points. Then A lies on the perpendicular bisector of B and C if and only if A is equidistant from B and C.

Proof: First assume that A lies on the perpendicular bisector l of B and C (Figure 25.8). The reflection σ_l maps B to C (by Theorem 25.4). Since σ_l fixes A, the distance from A to B equals the distance from A to C (by Axiom A5).

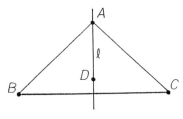

FIGURE 25.8.

Conversely, assume that A is equidistant from B and C. If A lies on line BC, then A is the midpoint of B and C (by Theorem 25.2vi), and so it lies on the perpendicular bisector of B and C. Thus, we can assume that A doesn't lie on BC (as in Figure 25.8). There is a ray $\overrightarrow{AD}$ that lies in the same halfplane of line AB as C and is such that $\angle BAD = \frac{1}{2}\angle BAC$ (by Axiom A4(i)). C doesn't lie on AB (since we're assuming that A doesn't lie on BC), and so we have $\angle BAC > 0$ (by the last sentence of Theorem 25.3). It follows that

$$0 < \angle BAD < \angle BAC,$$

and so $\overrightarrow{AD}$ doesn't equal $\overrightarrow{AC}$, and D doesn't lie on AB. Thus, Theorem 25.8(ii) shows that B and C lie on opposite sides of line AD and that

$$\angle BAC = \angle BAD + \angle DAC = \tfrac{1}{2}\angle BAC + \angle DAC.$$

Hence, we have

$$\angle DAC = \tfrac{1}{2}\angle BAC = \angle BAD. \tag{11}$$

Let E be the image of B under σ_{AD}. B and E lie on different sides of AD (by Axiom A5(ii)), and so do B and C (by the previous paragraph), and so C and E lie on the same side of AD. Since σ_{AD} maps B to E and fixes A and D, we have

$$\angle EAD = \angle BAD = \angle DAC$$

(by Axiom A5 and Equation 11). The last two sentences imply that $\overrightarrow{AE} = \overrightarrow{AC}$ (by Axiom A4(i)). Since it's also true that

$$d(A, E) = d(A, B) = d(A, C)$$

(by Axiom A5 and the assumption that A is equidistant from B and C), we have $E = C$ (by Theorem 25.2ii and iii). Thus, C is the image of B under σ_{AD}, and so AD is the perpendicular bisector of B and C (by Theorem 25.4). In particular, A lies on the perpendicular bisector of B and C, as desired. $\square$

In an absolute plane, we define *triangle ABC* to be the union of the segments $[A, B]$, $[B, C]$, $[C, A]$, where A, B, C are three points that don't lie on a line. We say that triangle ABC has *sides* $[A, B]$, $[B, C]$, $[C, A]$ and *angles* $\angle BAC$, $\angle ABC$, $\angle BCA$. The *length of a side* $[X, Y]$ of a triangle is the distance from X to Y. We say that triangles ABC and $A'B'C'$ are *congruent* if corresponding sides have equal lengths and corresponding angles are equal, that is, if $d(A, B) = d(A', B')$, $d(B, C) = d(B', C')$, $d(C, A) = d(C', A')$, $\angle BAC = \angle B'A'C'$, $\angle ABC = \angle A'B'C'$, and $\angle BCA = \angle B'C'A'$.

We can now extend the SAS congruence theorems for triangles to absolute planes. If two sides of one triangle have the same lengths as two sides of another triangle, and if the included angles are equal, we prove that the triangles are congruent by mapping one of the triangles to the other with a sequence of reflections.

THEOREM 25.10. In an absolute plane, let ABC and $A'B'C'$ be triangles such that $d(A, B) = d(A', B')$, $d(A, C) = d(A', C')$, and $\angle BAC = \angle B'A'C'$. Then the triangles are congruent.

Proof: If $A \neq A'$, let l be the perpendicular bisector of A and A' (Figure 25.9). If $A = A'$, let l be any line through A. In either case, σ_l maps A' to A (by Theorem 25.4 and Axiom A5(i)). Let B_1 and C_1 be the respective images of B' and C' under σ_l. Since σ_l maps A', B', C', to A, B_1, C_1, triangles $A'B'C'$ and AB_1C_1 are congruent (by Axiom A5(ii)). Together with the hypotheses of the theorem, this shows that $d(A, B) = d(A, B_1)$, $d(A, C) = d(A, C_1)$, and $\angle BAC = \angle B_1AC_1$.

If $B \neq B_1$, let m be the perpendicular bisector of B and B_1. If $B = B_1$, let m be line AB. In either case, m contains A (by Theorem 25.9 and the last sentence of the previous paragraph), and so σ_m fixes A and maps B_1 to B (by Theorem 25.4 and Axiom A5(i)). Let C_2 be the image of C_1 under σ_m. Since σ_m maps A, B_1, C_1 to A, B, C_2, triangles AB_1C_1 and ABC_2 are congruent (by Axiom A5(ii)). Together with the last sentence of the previous paragraph, this shows that $d(A, C) = d(A, C_2)$ and $\angle BAC = \angle BAC_2$.

If C and C_2 lie on opposite sides of line AB, set $C_3 = \sigma_{AB}(C_2)$. If C and C_2 lie on the same side of line AB, set $C_3 = C_2$. In either case, C and C_3 lie on the same side of line AB, triangles ABC_2 and ABC_3 are congruent, and we have $d(A, C) = d(A, C_3)$ and $\angle BAC = \angle BAC_3$ (by Axiom A5

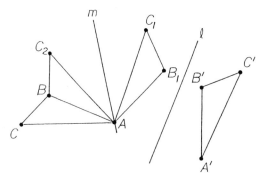

FIGURE 25.9.

and the last sentence of the previous paragraph). Then the rays $\overrightarrow{AC}$ and $\overrightarrow{AC_3}$ are equal (by Axiom A4(i)), and C equals C_3 (by Theorem 25.2ii and iii). The last two sentences show that triangles ABC_2 and ABC are congruent.

In short, we've proved that each of the triangles $A'B'C'$, AB_1C_1, ABC_2, ABC is congruent to the next, and so triangles $A'B'C'$ and ABC are congruent, as desired. □

In an absolute plane, the angles of a triangle can sum to 180 (by Theorem 0.8) or to less than 180 (by the discussion accompanying Figure 24.24). We end this section by showing that these are the only two possibilities. We define $\delta\triangle ABC$, the *defect* of triangle ABC, by setting

$$\delta\triangle ABC = 180 - \angle BAC - \angle ABC - \angle BCA. \qquad (12)$$

We prove that $\delta\triangle ABC \geq 0$, so the sum of the angles of any triangle in an absolute plane is at most 180. We require one preliminary result: If we divide a triangle into two subtriangles, the defect of the whole triangle is the sum of the defects of the two subtriangles.

THEOREM 25.11. In an absolute plane, let ABC be a triangle, and let D be a point on $[B, C]$ other than B and C. Then the equation

$$\delta\triangle ABC = \delta\triangle ABD + \delta\triangle ADC$$

holds (Figure 25.10).

Proof: $[C, D]$ doesn't contain B (by Theorem 25.2i), and B is the only point of CD that also lies on AB (by Axiom A1(i)). Thus, C and D lie on the same side of AB (by Axiom A3). Moreover, B and C lie on different sides of AD (by Axiom A3), and so we have

$$\angle BAC = \angle BAD + \angle DAC \qquad (13)$$

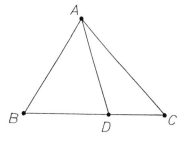

FIGURE 25.10.

(by Axiom A4(ii)). We also have

$$\angle BDA + \angle CDA = 180 \qquad\qquad (14)$$

(by Equation 3). It follows that

$$\delta \triangle ABD + \delta \triangle ADC = 180 - \angle BAD - \angle ABD - \angle BDA$$
$$+ 180 - \angle DAC - \angle ACD - \angle CDA$$
$$= 180 - \angle BAC - \angle ABD - \angle ACD$$

(by Equations 13 and 14)

$$= 180 - \angle BAC - \angle ABC - \angle ACB$$

(since $\overrightarrow{BD} = \overrightarrow{BC}$ and $\overrightarrow{CD} = \overrightarrow{CB}$, by Theorem 25.2v and iii)

$$= \delta \triangle ABC. \quad \square$$

Starting with any triangle in an absolute plane, we use the previous result to construct a sequence of triangles that have the same angle sum and two angles that approach zero. Since the third angle of each triangle is at most 180, it follows that the sum of the angles of the original triangle is at most 180. This is the main idea of the proof of the next theorem.

THEOREM 25.12. In an absolute plane, the sum of the angles of any triangle is at most 180.

Proof: Let ABC be a triangle (Figure 25.11). Let M be the midpoint of B and C (by Theorem 25.2vi), and let D be the point such that M is the midpoint of A and D (by Theorem 25.2vii). M is a point of $[B, C]$

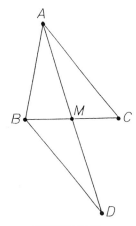

FIGURE 25.11.

other than B and C, and it is a point of $[A, D]$ other than A and D (by Theorem 25.2vi), and so we have $\angle AMC = \angle DMB$ (by Theorem 25.1 and Equation 4). Thus, since M is equidistant from A and D and from B and C, triangles AMC and DMB are congruent (by Theorem 25.10).

By Theorem 25.11 and the fact that triangles AMC and DMB are congruent, we have

$$\delta\triangle ABC = \delta\triangle ABM + \delta\triangle AMC = \delta\triangle ABM + \delta\triangle DMB = \delta\triangle ABD.$$

B doesn't lie on $[C, M]$ (by Theorem 25.2i), and it is the only point of CM on AB (by Axiom A1(i)). Thus, C and M lie on the same side of AB (by Axiom A3). Moreover, B and C lie on different sides of AM (by Axiom A3), and so Axiom A4(ii) shows that

$$\angle BAC = \angle BAM + \angle MAC$$

$$= \angle BAM + \angle MDB$$

(by the first paragraph of the proof)

$$= \angle BAD + \angle ADB$$

(since $\overrightarrow{AM} = \overrightarrow{AD}$ and $\overrightarrow{DM} = \overrightarrow{DA}$, by Theorem 25.2iii, v, and vi). It follows that the smaller of the two angles $\angle BAD$ and $\angle ADB$ is at most $\frac{1}{2}\angle BAC$ and that the larger is at most $\angle BAC$.

Accordingly, we can relabel A and D in some order as A_1 and C_1 so that

$$\delta\triangle A_1 BC_1 = \delta\triangle ABC,$$

$$\angle BA_1 C_1 \leq \tfrac{1}{2}\angle BAC,$$

$$\angle BC_1 A_1 \leq \angle BAC.$$

We now apply the same reasoning to triangle $A_1 BC_1$ instead of triangle ABC. This gives a triangle $A_2 BC_2$ such that

$$\delta\triangle A_2 BC_2 = \delta\triangle A_1 BC_1 = \delta\triangle ABC,$$

$$\angle BA_2 C_2 \leq \tfrac{1}{2}\angle BA_1 C_1 \leq \tfrac{1}{4}\angle BAC,$$

$$\angle BC_2 A_2 \leq \angle BA_1 C_1 \leq \tfrac{1}{2}\angle BAC.$$

Applying the same process to triangle $A_2 BC_2$ gives a triangle $A_3 BC_3$ such that

$$\delta\triangle A_3 BC_3 = \delta\triangle A_2 BC_2 = \delta\triangle ABC,$$

$$\angle BA_3 C_3 \leq \tfrac{1}{2}\angle BA_2 C_2 \leq \tfrac{1}{8}\angle BAC,$$

$$\angle BC_3 A_3 \leq \angle BA_2 C_2 \leq \tfrac{1}{4}\angle BAC.$$

Set $\alpha = \angle BAC$. Continuing as before, we obtain a triangle $A_n BC_n$ for every positive integer n such that

$$\delta\triangle A_n BC_n = \delta\triangle ABC, \tag{15}$$

$$\angle BA_n C_n \le \alpha/2^n, \tag{16}$$

$$\angle BC_n A_n \le \alpha/2^{n-1}. \tag{17}$$

Since triangles ABC and $A_n BC_n$ have the same defect (by Equation 15), their angles have the same sum. Thus, the sum of the angles of triangle ABC is

$$\angle A_n BC_n + \angle BA_n C_n + \angle BC_n A_n \le 180 + \alpha/2^n + \alpha/2^{n-1}$$

(by Axiom A4(i) and Equations 16 and 17). Since $\alpha/2^n$ and $\alpha/2^{n-1}$ approach zero as n becomes arbitrarily large, it follows that the sum of the angles of triangle ABC is less than or equal to 180. $\square$

EXERCISES

25.1. In an absolute plane, let ABC be a triangle. Prove that $d(A, C) = d(B, C)$ if and only if $\angle BAC = \angle ABC$.

25.2. In an absolute plane, let A and B be two points on a line n (Figure 25.12). Let l be a line on A other than n, and let m be a line on B other than n. Let C be a point of l other than A, and let D be a point of m that lies on the same side of n as C. Let E be a point of n that doesn't lie on $[A, B]$. We call $\angle CAE$ and $\angle DBE$ *corresponding angles* formed by l and m with n.

Prove the following result: In an absolute plane, if lines l and m form equal corresponding angles with a line n, then l and m are parallel.

23.3. In an absolute plane, let ABC be a triangle, and let l be a line. If l intersects a side of triangle ABC but doesn't contain any of the points A, B, C, prove that l intersects exactly two sides of the triangle.

(*Hint:* One possible approach is to consider how many of the points A, B, C lie on each side of l.)

25.4. In an absolute plane, let ABC be a triangle with a right angle at B (Figure 25.13). Let D be the point such that B is the midpoint of A and D (by Theorem 25.2vii). Assume that the line through D perpendicular to AB intersects AC at a point E.

(a) Prove that there is a point F on $\overrightarrow{DE}$ such that ABC and BDF are congruent triangles.
(b) Prove that AE and BF are parallel lines.

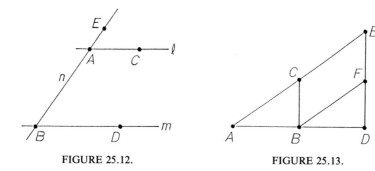

FIGURE 25.12.   FIGURE 25.13.

(c) Prove that F is a point on $[D, E]$ other than D and E and that C is a point on $[A, E]$ other than A and E.

(d) Use parts (a) and (c) and Theorems 25.11 and 25.12 to prove that

$$\delta \triangle ADE \geq 2\delta \triangle ABC.$$

(Exercises 26.6–26.11 depend on this exercise.)

25.5. In an absolute plane, let ABC be a triangle (Figure 25.10). Let D be a point on $[B, C]$ other than B and C. Prove that

$$\angle ADC \geq \angle BAD + \angle ABD \quad \text{and} \quad \angle ADC > \angle ABC.$$

25.6. In the absolute plane, let ABC and $A'B'C'$ be triangles such that $\angle ABC = \angle A'B'C'$, $d(B, C) = d(B', C')$, and $\angle ACB = \angle A'C'B'$. Prove that triangles ABC and $A'B'C'$ are congruent. (This result extends the ASA congruence property to absolute geometry. One possible approach is to use Theorem 25.10 to prove that there is a point D on $\overrightarrow{BA}$ such that triangles DBC and $A'B'C'$ are congruent. Then use the fact that $\angle ACB = \angle A'C'B'$ to prove that $D = A$.)

25.7. In an absolute plane, let ABC and $A'B'C'$ be triangles such that $\angle ABC = \angle A'B'C'$, $\angle BAC = \angle B'A'C'$, and $d(A, C) = d(A', C')$. Prove that triangles ABC and $A'B'C'$ are congruent.

(*Hint:* One possible approach to this exercise is to use Exercise 25.5 to adapt the hint to Exercise 25.6.)

25.8. In an absolute plane, let ABC and $A'B'C'$ be triangles such that corresponding sides have equal lengths. Prove that the triangles are congruent.

(This result extends the SSS congruence property to absolute geometry. One possible approach is as follows. First, prove that there is a point D such that D and C lie on different sides of AB and triangles ABD and $A'B'C'$ are congruent. Then, prove that $\angle ACD = \angle ADC$ and $\angle BCD = \angle BDC$.

FIGURE 25.14.

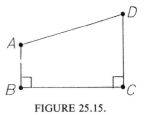

FIGURE 25.15.

Deduce that $\angle ACB = \angle ADB$ by considering the relative positions of A, B, and the point where AB are intersects CD. Conclude that triangles ABC and $A'B'C'$ are congruent.)

25.9. In an absolute plane, let A, B, C, D be four points such that AB and CD are perpendicular to BC and such that A and D lie on the same side of BC.

(a) If $d(A, B) = d(C, D)$, prove that B and C have the same perpendicular bisector as A and D and that $\angle BAD = \angle CDA$ (Figure 25.14).
(b) If $d(A, B) < d(C, D)$, prove that $\angle BAD > \angle CDA$. (See Figure 25.15. One possible approach is to let E be the point on $\overleftrightarrow{CD}$ such that $d(A, B) = d(C, E)$, by Theorem 25.2(ii). Then apply part (a) and Exercise 25.5.)
(c) Prove that $d(A, B) = d(C, D)$ if and only if $\angle BAD = \angle CDA$ (Figure 25.14).

25.10. In an absolute plane, the *distance from a point P to a line m* is $d(P, F)$, where F is the point of intersection of m and the line through P perpendicular to m. Two points are *equidistant* from m if their distances from m are equal.

(a) In an absolute plane, let l and m be parallel lines. Prove that l contains two points equidistant from m if and only if there is a line perpendicular to both l and m. (See Exercise 25.9 and Axiom A5.)
(b) In an absolute plane, let l and m be two lines. If three points on l are equidistant from m, prove that l and m are parallel.

25.11. In an absolute plane, let ABC be a triangle. Prove that $d(A, B) < d(A, C)$ if and only if $\angle ABC > \angle ACB$.

(*Hint:* If $d(A, B) < d(A, C)$, one possible approach is to let E be the point on $\overrightarrow{AC}$ such that $d(A, E) = d(A, B)$, and prove that $\angle ABC > \angle ABE = \angle AEB > \angle ACB$. Then deduce the converse result.)

25.12. In an absolute plane, let ABC be a triangle. Prove that $d(A, B) < d(A, C) + d(C, B)$.

(*Hint:* One possible approach is to let F be the point where AB intersects the line through C perpendicular to AB. Then use Equation 1 and Exercise 25.11 to prove that

$$d(A, B) \le d(A, F) + d(B, F) < d(A, C) + d(C, B).)$$

25.13. In an absolute plane, let ABC be a triangle with a right angle at B. Let D be a point of $[B, C]$ other than B and C.

(a) Prove that $d(A, D) < d(A, C)$. (See Exercise 25.11.)
(b) If E is a point of $[A, B]$ other than A and B, prove that $d(D, E) < d(A, C)$.

25.14. In an absolute plane, let ABC and $A'B'C'$ be triangles such that $\angle ABC = 90 = \angle A'B'C'$ and $d(A, C) = d(A', C')$. If $d(A, B) < d(A', B')$, prove that $d(B, C) > d(B', C')$, $\angle BAC > \angle B'A'C'$, and $\angle BCA < \angle B'C'A'$. (See Exercise 25.13.)

25.15. In an absolute plane, let ABC and $A'B'C'$ be triangles such that $\angle ABC = \angle A'B'C'$, $d(A, B) = d(A', B')$, $d(A, C) = d(A', C')$, and $\angle BCA$ and $\angle B'C'A'$ are either both less than or equal to 90 or both greater than or equal to 90. Prove that triangles ABC and $A'B'C'$ are congruent.

Section 26.

Axiomatic Euclidean and Hyperbolic Geometry

We used the axiomatic method in the last section to study properties shared by the Euclidean and the hyperbolic planes. We now study differences between the planes. We've seen that two differences are whether the angles of triangles sum to 180 and whether a point lies on at most one line parallel to a given line (by the discussion accompanying Figures 24.23 and 24.24). We now consider two more differences: whether rectangles exist and whether there are similar triangles that aren't congruent. We use the axiomatic approach to show that all four of these differences are related.

A *rectangle ABCD* is the union of the segments $[A, B]$, $[B, C]$, $[C, D]$, and $[D, A]$, where A, B, C, D are four points such that

$$\angle ABC = 90 = \angle BCD = \angle CDA = \angle DAB$$

(Figure 26.1). In the Euclidean plane, if A, B, C are three points such that $\angle ABC = 90$, there is a point D such that $ABCD$ is a rectangle: As in Figure 26.1, D is the point where the line l through A perpendicular to AB intersects the line m through C perpendicular to BC. This result does not extend to the hyperbolic plane. For example, we have h$\angle ABC = 90$ in

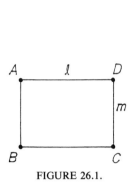

FIGURE 26.1.

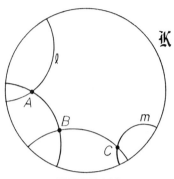

FIGURE 26.2.

Figure 26.2, but the line l through A perpendicular to AB doesn't intersect the line m through C perpendicular to BC, and so there is no point D such that $ABCD$ is a rectangle. The next result shows that, in any absolute plane, the existence of a rectangle is linked to the defect of a triangle. Of course, a triangle has defect zero if and only if its angles sum to 180 (by Equation 12 of Section 25). We say that triangle ABC has a *right angle* at B if $\angle ABC = 90$.

THEOREM 26.1. In an absolute plane, if ABC is a triangle that has a right angle at B and defect zero, then there is a point D such that $ABCD$ is a rectangle. Conversely, if $ABCD$ is any rectangle, then ABC is a triangle with a right angle at B and defect zero (Figure 26.3).

Proof: First assume that ABC is a triangle with a right angle at B and defect zero. The angles of the triangle sum to 180, and $\angle ABC$ is 90, and so we have

$$\angle BAC + \angle ACB = 90. \tag{1}$$

Let m be the line through A perpendicular to AB (by Theorem 25.5) (Figure 26.4). If $\overrightarrow{AE}$ and $\overrightarrow{AF}$ are the two rays on m that originate at A, then A is a point of $[E, F]$ other than E and F (by Theorem 25.1), and E and F lie on different sides of AB (by Axiom A3). We can assume that E lies on the same side of AB as C (by interchanging E and F, if necessary). There is a point D of $\overrightarrow{AE}$ other than A such that

$$d(A, D) = d(B, C) \tag{2}$$

(by Theorem 25.2ii). Since $\overrightarrow{AD} = \overrightarrow{AE}$ (by Theorem 25.2iii), either D equals E or $[D, E]$ doesn't contain A (by Theorem 25.1). In either case D lies on the same side of AB as E (by Axiom A3). Thus, since E and C lie on the same side of AB, so so D and C.

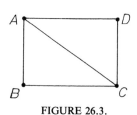

FIGURE 26.3.

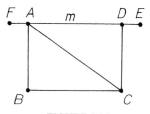

FIGURE 26.4.

Since AD and BC are perpendicular to AB at distinct points A and B, they are parallel (by Theorem 25.6). Thus, B and C lie on the same side of AD (by Axiom A3), and $\overleftrightarrow{AC} \neq \overleftrightarrow{AD}$. Hence, B and D lie on different sides of AC (by Theorem 25.8i and the last sentence of the previous paragraph). Thus, we have

$$\angle BAC + \angle CAD = \angle BAD = 90 \qquad (3)$$

(by Axiom A4(ii)). Combining Equations 1 and 3 gives

$$\angle ACB = \angle CAD. \qquad (4)$$

Equations 2 and 4 imply that triangles ACB and CAD are congruent (by Theorem 25.10, since $[A, C]$ is a side of both triangles). Thus, we have

$$\angle CDA = \angle ABC = 90 \qquad (5)$$

and

$$\angle DCA = \angle BAC. \qquad (6)$$

A and D lie on the same side of BC (by Axiom A3 and the first sentence of the previous paragraph), and B and D lie on different sides of AC (by the third sentence of the previous paragraph). Thus, we have

$$\angle DCB = \angle DCA + \angle ACB \qquad \text{(by Axiom A4(ii))}$$

$$= \angle BAC + \angle ACB \qquad \text{(by Equation 6)}$$

$$= 90 \qquad \text{(by Equation 1).} \qquad (7)$$

Equations 3, 5, and 7 show that $ABCD$ is a rectangle, as desired.

Conversely, assume that $ABCD$ is a rectangle. AB and CD are perpendicular to BC at distinct points B and C (by the last sentence of Theorem 25.3), and so they are parallel (by Theorem 25.6). Accordingly, ABC is a triangle with a right angle at B, $\overleftrightarrow{AC}$ doesn't equal $\overrightarrow{AD}$, and C and D lie on the same side of AB (by Axiom A3). By symmetry, B and C lie on the same side of AD, and so B and D lie on different sides of AC (by Theorem 25.8i). Thus, we have

$$\angle BAC + \angle CAD = \angle BAD = 90 \qquad (8)$$

(by Axiom A4(ii)). By symmetry, we also have

$$\angle BCA + \angle ACD = \angle BCD = 90. \tag{9}$$

It follows that

$$\delta\triangle ABC + \delta\triangle ADC = (90 - \angle BAC - \angle BCA)$$
$$+ (90 - \angle CAD - \angle ACD)$$

(since $\angle ABC = 90 = \angle CDA$)

$$= 0$$

(by Equations 8 and 9). Since the defect of every triangle is nonnegative (by Theorem 25.12), the last sentence implies that triangle ABC has defect zero, as desired. □

If one triangle in absolute plane has defect zero, we want to prove that all triangles in the plane have defect zero. We start by deducing a special case of this result from the previous theorem.

THEOREM 26.2. In an absolute plane, let AFC be a triangle that has a right angle at F and defect zero. Let E be any point on $\overrightarrow{FC}$ other than F. Then triangle AFE has defect zero (Figure 26.5).

Proof: There is a point D such that $AFCD$ is a rectangle, by Theorem 26.1 (Figure 26.6).

Let C_1 be the point such that C is the midpoint of C_1 and F, and let D_1 be the point such that D is the midpoint of D_1 and A (by Theorem 25.2vii). We have

$$d(F, C_1) = 2d(F, C)$$

(by Theorem 25.2vi). Since $[F, C_1]$ contains C (by Theorem 25.2vi), so does $\overrightarrow{FC_1}$ (by Theorem 25.2v), and we have $\overrightarrow{FC_1} = \overrightarrow{FC}$ (by Theorem 25.2iii). It follows that

$$\angle AFC_1 = \angle AFC = 90. \tag{10}$$

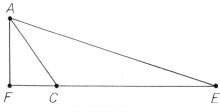

FIGURE 26.5.

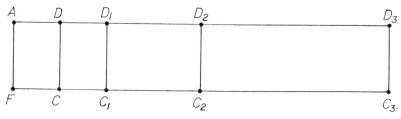

FIGURE 26.6.

CD is the perpendicular bisector of F and C_1 and of A and D_1 (since $AFCD$ is a rectangle, C is the midpoint of F and C_1, and D is the midpoint of A and D_1). Thus, the reflection σ_{CD} interchanges F with C_1 and A with D_1 (by Theorem 25.4). Hence, if we apply σ_{CD} to A, F, and C_1, Equation 10 becomes

$$\angle D_1 C_1 F = 90 \tag{11}$$

(by Axiom A5). By symmetry with Equations 10 and 11, we have

$$\angle FAD_1 = 90 = \angle C_1 D_1 A.$$

Together with Equations 10 and 11, this shows that $AFC_1 D_1$ is a rectangle.

If we replace C and D with C_1 and D_1 in the last two paragraphs, let C_2 and D_2 be the points that replace C_1 and D_1. Then $AFC_2 D_2$ is a rectangle, and we have

$$d(F, C_2) = 2d(F, C_1) = 4d(F, C)$$

and $\overrightarrow{FC_2} = \overrightarrow{FC_1} = \overrightarrow{FC}$.

Similarly, if we replace C and D with C_2 and D_2 in the second and third paragraphs of the proof, let C_3 and D_3 be the points that replace C_1 and D_1. Then $AFC_3 D_3$ is a rectangle, and we have

$$d(F, C_3) = 2d(F, C_2) = 8d(F, C)$$

and $\overrightarrow{FC_3} = \overrightarrow{FC_2} = \overrightarrow{FC}$.

We continue in this way, successively obtaining points C_i and D_i for every positive integer i such that $AFC_i D_i$ is a rectangle,

$$d(F, C_i) = 2^i d(F, C), \tag{12}$$

and $\overrightarrow{FC_i} = \overrightarrow{FC}$.

Since $d(F, C) > 0$ (by Theorem 25.2ii), there is a positive integer k such that

$$2^k d(F, C) > d(F, E).$$

Then $d(F, E)$ is less than $d(F, C_k)$ (by Equation 12). Therefore, since E is a point of $\overrightarrow{FC} = \overrightarrow{FC_k}$ other than F, E is a point of $[F, C_k]$ other than F

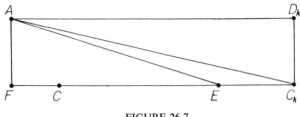

FIGURE 26.7.

and C_k (by Theorem 25.2iv) (Figure 26.7). Since AFC_kD_k is a rectangle, we have

$$0 = \delta\triangle AFC_k = \delta\triangle AFE + \delta\triangle AEC_k \tag{13}$$

(by Theorems 25.11 and 26.1). Since the defect of every triangle is non-negative (by Theorem 25.12), (13) implies that the defect of triangle AFE is zero, as desired. □

We use the next result to generalize the previous theorem. If ABC is a triangle in an absolute plane, the *foot of the perpendicular from A to BC* is the point F where BC intersects the line through A perpendicular to BC (Figure 26.8). Theorem 25.5 and Axiom A1(i) show that the point F exists and is unique.

THEOREM 26.3. In an absolute plane, let ABC be a triangle. Let F be the foot of the perpendicular from A to BC (Figure 26.8).

(i) if $\angle ABC < 90$, then F is a point of $\overrightarrow{BC}$ other than B.
(ii) If $\angle ABC < 90$ and $\angle ACB < 90$, then F is a point of $[B, C]$ other than B and C.

Proof:
(i) if $\overrightarrow{BE}$ is the ray opposite $\overrightarrow{BC}$, then B is a point of $[C, E]$ other than C and E (by Theorem 25.1). Thus, we have

$$\angle ABE = 180 - \angle ABC > 90$$

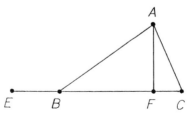

FIGURE 26.8.

(by Equation 3 of Section 25 and the assumption that $\angle ABC < 90$). Because AF is perpendicular to BC, we have $F \neq B$ (since $\angle ABC < 90$) and $\angle ABF < 90$ (by Theorem 25.12 applied to triangle ABF). The last two sentences show that $\overrightarrow{BF} \neq \overrightarrow{BE}$, and so we have $\overrightarrow{BF} = \overrightarrow{BC}$ (by Theorem 25.1). Thus, F is a point on $\overrightarrow{BC}$ other than B (by Theorem 25.2iii).

(ii) By part (i), F is point on $\overrightarrow{BC}$ other than B, and it is a point on $\overrightarrow{CB}$ other than C. Thus, F is a point on $[B, C]$ other than B and C (by Theorem 25.2v). □

We can now generalize Theorem 26.2 by proving that, if one triangle in an absolute plane has defect zero, then all triangles in the plane have defect zero.

THEOREM 26.4. In an absolute plane, if the angles of one triangle sum to 180, then the angles of every triangle sum to 180.

Proof: Let ABC be the given triangle whose angles sum to 180. At most one angle of the triangle is greater than or equal to 90 (by Theorem 25.12 and the last sentence of Theorem 25.3), and so we can assume that $\angle ABC < 90$ and $\angle ACB < 90$ (by relabeling A, B, C, if necessary). Let F be the foot of the perpendicular from A to BC (Figure 26.8). F is a point of $[B, C]$ other than B and C (by Theorem 26.3ii), and so we have

$$\delta \triangle ABF + \delta \triangle AFC = \delta \triangle ABC = 0 \qquad (14)$$

(by Theorem 25.11). Since the defect of every triangle is nonnegative (by Theorem 25.12), the equations in (14) imply that the defect of triangle AFC is zero.

Let GHI be any triangle in the plane with a right angle at H (Figure 26.9). By Theorem 25.2(ii), there is a point G' on $\overrightarrow{FA}$ other than F such that

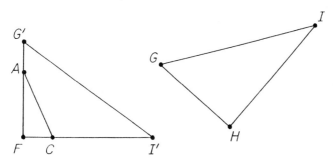

FIGURE 26.9.

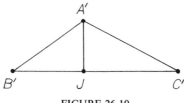

FIGURE 26.10.

$d(F, G') = d(H, G)$, and there is a point I' on $\overrightarrow{FC}$ other than F such that $d(F, I') = d(H, I)$. Since triangle AFC has defect zero, so does triangle AFI' (by Theorem 26.2). Then since triangle AFI' has defect zero, so does triangle $G'FI'$ (by Theorem 26.2). Moreover, triangles GHI and $G'FI'$ are congruent (by Theorem 25.10), and so triangle GHI has defect zero. In short, every triangle in the plane with a right angle has defect zero.

Finally, let $A'B'C'$ be any triangle in the plane (Figure 26.10). As in the first paragraph of the proof, we can ensure that $\angle A'B'C' < 90$ and $\angle A'C'B' < 90$ by relabeling A', B', C', if necessary. Let J be the foot of the perpendicular from A' to $B'C'$. J is a point of $[B', C']$ other than B' and C' (by Theorem 26.3ii). Thus, we have

$$\delta\triangle A'B'C' = \delta\triangle A'B'J + \delta\triangle A'JC' = 0$$

(by Theorem 25.11 and the previous paragraph), and so the angles of triangle $A'B'C'$ sum to 180. $\square$

Let P be a point that doesn't lie on a line l in an absolute plane. We want to prove that P lies on a unique line parallel to l if the plane contains enough rectangles. The proof uses the following technical result, which we break off as a separate theorem for ease of reference.

THEOREM 26.5. In an absolute plane, let PBC be a triangle with a right angle at B. Let A be a point such that B is the midpoint of A and P. If there is a point D such that $ABCD$ is a rectangle, then there is a point E on line PC such that A is the foot of the perpendicular from E to line PB (Figure 26.11).

Proof: Let E be the point such that D is the midpoint of E and A (by Theorem 25.2vii). Since $\angle BAD = 90$, A is the foot of the perpendicular from E to PB (by the last sentence of Theorem 25.3). Accordingly, if we prove that E lies on PC, we're done.

D is a point of $[A, E]$ other than A and E (by Theorem 25.2vi), and so $[D, E]$ doesn't contain A (by Theorem 25.2i). Thus, since A is the only point of DE that lies on AC (by Axiom A1(i)), D and E lie on the same side

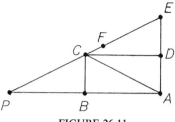

FIGURE 26.11.

of AC (by Axiom A3). Moreover, A and E lie on different sides of CD (by Axiom A3 and Theorem 25.2vi), and so we have

$$\angle ACE = \angle ACD + \angle ECD$$

(by Axiom A4(ii)). We have $\angle ACD = \angle ECD$, since σ_{CD} interchanges A and E, fixes C and D, and preserves angles (by Theorem 25.4 and Axiom A5). The last two sentences give

$$\angle ACE = 2\angle ACD. \tag{15}$$

Since B is the midpoint of A and P, we obtain the following analogue of Equation 15:

$$\angle PCA = 2\angle BCA. \tag{16}$$

Since $ABCD$ is a rectangle, we have

$$\angle BCA + \angle ACD = 90 \tag{17}$$

(by Equation 9). Combining Equations 15–17 gives

$$\angle PCA + \angle ACE = 2\angle BCA + 2\angle ACD = 180.$$

D and B lie on different sides of AC (by the last paragraph of the proof of Theorem 26.1). D and E lie on the same side of AC (by the second-to-last paragraph), and, by symmetry, B and P lie on the same side of AC. Hence, E and P lie on different sides of AC.

On the other hand, if $\overrightarrow{CF}$ is the ray opposite $\overrightarrow{CP}$, C is a point of $[P, F]$ other than P and F (by Theorem 25.1). P and F lie on opposite sides of AC (by Axiom A3), and the equation

$$\angle PCA + \angle ACF = 180$$

holds (by Equation 3 of Section 25). The two properties in the last sentence determine the ray $\overrightarrow{CF}$ uniquely (by Axiom A4(i)). Since $\overrightarrow{CE}$ satisfies the same two properties (by the last two paragraphs), $\overrightarrow{CE}$ equals $\overrightarrow{CF}$. Thus, E lies on line $CF = PC$ (by Theorem 25.2iii and Axiom A1(i)), as desired. $\square$

We say that triangles ABC and $A'B'C'$ are *similar* if corresponding angles are equal: $\angle BAC = \angle B'A'C'$, $\angle ABC = \angle A'B'C'$, and $\angle ACB = \angle A'C'B'$.

We can now prove the equivalence of a number of conditions about the angle sum of triangles, the existence of rectangles, the uniqueness of parallel lines, and the existence of similar triangles in an absolute plane. Conditions are called *equivalent* when each of the conditions implies all of the others, and so either all or none of the conditions hold in any given case. All of the conditions in the next theorem hold in the Euclidean plane. None of them holds in the hyperbolic plane, since the theorem shows that the conditions are equivalent, and since the discussions accompanying Figures 24.24, 26.2, and 24.23 show that conditions (ii), (iii), and (v) don't hold in the hyperbolic plane.

THEOREM 26.6. In an absolute plane, the following conditions are equivalent.

   (i) The angles of one triangle sum to 180.
  (ii) The angles of every triangle sum to 180.
 (iii) For every triangle ABC with a right angle at B, there is a point D such that $ABCD$ is a rectangle.
 (iv) There is a rectangle.
  (v) For every point P and every line l that doesn't contain P, P lies on a unique line parallel to l.
 (vi) There is a point P and there is a line l that doesn't contain P such that P lies on a unique line parallel to l.
(vii) There exist triangles that are similar but not congruent.

Proof: We prove the implications in Figure 26.12. One can follow arrows around this figure from any of the seven conditions to all of the others. Thus, each condition implies all of the others, and the seven conditions are equivalent.

(i) implies (ii), by Theorem 26.4. (ii) implies (iii), by Theorem 26.1.

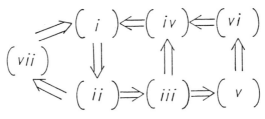

FIGURE 26.12.

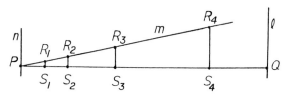

FIGURE 26.13.

(iii) implies (iv): There are two points A and B (by Axiom A1(ii)). There is a point C such that ABC is a triangle with a right angle at B (by Axiom A4(i) and the last sentence of Theorem 25.3). Thus, by condition (iii), there is a point D such that $ABCD$ is a rectangle.

(iv) implies (i), by Theorem 26.1. We've now established the implications in the middle square of Figure 26.12, and so conditions (i)–(iv) are equivalent.

(iii) implies (v): Let Q be the foot of the perpendicular from P to l (Figure 26.13). Q doesn't equal P (since l contains Q but not P). Let n be the line through P perpendicular to PQ (by Theorem 25.5). The lines l and n are parallel (by Theorem 25.6). Let m be a line through P other than n and PQ. If we prove that m intersects l, then n is the unique line through P parallel to l, as desired.

The two rays on m originating at P form angles with $\overrightarrow{PQ}$ that sum to 180 (by Theorem 25.1 and Equation 3 of Section 25). These angles don't equal 90 (by Axiom A4(i)), and so one is less than 90 and the other is greater than 90. Accordingly, let R_1 be a point on m such that $\angle QPR_1 < 90$. Let S_1 be the foot of the perpendicular from R_1 to PQ. S_1 is a point of $\overrightarrow{PQ}$ other than P (by Theorem 26.3i), and so we have

$$\overrightarrow{PS_1} = \overrightarrow{PQ} \tag{18}$$

(by Theorem 25.2iii).

Let S_2 be the point such that S_1 is the midpoint of S_2 and P (by Theorem 25.2vii). Since R_1 doesn't lie on PS_1 (by the last sentence of Theorem 25.3), $R_1 S_1 S_2$ is a triangle with a right angle at S_1. Thus, there is a point T such that $S_2 S_1 R_1 T$ is a rectangle (by the assumption that (iii) holds). Hence, there is a point R_2 on m such that S_2 is the foot of the perpendicular from R_2 to PQ (by Theorem 26.5).

If we replace S_1 and R_1 in the previous paragraph with S_2 and R_2, we see that there are points S_3 and R_3 such that S_2 is the midpoint of S_3 and P, R_3 lies on m, and S_3 is the foot of the perpendicular from R_3 to PQ. Proceeding in this way, we define points S_i and R_i for every positive integer i such that S_i is the midpoint of P and S_{i+1}, R_i lies on m, and S_i is the foot of the perpendicular from R_i to PQ.

Since S_i is the midpoint of P and S_{i+1}, S_i is a point on $\overrightarrow{PS_{i+1}}$ other than P (by Theorem 25.2vi and v), and so we have

$$\overrightarrow{PS_i} = \overrightarrow{PS_{i+1}} \tag{19}$$

for every positive integer i (by Theorem 25.2iii). Equations 18 and 19 give

$$\overrightarrow{PQ} = \overrightarrow{PS_1} = \overrightarrow{PS_2} = \overrightarrow{PS_3} = \cdots . \tag{20}$$

The fact that S_i is the midpoint of P and S_{i+1} for every positive integer i implies that

$$d(P, S_1) = \tfrac{1}{2}d(P, S_2) = \tfrac{1}{4}d(P, S_3) = \cdots$$

(by Theorem 25.2vi). Thus, we have

$$d(P, S_i) = 2^{i-1}d(P, S_1) \tag{21}$$

for every positive integer i.

Since $d(P, S_1) > 0$ (by Axiom A2 and Equation 1 of Section 25), there is a positive integer k such that

$$2^{k-1}d(P, S_1) > d(P, Q). \tag{22}$$

Q is a point on $\overrightarrow{PS_k}$ such that

$$d(P, Q) < d(P, S_k),$$

by (20)–(22) and Theorem 25.2(iii). Thus, Q is a point on $[P, S_k]$ other than P and S_k (by Theorem 25.2iv) (Figure 26.14).

Since l and $R_k S_k$ are perpendicular to PQ at distinct points Q and S_k, they're parallel (by Theorem 25.6). Thus, R_k and S_k lie on the same side of l (by Axiom A3). Since $[P, S_k]$ contains the point Q of l, P and S_k lie on opposite sides of l (by Axiom A3). The last two sentences show that P and R_k lie on opposite sides of l. Thus, l intersects $[P, R_k]$ (by Axiom A3), and so l intersects m, as desired.

(v) implies (vi) because there are two points P and Q (by Axiom A1(ii)) and a line l through Q that doesn't contain P (by Axiom A4(i) and the last sentence of Theorem 25.3).

(vi) implies (iv): Let P and l be as in part (vi). (See Figure 26.15, where, as always, small squares mark perpendicular lines.) Let Q be the foot

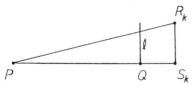

FIGURE 26.14.

FIGURE 26.15.

of the perpendicular from P to l. Let R be a point on l other than Q (by Axiom A2). Let m be the line through R perpendicular to l (by Theorem 25.5). Let S be the foot of the perpendicular from P to m.

Since P doesn't lie on l, P doesn't equal Q, and S doesn't equal R. Q doesn't equal R (by the choice of R), and l is perpendicular to PQ and SR, and so PQ and SR are parallel (by Theorem 25.6). Thus, neither P nor Q equals either R or S. The last three sentences show that P, Q, R, S are four distinct points. We have

$$\angle PQR = 90 = \angle QRS = \angle RSP,$$

by the choices of Q, R, and S.

Let n be the line through P perpendicular to PQ (by Theorem 25.5). Since n and l are perpendicular to PQ at distinct points P and Q, they are parallel (by Theorem 25.6). Moreover, PS and l are perpendicular to SR at distinct points S and R, and so they are parallel (by Theorem 25.6). Since P lies on a unique line parallel to l (by the assumption that (vi) holds), the last two sentences show that PS is the line n through P perpendicular to PQ. Thus, we have $\angle SPQ = 90$. Together with the previous paragraph, this shows that $PQRS$ is a rectangle.

We've now established the implications in the two squares on the right of Figure 26.12, and so conditions (i)–(vi) are equivalent.

(ii) implies (vii): There are two points A and B by Axiom A1(ii), and there is a point C such that ABC is a triangle with a right angle at B (by Axiom A4(i) and the last sentence of Theorem 25.3) (Figure 26.16). $\overleftrightarrow{AC}$ contains a point P other than A such that

$$d(A, P) \neq d(A, C) \tag{23}$$

(by Theorem 25.2ii). Let Q be the foot of the perpendicular from P to line AB. We have

$$\overrightarrow{AP} = \overrightarrow{AC} \tag{24}$$

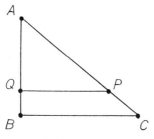

FIGURE 26.16.

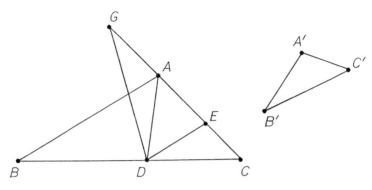

FIGURE 26.17.

(by Theorem 25.2iii), which implies that

$$\angle BAP = \angle BAC < 90$$

(by Theorem 25.12 and the fact that $\angle ABC = 90$). Then Q is a point $\overleftrightarrow{AB}$ other than A (by Theorem 26.3i), and so we have $\overrightarrow{AQ} = \overrightarrow{AB}$ (by Theorem 25.2iii). Thus, by Equation 24, we have

$$\angle PAQ = \angle BAC. \tag{25}$$

Since $\angle AQP = 90 = \angle ABC$, Equation 25 and condition (ii) imply that $\angle APQ = \angle ACB$, and so triangles ABC and AQP are similar. These triangles aren't congruent, by (23).

(vii) implies (i): Let triangles ABC and $A'B'C'$ be similar but not congruent (Figure 26.17). Corresponding angles are equal, and corresponding sides are not all of equal length. By symmetry, we can assume that

$$d(C', B') < d(C, B). \tag{26}$$

By Theorem 25.2(ii), there is a point D on $\overrightarrow{CB}$ other than C such that

$$d(C, D) = d(C', B'), \tag{27}$$

and there is a point E on $\overrightarrow{CA}$ other than C such that $d(C, E) = d(C', A')$. We have $\overrightarrow{CD} = \overrightarrow{CB}$ and $\overrightarrow{CE} = \overrightarrow{CA}$ (by Theorem 25.2iii), and so we have $\angle DCE = \angle BCA = \angle B'C'A'$. The last two sentences imply that triangles EDC and $A'B'C'$ are congruent (by Theorem 25.10), and so triangles ABC and EDC are similar.

D is a point on $\overrightarrow{CB}$ other than C such that $d(C, D) < d(C, B)$ (by (26) and (27)), and so D lies on $[B, C]$ and doesn't equal B or C (by Theorem 25.2iv). Thus, we have

$$\angle BAD + \angle DAC = \angle BAC$$

(by Equation 13 of Section 25), and it follows that

$$\angle DAC < \angle BAC \tag{28}$$

(by the last sentence of Theorem 25.3). Let G be any point on $\overrightarrow{CA}$ that doesn't lie on $[C, A]$. Theorem 25.2(iv) shows that

$$d(C, A) < d(C, G)$$

and lets us conclude that A is a point of $[C, G]$ other than C and G. Thus, A lies on $\overrightarrow{GC}$ (by Theorem 25.2v), and so we have $\overrightarrow{GA} = \overrightarrow{GC}$ (by Theorem 25.2iii). Accordingly, we have

$$\angle DGC = \angle DGA \le 180 - \angle DAG - \angle ADG$$

(by Theorem 25.12)

$$< 180 - \angle DAG$$

(by the last sentence of Theorem 25.3)

$$= \angle DAC$$

(by Equation 3 of Section 25)

$$< \angle BAC \tag{29}$$

(by (28)).

E is a point on $\overrightarrow{CA}$ such that

$$\angle DEC = \angle BAC \tag{30}$$

(by the last sentence of the second-to-last paragraph). Comparing (28) and (30) shows that $E \neq A$, and comparing (29) and (30) shows that E doesn't lie outside of $[C, A]$. In short, E is a point of $[C, A]$ other than C and A. Therefore, since D is a point of $[B, C]$ other than B and C (by the first sentence of the previous paragraph), Theorem 25.11 shows that

$$\delta \triangle ABC = \delta \triangle ABD + \delta \triangle ADC$$

$$= \delta \triangle ABD + \delta \triangle ADE + \delta \triangle EDC. \tag{31}$$

Since triangles ABC and EDC are similar (by the second-to-last paragraph), $\delta \triangle ABC$ equals $\delta \triangle EDC$. Thus, (31) implies that $\delta \triangle ABD$ is zero (since the defect of every triangle is nonnegative, by Theorem 25.12). Hence, the angles of triangle ABD sum to 180, and condition (i) holds.

We've proved every implication in Figure 26.12, and so conditions (i)–(vii) are equivalent. $\square$

The *axioms of Euclidean geometry* are Axioms A1–A5 of absolute geometry and any one of the conditions in Theorem 26.6. *Axiomatic Euclidean geometry* is the study of the logical consequences of these axioms and the accompanying definitions. Because the conditions in Theorem 26.6 are equivalent, we can take any one of them as an axiom of Euclidean geometry in addition to Axioms A1–A5 and deduce all of the other conditions in Theorem 26.6.

The axioms of Euclidean geometry and the accompanying definitions hold in the familiar Euclidean plane constructed from ordered pairs of real numbers. Conversely, let $\mathcal{P}$ be any plane that satisfies the axioms of Euclidean geometry and the accompanying definitions. It can be shown that the points of $\mathcal{P}$ can be matched up with ordered pairs of real numbers so that lines, coordinate systems on lines, and angles are determined as usual in the Euclidean plane. In effect, the familiar Euclidean plane is the *only* plane that satisfies the axioms of Euclidean geometry and the accompanying definitions. These axioms and definitions characterize the Euclidean plane completely.

Because the conditions of Theorem 26.6 are equivalent, they are either all true or all false in an absolute plane. If any one of them is false, they all are. Thus, the negations of the conditions in Theorem 26.6 are also equivalent.

The sum of the angles of a triangle in an absolute plane is at most 180 (by Theorem 25.12). Accordingly, if the angles of a triangle don't sum to 180, their sum is less than 180.

Let P be a point that doesn't lie on a line l in an absolute plane. P lies on at least one line parallel to l (by Theorem 25.7). Thus, if P doesn't lie on a unique line parallel to l, it lies on more than one line parallel to l.

The last three paragraphs show that Theorem 26.6 implies the following result, where each condition of Theorem 26.6 is negated. Note that the negation of a "for every" statement is a "there exists" statement: if something doesn't always happen, then it fails to happen at least once. Likewise, the negation of a "there exists" statement is a "for every" statement.

THEOREM 26.7. In an absolute plane, the following conditions are equivalent.

    (i) The sum of the angles of every triangle is less than 180.
    (ii) The sum of the angles of one triangle is less than 180.
    (iii) There is a triangle ABC with a right angle at B such that no point D forms a rectangle $ABCD$.
    (iv) There are no rectangles.

(v) There is a point P and there is a line l that doesn't contain P such that P lies on more than one line parallel to l.

(vi) For every point P and every line l that doesn't contain P, P lies on more than one line parallel to l.

(vii) Any two similar triangles are congruent. □

The conditions in Theorem 26.7 all hold in the hyperbolic plane (by the discussion before Theorem 26.6). Condition (vii) is particularly notable: Two triangles in the hyperbolic plane can only have the same angles if they are the same size. Accordingly, if two figures in the hyperbolic plane have exactly the same shape, they must have the same size (unless each of the figures lies on a line). In the hyperbolic plane, there is no scale model of a figure that doesn't lie on a line.

The *axioms of hyperbolic geometry* are Axioms A1–A5 of absolute geometry and any one of the conditions in Theorem 26.7. *Axiomatic hyperbolic geometry* is the study of the logical consequences of these axioms and the accompanying definitions. Because the conditions in Theorem 26.7 are equivalent, we can take any one of them as an axiom of hyperbolic geometry in addition to Axioms A1–A5 and deduce all of the other conditions in Theorem 26.7.

Let $\mathcal{P}$ be any plane that satisfies the axioms of hyperbolic geometry and the accompanying definitions. It can be shown with some effort that the points of $\mathcal{P}$ can be matched up with the points of the hyperbolic plane constructed from a circle $\mathcal{K}$ in the Euclidean plane. Thus, the axioms of hyperbolic geometry and the accompanying definitions characterize hyperbolic planes completely.

It can be shown that the plane $\mathcal{P}$ uniquely determines the radius of the circle $\mathcal{K}$ in the previous paragraph. Thus, hyperbolic planes determined by circles of different radii are essentially different. Rather than one hyperbolic plane, there are infinitely many, constructed from circles of different radii.

Instead of using circles of different radii to construct all hyperbolic planes, we could use just one circle and replace the natural logarithm in Theorem 24.3 with logarithms to any base greater than one. This multiplies coordinates on lines and distances between points by a constant positive factor. The fact that similar triangles in the hyperbolic plane must be congruent explains why the hyperbolic plane is changed in an essential way when distances are multiplied by a constant factor. This is very different from the Euclidean plane, where the distance scale is arbitrary and distances can be multiplied by a positive factor without fundamentally changing the plane.

Finally, every absolute plane $\mathcal{P}$ satisfies the axioms of either Euclidean or hyperbolic geometry and the accompanying definitions (since the conditions in Theorem 26.7 are the negations of those in Theorem 26.6). Accordingly, the discussions after Theorems 26.6 and 26.7 show that $\mathcal{P}$ can be identified with either the familiar Euclidean plane or a hyperbolic plane constructed from a circle. Thus, the Axioms A1–A5 of absolute geometry characterize the Euclidean and hyperbolic planes.

EXERCISES

26.1. Prove that the following condition in an absolute plane is equivalent to the conditions in Theorem 26.6: If l, m, n are three lines such that l is parallel to m and m is parallel to n, then l is parallel to n.

26.2. Prove that the following condition in an absolute plane is equivalent to the conditions in Theorem 26.7: For every point P and every line l that doesn't contain P, P lies on infinitely many lines parallel to l.

26.3. Let corresponding angles be defined as in Exercise 25.2. Prove that the following condition in an absolute plane is equivalent to the conditions in Theorem 26.6: Any two parallel lines l and m form equal corresponding angles with any line that intersects both l and m.

26.4. In an absolute plane that satisfies the conditions of Theorem 26.7, let A be a point that doesn't lie on a line m. Prove that A lies on infinitely many lines l such that there is a line perpendicular to both l and m.

26.5. Prove that the following condition in an absolute plane is equivalent to the conditions in Theorem 26.6: If A, B, C, D are four points such that A and D lie on the same side of line BC, and if $\angle ABC$ and $\angle BCD$ are less than $90°$, then the rays $\overrightarrow{BA}$ and $\overrightarrow{CD}$ intersect. (See Figure V.1 and the accompanying discussion in the chapter introduction.)

26.6. In an absolute plane, let PS_1R_1 be a triangle with a right angle at S_1 (Figure 26.13). Let $S_2, S_3, \ldots$ be points such that S_i is the midpoint of P and S_{i+1} for each positive integer i. Let t be a positive integer such that the line through S_t perpendicular to PS_1 intersects PR_1 at a point R_t. Use Exercise 25.4 to prove that

$$\delta \triangle PR_1S_1 < 180/2^{t-1}.$$

26.7. In an absolute plane that satisfies the conditions of Theorem 26.7, let n be a line. Let r be a ray that originates at a point P of n and doesn't lie on n.

(a) Prove that there is a point S on r other than P such that n is parallel to the line through S perpendicular to PS. (See Exercise 26.6.)
(b) If T is a point such that $[P, T]$ contains S, prove than n is parallel to the line through T perpendicular to PT.

26.8. In an absolute that satisfies the conditions of Theorem 26.7, let A and B be two points. Let C, D, E be points on one side of AB such that

$$\angle BAC < \angle BAD < \angle BAE.$$

(a) Prove that there is a line p that intersects $\overleftrightarrow{AC}$ and is perpendicular to AC and parallel to AB and AD. Prove that there is a line q that intersects $\overleftrightarrow{AE}$ and is perpendicular to AE and parallel to AB and AD. (See Exercise 26.7.)
(b) Prove that p and q are parallel.
(c) Prove that any two of the lines p, q, and AB lie on the same side of the third.

26.9. Prove that each of the following conditions in an absolute plane is equivalent to the conditions in Theorem 26.7.

(a) There are three parallel lines such that any two of the lines lie on the same side of the third.
(b) There are three lines such that every line in the plane is parallel to at least one of the three lines.

(*Hint:* See Exercise 26.8. Axioms A2 and A3 can be used to prove that three lines as in (a) do not all intersect any one line.)

26.10. In an absolute plane, prove that the following condition is equivalent to the conditions in Theorem 26.7: If l and m are any two lines, there is a line n perpendicular to m and parallel to l.

(*Hint:* To obtain the line n when the conditions of Theorem 26.7 hold and l and m are parallel, one possible approach is to combine Theorem 26.7vi with Exercise 26.7.)

26.11. In an absolute plane, prove that the following condition is equivalent to the conditions in Theorem 26.7: If l and m are any two lines, there is a line parallel to both l and m. (See Exercise 26.8a.)

26.12. In an absolute plane, prove that the following condition is equivalent to the conditions of Theorem 26.7: No line contains three points equidistant from another line. (See Exercises 25.9 and 25.10.)

26.13. In an absolute plane that satisfies the conditions of Theorem 26.7, let G and H be two points (Figure 26.18). Let l and m be the lines perpendicular to GH through G and H, respectively. Let R be a point on l other than G, and let S be a point on $[G, R]$ other than G and R. Let U and V be the feet of the perpendiculars to m from R and S, respectively. Prove that $90 > \angle GSV > \angle GRU$ and $d(G, H) < d(S, V) < d(R, U)$. (See Exercise 25.9.)

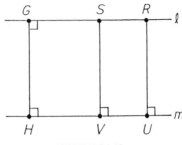

FIGURE 26.18.

26.14. Let the notation be as in Exercise 26.13.

(a) Prove that $d(G, R) < d(G, H) + d(H, U) + d(R, U)$. (See Exercise 25.12.)

(b) Prove that there is a point W on m such that $[H, W]$ contains the feet of the perpendiculars to m from all points on $\overrightarrow{GR}$. (See Exercise 26.7.)

(c) Let the distance from a point to a line be defined as in Exercise 25.10. For any number N, prove that there is a point on $\overrightarrow{GR}$ whose distance from m is greater than N.

26.15. In an absolute plane that satisfies the conditions of Theorem 26.7, let G and H be two points (Figure 26.19). Let l and m be the lines perpendicular to GH through G and H, respectively. Let A be a point on l, and let F be the foot of the perpendicular from A to m. Prove that there are points P and Q that lie on different sides of AF, lie on the same side of l as F, and are such that AP and AQ are parallel to m.

26.16. In an absolute plane that satisfies the conditions of Theorem 26.7, let A be a point that doesn't lie on a line m. Let F be the foot of the perpendicular from A to m, and let S be a side of line AF. Prove that S contains a point B that has the following two properties (Figure 26.20):

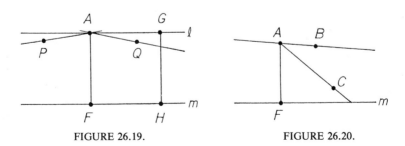

FIGURE 26.19.   FIGURE 26.20.

(i) AB and m are parallel lines.
(ii) If C is any point of S that lies on the same side of AB as F, then AC intersects m.

(*Hint:* If $\mathfrak{I}$ is a nonempty set of positive numbers, there is a nonnegative number b that satisfies the following two conditions.

(i) b is less then or equal to every element of $\mathfrak{I}$.
(ii) No real number greater than b is less than or equal to every element of $\mathfrak{I}$.

The number b is called the *greatest lower bound* of $\mathfrak{I}$. One possible approach to the exercise is to consider the greatest lower bound of the set of all $\angle FAD$ as D varies over all points of S such that AD is parallel to m.)

26.17. In the notation of Exercise 26.16, prove that no line is perpendicular to both AB and m. (See Exercise 26.15.)

26.18. In an absolute plane, prove that the conditions of Theorem 26.7 are equivalent to the following condition: There are parallel lines l and m such that no two points of l are equidistant from m. (See Exercises 25.10 and 26.17. Note that an absolute plane satisfying the conditions of Theorem 26.7 also contains parallel lines l and m such that l contains two but not three points equidistant from m, by Exercises 25.10 and 26.12.)

26.19. We say that AB is *horoparallel* to m at A if these points and lines are as in Exercise 26.16.
  In an absolute plane that satisfies the conditions of Theorem 26.7, let A and A' be two points on a line l, and let A'' be a point on a line m.

(a) If l is horoparallel to m at A, prove that l is horoparallel to m at A'.
(b) If l is horoparallel to m at A, prove that m is horoparallel to l at A''.

(Accordingly, horoparallelism is a symmetric property of two lines that doesn't depend on choosing a point on either line, and so we can say that "two lines are horoparallel.")

Index